Lecture Notes in Computer Science 965
Edited by G. Goos, J. Hartmanis and J. van Leeuwen

Springer

Berlin
Heidelberg
New York
Barcelona
Budapest
Hong Kong
London
Milan
Paris
Tokyo

Horst Reichel (Ed.)

Fundamentals of Computation Theory

10th International Conference, FCT '95
Dresden, Germany, August 22-25, 1995
Proceedings

Springer

Series Editors

Gerhard Goos
Universität Karlsruhe
Vincenz-Priessnitz-Straße 3, D-76128 Karlsruhe, Germany

Juris Hartmanis
Department of Computer Science, Cornell University
4130 Upson Hall, Ithaca, NY 14853, USA

Jan van Leeuwen
Department of Computer Science, Utrecht University
Padualaan 14, 3584 CH Utrecht, The Netherlands

Volume Editor

Horst Reichel
Fakultät Informatik, Technische Universität Dresden
Mommsenstraße 13, D-01069 Dresden

Cataloging-in-Publication data applied for

Die Deutsche Bibliothek - CIP-Einheitsaufnahme

Fundamentals of computation theory : 10th international conference ; proceedings / FCT '95, Dresden, Germany, August 22 - 25, 1995. Horst Reichel (ed.). - Berlin ; Heidelberg ; New York ; Barcelona ; Budapest ; Hong Kong ; London ; Milan ; Paris ; Tokyo : Springer, 1995
(Lecture notes in computer science ; Vol. 965)
ISBN 3-540-60249-6
NE: Reichel, Horst [Hrsg.]; FCT <10, 1995, Dresden>; GT

CR Subject Classification (1991): F.1-4, G.2, I.3.5, E.1

ISBN 3-540-60249-6 Springer-Verlag Berlin Heidelberg New York

Typesetting: Camera-ready by author
SPIN 10486648 06/3142 - 5 4 3 2 1 0 Printed on acid-free paper

Foreword

This volume contains the Proceedings of the Tenth Conference on Fundamentals of Computation Theory, held in Dresden, August 22–25, 1995. The previous conferences of the FCT series took place in Poznan–Kornik (1977), Wendisch–Rietz (1979), Szeged (1981), Borgholm (1983), Cottbus (1985), Kazan (1987), Szeged (1989), Berlin (1991) and Szeged (1993). Following the tradition of the predecessors the conference covered a broad range of topics of theoretical computer science:

- Algorithms and data structures
- Automata and formal languages
- Categories and types
- Computability and complexity
- Computational logics
- Computational geometry
- Foundations of system specifications
- Learning theory
- Parallelism and concurrency
- Rewriting and high-level replacement systems
- Semantics

As a novelty, in addition to three days of invited and selected lectures there was a one–day minisymposium devoted to *Specification of time–critical systems* and organized by E.-R. Olderog.

The proceedings present five invited lectures, including the invited lectures of the minisymposium, and 32 short communications selected by the international program committee consisting of J. Balcazar (Barcelona), R.G. Bukharajev (Kazan), Z. Esik (Szeged), J. Gruska (Bratislava), T. Hagerup (Saarbrücken), H. Jürgensen (London, Canada), M. Main (Boulder), U. Montanari (Pisa), E.-R. Olderog (Oldenburg), T. Ottmann (Freiburg), G. Paun (Bucharest), H. Reichel (Dresden), A. Slissenko (Paris), J. Tiuryn (Warsaw), P. Vitanyi (Amsterdam), I. Wegener (Dortmund), P. Widmayer (Zürich).

We would like to extend our sincer thanks to all program committee members and all referees listed below for their care in reviewing and selecting the submitted papers: D. Beauquier, M. Bellia, M. Bertol, L. Bernátsky, C. Bertram-Kretzberg, B. Bollig, J. Bröcker, V. Bruyére, A. Bucciarelli, L. Budach, H. Buhrman, A. Carbone, Ch. Choffrut, H. Comon, A. Corradini, L. Csirmaz, W.F. De-La-Vega, M. Dietzfelbinger, K. Diks, J. Desel, P. Duris, J. Eckerle, J. Esparza, M. Fernandez, G. Ferrari, P. Fischer, G. Ghelli, J. Gabarro, R. Gavaldá, D. Gomm, A. Heinz, Ch. Hipke, Y. Hirshfeld, F. Huber-Waeschle, D. Kesner, M. Kochol, M. Krause, B. Kroell, A. Labella, P. Laeunchli, H. Lefmann, H.-P. Lenhof, P. Lescanne, G. Longo, J. Magun, B. Mahr, V. Manca, E. Manes, Y. Manoussakis, A. Marzetta, G. Mauri, E. Mayordomo, F. Moller, P.D. Mosses, A. Muscholl, J. Ohlbach, R. Pajarola, D. Pardubska, V. Priebe, R. Raman, Th. Roos, P. Rosolini, M. Sauerhoff, St. Schrodl, H.U. Simon, M. Smid, R. Szelepcsenyi, S. Vágvólgyi, B. Velickovic, H. Vogler, H.-J. Voss, P.-A. Wacrenier,

We gratefully acknowledge the financial support provided by the following institutions:

- Deutsche Forschungsgemeinschaft
- Sächsisches Staatsministerium für Wissenschaft und Kunst
- Gesellschaft von Freunden und Förderern der TU Dresden e.V.

The conference was organized by the Department of Theoretical Computer Science of the Technical University of Dresden. I would like to thank my colleagues H. Findeisen, R. Hebenstreit and M. Posegga for their support. Without their help the conference would have not been possible.

Finally , I would like to thank all authors and Springer Verlag, in particular Alfred Hofmann, for their excellent cooperation in the publication of the proceedings.

Dresden, June 1995 Horst Reichel

Table of Contents

Invited Lectures

Communications

Discrete Time Process Algebra with Abstraction

J.C.M. Baeten
Department of Computing Science, Eindhoven University of Technology, P.O.Box 513, 5600 MB Eindhoven, The Netherlands

J.A. Bergstra
Programming Research Group, University of Amsterdam, Kruislaan 403, 1098 SJ Amsterdam, The Netherlands
and
Department of Philosophy, Utrecht University, Heidelberglaan 8, 3584 CS Utrecht, The Netherlands

The axiom system ACP of [BEK84a] was extended to discrete time in [BAB95]. Here we proceed to define the silent step in this theory in branching bisimulation semantics [GLW91, BAW90] rather than weak bisimulation semantics [MIL89, BEK85]. We present versions based on relative timing and on absolute timing. Both approaches are integrated using parametric timing. The time free ACP theory is embedded in the discrete time theory.
Note: Partial support received from ESPRIT Basic Research Action 7166, CONCUR2.

1. INTRODUCTION.

Process algebra in the form of ACP [BEK85, BAW90], CCS [MIL89] or CSP [BRHR84] involves no explicit mention of time. Time is present in the interpretation of sequential composition: in p·q (ACP notation) the process p should be executed before q. Process algebras can be introduced that support standardised features to incorporate a quantative view on time. Time may be represented by means of non-negative reals, and actions can be given time stamps. This line is followed in [BAB91] for ACP, in [MOT90] for CCS and in [RER88] for CSP.

A second option is to divide time in slices indexed by natural numbers, to have an implicit or explicit time stamping mechanism that determines for each action the time slice in which it occurs and to have a time order within each slice only. This line has been followed in ATP [NSVR90], [NIS94], a process algebra that adds time slicing to a version of ACP based on action prefixing rather than sequential composition. Further, [GRO90] has extended ACP with time slices whereas [MOT89] have added these features to CCS. Following [BAB95], we use the phrase discrete time process algebra if an enumeration of time slices is used.

The objective of this paper is to extend the discrete time process algebra of ACP as given in [BAB95] with the silent step τ. This will allow a notion of abstraction. We base our work on the branching bisimulation of [GLW91]. We mention that [KLU93] has extended the real time ACP of [BAB91] with silent steps. We present three views on discrete time process algebra with silent step. In section 2, we consider discrete time process algebra with abstraction in *relative timing*, where timing refers to the execution of the previous action. In section 3, we have discrete time process algebra with abstraction in *absolute timing*, where all timing refers to an absolute clock. Here again, we only consider the two-phase version. In section 4, we have discrete time process algebra with *parametric timing*, where absolute and relative timing are integrated. For parametric timing, we introduce a model based on time spectrum sequences.

An underlying viewpoint of the present paper is that for a given time free atomic action, there may be different timed versions. We mention some: fts(a) stands for action a in the first time slice, followed by immediate termination, cts(a) is a in the current time slice with immediate termination, ats(a) is a in any time slice with immediate termination, atstau(a) = ats(a)·ats(τ) is a in any time slice followed by silent termination in any subsequent slice. It turns out that for a an atomic action or τ, the interpretation of a as atstau(a) is the appropriate (homomorphic) embedding of time free process algebra into timed process algebra.

There are many practical uses conceivable for timed process algebras. In particular, we mention the TOOLBUS (see [BEK94, 95]). This TOOLBUS contains a program notation called T which is syntactically sugared discrete time process algebra. Programs in T are called T-scripts. The runtime system is also described in terms of discrete time process algebra. By using randomised symbolic execution the TOOLBUS implementation enacts that the axioms of process algebra can be viewed as correctness preserving transformations of T-scripts. A comparable part of disrete time process algebra that is used to describe T-scripts has also been used for the description of φSDL, flat SDL, a subset of SDL that leaves out modularisation and concentrates on timing aspects (see [BEM95]).

2. DISCRETE TIME PROCESS ALGEBRA WITH RELATIVE TIMING.

We start out from the relative discrete time process algebra of [BAB95]. First, we consider the theory with only nondelayable actions, next we add the delayable or time free actions.

2.1 BASIC PROCESS ALGEBRA WITHOUT TIME FREE ACTIONS.

The signature of $BPA_{\bar{drt}}$ has constants cts(a) (for a ∈ A), denoting a in the current time slice, and cts(δ), denoting a deadlock at the end of the current time slice. The superscript $^{-}$ denotes that time free atoms are not part of the signature. Also, we have the immediate deadlock constant δ introduced in [BAB95]. This constant denotes an immediate and catastrophic deadlock. Within a time slice, there is no explicit mention of the passage of time, we can see the passage to the next time slice as a clock tick. Thus, the cts(a) can be called nondelayable actions: the action must occur before the next clock tick. The operators are alternative and sequential composition, and the relative discrete time unit delay σ_{rel} (the notation σ taken from [HER90]). The process $\sigma_{rel}(x)$ will start x after one clock tick, i.e. in the next time slice. In addition, we add the auxiliary operator ν_{rel}. This operator disallows an initial time step, it gives the part of a process that starts with an action in the current time slice. It was called a time out at the end of the current time slice in [BAB92], there, the notation $x \gg_{dt} 1$ was used for $\nu_{rel}(x)$. (The greek letter ν sounds like "now"; this correspondence is even stronger in Dutch.)

The axiom DRT1 is the time factorization axiom: it says that the passage of time by itself cannot determine a choice. The addition of a silent step in strong bisimulation semantics now just amounts to the presence of a new constant $cts(\tau)$, with the same axioms as the $cts(a)$ constants. We write $A_\tau = A \cup \{\tau\}, A_\delta = A \cup \{\delta\}$ etc.

The standard process algebra BPA_δ can be considered as an SRM specification (Subalgebra of Reduced Model, in the terminology of [BAB94]) of the present theory: consider the initial algebra of $BPA_{drt}^{-} + DCS$, reduce the signature by omitting $\dot{\delta}$, σ_{rel}, ν_{rel}, then BPA_δ is a complete axiomatisation of the reduced model, under the interpretation of a, δ by $cts(a)$, $cts(\delta)$ (note that $x + cts(\delta) = x$ for all closed terms x except $\dot{\delta}$).

$x + y = y + x$	A1	$\sigma_{rel}(x) + \sigma_{rel}(y) = \sigma_{rel}(x + y)$	DRT1
$(x + y) + z = x + (y + z)$	A2	$\sigma_{rel}(x) \cdot y = \sigma_{rel}(x \cdot y)$	DRT2
$x + x = x$	A3	$\sigma_{rel}(\dot{\delta}) = cts(\delta)$	DRT3
$(x + y) \cdot z = x \cdot z + y \cdot z$	A4	$cts(a) + cts(\delta) = cts(a)$	DRT4
$(x \cdot y) \cdot z = x \cdot (y \cdot z)$	A5	$\nu_{rel}(\dot{\delta}) = \dot{\delta}$	DCS1
		$\nu_{rel}(cts(a)) = cts(a)$	DCS2
$x + \dot{\delta} = x$	A6ID	$\nu_{rel}(\sigma_{rel}(x)) = cts(\delta)$	DCS3
$\dot{\delta} \cdot x = \dot{\delta}$	A7ID	$\nu_{rel}(x + y) = \nu_{rel}(x) + \nu_{rel}(y)$	DCS4
		$\nu_{rel}(x \cdot y) = \nu_{rel}(x) \cdot y$	DCS5

TABLE 1. $BPA_{drt}^{-} + DCS$.

2.2 STRUCTURED OPERATIONAL SEMANTICS.

We give a semantics in terms of operational rules. We have the following relations on the set of closed process expressions P:

- action step $\subseteq P \times A_\tau \times P$, notation $p \xrightarrow{a} p'$ (denotes action execution)
- action termination $\subseteq S \times A_\tau$, notation $p \xrightarrow{a} \surd$ (execution of a terminating action).
- time step $\subseteq P \times P$, notation $p \xrightarrow{\sigma} p'$ (denotes passage to the next time slice)
- immediate deadlock $\subseteq P$, notation $ID(p)$ (immediate deadlock, holds only for process expressions equal to $\dot{\delta}$).

We enforce the time factorization axiom DRT1 by phrasing the rules so that each process expression has at most one σ-step: in a transition system, each node has at most one outgoing σ-edge. This operational semantics uses predicates and negative premises. Still, using terminology and results of [VER94], the rules satisfy the *panth* format, and determine a unique transition relation on closed process expressions. Strong bisimulation is defined as usual, so a binary relation R on P is a *strong bisimulation* iff the following conditions hold:

i. if $R(p,q)$ and $p \xrightarrow{u} p'$ $(u \in A_{\tau\sigma})$, then there is q' such that $q \xrightarrow{u} q'$ and $R(p',q')$
ii. if $R(p,q)$ and $q \xrightarrow{u} q'$ $(u \in A_{\tau\sigma})$, then there is p' such that $p \xrightarrow{u} p'$ and $R(p',q')$
iii. if $R(p,q)$ then $p \xrightarrow{a} \surd$ iff $q \xrightarrow{a} \surd$ $(a \in A_\tau)$ and $ID(p)$ iff $ID(q)$.

Two terms p,q are *strong bisimulation equivalent*, $p \leftrightarrow q$, if there exists a strong bisimulation relating them. From [BAB95] we know that the axiomatisation in table 1 is sound and complete for the model of closed process expressions modulo strong bisimulation.

$$\mathrm{cts}(a) \xrightarrow{a} \surd \qquad \mathrm{ID}(\dot{\delta}) \qquad \frac{\neg \mathrm{ID}(x)}{\sigma_{rel}(x) \xrightarrow{\sigma} x}$$

$$\frac{x \xrightarrow{a} x'}{x \cdot y \xrightarrow{a} x' \cdot y} \qquad \frac{x \xrightarrow{a} \surd}{x \cdot y \xrightarrow{a} y} \qquad \frac{x \xrightarrow{\sigma} x'}{x \cdot y \xrightarrow{\sigma} x' \cdot y} \qquad \frac{\mathrm{ID}(x)}{\mathrm{ID}(x \cdot y)}$$

$$\frac{x \xrightarrow{a} x'}{x+y \xrightarrow{a} x', y+x \xrightarrow{a} x'} \qquad \frac{x \xrightarrow{a} \surd}{x+y \xrightarrow{a} \surd, y+x \xrightarrow{a} \surd} \qquad \frac{\mathrm{ID}(x), \mathrm{ID}(y)}{\mathrm{ID}(x+y)}$$

$$\frac{x \xrightarrow{\sigma} x', y \xrightarrow{\sigma} y'}{x+y \xrightarrow{\sigma} x'+y'} \qquad \frac{x \xrightarrow{\sigma} x', y \not\xrightarrow{\sigma}}{x+y \xrightarrow{\sigma} x', y+x \xrightarrow{\sigma} x'}$$

$$\frac{x \xrightarrow{a} x'}{\nu_{rel}(x) \xrightarrow{a} x'} \qquad \frac{x \xrightarrow{a} \surd}{\nu_{rel}(x) \xrightarrow{a} \surd} \qquad \frac{\mathrm{ID}(x)}{\mathrm{ID}(\nu_{rel}(x))}$$

TABLE 2. Operational rules for BPA_{drt}^{-} + DCS ($a \in A_\tau$).

2.3 BRANCHING BISIMULATION.

Now, we want to define a notion of bisimulation that takes into account the special status of the silent step. We start from the definition of branching bisimulation in the time free case, and adapt this to our timed setting. An immediate deadlock term can only be related to another immediate deadlock term. As a consequence, we can have no term that is related to $\surd$. This is different from the usual definition of branching bisimulation of [GLW91, BAW90]. In order to emphasise this fact, we call the relations to be defined branching tail bisimulations. A useful notation is to write $p \Rightarrow q$ if it is possible to reach q from p by executing a number of τ-steps (0 or more).

A binary relation R on P is a *branching tail bisimulation* if:

i. if $R(p,q)$ and $p \xrightarrow{u} p'$ ($u \in A_{\tau\sigma}$) then either:
   a. $u \equiv \tau$ and $R(p',q)$, or
   b. there are $q^*, q' \in P$ such that $q \Rightarrow q^* \xrightarrow{u} q'$ and $R(p,q^*)$, $R(p',q')$.

ii. if $R(p,q)$ and $q \xrightarrow{u} q'$ ($u \in A_{\tau\sigma}$), then either:
   a. $u \equiv \tau$ and $R(p,q')$, or
   b. there are $p^*, p' \in P$ such that $p \Rightarrow p^* \xrightarrow{u} p'$ and $R(p^*,q)$, $R(p',q')$.

iii. if $R(p,q)$ and $p \xrightarrow{a} \surd$ ($a \in A_\tau$) then there is $q' \in P$ such that $q \Rightarrow q' \xrightarrow{a} \surd$ and $R(p,q')$.

iv. if $R(p,q)$ and $q \xrightarrow{a} \surd$ ($a \in A_\tau$) then there is $p' \in P$ such that $p \Rightarrow p' \xrightarrow{a} \surd$ and $R(p',q)$.

v. if $R(p,q)$ then $\mathrm{ID}(p)$ iff $\mathrm{ID}(q)$.

Two terms p,q are *branching tail bisimulation equivalent*, $p \leftrightarrow_{bt} q$, if there is a branching tail bisimulation relating p and q.

If R is a branching tail bisimulation, then we say that the related pair p,q satisfies the *root condition* if:

vi. if $p \xrightarrow{u} p'$ ($u \in A_{\tau\sigma}$), then there is q' such that $q \xrightarrow{u} q'$ and $R(p',q')$

vii. if $q \xrightarrow{u} q'$ ($u \in A_{\tau\sigma}$), then there is p' such that $p \xrightarrow{u} p'$ and $R(p',q')$
viii. $p \xrightarrow{a} \surd$ ($a \in A_\tau$) iff $q \xrightarrow{a} \surd$.

Two terms p,q are *rooted branching tail bisimulation equivalent*, $g \underline{\leftrightarrow}_{rbt}\, h$, if there is a branching bisimulation R relating p and q, such that the pair p,q satisfies the root condition, and all related pairs that can be reached from p and q by only performing σ-steps also satisfy the root condition.
Thus the root condition amounts to the condition that the bisimulation is strong as long as no action is executed.

We can prove that rooted branching tail bisimulation is a congruence relation, and thus we obtain the algebra $P/\underline{\leftrightarrow}_{rbt}$ of closed process expressions modulo rooted branching tail bisimulation. We can establish that the algebra $P/\underline{\leftrightarrow}_{rbt}$ satisfies all laws of BPA_{drt}^{-}.

Note that if we take the reduced model obtained by omitting the immediate deadlock constant, then we can define a notion of branching bisimulation without the tail condition, where a term can be related to $\surd$. Using the embedding of discrete time process algebra into real time process algebra given in [BAB92, 95] we find that our notion of silent step in time is in line with the notion of branching bisimulation in time of [KLU93] (albeit that he has no tail condition).

2.4 AXIOMATISATION FOR SILENT STEP.

Now we want to formulate algebraic laws for the silent step, that hold in the algebra $P/\underline{\leftrightarrow}_{rbt}$. We cannot just transpose the well-known laws, because of the special status of σ-steps. An example: $cts(a)\cdot(cts(\tau)\cdot(\sigma_{rel}(cts(b)) + \sigma_{rel}(cts(c))) + \sigma_{rel}(cts(b)))$ is not rooted branching tail bisimilation equivalent to $cts(a)\cdot(\sigma_{rel}(cts(b)) + \sigma_{rel}(cts(c)))$, as in the first term, the choice for b can be made by the execution of the σ-step, and in the second term, the choice must be made after the σ-step (it is equal to $cts(a)\cdot\sigma_{rel}(cts(b) + cts(c))$). In order to ensure that we do not split terms that both have an initial σ-step, we use the time out operator ν_{rel}.

In table 3, DRTB1 and DRTB2 are variants of the branching law B2 ($x\cdot(\tau\cdot(y + z) + y) = x\cdot(y + z)$). In DRTB1, we have the case where y does not have an initial time step, in DRTB2 this is the case for z. We add $cts(\delta)$ to ensure that the expression following $cts(\tau)$ does not equal $\dot{\delta}$. A simple instance is the identity $x\cdot cts(\tau)\cdot cts(\tau) = x\cdot cts(\tau)$. However, we do not have the law $x\cdot cts(\tau) = x$ (the counterpart of the branching law B1), as $cts(a)\cdot cts(\tau)\cdot\dot{\delta}$ must be distinguished from $cts(a)\cdot\dot{\delta}$ (this can be appreciated in a setting with parallel composition: the first term allows execution of actions in the current time slice from a parallel component after the execution of a, the second term does not).

The law DRTB3 allows to omit a τ-step following one or more time steps. A simple instance is $cts(a)\cdot\sigma_{rel}(cts(\tau)\cdot cts(b)) = cts(a)\cdot\sigma_{rel}(cts(b))$.

$x\cdot(cts(\tau)\cdot(\nu_{rel}(y) + z + cts(\delta)) + \nu_{rel}(y)) = x\cdot(\nu_{rel}(y) + z + cts(\delta))$	DRTB1
$x\cdot(cts(\tau)\cdot(y + \nu_{rel}(z) + cts(\delta)) + y) = x\cdot(y + \nu_{rel}(z) + cts(\delta))$	DRTB2
$cts(a)\cdot x = cts(a)\cdot y \Rightarrow cts(a)\cdot(\sigma_{rel}(x) + \nu_{rel}(z)) = cts(a)\cdot(\sigma_{rel}(y) + \nu_{rel}(z))$	DRTB3

TABLE 3. $BPA_{drt}^{-}\tau = BPA_{drt}^{-} + DCS + DRTB1\text{-}3$ ($a \in A_\tau$).

2.5 PROPOSITION: The model $P/\underline{\leftrightarrow}_{rbt}$ satisfies the axioms DRTB1-3. We leave it as an open problem, whether the axiomatisation of $BPA_{\bar{drt}}\tau$ is complete for the model $P/\underline{\leftrightarrow}_{rbt}$.

Using the interpretation of a,δ,τ by cts(a), cts(δ), cts(τ), we do not obtain the time free theory $BPA_\delta{}^\tau$ of [GLW91, BAW90] as an SRM specification of $BPA_{\bar{drt}}\tau$. If we reduce the initial algebra by omitting $\dot{\delta}$ and the σ_{rel} and ν_{rel} operators, then the first branching law $x \cdot \tau = x$ will not hold, but instead $x \cdot \tau \cdot y = x \cdot y$. We can nevertheless obtain $BPA_\delta{}^\tau$ as an SRM specifications, if we use a different interpretation of the constants: define $ctstau(a) = cts(a) \cdot cts(\tau)$ ($a \in A_\tau$), then $BPA_\delta{}^\tau$ becomes an SRM specification of $BPA_{\bar{drt}}\tau$ with the interpretation of a,τ,δ as ctstau(a), ctstau(τ), cts(δ).

2.6 PROPOSITION. The following laws are derivable from $BPA_{\bar{drt}}\tau$:
1. $ctstau(a) \cdot ctstau(\tau) = ctstau(a)$
2. $x \cdot (ctstau(\tau) \cdot (\nu_{rel}(y) + z + cts(\delta)) + \nu_{rel}(y)) = x \cdot (\nu_{rel}(y) + z + cts(\delta))$
3. $x \cdot (ctstau(\tau) \cdot (y + \nu_{rel}(z) + cts(\delta)) + y) = x \cdot (y + \nu_{rel}(z) + cts(\delta))$

PROOF: 1 is straightforward; for 2, we calculate:
$x \cdot (ctstau(\tau) \cdot (\nu_{rel}(y) + z + cts(\delta)) + \nu_{rel}(y)) = x \cdot (cts(\tau) \cdot (cts(\tau) \cdot (\nu_{rel}(y) + z + cts(\delta)) + cts(\delta)) + {}$
$+ \nu_{rel}(y)) = x \cdot (cts(\tau) \cdot (\nu_{rel}(y) + z + cts(\delta)) + \nu_{rel}(y)) = x \cdot (y + \nu_{rel}(z) + cts(\delta))$. 3 goes similarly.

As a consequence, if we take the SRM obtained by omitting the cts(a) constants for $a \in A_\tau$, then we can prove $x \cdot ctstau(\tau) = x$ for all closed terms. Thus, we have embedded the time free theory into the relative discrete time theory.

2.7 DELAYABLE ACTIONS.
However, this is not the embedding of the time free theory into the discrete time theory that we want: it reduces the whole world to one time slice. Rather, we want to interpret the time free actions as actions that can occur in any time slice. Following [BAB95], we extend $BPA_{\bar{drt}}$ to BPA_{drt} by introducing constants ats(a) (for $a \in A_{\tau\delta}$). ats(a) executes a in an arbitrary time slice, followed by immediate termination. We define these constants using the operator $\lfloor\_\rfloor^\omega$, the *unbounded start delay* operator of [NIS94]: $\lfloor x\rfloor^\omega$ can start the execution of x in the current time slice, or delay unchanged to the next time slice. The defining axiom for this operator takes the form of a recursive equation. In order to prove identities for the unbounded start delay operator, we need a restricted form of the Recursive Specification Principle RSP of [BEK86]. We call this RSP(USD). We give operational rules for the new operator in table 5. Following [BAB95] we can prove that BPA_{drt} is a complete axiomatisation of the model of closed process expressions modulo strong bisimulation equivalence. We can also prove that the model $P/\underline{\leftrightarrow}_{rbt}$ satisfies the axioms of BPA_{drt}.

$\lfloor x\rfloor^\omega = \nu_{rel}(x) + \sigma_{rel}(\lfloor x\rfloor^\omega)$	USD
$ats(a) = \lfloor cts(a)\rfloor^\omega$	ATS
$y = \nu_{rel}(x) + \sigma_{rel}(y) \Rightarrow y = \lfloor x\rfloor^\omega$	RSP(USD)

TABLE 4. $BPA_{drt} = BPA_{\bar{drt}}$ + USD, ATS ($a \in A_{\tau\delta}$).

$$\frac{x \xrightarrow{a} x'}{\lfloor x \rfloor^{\omega} \xrightarrow{a} x'} \qquad \frac{x \xrightarrow{a} \surd}{\lfloor x \rfloor^{\omega} \xrightarrow{a} \surd} \qquad \lfloor x \rfloor^{\omega} \xrightarrow{\sigma} \lfloor x \rfloor^{\omega}$$

TABLE 5. Operational rules for unbounded start delay ($a \in A_\tau$).

2.8 PROPOSITION. The following identities are derivable from BPA_{drt} + RSP(USD):

1. $\lfloor \dot{\delta} \rfloor^{\omega} = ats(\delta)$	5. $ats(a) = cts(a) + \sigma_{rel}(ats(a))$	(ADRT)
2. $\lfloor x + y \rfloor^{\omega} = \lfloor x \rfloor^{\omega} + \lfloor y \rfloor^{\omega}$	6. $ats(a) + ats(\delta) = ats(a)$	(A6A)
3. $\lfloor x \cdot y \rfloor^{\omega} = \lfloor x \rfloor^{\omega} \cdot y$	7. $ats(\delta) \cdot x = ats(\delta)$	(A7)
4. $\lfloor \sigma_{rel}(x) \rfloor^{\omega} = ats(\delta)$.		

2.9 AXIOMATISATION FOR SILENT STEP.

We have one additional axiom for the constant $ats(\tau)$, DTB4.

$$x \cdot (ats(\tau) \cdot \lfloor y + z \rfloor^{\omega} + \lfloor y \rfloor^{\omega}) = x \cdot \lfloor y + z \rfloor^{\omega} \qquad \text{DTB4}$$

TABLE 6. $BPA_{drt}\tau = BPA_{drt}$ + DTB4.

2.10 PROPOSITION: The model $P/\underline{\leftrightarrow}_{rbt}$ satisfies the axioms of $BPA_{drt}\tau$. Again, completeness is an open problem.

Again, we cannot obtain the time free theory $BPA_\delta{}^\tau$ as an SRM specification, interpreting a by $ats(a)$, since $ats(a) \cdot ats(\tau)$ is not branching tail bisimilar to $ats(a)$: the first term can still perform time steps after executing the action. Note that the second branching law is valid, as the unbounded start delay is the identity on all time free processes. In order to achieve an SRM specification, nonetheless, we will define a different interpretation of time free atoms in our timed setting. We define constants $atstau(a)$ as follows ($a \in A_\tau$): $\quad atstau(a) = ats(a) \cdot ats(\tau)$.

Interpreting a, τ, δ by $atstau(a)$, $atstau(\tau)$, $ats(\delta)$, we do obtain $BPA_\delta{}^\tau$ as an SRM specification.

2.11 PROPOSITION. The following laws are derivable from $BPA_{drt}\tau$:

1. $atstau(a) \cdot atstau(\tau) = atstau(a)$
2. $x \cdot (atstau(\tau) \cdot \lfloor y + z \rfloor^{\omega} + \lfloor y \rfloor^{\omega}) = x \cdot \lfloor y + z \rfloor^{\omega}$

PROOF: as before.

2.12 ADDITIONAL OPERATOR.

We explore the theory $BPA_{drt}\tau$ a bit further by adding an extra operator: $\bar{\nu}_{rel}$ is the counterpart of the operator ν_{rel}, it gives the part of the process that starts with an initial time step. $\bar{\nu}_{rel}$ was called initialisation in the following time slice in [BAB92], $\bar{\nu}_{rel}(x)$ was notated $1 \gg_{dt} x$. The axiom DRTSA, the Discrete Relative Time Slice Axiom, allows to split each term into two components. Both models $P/\underline{\leftrightarrow}$, $P/\underline{\leftrightarrow}_{rbt}$ satisfy the axioms DRTSA, DNS1-4, if we add the operational rules in table 8.

$x = \nu_{rel}(x) + \overline{\nu}_{rel}(x)$	DRTSA		
$\overline{\nu}_{rel}(\dot{\delta}) = \dot{\delta}$	DNS1	$\overline{\nu}_{rel}(x + y) = \overline{\nu}_{rel}(x) + \overline{\nu}_{rel}(y)$	DNS3
$\overline{\nu}_{rel}(cts(a)) = cts(\delta)$	DNS2	$\overline{\nu}_{rel}(x \cdot y) = \overline{\nu}_{rel}(x) \cdot y$	DNS4

TABLE 7. Axioms for the additional operator ($a \in A_\tau$).

$$\frac{x \xrightarrow{\sigma} x'}{\overline{\nu}_{rel}(x) \xrightarrow{\sigma} x'} \qquad \frac{ID(x)}{ID(\overline{\nu}_{rel}(x))}$$

TABLE 8. Operational rules for the additional operator ($a \in A_\tau$).

2.13 PARALLEL COMPOSITION.

The extension of the theory $BPA_{drt}\tau$ with parallel composition, with or without communication, can be done along the lines of [BAB95]. The additional syntax has binary operators $\parallel$ (merge), $\mathbin{|\!\!\lfloor}$ (left merge), $\mid$ (communication merge), unary operators ∂_H (encapsulation operator, for $H \subseteq A$), τ_I (abstraction operator, for $I \subseteq A$). We present axioms for $ACP_{drt}\tau$ in table 9, operational rules in table 10. We assume given a partial commutative associative communication function $\gamma: A \times A \to A$.

$cts(a) \mid cts(b) = cts(c)$	if $\gamma(a,b)=c$	DRTCF1	
$cts(a) \mid cts(b) = cts(\delta)$	otherwise	DRTCF2	
$x \parallel y = x \mathbin{|\!\!\lfloor} y + y \mathbin{|\!\!\lfloor} x + x \mid y$		CM1	
$\dot{\delta} \mathbin{|\!\!\lfloor} x = \dot{\delta}$		LMID1	
$x \mathbin{|\!\!\lfloor} \dot{\delta} = \dot{\delta}$		LMID2	
$(x + y) \mathbin{|\!\!\lfloor} z = x \mathbin{|\!\!\lfloor} z + y \mathbin{|\!\!\lfloor} z$		CM4	
$cts(a) \mathbin{|\!\!\lfloor} (x + cts(\delta)) = cts(a) \cdot (x + cts(\delta))$		DRTCM2	
$cts(a) \cdot x \mathbin{|\!\!\lfloor} (y + cts(\delta)) = cts(a) \cdot (x \parallel (y + cts(\delta)))$		DRTCM3	
$\sigma_{rel}(x) \mathbin{|\!\!\lfloor} (\sigma_{rel}(y) + \nu_{rel}(z)) = \sigma_{rel}(x \mathbin{|\!\!\lfloor} y)$		DRT5	

$\dot{\delta} \mid x = \dot{\delta}$	CMID1	$cts(a) \cdot x \mid cts(b) = cts(a \mid b) \cdot x$	DRTCM5
$x \mid \dot{\delta} = \dot{\delta}$	CMID2	$cts(a) \mid cts(b) \cdot x = cts(a \mid b) \cdot x$	DRTCM6
$(x + y) \mid z = x \mid z + y \mid z$	CM8	$cts(a) \cdot x \mid cts(b) \cdot y = cts(a \mid b) \cdot (x \parallel y)$	DRTCM7
$x \mid (y + z) = x \mid y + x \mid z$	CM9	$\nu_{rel}(x) \mid \sigma_{rel}(y) = cts(\delta)$	DRT6
$\sigma_{rel}(x) \mid \sigma_{rel}(y) = \sigma_{rel}(x \mid y)$	DRT8	$\sigma_{rel}(x) \mid \nu_{rel}(y) = cts(\delta)$	DRT7

$\partial_H(\dot{\delta}) = \dot{\delta}$	DID	$\partial_H(cts(a)) = cts(a)$	if $a \notin H$	DRTD1
$\partial_H(x + y) = \partial_H(x) + \partial_H(y)$	D3	$\partial_H(cts(a)) = cts(\delta)$	if $a \in H$	DRTD2
$\partial_H(x \cdot y) = \partial_H(x) \cdot \partial_H(y)$	D4	$\partial_H(\sigma_{rel}(x)) = \sigma_{rel}(\partial_H(x))$		DRT9
$\tau_I(\dot{\delta}) = \dot{\delta}$	TIID	$\tau_I(cts(a)) = cts(a)$	if $a \notin I$	DRTTI1
$\tau_I(x + y) = \tau_I(x) + \tau_I(y)$	TI3	$\tau_I(cts(a)) = cts(\delta)$	if $a \in I$	DRTTI2
$\tau_I(x \cdot y) = \tau_I(x) \cdot \tau_I(y)$	TI4	$\tau_I(\sigma_{rel}(x)) = \sigma_{rel}(\tau_I(x))$		DRT10

TABLE 9. Additional axioms for $ACP_{drt}\tau$ ($a \in A_{\delta\tau}$).

$$\frac{x \xrightarrow{a} x',\ \neg\, ID(y)}{x \| y \xrightarrow{a} x' \| y,\ y \| x \xrightarrow{a} y \| x',\ x \mathbin{\underline{\|}} y \xrightarrow{a} x' \| y} \qquad \frac{x \xrightarrow{a} \surd,\ \neg\, ID(y)}{x \| y \xrightarrow{a} y,\ y \| x \xrightarrow{a} y,\ x \mathbin{\underline{\|}} y \xrightarrow{a} y}$$

$$\frac{x \xrightarrow{\sigma} x',\ y \xrightarrow{\sigma} y'}{x \| y \xrightarrow{\sigma} x' \| y',\ x \mathbin{\underline{\|}} y \xrightarrow{\sigma} x' \mathbin{\underline{\|}} y',\ x \mid y \xrightarrow{\sigma} x' \mid y'}$$

$$\frac{ID(x)}{ID(x \| y),\ ID(y \| x),\ ID(x \mathbin{\underline{\|}} y),\ ID(y \mathbin{\underline{\|}} x),\ ID(x \mid y),\ ID(y \mid x)}$$

$$\frac{x \xrightarrow{a} x',\ y \xrightarrow{b} y',\ a \mid b = c}{x \| y \xrightarrow{c} x' \| y',\ x \mid y \xrightarrow{c} x' \| y'} \qquad \frac{x \xrightarrow{a} x',\ y \xrightarrow{b} \surd,\ a \mid b = c}{x \| y \xrightarrow{c} x',\ x \mid y \xrightarrow{c} x',\ y \| x \xrightarrow{c} x',\ y \mid x \xrightarrow{c} x'}$$

$$\frac{x \xrightarrow{a} \surd,\ y \xrightarrow{b} \surd,\ a \mid b = c}{x \| y \xrightarrow{c} \surd,\ x \mid y \xrightarrow{c} \surd}$$

$$\frac{x \xrightarrow{a} x',\ a \notin H}{\partial_H(x) \xrightarrow{a} \partial_H(x')} \qquad \frac{x \xrightarrow{\sigma} x'}{\partial_H(x) \xrightarrow{\sigma} \partial_H(x')} \qquad \frac{x \xrightarrow{a} \surd,\ a \notin H}{\partial_H(x) \xrightarrow{a} \surd} \qquad \frac{ID(x)}{ID(\partial_H(x))}$$

$$\frac{x \xrightarrow{a} x',\ a \notin I}{\tau_I(x) \xrightarrow{a} \tau_I(x')} \qquad \frac{x \xrightarrow{a} x',\ a \in I}{\tau_I(x) \xrightarrow{\tau} \tau_I(x')} \qquad \frac{x \xrightarrow{a} \surd,\ a \notin I}{\tau_I(x) \xrightarrow{a} \surd} \qquad \frac{x \xrightarrow{a} \surd,\ a \in I}{\tau_I(x) \xrightarrow{\tau} \surd}$$

$$\frac{x \xrightarrow{\sigma} x'}{\tau_I(x) \xrightarrow{\sigma} \tau_I(x')} \qquad \frac{ID(x)}{ID(\tau_I(x))}$$

TABLE 10. Additional operational rules for $ACP_{drt}\tau$.

2.14 PROPOSITION. We derive the following identities for time free atoms in $ACP_{drt}\tau + RSP(USD)$:

1. $ats(a) \mathbin{\underline{\|}} \lfloor x \rfloor^{\omega} = ats(a) \cdot \lfloor x \rfloor^{\omega}$ (CM2)
2. $ats(a) \cdot x \mathbin{\underline{\|}} \lfloor y \rfloor^{\omega} = ats(a) \cdot (x \| \lfloor y \rfloor^{\omega})$ (CM3)
3. $ats(a) \cdot x \mid ats(b) = ats(a \mid b) \cdot x$ (CM5)
4. $ats(a) \mid ats(b) \cdot x = ats(a \mid b) \cdot x$ (CM6)
5. $ats(a) \cdot x \mid ats(b) \cdot y = ats(a \mid b) \cdot (x \| y)$ (CM7)
6. $\partial_H(ats(a)) = ats(a)$ if $a \notin H$ (D1)
7. $\partial_H(ats(a)) = ats(\delta)$ if $a \in H$ (D2)
8. $\tau_I(ats(a)) = ats(a)$ if $a \notin I$ (TI1)
9. $\tau_I(ats(a)) = ats(\tau)$ if $a \in I$ (TI2)

The following identity, useful in verifications, can be proved for all closed terms x,y:

10. $cts(a) \cdot (cts(\tau) \cdot (x + cts(\delta)) \| y) = cts(a) \cdot ((x + cts(\delta)) \| y)$.

3. DISCRETE TIME PROCESS ALGEBRA WITH ABSOLUTE TIMING.

We present a version of the theory in section 2 using absolute timing, where all timing is related to a global clock.

3.1 BASIC PROCESS ALGEBRA.

We start with constants fts(a), denoting a in the first time slice ($a \in A_{\tau\delta}$), followed by immediate termination. Besides, we have operators +,· as before. In addition, we have the absolute discrete time unit delay σ_{abs}. Axiom DAT4 uses the absolute value operator. This operator turns out to be the identity for all processes using absolute timing only: it initialises a process in the first time slice (the numbering of the axioms is taken from [BAB95]). Note that in the term $\sigma_{abs}(\mathrm{fts}(a))\cdot\mathrm{fts}(\delta)$, after execution of the a in slice 2, an immediate deadlock will occur. This term is different from $\sigma_{abs}(\mathrm{fts}(a)\cdot\mathrm{fts}(\delta))$, where after the execution of a in slice 2, further activity in this slice can take place (of a parallel process). We conclude that in the absolute time theory, the immediate deadlock constant $\dot{\delta}$ is necessary. The axiomatisation of BPA^{-}_{dat} + DFS adds the axioms of table 11 to the axioms A1-5, A6ID, A7ID (see the left hand side of table 1).

$\sigma_{abs}(\dot{\delta}) = \mathrm{fts}(\delta)$	DAT1	$\lvert\dot{\delta}\rvert = \dot{\delta}$	AV1
$\mathrm{fts}(a) + \mathrm{fts}(\delta) = \mathrm{fts}(a)$	DAT2	$\lvert x + y\rvert = \lvert x\rvert + \lvert y\rvert$	AV3
$\sigma_{abs}(x) + \sigma_{abs}(y) = \sigma_{abs}(x + y)$	DAT3	$\lvert x\cdot y\rvert = \lvert x\rvert\cdot y$	AV4
$\sigma_{abs}(x)\cdot\dot{\delta} = \sigma_{abs}(x\cdot\dot{\delta})$	DAT4	$\lvert\mathrm{fts}(a)\rvert = \mathrm{fts}(a)$	AV8
$\sigma_{abs}(x)\cdot(\mathrm{fts}(a) + y) = \sigma_{abs}(x)\cdot y$	DAT5	$\lvert\sigma_{abs}(x)\rvert = \sigma_{abs}(\lvert x\rvert)$	AV9
$\sigma_{abs}(x)\cdot(\mathrm{fts}(a)\cdot y + z) = \sigma_{abs}(x)\cdot z$	DAT6	$\lvert\nu_{abs}(x)\rvert = \nu_{abs}(\lvert x\rvert)$	AV10
$\sigma_{abs}(x)\cdot\sigma_{abs}(y) = \sigma_{abs}(x\cdot\lvert y\rvert)$	DAT7		
$\nu_{abs}(\dot{\delta}) = \dot{\delta}$	DFS1	$\nu_{abs}(x\cdot y) = \nu_{abs}(x)\cdot y$	DFS4
$\nu_{abs}(\mathrm{fts}(a)) = \mathrm{fts}(a)$	DFS2	$\nu_{abs}(x + y) = \nu_{abs}(x) + \nu_{abs}(y)$	DFS5
$\nu_{abs}(\sigma_{abs}(x)) = \mathrm{fts}(\delta)$	DFS3		

TABLE 11. BPA^{-}_{dat} + DFS ($a \in A_{\delta\tau}$).

3.2 STRUCTURED OPERATIONAL SEMANTICS.

The operational rules are more complicated in this case, as we have to keep track of which time slice we are in, we have to keep track of the global clock. $\langle x,n\rangle$ denotes x in the (n+1)st time slice. We have:

- if $\langle x,n\rangle \xrightarrow{\sigma} \langle x',n'\rangle$, then $x \equiv x'$, $n' = n+1$
- if $\langle x,n\rangle \xrightarrow{a} \langle x',n'\rangle$ or $\langle x,n\rangle \xrightarrow{a} \langle \surd,n'\rangle$, then $n' = n$.

The operational rules for the absolute value operator are trivial.

$$\langle \mathrm{fts}(a), 0\rangle \xrightarrow{a} \langle \surd, 0\rangle \qquad \mathrm{ID}(\langle \dot{\delta}, n\rangle) \qquad \mathrm{ID}(\langle \mathrm{fts}(a), n+1\rangle)$$

$$\frac{\neg \mathrm{ID}(\langle x,0\rangle)}{\langle \sigma_{abs}(x),0\rangle \xrightarrow{\sigma} \langle \sigma_{abs}(x),1\rangle} \qquad \frac{\mathrm{ID}(\langle x,n\rangle)}{\mathrm{ID}(\langle \sigma_{abs}(x),n+1\rangle)} \qquad \frac{\langle x,n\rangle \xrightarrow{\sigma} \langle x,n+1\rangle}{\langle \sigma_{abs}(x),n+1\rangle \xrightarrow{\sigma} \langle \sigma_{abs}(x),n+2\rangle}$$

$$\frac{\langle x,n\rangle \xrightarrow{a} \langle x',n\rangle}{\langle \sigma_{abs}(x),n+1\rangle \xrightarrow{a} \langle \sigma_{abs}(x'),n+1\rangle} \qquad \frac{\langle x,n\rangle \xrightarrow{a} \langle \surd,n\rangle}{\langle \sigma_{abs}(x),n+1\rangle \xrightarrow{a} \langle \surd,n+1\rangle}$$

$$\frac{\langle x,n\rangle \xrightarrow{a} \langle x',n\rangle}{\langle x+y,n\rangle \xrightarrow{a} \langle x',n\rangle, \langle y+x,n\rangle \xrightarrow{a} \langle x',n\rangle} \qquad \frac{\langle x,n\rangle \xrightarrow{a} \langle \surd,n\rangle}{\langle x+y,n\rangle \xrightarrow{a} \langle \surd,n\rangle, \langle y+x,n\rangle \xrightarrow{a} \langle \surd,n\rangle}$$

$$\frac{\langle x,n\rangle \xrightarrow{\sigma} \langle x,n+1\rangle}{\langle x+y,n\rangle \xrightarrow{\sigma} \langle x+y,n+1\rangle, \langle y+x,n\rangle \xrightarrow{\sigma} \langle y+x,n+1\rangle} \qquad \frac{\mathrm{ID}(\langle x,n\rangle), \mathrm{ID}(\langle y,n\rangle)}{\mathrm{ID}(\langle x+y,n\rangle)}$$

$$\frac{\langle x,n\rangle \xrightarrow{a} \langle x',n\rangle}{\langle x\cdot y,n\rangle \xrightarrow{a} \langle x'\cdot y,n\rangle} \qquad \frac{\langle x,n\rangle \xrightarrow{\sigma} \langle x,n+1\rangle}{\langle x\cdot y,n\rangle \xrightarrow{\sigma} \langle x\cdot y,n+1\rangle} \qquad \frac{\langle x,n\rangle \xrightarrow{a} \langle \surd,n\rangle}{\langle x\cdot y,n\rangle \xrightarrow{a} \langle y,n\rangle} \qquad \frac{\mathrm{ID}(\langle x,n\rangle)}{\mathrm{ID}(\langle x\cdot y,n\rangle)}$$

$$\frac{\langle x,0\rangle \xrightarrow{a} \langle x',0\rangle}{\langle \nu_{abs}(x),0\rangle \xrightarrow{a} \langle x',0\rangle} \qquad \frac{\langle x,0\rangle \xrightarrow{a} \langle \surd,0\rangle}{\langle \nu_{abs}(x),0\rangle \xrightarrow{a} \langle \surd,0\rangle} \qquad \frac{\mathrm{ID}(\langle x,0\rangle)}{\mathrm{ID}(\langle \nu_{abs}(x),0\rangle)} \qquad \mathrm{ID}(\langle \nu_{abs}(x),n+1\rangle)$$

TABLE 12. Operational semantics of $\mathrm{BPA}_{\bar{\mathrm{dat}}}$ + DFS ($a \in A_\tau$).

3.3 BISIMULATION.

We also have to adapt the definition of bisimulation. A *strong bisimulation* is a binary relation R on $P \times \mathbb{N}$ such that ($u \in A_{\tau\sigma}$):

i. whenever $R(\langle s, n\rangle, \langle s', n'\rangle)$ then $n = n'$ and $\mathrm{ID}(\langle s, n\rangle)$ iff $\mathrm{ID}(\langle s', n'\rangle)$.

ii. whenever $R(\langle s, n\rangle, \langle t, n\rangle)$ and $\langle s,n\rangle \xrightarrow{u} \langle s',n'\rangle$, then there is a term t' such that $\langle t,n\rangle \xrightarrow{u} \langle t',n'\rangle$ and $R(\langle s', n'\rangle, \langle t', n'\rangle)$

iii. whenever $R(\langle s, n\rangle, \langle t, n\rangle)$ and $\langle t,n\rangle \xrightarrow{u} \langle t',n'\rangle$, then there is a term s' such that $\langle s,n\rangle \xrightarrow{u} \langle s',n'\rangle$ and $R(\langle s', n'\rangle, \langle t', n'\rangle)$

iv. whenever $R(\langle s, n\rangle, \langle t, n\rangle)$, then $\langle s,n\rangle \xrightarrow{a} \langle \surd,n\rangle$ iff $\langle t,n\rangle \xrightarrow{a} \langle \surd,n\rangle$ ($a \in A_\tau$).

We say process expressions x and y are *strong bisimulation equivalent*, denoted $x \leftrightarroweq y$, if there exists a strong bisimulation with $R(\langle x,0\rangle, \langle y,0\rangle)$. From [BAB95] we know that the axiomatisation in table 9 is complete for the model of closed process expressions modulo strong bisimulation.

A *branching tail bisimulation* is a binary relation R on $P \times \mathbb{N}$ such that:

i. whenever $R(\langle s, n\rangle, \langle s', n'\rangle)$ then $n = n'$ and $\mathrm{ID}(\langle s, n\rangle)$ iff $\mathrm{ID}(\langle s', n'\rangle)$.

ii. whenever $R(\langle s, n\rangle, \langle t, n\rangle)$ and $\langle s,n\rangle \xrightarrow{u} \langle s',n'\rangle$ ($u \in A_{\tau\sigma}$), then either:

a. $u \equiv \tau$ and $R(\langle s', n\rangle, \langle t, n\rangle)$, or

b. there are $t^*, t' \in P$ such that $\langle t,n\rangle \Rightarrow \langle t^*,n\rangle \xrightarrow{u} \langle t',n'\rangle$ and $R(\langle s,n\rangle, \langle t^*,n\rangle)$, $R(\langle s',n'\rangle, \langle t',n'\rangle)$

iii. whenever $R(\langle s,n\rangle, \langle t,n\rangle)$ and $\langle t,n\rangle \xrightarrow{u} \langle t',n'\rangle$ ($u \in A_{\tau\sigma}$), then either:

a. $u \equiv \tau$ and $R(\langle s,n\rangle, \langle t',n\rangle)$, or

b. there are $s^*, s' \in P$ such that $\langle s,n\rangle \Rightarrow \langle s^*,n\rangle \xrightarrow{u} \langle s',n'\rangle$ and $R(\langle s^*,n\rangle, \langle t,n\rangle)$, $R(\langle s',n'\rangle, \langle t',n'\rangle)$

iv. if $R(\langle s,n\rangle, \langle t,n\rangle)$ and $\langle s,n\rangle \xrightarrow{a} \langle \surd,n\rangle$ ($a \in A_\tau$) then there is $t' \in P$ such that $\langle t,n\rangle \Rightarrow \langle t',n\rangle \xrightarrow{a} \langle \surd,n\rangle$ and $R(\langle s,n\rangle, \langle t',n\rangle)$.

v. if $R(\langle s,n\rangle, \langle t,n\rangle)$ and $\langle t,n\rangle \xrightarrow{a} \langle \surd,n\rangle$ ($a \in A_\tau$) then there is $s' \in P$ such that $\langle s,n\rangle \Rightarrow \langle s',n\rangle \xrightarrow{a} \langle \surd,n\rangle$ and $R(\langle s',n\rangle, \langle t,n\rangle)$.

We say process expressions x and y are *branching tail bisimilation equivalent*, denoted $x \leftrightarrow_{bt} y$, if there exists a branching tail bisimulation with $R(\langle x,0\rangle, \langle y,0\rangle)$. Process expressions x and y are *rooted branching tail bisimilation equivalent*, denoted $x \leftrightarrow_{rbt} y$, if there exists a branching tail bisimulation with $R(\langle x,0\rangle,\langle y,0\rangle)$, that is strong for all pairs that can be reached from $\langle x,0\rangle,\langle y,0\rangle$ by just performing time steps.

3.4 AXIOMATISATION FOR SILENT STEP.

Axioms for the silent step are comparable to the ones for relative time. The model of closed process expressions modulo rooted branching tail bisimulation satisfies the following axioms. Again, we can find the time free theory $BPA_\delta{}^\tau$ as an SRM specification by defining the constants $ftstau(a) = fts(a)\cdot fts(\tau)$.

$fts(a)\cdot(fts(\tau)\cdot(v_{abs}(x) + y + fts(\delta)) + v_{abs}(x)) = fts(a)\cdot(v_{abs}(x) + y + fts(\delta))$	DATB1
$fts(a)\cdot(fts(\tau)\cdot(x + v_{abs}(y) + fts(\delta)) + x) = fts(a)\cdot(x + v_{abs}(y) + fts(\delta))$	DATB2
$fts(a)\cdot x = fts(a)\cdot y \Rightarrow fts(a)\cdot(\sigma_{abs}(x) + v_{abs}(z)) = fts(a)\cdot(\sigma_{abs}(y) + v_{abs}(z))$	DATB3

TABLE 13. Axioms for $BPA_{\overline{dat}}\tau$.

3.5 DELAYABLE ACTIONS.

The extension with delayable actions is not so smooth as in the relative time case. We can define a recursive equation for the unbounded start delay operator, but it uses the operator μ that we will only define in section 4. Instead, we define the unbounded start delay operator by structural induction. Also, time free atoms cannot simply be defined using the unbounded start delay operator. Instead, we will define them by a recursive equation. The theory BPA_{dat} adds the axioms in table 14 to $BPA_{\overline{dat}}$ + DFS. Operational rules are presented in table 15.

$\lfloor \dot{\delta} \rfloor^\omega = ats(\delta)$	USD1	$ats(a) = fts(a) + \sigma_{abs}(ats(a))$	ADAT
$\lfloor fts(a) \rfloor^\omega = fts(a) + ats(\delta)$	USD2	$\lvert ats(a) \rvert = ats(a)$	AV2
$\lfloor \sigma_{abs}(x) \rfloor^\omega = \sigma_{abs}(\lfloor x \rfloor^\omega)$	USD3	$v_{abs}(\lfloor x \rfloor^\omega) = v_{abs}(x)$	DFS6
$\lfloor x + y \rfloor^\omega = \lfloor x \rfloor^\omega + \lfloor y \rfloor^\omega$	USD4	$\lvert \lfloor x \rfloor^\omega \rvert = \lfloor x \rfloor^\omega$	AV11
$\lfloor x\cdot y \rfloor^\omega = \lfloor x \rfloor^\omega\cdot y$	USD5		

TABLE 14. Axioms for BPA_{dat}.

$$\langle ats(a), n\rangle \xrightarrow{a} \langle \surd, n\rangle \qquad \langle ats(a), n\rangle \xrightarrow{\sigma} \langle ats(a), n+1\rangle \qquad \langle ats(\delta), n\rangle \xrightarrow{\sigma} \langle ats(\delta), n+1\rangle$$

$$\langle \lfloor x\rfloor^{\omega}, n\rangle \xrightarrow{\sigma} \langle \lfloor x\rfloor^{\omega}, n+1\rangle \qquad \frac{\langle x,n\rangle \xrightarrow{a} \langle x',n\rangle}{\langle \lfloor x\rfloor^{\omega},n\rangle \xrightarrow{a} \langle x',n\rangle} \qquad \frac{\langle x,n\rangle \xrightarrow{a} \langle \surd,n\rangle}{\langle \lfloor x\rfloor^{\omega},n\rangle \xrightarrow{a} \langle \surd,n\rangle}$$

TABLE 15. Operational rules for time free theory ($a \in A_\tau$).

3.6 AXIOMATISATION FOR SILENT STEP.

The axiomatisation for silent step has the same rule, DTB4, as in the relative time case. Again, we can find $BPA_\delta{}^\tau$ as an SRM specification by using the interpretation of a as atstau(a).
The extension with parallel composition can be found along the same lines as for the relative time case (see [BAB95]).

4. PARAMETRIC TIME.

In this section we integrate the absolute time and the relative time approach. All axioms presented in the previous sections are still valid for all parametric time processes. We obtain a finite axiomatization, that allows an elimination theorem. As a consequence, we can expand expressions like cts(a) || fts(b), cts(a) || (fts(b) + ats(δ)).

In [BAB95], we introduced the operators ⊙, the (relative) time spectrum combinator, and μ, the spectrum tail operator. The absolute value operator introduced in 3.1 can also be called the spectrum head operator. P ⊙ Q is a process that when initialised in the first time slice behaves as |P|; when initialised in slice n+1 its behaviour is determined by Q as follows: initialise in slice n thereafter apply σ_{abs}. μ(X) computes a process such that X = |X| ⊙ μ(X). For a parametric discrete time process we have the time spectrum sequence |X|, |μ(X)|, $|\mu^2(X)|$, ... For each infinite sequence $(P_n)_{n\in N}$ one may imagine a process P with $|\mu^n(P)| = P_n$ though not all such P can be finitely expressed.

4.1 BASIC PROCESS ALGEBRA.

$\lvert cts(a)\rvert = fts(a)$	AV10	$\mu(\dot\delta) = \dot\delta$	ST1
$\lvert\sigma_{rel}(X)\rvert = \sigma_{abs}(\lvert\mu(X)\rvert)$	AV11	$\mu(cts(a)) = cts(a)$	ST2
$\lvert\nu_{rel}(X)\rvert = \nu_{abs}(\lvert X\rvert)$	AV12	$\mu(X+Y) = \mu(X) + \mu(Y)$	ST3
$\sigma_{abs}(X) = \sigma_{abs}(\lvert X\rvert)$	AV13	$\mu(X\cdot Y) = \mu(X)\cdot\mu(Y)$	ST4
$\nu_{abs}(X) = \nu_{abs}(\lvert X\rvert)$	AV14	$\mu(\sigma_{rel}(X) = \sigma_{rel}(\mu(X))$	ST5
		$\mu(\nu_{rel}(x)) = \nu_{rel}(\mu(x))$	ST6
$\lvert X \odot Y\rvert = \lvert X\rvert$	SC1	$\mu(fts(a)) = \dot\delta$	ST7
$\mu(X \odot Y) = Y$	SC2	$\mu(\sigma_{abs}(X)) = \lvert X\rvert$	ST8
$X = \lvert X\rvert \odot \mu(X)$	SC3	$\mu(\nu_{abs}(x)) = \dot\delta$	ST9
$\lfloor x\rfloor^{\omega} = \nu_{abs}(x) + \sigma_{abs}(\lfloor\mu(x)\rfloor^{\omega})$	USDA	$\mu(\lfloor x\rfloor^{\omega}) = \lfloor\mu(x)\rfloor^{\omega}$	ST10

TABLE 17. $BPA_{dpt} = BPA_{drt} + BPA_{dat}$ + AV10-14, ST1-10, SC1-3, USDA.

Note that the three axioms DAT5-7 can be replaced by $\sigma_{abs}(X)\cdot Y = \sigma_{abs}(X\cdot\mu(Y))$. We can define:

- x is an absolute time process iff $BPA_{dpt} \vdash x = |x|$
- x is a relative time process iff $BPA_{dpt} \vdash x = \mu(x)$.

4.2 BASIC FORM.

We claim that each BPA_{dpt} process expression can be written in the form

$$P_1 \odot P_2 \odot \ldots P_n \odot Q$$

(we omit brackets, using the convention that $\odot$ associates to the right), such that each P_i is a BPA_{dat}-term and Q is a BPA_{drt}-term.

The way we achieve this is by writing

$$X = |X| \odot |\mu(X)| \odot \ldots |\mu^n(X)| \odot \mu^{n+1}(X).$$

Now one can reduce each $|\mu^n(X)|$ to a BPA_{dat}-term and if n is sufficiently large, we can write $\mu^{n+1}(X))$ without any σ_{abs} or $fts(a)$ using ST1-7, so will be in the relative time signature.

We call $(|X|, |\mu(X)|, |\mu^2(X)|, \ldots)$ the time spectrum expansion sequence (TSS) of X.

Note that we obtain the following spectrum expansion for the v_{rel} operator:

$$v_{rel}(x) = v_{abs}(|x|) \odot v_{rel}(\mu(x)).$$

For further details, we refer to [BAB95].

REFERENCES.

[BAB91] J.C.M. BAETEN & J.A. BERGSTRA, *Real time process algebra,* Formal Aspects of Computing 3 (2), 1991, pp. 142-188.

[BAB92] J.C.M. BAETEN & J.A. BERGSTRA, *Discrete time process algebra (extended abstract),* in: Proc. CONCUR'92, Stony Brook (W.R. Cleaveland, ed.), LNCS 630, Springer 1992, pp. 401-420.

[BAB93] J.C.M. BAETEN & J.A. BERGSTRA, *Real space process algebra,* Formal Aspects of Computing 5 (6), 1993, pp. 481-529.

[BAB94] J.C.M. BAETEN & J.A. BERGSTRA, *On sequential composition, action prefixes and process prefix,* Formal Aspects of Computing 6 (3), 1994, pp. 250-268.

[BAB95] J.C.M. BAETEN & J.A. BERGSTRA, *Discrete time process algebra,* report P9208c, Programming Research Group, University of Amsterdam 1995. To appear in Formal Aspects of Computing.

[BAW90] J.C.M. BAETEN & W.P. WEIJLAND, *Process algebra,* Cambridge Tracts in Theor. Comp. Sci. 18, Cambridge University Press 1990.

[BEK85] J.A. BERGSTRA & J.W. KLOP, *Algebra of communicating processes with abstraction,* TCS 37 (1), 1985, pp. 77-121.

[BEK86] J.A. BERGSTRA & J.W. KLOP, *Verification of an alternating bit protocol by means of process algebra,* in: Math. Methods of Spec. and Synthesis of Software Systems '85 (W. Bibel & K.P. Jantke, eds.), Springer LNCS 215, 1986, pp. 9-23.

[BEK94] J.A. BERGSTRA & P. KLINT, *The toolbus - a component interconnection architecture -*, report P9408, Programming Research Group, University of Amsterdam 1994.

[BEK95] J.A. BERGSTRA & P. KLINT, *The discrete time toolbus*, report P9502, Programming Research Group, University of Amsterdam 1995.

[BEM95] J.A. BERGSTRA & C.A. MIDDELBURG, *A process algebra semantics for φSDL*, report LGPS 129, Dept. of Philosophy, Utrecht University 1995.

[BRHR84] S.D. BROOKES, C.A.R. HOARE & A.W. ROSCOE, *A theory of communicating sequential processes*, Journal of the ACM 31 (3), 1984, pp. 560-599.

[GLW91] R.J. VAN GLABBEEK & W.P. WEIJLAND, *Branching time and abstraction in bisimulation semantics*, CWI report CS-R9120, 1991.

[GRO90a] J.F. GROOTE, *Specification and verification of real time systems in ACP*, in: Proc. 10th Symp. on Protocol Specification, Testing and Verification, Ottawa (L. Logrippo, R.L. Probert & H. Ural, eds.), North-Holland, Amsterdam 1990, pp. 261-274.

[HER90] M. HENNESSY & T. REGAN, *A temporal process algebra*, report 2/90, University of Sussex 1990.

[KLU93] A.S. KLUSENER, *Models and axioms for a fragment of real time process algebra*, Ph.D. Thesis, Eindhoven University of Technology 1993.

[MIL89] R. MILNER, *Communication and concurrency*, Prentice-Hall 1989.

[MOT89] F. MOLLER & C. TOFTS, *A temporal calculus of communicating systems*, report LFCS-89-104, University of Edinburgh 1989.

[MOT90] F. MOLLER & C. TOFTS, *A temporal calculus of communicating systems*, in: Proc. CONCUR'90, Amsterdam (J.C.M. Baeten & J.W. Klop, eds.), LNCS 458, Springer 1990, pp. 401-415.

[NIS94] X. NICOLLIN & J. SIFAKIS, *The algebra of timed processes ATP: theory and application*, Information & Computation 114, 1994, pp. 131-178.

[NSVR90] X. NICOLLIN, J.-L. RICHIER, J. SIFAKIS & J. VOIRON, *ATP: an algebra for timed processes*, in Proc. IFIP TC2 Conf. on Progr. Concepts & Methods, Sea of Gallilee, Israel 1990.

[RER88] G.M. REED & A.W. ROSCOE, *A timed model for communicating sequential processes*, TCS 58, 1988, pp. 249-261.

[VER94] C. VERHOEF, *A congruence theorem for structured operational semantics with predicates and negative premises*, in: Proc. CONCUR'94, Uppsala (B. Jonsson & J. Parrow, eds.), Springer LNCS 836, 1994, pp. 433-448.

A Duration Calculus with Infinite Intervals

Zhou Chaochen*, Dang Van Hung** and Li Xiaoshan***

UNU/IIST, P.O. Box 3058, Macau

Abstract. This paper introduces infinite intervals into the Duration Calculus [33]. The extended calculus defines a state duration over an infinite interval by a property which specifies the limit of the state duration over finite intervals, and excludes the description operator. Thus the calculus can be established without involvement of unpleasant calculation of infinity. With limits of state durations, one can treat conventional liveness and fairness, and can also measure liveness and fairness through properties of limits. Including both finite and infinite intervals, the calculus can, in a simple manner, distinguish between terminating behaviour and nonterminating behaviour, and therefore directly specify and reason about sequentiality.

1 Introduction

The Duration Calculus (abbr. DC) [33] is an extension of the Interval Temporal Logic [15]. It is restricted to *finite* intervals, and uses *chop* ($\frown$) as the only modality. *Chop* is a *contracting* operator, by which, from a given interval, we can reach its subintervals. This restriction prohibits the DC from specifying *unbounded* liveness and fairness properties of computing systems, such as a circuit which oscillates *forever*, or two users which are served so fairly that they have equal service durations *at last*.

In order to cope with unbounded liveness and fairness, [18], [3] and [22] introduce *expanding* modalities, while keep the restriction to finite intervals. [18] defines two *weakest inverses* of the chop. [3] generalizes the chop by introducing *backward* intervals. Inspired by [26], [22] introduces two *expanding* modalities into the DC. They are designated T and D in [26], and $\rhd$ and $\lhd$ in [22]. By $\rhd$ and $\lhd$, from a given interval one can refer to its superintervals: an interval $[a, b]$ satisfies

$$D_1 \rhd D_2$$

iff there exists c such that $c \geq b$, $[a, c]$ satisfies D_1, and $[b, c]$ satisfies D_2. $\lhd$ is defined symmetrically. With $\rhd$ and $\lhd$, one can specify unbounded liveness and fairness. Let Boolean function W model the output of an oscillator.[4] The oscillator can therefore be specified by

---

* On leave of absence from the Software Institute, Academia Sinica

** On leave of absence from the Institute of Information Technology of Vietnam

*** On leave of absence from the Software Institute, Academia Sinica

[4] By $W(t) = 1$ (0), we mean that W is connected to power (ground) at time t.

$$\neg((\neg(true \rhd (\neg\lceil W \rceil \wedge \neg\lceil \neg W \rceil))) \rhd true)$$

where $\lceil W \rceil$ ($\lceil \neg W \rceil$) means that W has value 1 (0) everywhere inside an interval. The formula can be read as:

There is no such a right expansion that one cannot find out an interval right of the expansion, inside which W is neither everywhere 1 *nor everywhere* 0.

Let Boolean functions S_1 and S_2 model system service for the two users respectively. The specification that S_1 and S_2 at last have equal duration can be formulated by

$$\forall \epsilon > 0.(\neg(|\textstyle\int S_1 - \int S_2| \geq \epsilon \rhd true)) \rhd true$$

where $\int S$ is a duration expression of the DC. It is a function from intervals to real numbers. Given interval $[a, b]$, the value of $\int S$ is defined as

$$\int_a^b S(t)dt$$

i.e. the presence duration of state S in the interval. The above formula can be read as:

For any $\epsilon > 0$, one can find out a right expansion, such that no further right expansion will make a difference between presence durations of S_1 and S_2 greater than or equal to ϵ.

Although the DC with the two additional modalities can express liveness and fairness properties of computing systems, it still has problems in differentiating finite system behaviour from infinite one *syntactically*. An infinite behaviour determines system states eternally, while a finite behaviour represents a termination, which determines system states up to some moment in time, and allows arbitrary continuation. It seems that, in order to define sequential composition (;) in a finite interval based DC, an extra state (like $\surd$ in CSP [10]) might be necessary, which syntactically indicates a termination. The CSP school introduces not only $\surd$ state, but also *refusals* (or *ready sets*), in order to deal with liveness properties in a *finite trace* based language. It therefore exhibits another possible approach to extending the expressiveness of the DC, which introduces new states instead of new modalities.

The third approach to extend the DC for specifying and reasoning about unbounded liveness and fairness properties is to remove the restriction of finite intervals by introducing infinite intervals into the calculus. In [21], [16] and [17], a kind of infinite interval has been introduced. [21] is an extension of Temporal Logic, and [16] and [17] extend Interval Temporal Logic with infinite intervals. [16] and [17] let inf stand for infinite intervals, and includes an axiom

$$(D \wedge inf); C \;\equiv\; D \wedge inf$$

which defines nicely the sequential composition of a nonterminate (infinite) behaviour: $(D \wedge inf)$. Unfortunately both of them have not proposed a way to deal with values of temporal variables (e.g. interval length) over infinite intervals yet.

Since the length of an infinite interval is *infinity*: ∞, the treatment of infinity becomes an obstacle to the development of an infinite Duration Calculus. Many of the textbooks of mathematical analysis include algebraic laws of infinity, such as

$$\infty + \infty = \infty,$$
$$\infty \cdot \infty \ = \infty.$$

However those laws are far from complete. For example, they do not provide laws for subtraction: $(\infty - \infty)$.

[36] attempts the problem by assigning $\perp$ (*'undefined'*) to the length of infinite intervals, and assigning false to atomic formulas with occurrence of $\perp$. Therefore

$$\ell = \ell$$

becomes false for infinite intervals, where ℓ designates interval length. It unfortunately opposes mathematical common sense.

In the foundations of mathematics, the Intuitionism denies the existence of infinity, but recognizes a 'manifold of possibilities open towards infinity' [27]. Inspired by the Intuitionism, this paper will establish a Duration Calculus of both finite and infinite intervals (called DC^i), which treats ∞ as a property rather than an entity. DC^i is able to formulate and reason about properties which characterize infinity, but rejects calculations of infinity by excluding the description operator from the calculus.

In DC^i, a DC formula D is satisfied by infinite interval $[a, \infty)$, iff for any b $(\geq a)$, $[a, b]$ satisfies D. In other words, D is an invariant for all finite prefixes of the infinite interval. That D is satisfied by an infinite interval is designated as D^i, and called *infinite satisfaction*. Similarly the satisfaction of D by a finite interval is designated as D^f, and called *finite satisfaction*. DC^i is a first order logic of the infinite and finite satisfactions of DC formulas.

The unbounded liveness of system state S can be represented by infinite presence of S. It can be formulated in DC^i by

$$\forall x \exists y > x.(\ell > y \Rightarrow (\ell = y) \frown \lceil S \rceil \frown true)^i$$

where $\frown$ designates the *chop* operator. An interval satisfies $(D_1 \frown D_2)$, iff the interval can be chopped into two subintervals such that the left subinterval satisfies D_1, and the right one satisfies D_2. The plain meaning of the previous formula is that, after any time x, one can always find a time y such that S appears right after y. A formulation of the oscillator can be obtained by postulating infinite presence of both W and $\neg W$.

One can also measure unbounded liveness of state S through its duration over an infinite time interval.[5] An infinite duration of state S over an infinite interval can be specified in DC^i

$$\forall x \exists y.(\ell \geq y \;\Rightarrow\; \textstyle\int S > x)^i.$$

This formula is almost a direct translation of the Cauchy definition of the limit of infinity. Similarly we can specify finite limits such as v is the value of $\int S$ over an infinite interval:

$$\forall \epsilon > 0 \exists x.(\ell \geq x \;\Rightarrow\; |v - \textstyle\int S| < \epsilon)^i.$$

The unbounded fairness between states S_1 and S_2 can be measured by the ratio of their durations over an infinite interval. For example

$$\forall \epsilon > 0 \exists x.(\ell \geq x \;\Rightarrow\; |\textstyle\int S_1 - r \cdot \int S_2| < \epsilon)^i.$$

specifies that the limit of the ratio of the duration of S_1 to the duration of S_2 is r.

With infinite intervals, we can establish a theory of limits of state durations. Unbounded liveness and fairness properties are essentially kinds of limit properties of state durations. A limit theory can facilitate specifications and verifications of liveness and fairness properties. It can also help us specify properties of hybrid systems, e.g. system stability.

All behaviour defined by formula $\mathcal{G}$ of DC^i terminate, iff

$$\mathcal{G} \;\Rightarrow\; fin$$

where fin specifies finite intervals, and

$$fin \;\widehat{=}\; true^f$$

Similarly, $\mathcal{G}$ defines non-terminating behaviour, iff

$$\mathcal{G} \;\Rightarrow\; inf$$

where inf specifies infinite intervals, and

$$inf \;\widehat{=}\; true^i$$

Therefore an operator of sequential composition can be simply defined in DC^i.

The syntax and semantics of DC^i are explained in Section 3 in detail. In Section 2, the DC is briefly reviewed. In Section 5, we list various examples of DC^i specifications, which include specifications of duration limit, liveness and fairness properties, and a definition of the sequential composition operator.

[5] That is why we choose the term – *unbounded* liveness and fairness, instead of *qualitative* liveness and fairness.

Regarding inference rules of DC^i, we will of course adopt the rules for first order predicate calculus. Besides, by the definition of finite and infinite satisfactions, we can derive DC^i theorems from DC theorems. For example, if

$$\vdash_{DC} D$$

then all finite intervals will satisfy D^f, and all infinite intervals will satisfy D^i. Hence

$$\vdash_i D^f \vee D^i$$

where $\vdash_i$ is an abbreviation of $\vdash_{DC^i}$. An inference system is given in Section 4. We cannot conclude the completeness of the inference system in the paper.

2 Duration Calculus (DC)

In this section, the Duration Calculus is briefly reviewed [32].

Research into the Duration Calculus was started by the ProCoS project (Provably Correct Systems: Esprit BRA 3104) in 1989, when the project was developing formal techniques for designing real-time safety critical systems. Several calculi have been developed since then. They are the Duration Calculus, the Extended Duration Calculus, the Mean Value Calculus, the Probabilistic Duration Calculus and the three calculi as mentioned in Section 1.

The Duration Calculus is a real-time interval logic [33]. It formalizes integrals of Boolean functions over finite intervals, and can be used to specify and reason about timing and logical constraints on discrete states of a system. All the other calculi are extensions of the Duration Calculus. The Extended Duration Calculus [38] extends the Duration Calculus with piecewise continuity/differentiability of functions. It can capture properties of continuous states, and can be used for designing hybrid systems. The Mean Value Calculus [37] extends the Duration Calculus by replacing integrals of Boolean functions with their mean values, so that it can use δ-functions to represent instant actions such as communications and events. The Mean Value Calculus can be used to refine from state based specifications via mixed state and event specifications to event based specifications. The Probabilistic Duration Calculus [13, 14, 1] provides designers with a set of rules to reason about and calculate dependability of a system with respect to its components. As explained in Section 1, the other three calculi introduce more modalities (or inverse intervals), so that they can deal with unbounded liveness and fairness properties.

The Duration Calculi have been used to specify a number of examples of hybrid systems [19, 20, 23, 2, 29, 30, 28, 9]. The Calculi have also been used to define real-time semantics for Occam-like languages [34, 8], and to specify real-time behaviour of schedulers [34, 31] and circuits [7].

As to mechanical support tool for the Duration Calculi, the decidability and undecidability results of the Duration Calculus have been published [35, 4], and an automatic model checker for a decidable subclass of the Duration Calculus has been implemented in Standard ML [25]. Efficient model checking algorithms for Linear Duration Invariants have been discovered [11, 39]. They employ the technique of Linear Programming. A tool for constructing DC specifications and checking DC proofs has been implemented by using PVS [24].

We present below some of the main features of DC.

In DC, a system state represents a logical property of the system. Presence of a state means that the property holds. Absence means that the property does not hold. State is modelled by Boolean functions over time: $\mathbf{R} \rightarrow \{0,1\}$, where time is modelled by real numbers: $\mathbf{R}$. When the function has value 1 at time t, it represents presence of the state at t. Symmetrically, value 0 represents absence of the state at a time.

For an arbitrary state S and an arbitrary finite interval $[a,b]$, the duration of S, designated $\int S$, is defined by the value of the integral of S over the interval $[a,b]$

$$\int_a^b S(t)dt$$

It follows that $\int 1$ equals $(b-a)$, i.e. the length of the finite interval. Thus we introduce an abbreviation to designate interval length

$$\ell \mathrel{\widehat{=}} \int 1$$

The following BNFs review the inductive definitions of the DC syntax. Let S stand for states, τ for terms, A for atomic formulas, and D for duration formulas.

$$S ::= \; P \mid \neg S \mid S \vee S$$

where P stands for primitive states.

$$\tau ::= \; \int S \mid r \mid x \mid f(\tau, ..., \tau) \mid \tau + \tau \mid \tau - \tau \mid ...$$

where r stands for constants, x for global variables, and f for function symbols.

According to the meaning of $\int S$ presented above, the denotations of terms are functions from finite intervals to reals, called *interval functions*: $\mathbf{I} \rightarrow \mathbf{R}$, where

$$\mathbf{I} \mathrel{\widehat{=}} \{[a,b] \mid a, b \in \mathbf{R} \;\&\; b \geq a\}$$

Therefore a model of DC formula, designated Π, consists of interpretations for primitive states P, global variables x, and function symbols f. State P is interpreted as a Boolean function in $(\mathbf{R} \rightarrow \{0,1\})$, x as a real number (a constant interval function), and f as a function in $(\mathbf{R}^n \rightarrow \mathbf{R})$ (a constant functional of interval functions), where n is the arity of f. The arithmetical operators and Boolean operators are here applied to functions, and their interpretations are pointwise extensions of the standard one.

The atomic formulas are

$$A ::= true \mid false \mid \tau = \tau \mid \tau > \tau \mid \ldots$$

and the formulas are

$$D ::= A \mid \neg A \mid D \vee D \mid D \frown D \mid \exists x.D$$

where x is a global variable, and $\frown$ designates the *chop* operator.[6]

The semantics of the formulas can be defined by formula satisfactions. Given model Π, a finite interval $[a, b]$ satisfies formula D under model Π, written as

$$\Pi, [a, b] \models_{DC} D$$

iff the values of the terms over $[a, b]$ satisfy D, where the meaning of operators $=$ and $>$, connectives $\neg$ and $\vee$ and quantifier $\exists x$ is standard, and the satisfaction of $B \frown C$ by finite interval $[a, b]$ is defined as that there exists m $(a \leq m \leq b)$, such that B is satisfied by $[a, m]$, and C is satisfied by $[m, b]$. When formula D is satisfied by all finite intervals under model Π, we say that D is satisfied by Π, written as

$$\Pi \models_{DC} D$$

If D is satisfied by any models, D is called *valid*, written as

$$\models_{DC} D$$

In DC, we also use the following abbreviations.

$$\lceil S \rceil \widehat{=} (\textstyle\int S = \ell) \wedge (\ell > 0)$$

$\lceil S \rceil$ means that state S is present (almost) everywhere in a non-point interval.

$$\lceil\, \rceil \widehat{=} (\ell = 0)$$

$\lceil\, \rceil$ defines point intervals.

We also use D^* as an abbreviation of $(\lceil\, \rceil \vee D)$.

For a formula D

$$\Diamond D \widehat{=} true \frown D \frown true$$

Thus $\Diamond D$ is satisfied by a finite interval in which D holds for some subinterval, and similarly

[6] We overload the Boolean operator ($\neg$ and $\vee$). They apply to states and also formulas.

$$\Diamond D \mathrel{\hat{=}} D \frown true$$

$\Diamond D$ holds for a finite interval, iff D holds for a prefix of the interval.

The dualities are

$$\Box D \mathrel{\hat{=}} \neg\Diamond\neg D$$
$$\Box D \mathrel{\hat{=}} \neg\Diamond\neg D$$

They are true of a finite interval, in which D holds in every subinterval or prefix respectively.

DC is an extension of Interval Temporal Logic (ITL). It therefore employs all the axioms and rules of first order ITL, but also has a small set of additional axiom-s and rules, which constitute a relatively complete inference system [6]. They are

Axiom 1: $\int 0 = 0$

Axiom 2: For an arbitrary state S

$$\int S \geq 0$$

Axiom 3: For arbitrary states S_1 and S_2

$$\int S_1 + \int S_2 = \int(S_1 \vee S_2) + \int(S_1 \wedge S_2)$$

Axiom 4: Let S be a state and r, s non-negative reals

$$(\int S = r + s) \Leftrightarrow (\int S = r) \frown (\int S = s)$$

States are assumed *finitely variable.* That is, a state can have only a finite number of alternations of its presence and absence in a finite interval. DC establishes two induction rules which axiomatize finite variability. Let X denote a formula letter occurring in the formula $R(X)$ and let S be a state.

Forward Induction Rule:
If $R(\lceil\,\rceil)$ holds, and $R(X \vee (X \frown \lceil S \rceil)) \wedge R(X \vee (X \frown \lceil \neg S \rceil))$ is provable from $R(X)$, then $R(true)$ holds.

Backward Induction Rule:
If $R(\lceil\,\rceil)$ holds, and $R(X \vee (\lceil S \rceil \frown X)) \wedge R(X \vee (\lceil \neg S \rceil \frown X))$ is provable from $R(X)$, then $R(true)$ holds.

3 Duration Calculus with Infinite Intervals (DC^i)

DC^i is a first order logic of finite and infinite satisfactions of DC.

DC^i designates finite satisfaction of DC formula D by D^f, and infinite satisfaction of D by D^i. D^f holds for finite intervals only, and D^i for infinite intervals only. DC^i shares models with DC. A finite interval satisfies D^f under a model, iff the interval satisfies D in terms of the semantics of DC. An infinite interval satisfies D^i under a model, iff all its *finite* prefixes satisfy D in terms of the semantics of DC. Let $[a, b]$ stand for finite intervals and $[a, \infty)$ for infinite intervals henceforth. We define

1. $\Pi, [a,b] \models_i D^f$ iff $\Pi, [a,b] \models_{DC} D$
   where $\models_i$ designates the satisfaction relation of DC^i.
2. $\Pi, [a,\infty) \not\models_i D^f$
3. $\Pi, [a,b] \not\models_i D^i$
4. $\Pi, [a,\infty) \models_i D^i$ iff $\Pi, [a,b] \models_{DC} D$ for all b $(\geq a)$.

D^f and D^i constitute atomic formulas of DC^i. Let $\mathcal{G}$ stand for formulas of DC^i. The BNF that defines syntax of DC^i can be given by

$$\mathcal{G} ::= D^f \mid D^i \mid \neg\mathcal{G} \mid \mathcal{G} \vee \mathcal{G} \mid \exists x \mathcal{G}$$

where D stands for formulas of DC, and x for global variables of DC. We here adopt the standard semantics for $\neg$, $\vee$ and $\exists x$, and the conventional way of introducing $\wedge$, $\Rightarrow$, $\Leftrightarrow$ and $\forall x$.

Formula $\mathcal{G}$ is satisfied by model Π, iff $\mathcal{G}$ is satisfied by any interval (finite or infinite) under model Π. That is,

$$\Pi \models_i \mathcal{G}$$

iff for any a and b $(b \geq a)$

$$\Pi, [a,b] \models_i \mathcal{G},$$

and

$$\Pi, [a,\infty) \models_i \mathcal{G}$$

Formula $\mathcal{G}$ is valid, iff $\mathcal{G}$ is satisfied by any models. That is,

$$\models_i \mathcal{G}$$

iff for any model Π

$$\Pi \models_i \mathcal{G}$$

Example 1: For any Π, a and b $(\geq a)$

$$\Pi, [a,b] \models_i fin,$$

and

$$\Pi, [a,\infty) \models_i inf$$

where

$$fin \mathrel{\widehat{=}} true^f$$
$$inf \mathrel{\widehat{=}} true^i$$

fin represents all finite intervals, and inf all infinite intervals. Thus, for any Π,

$$\Pi \models_i (fin \vee inf)$$

So

$$\models_i (fin \vee inf)$$

We let

$$True \mathrel{\widehat{=}} (fin \vee inf)$$
$$False \mathrel{\widehat{=}} \neg True$$

They represent the *truth* and *falsehood* of DC^i.

Example 2: For Π which interprets P with constant function 1, we have

$$\Pi, [a, b] \models_i \lceil P \rceil^{*f}$$

for all a and b ($\geq a$), and also

$$\Pi, [a, \infty) \models_i \lceil P \rceil^{*i}$$

Hence

$$\Pi \models_i (\lceil P \rceil^{*f} \vee \lceil P \rceil^{*i})$$

However $(\lceil P \rceil^{*f} \vee \lceil P \rceil^{*i})$ is not valid

$$\not\models_i (\lceil P \rceil^{*f} \vee \lceil P \rceil^{*i})$$

since all other models do not satisfy the formula.

Example 3:

$$\Pi, [a, \infty) \models_i \exists x.(\ell > x \Rightarrow (\ell = x) \frown \lceil P \rceil)^i$$

iff Π assigns 1 to P from some time to eternity. However, for any Π and a,

$$\Pi, [a, \infty) \models_i (\exists x(\ell > x \Rightarrow (\ell = x) \frown \lceil P \rceil))^i$$

since for any Π and $[a, b]$, when $x > (b - a)$, we have

$$\Pi, [a, b] \models_{DC} \ell > x \Rightarrow (\ell = x) \frown \lceil P \rceil$$

Note that the Example shows that

$$(\exists x D)^i \Rightarrow \exists x.D^i$$

is not valid. So $\exists$ cannot be distributed over i.

The followings are useful properties of satisfaction and validity of DC^i formula. They can be easily derived from the definitions above.

Monotonicity: For any Π, if

$$\Pi \models_{DC} (D_1 \Rightarrow D_2)$$

then

$$\Pi \models_i (D_1^f \Rightarrow D_2^f) \text{ and}$$
$$\Pi \models_i (D_1^i \Rightarrow D_2^i)$$

***f&i*-Exclusion:** The finite and infinite satisfactions are mutually excluded.

1. $(fin \Leftrightarrow \neg inf)$
2. For any Π, if
$$\Pi \models_i (\mathcal{G}_1 \Rightarrow fin)$$
and
$$\Pi \models_i (\mathcal{G}_2 \Rightarrow inf)$$
then
$$\Pi \models_i \forall x.(\mathcal{G}_1 \lor \mathcal{G}_2) \Rightarrow (\forall x.\mathcal{G}_1 \lor \forall x.\mathcal{G}_2)$$

Proof. We only give proof for the second property, since the mutual exclusion of fin and inf is clear. Suppose that
$$\Pi, [a,b] \models_i \forall x.(\mathcal{G}_1 \lor \mathcal{G}_2)$$
Then for any x
$$\Pi, [a,b] \models_i (\mathcal{G}_1 \lor \mathcal{G}_2)$$
By the assumption
$$\Pi \models_i (\mathcal{G}_2 \Rightarrow inf)$$
we have
$$\Pi, [a,b] \not\models_i \mathcal{G}_2$$
Hence for any x
$$\Pi, [a,b] \models_i \mathcal{G}_1$$
That is
$$\Pi, [a,b] \models_i \forall x.\mathcal{G}_1$$
So
$$\Pi, [a,b] \models_i \forall x.(\mathcal{G}_1 \lor \mathcal{G}_2) \Rightarrow (\forall x.\mathcal{G}_1 \lor \forall x.\mathcal{G}_2)$$
Similarly we can prove
$$\Pi, [a,\infty) \models_i \forall x.(\mathcal{G}_1 \lor \mathcal{G}_2) \Rightarrow (\forall x.\mathcal{G}_1 \lor \forall x.\mathcal{G}_2)$$

***f*-Distributivity:** $\neg$, $\lor$ and $\exists$ distribute over f.

1. $\models_i \neg D^f \Leftrightarrow (inf \lor (\neg D)^f)$
2. $\models_i D_1^f \lor D_2^f \Leftrightarrow (D_1 \lor D_2)^f$
3. $\models_i \exists x.D^f \Leftrightarrow (\exists x D)^f$

***i*-Closure:** The infinite satisfaction implies a property of prefix closure.

$$\models_i D^i \Leftrightarrow (\Box D)^i$$

Proof. Suppose that there are Π and $[a,\infty)$ which satisfy D^i

$$\Pi, [a,\infty) \models_i D^i$$

Then for any b $(\geq a)$

$$\Pi, [a,b] \models_{DC} D$$

This implies that for arbitrary given b ($\geq a$) and any c ($b \geq c \geq a$)

$$\Pi, [a, c] \models_{DC} D$$

Hence by the definition of $\Box D$

$$\Pi, [a, b] \models_{DC} \Box D$$

Therefore

$$\Pi, [a, \infty) \models_i (\Box D)^i$$

So the proof of the first half of the equivalence is completed. The second half can be derived from Monotonicity, since

$$\models_{DC} \Box D \Rightarrow D$$

***i*-Distributivity:** *i*-Distributivity is more complicated than *f*-Distributivity.

1. $\models_i \neg D^i \Leftrightarrow (fin \vee \exists x.(\ell = x \Rightarrow \neg D)^i)$
   Proof. By the satisfaction definition, for any Π and $[a, b]$
   $$\Pi, [a, b] \models_i \neg D^i$$
   Thus
   $$\Pi, [a, b] \models_i \neg D^i \Leftrightarrow (fin \vee \exists x.(\ell = x \Rightarrow \neg D)^i)$$
   It remains to show the equivalence with respect to the infinite satisfactions. Suppose that there are Π and $[a, \infty)$ such that
   $$\Pi, [a, \infty) \models_i \neg D^i$$
   This means
   $$\Pi, [a, \infty) \not\models_i D^i$$
   By the satisfaction definition, there must exist c ($\geq a$)
   $$\Pi, [a, c] \not\models_{DC} D$$
   That is,
   $$\Pi, [a, c] \models_{DC} \neg D$$
   Then we let $x = (c - a)$, and for any b ($\geq a$) we have
   $$\Pi, [a, b] \models_{DC} (\ell = x \Rightarrow \neg D)$$
   Hence by the satisfaction definition again
   $$\Pi, [a, \infty) \models_i (\ell = x \Rightarrow \neg D)^i$$
   By the rule $\exists_+$ of first order logic
   $$\Pi, [a, \infty) \models_i \exists x.(\ell = x \Rightarrow \neg D)^i$$
   Thus we complete the proof of $\Rightarrow$ of the equivalence. The proof of $\Leftarrow$ can be presented in a similar way. We leave it out.
2. $\models_i (D_1^i \vee D_2^i) \Leftrightarrow (\Box D_1 \vee \Box D_2)^i$
   Proof. The first half of the equivalence can be proved as follows.
   $$\begin{array}{llll} \models_i & D_1^i & \Rightarrow (\Box D_1)^i & \text{(by } i\text{-Closure)} \\ \models_{DC} & \Box D_1 & \Rightarrow (\Box D_1 \vee \Box D_2) & \text{(by } \vee_+) \\ \models_i & (\Box D_1)^i & \Rightarrow (\Box D_1 \vee \Box D_2)^i & \text{(by Monotonicity)} \\ \models_i & D_1^i & \Rightarrow (\Box D_1 \vee \Box D_2)^i & \text{(by Transitivity of } \Rightarrow) \end{array}$$
   Similarly, we can prove
   $$\models_i D_2^i \Rightarrow (\Box D_1 \vee \Box D_2)^i$$
   Thus we conclude

$\models_i (D_1^i \vee D_2^i) \Rightarrow (\Box D_1 \vee \Box D_2)^i$

It remains to show the second half of the equivalence. Suppose

$\Pi, [a, \infty) \models_i (\Box D_1 \vee \Box D_2)^i$

Then for any b ($\geq a$) by the satisfaction definition

$\Pi, [a, b] \models_{DC} (\Box D_1 \vee \Box D_2)$

That is,

$\Pi, [a, b] \models_{DC} \Box D_1$

or

$\Pi, [a, b] \models_{DC} \Box D_2$

Therefore at least one of $\Box D_1$ and $\Box D_2$ must be satisfied by infinitely many intervals under model Π. They all start from a, and cover $[a, \infty)$. Suppose that there are b_j ($\geq a$) ($j = 1, 2, ...$)

$\Pi, [a, b_j] \models_{DC} \Box D_1$

and

$\lim_{j \to \infty} b_j = \infty.$

Then, for any b ($\geq a$), we can find a b_n such that ($b_n \geq b$). Hence

$\Pi, [a, b] \models_{DC} \Box D_1$

by the definition of $\Box$ and

$\Pi, [a, b_n] \models_{DC} \Box D_1.$

Thus by the satisfaction definition

$\Pi, [a, \infty) \models_i (\Box D_1)^i.$

So by Monotonicity we conclude

$\Pi, [a, \infty) \models_i D_1^i,$

and then the second half of the equivalence.

3. $\models_i \forall x.D^i \Leftrightarrow (\forall x D)^i$

Proof.

$\models_i (\forall x D)^i \Rightarrow \forall x.D^i$

can be derived from Monotonicity and $\forall_+$. Conversely, suppose that

$\Pi, [a, \infty) \models_i \forall x.D^i$

That is, for any x

$\Pi, [a, \infty) \models_i D^i$

Then by the satisfaction definition, for any b ($\geq a$)

$\Pi, [a, b] \models_i D$

Hence

$\Pi, [a, b] \models_i \forall x D$

Therefore

$\Pi, [a, \infty) \models_i (\forall x D)^i$

The proof is completed.

4 Inference Rules of DC^i

In this section, we establish an inference system for DC^i. Since DC^i is a first order logic, the inference system of DC^i will adopt all the axioms and rules of first order predicate calculus. DC^i also includes real numbers and their arithmetical operations, so the DC^i inference system contains real arithmetic. Here we do not repeat the rules taken from first order predicate calculus and real arithmetic. Of course, DC^i has its own rules for inferring the finite and infinite satisfactions of DC formulas. Those inferences in DC^i shall very much involve DC inferences. They may take DC theorems as their premises. According to the properties of satisfaction and validity of DC^i formulas which are listed in the previous section, we can introduce the following four groups of inference rules of DC^i. (i-Closure is implied by i-Distributivity of $\vee$.)

Monotonicity: If

$$\vdash_{DC} (D_1 \Rightarrow D_2)$$

then

$$\vdash_i (D_1^f \Rightarrow D_2^f)$$

and

$$\vdash_i (D_1^i \Rightarrow D_2^i)$$

***f&i*-Exclusion:** There are two rules, regarding mutual exclusion of the finite and infinite satisfactions.

1. $\vdash_i (fin \Leftrightarrow \neg inf)$
2. If
   $$\vdash_i \mathcal{G}_1 \Rightarrow fin$$
   and
   $$\vdash_i \mathcal{G}_2 \Rightarrow inf$$
   then
   $$\vdash_i \forall x.(\mathcal{G}_1 \vee \mathcal{G}_2) \Rightarrow (\forall x.\mathcal{G}_1 \vee \forall x.\mathcal{G}_2)$$

***f*-Distributivity:** It contains three rules.

1. $\vdash_i \neg D^f \Leftrightarrow (inf \vee (\neg D)^f)$
2. $\vdash_i (D_1^f \vee D_2^f) \Leftrightarrow (D_1 \vee D_2)^f$
3. $\vdash_i (\exists x.D^f) \Leftrightarrow (\exists x D)^f$

***i*-Distributivity:** It also contains three rules.

1. $\vdash_i \neg D^i \Leftrightarrow (fin \vee \exists x.(\ell = x \Rightarrow \neg D)^i)$
2. $\vdash_i (D_1^i \vee D_2^i) \Leftrightarrow (\Box D_1 \vee \Box D_2)^i$
3. $\vdash_i (\forall x.D^i) \Leftrightarrow (\forall x D)^i$

With the rules, we can prove

Theorem 1.

1. $\vdash_i (D^f \Rightarrow fin)$

2. $\vdash_i (D^i \Rightarrow inf)$
3. If

$$\vdash_{DC} D,$$

then

$$\begin{aligned} \vdash_i \ & (fin \Rightarrow D^f) \\ & \wedge (inf \Rightarrow D^i) \\ & \wedge (D^f \vee D^i) \end{aligned}$$

Proof. The proof can be easily obtained by applying the rules of Monotonicity and the (mutual) Exclusion of fin and inf. We omit the proof.

Theorem 2. (i-Closure)

$$\vdash_i D^i \Leftrightarrow (\Box D)^i$$

Proof. Let D_1 be D and D_2 be $true$ in the i-Distributivity of $\vee$. We have

$$(D^i \vee inf) \Leftrightarrow (\Box D \vee \Box true)^i$$

By ($\Box true \Leftrightarrow true$) in DC, Theorem 1(2) and Monotonicity, we can derive the theorem from the previous equivalence.

Theorem 3.

1. $\vdash_i \neg false^i$
2. $\vdash_i \forall x.\neg(\ell = x)^i$

Proof. The proof of the first statement:

$$\begin{aligned} \neg false^i & \Leftrightarrow (fin \vee \exists x.(\ell = x \Rightarrow \neg false)^i) && (i\text{-Distributivity of } \neg) \\ & \Leftrightarrow (fin \vee \exists x.inf) && (\text{Monotonicity}) \\ & \Leftrightarrow (fin \vee inf) && (\text{first order logic}) \\ & \Leftrightarrow True && (f\&i\text{-Exclusion}) \end{aligned}$$

The proof of the second statement: for any x,

$$\neg(\ell = x)^i \Leftrightarrow (fin \vee \exists y.(\ell = y \Rightarrow \ell \neq x))^i \ (i\text{-Distributivity of } \neg)$$

Let $f(x) = x + 1$.

$$(\ell = f(x) \Rightarrow \ell \neq x)^i \Leftrightarrow inf \ (\text{arithmetic \& Monotonicity})$$
$$\exists y.(\ell = y \Rightarrow \ell \neq x)^i \Leftrightarrow inf \ (\exists_+)$$

Hence

$$\neg(\ell = x)^i \Leftrightarrow (fin \vee inf)$$

and the proof can be completed by using $\forall_+$.

In the proof of Theorem 3.2, we apply Skolemisation to DC^i formulas. We can prove a general theorem of application of Skolemisation. For example, let $\mathcal{G}_1$ be

$$\exists x_1 \forall y_1 \exists z_1 . D_1(x_1, y_1, z_1)^i$$

and $\mathcal{G}_2$ be

$$\forall y_2 \exists z_2 . D_2(y_2, z_2)^i$$

Theorem 4. If for any x_1 and f_1 there exists f_2 such that

$$\vdash_{DC} \ (\Box \forall y_1 . D_1(x_1, y_1, f_1(y_1))) \ \Rightarrow \ \forall y_2 . D_2(y_2, f_2(y_2))$$

then

$$\mathcal{G}_1 \ \Rightarrow \ \mathcal{G}_2$$

Proof. A proof can be derived from i-Closure (Theorem 2), i-Distributivity of $\forall$ and rules of first order logic. We omit here the proof details.

Theorem 5. (i-Distributivity of $\wedge$)

$$\vdash_i \ (D_1^i \ \wedge \ D_2^i) \ \Leftrightarrow \ (D_1 \ \wedge \ D_2)^i$$

Proof. By Monotonicity, we can easily prove the $\Rightarrow$ part of the equivalence. Conversely, we first distribute $\wedge$ over i by i-Distributivity of $\neg$ and $\vee$, and rules of first order logic

$$\begin{aligned}(D_1^i \ \wedge \ D_2^i) &\Leftrightarrow \neg(\neg D_1^i \ \vee \ \neg D_2^i) \\ &\Leftrightarrow \neg(\exists x.(\ell = x \ \Rightarrow \ \neg D_1)^i \ \vee \ \exists y.(\ell = y \ \Rightarrow \ \neg D_2)^i) \\ &\Leftrightarrow \neg \exists x, y.(\Box(\ell = x \ \Rightarrow \ \neg D_1) \ \vee \ \Box(\ell = y \ \Rightarrow \ \neg D_2))^i \\ &\Leftrightarrow \forall x, y \exists z.(\ell = z \ \Rightarrow \ \neg(\Box(\ell = x \ \Rightarrow \ \neg D_1) \ \vee \ \Box(\ell = y \ \Rightarrow \ \neg D_2)))^i \\ &\Leftrightarrow \forall x, y \exists z.(\ell = z \ \Rightarrow \ (\Diamond(\ell = x \ \wedge \ D_1) \ \wedge \ \Diamond(\ell = y \ \wedge \ D_2)))^i\end{aligned}$$

We now apply *reductio ad absurdum* to prove the conclusion. By i-Distributivity of $\neg$ and i-Closure

$$\neg(D_1 \ \wedge \ D_2)^i \ \Leftrightarrow \exists u.\Box(\ell = u \ \Rightarrow \ \neg(D_1 \wedge D_2))$$

For any u and f, let $g = f(u, u)$. We can prove

$$\begin{aligned}\vdash_{DC} \quad &((\ell = f(u, u) \ \Rightarrow \ (\Diamond(\ell = u \ \wedge \ D_1) \ \wedge \ \Diamond(\ell = u \ \wedge \ D_2))) \\ &\wedge \Box(\ell = u \ \Rightarrow \ \neg(D_1 \wedge D_2)) \ \wedge \ \ell = g) \qquad \Rightarrow \ false\end{aligned}$$

Therefore by $\forall_-$

$$\begin{aligned}\vdash_{DC} \quad &(\forall x, y.(\ell = f(x, y) \ \Rightarrow \ (\Diamond(\ell = x \ \wedge \ D_1) \ \wedge \ \Diamond(\ell = y \ \wedge \ D_2))) \\ &\wedge \Box(\ell = u \ \Rightarrow \ \neg(D_1 \wedge D_2)) \ \wedge \ \ell = g) \qquad \Rightarrow \ false\end{aligned}$$

Hence by *reductio ad absurdum*

$$\begin{aligned}\vdash_{DC} \quad &\forall x, y.(\ell = f(x, y) \ \Rightarrow \ (\Diamond(\ell = x \ \wedge \ D_1) \ \wedge \ \Diamond(\ell = y \ \wedge \ D_2))) \\ &\Rightarrow (\ell = g \ \Rightarrow \ \neg \Box(\ell = u \ \Rightarrow \ \neg(D_1 \wedge D_2)))\end{aligned}$$

Then by Monotonicity and i-Distributivity of $\forall$

$$\vdash_i \quad \forall x, y.(\ell = f(x,y) \Rightarrow (\Diamond(\ell = x \wedge D_1) \wedge \Diamond(\ell = y \wedge D_2)))^i$$
$$\Rightarrow (\ell = g \Rightarrow \neg\Box(\ell = u \Rightarrow \neg(D_1 \wedge D_2)))^i$$

By $\exists_+$,

$$\vdash_i \quad \forall x, y \exists z.(\ell = z \Rightarrow (\Diamond(\ell = x \wedge D_1) \wedge \Diamond(\ell = y \wedge D_2)))^i$$
$$\Rightarrow \exists v.(\ell = v \Rightarrow \neg\Box(\ell = u \Rightarrow \neg(D_1 \wedge D_2)))^i$$

By i-Distributivity of $\neg$

$$(D_1^i \wedge D_2^i) \Rightarrow \neg(\Box(\ell = u \Rightarrow \neg(D_1 \wedge D_2)))^i$$

By $\forall_+$

$$(D_1^i \wedge D_2^i) \Rightarrow \forall u.\neg(\Box(\ell = u \Rightarrow \neg(D_1 \wedge D_2)))^i$$

That is

$$(D_1^i \wedge D_2^i) \Rightarrow \neg\exists u.(\Box(\ell = u \Rightarrow \neg(D_1 \wedge D_2)))^i$$

So

$$(D_1^i \wedge D_2^i) \Rightarrow \neg\neg(D_1 \wedge D_2)^i$$
$$\Rightarrow (D_1 \wedge D_2)^i$$

We can establish a general theorem about *reductio ad absurdum* in DC^i. Let $\mathcal{G}_1$ and $\mathcal{G}_2$ be introduced as above. We can prove

Theorem 6. If for any x_1, f_1 and f_2 there exists g such that

$$\vdash_{DC} \quad (\Box\forall y_1.D_1(x_1, y_1, f_1(y_1)) \wedge \Box\forall y_2.D_2(y_2, f_2(y_2)) \wedge \ell = g) \Rightarrow \mathit{false}$$

then

$$\vdash_i \mathcal{G}_1 \Rightarrow \neg\mathcal{G}_2$$

Proof. Similar to the proof given in Theorem 5. The proof is omitted here.

By f- and i-Distributivity, we can reduce DC^i formulas to a kind of normal form, called DC^i *prenix normal form*:

Theorem 7. For any DC^i formula $\mathcal{G}$, there exists DC formulas D_1 and D_2 and a prenix (a sequence of quantifiers), designated α, such that

$$\vdash_i \mathcal{G} \Leftrightarrow (D_1^f \vee \alpha.D_2^i)$$

Proof. By applying rules of first order logic, one can reduce $\mathcal{G}$ to prenix form with a matrix of disjunctive normal form. By f- and i-Distributivity of $\neg$, we can transform the matrix into a disjunctive normal form without negations of atomic formulas of DC^i. Of course, when the i-Distributivity of $\neg$ is applied during the transformation, the prenix will be augmented. Suppose that the prenix after the augmentation is α.

Then applying f- and i-Distributivity of $\wedge$, one can obtain a matrix of

$$D^f_{k_1} \vee \ldots \vee D^f_{k_m} \vee D^i_{k_{m+1}} \vee \ldots \vee D^i_{k_p} \bigvee_j (D^f_{j_1} \wedge D^i_{j_2})$$

By the first rule of $f\&i$-Exclusion

$$(D^f_{j_1} \wedge D^i_{j_2}) \Rightarrow \mathit{False}$$

Therefore the matrix can be reduced to

$$D^f_{k_1} \vee \ldots \vee D^f_{k_m} \vee D^i_{k_{m+1}} \vee \ldots \vee D^i_{k_p}$$

Applying f- and i-Distributivity of $\vee$, we can transform the matrix into

$$(D^f_q \vee D^i_2)$$

By the distributivity of $\exists$ over $\vee$ and the distributivity of $\forall$ over $\vee$ with formulas of mutual exclusion (stated as the second rule of $f\&i$-Exclusion), we can move the prenix into the matrix, and obtain

$$(\alpha.D^f_q \;\vee\; \alpha.D^i_2)$$

Applying the f-Distributivity of $\exists$ and $\forall$, one can reduce

$$\alpha.D^f_q$$

to formula

$$(\alpha.D_q)^f$$

designated D^f_1. Therefore formula $\mathcal{G}$ is at last reduced to its equivalent prenix normal form

$$D^f_1 \;\vee\; \alpha.D^i_2$$

Example. Reduce $\mathcal{G}$

$$\forall x(D^f_1 \;\vee\; \neg\forall y\exists z.D^i_2) \;\wedge\; (D^f_3 \;\vee\; \exists u.D^i_4)$$

to prenix normal form.

Following the procedure presented in the proof of Theorem 7:

$$\begin{aligned}
\mathcal{G} \Leftrightarrow\; & \exists u\forall x\exists y\forall z.((D^f_1 \;\wedge\; D^f_3) \vee (D^f_1 \;\wedge\; D^i_4) \\
& \vee(\neg D^i_2 \;\wedge\; D^f_3) \vee (\neg D^i_2 \;\wedge\; D^i_4)) \\
\Leftrightarrow\; & \exists u\forall x\exists y\forall z\exists v\exists w.((D^f_1 \;\wedge\; D^f_3) \vee (D^f_1 \;\wedge\; D^i_4) \\
& \vee((\ell = v \;\Rightarrow\; \neg D_2)^i \;\wedge\; D^f_3) \vee ((\ell = w \;\Rightarrow\; \neg D_2)^i \;\wedge\; D^i_4)) \\
\Leftrightarrow\; & \exists u\forall x\exists y\forall z\exists v\exists w.((D_1 \wedge D_3)^f \vee (D^f_1 \;\wedge\; D^i_4) \\
& \vee((\ell = v \Rightarrow \neg D_2)^i \;\wedge\; D^f_3) \vee ((\ell = w \;\Rightarrow\; \neg D_2) \wedge D_4)^i) \\
\Leftrightarrow\; & \exists u\forall x\exists y\forall z\exists v\exists w.((D_1 \wedge D_3)^f \;\vee\; ((\ell = v \;\Rightarrow\; \neg D_2) \wedge D_4)^i) \\
\Leftrightarrow\; & (\forall x.(D_1 \wedge D_3)^f) \;\vee\; \exists u\forall x\exists y\forall z\exists w.((\ell = w \;\Rightarrow\; \neg D_2) \wedge D_4)^i \\
\Leftrightarrow\; & (\forall x.(D_1 \wedge D_3))^f \;\vee\; \exists u\forall x\exists y\forall z\exists w.((\ell = w \;\Rightarrow\; \neg D_2) \wedge D_4)^i
\end{aligned}$$

5 DC^i Specifications

This section explains how to use DC^i to specify duration limits, liveness and fairness of states, and program semantics, in particular the semantics of the sequential composition.

5.1 Limit, Liveness and Fairness

Due to the assumption of the finite variability, a model of DC^i can be regarded as a countable sequence of states.

A sequence of states has a limit, iff one of the states appears constantly from some time forever. It can be specified by

$$\exists x.(\ell > x \Rightarrow (\ell = x) \frown \lceil S \rceil)^i$$

specifies models, which take state S as limit, abbreviated $\lceil S \rceil^\infty$. If S appears everywhere in a model, then the model satisfies

$$\lceil S \rceil^{*i}$$

State S is a limit of a subsequence of a model, iff

$$\forall x \exists y.(y > x \wedge (\ell > y \Rightarrow (\ell = y) \frown \Diamond \lceil S \rceil))^i$$

That is, for any x there exists y ($> x$) such that S appears right after y. The formula is abbreviated as $\lceil S \rceil^\omega$. $\lceil S \rceil^\omega$ is also a specification of the conventional liveness of state S.

An oscillator with W as its output can be specified by

$$\lceil W \rceil^\omega \wedge \lceil \neg W \rceil^\omega$$

With the inference system of DC^i, we can prove

$$(\lceil W \rceil^\omega \wedge \lceil \neg W \rceil^\omega) \Rightarrow \neg(\lceil W \rceil^\infty \vee \lceil \neg W \rceil^\infty)$$

Fairness can be regarded as relations between live states. Let S_1 stand for request, and S_2 for response. A *strong* fairness for a request can be specified as

$$\lceil S_1 \rceil^\omega \Rightarrow \lceil S_2 \rceil^\omega$$

and a *weak* fairness as

$$\lceil S_1 \rceil^\infty \Rightarrow \lceil S_2 \rceil^\omega$$

Trivially one can prove that the strong fairness implies the weak fairness.

A model can also be regarded as a set of Boolean valued functions, which interpret states. Therefore we can investigate limits of state durations over infinite intervals.

That state S has infinite duration over an infinite interval can be specified by

$$\forall x \exists y.(\ell > y \Rightarrow \textstyle\int S > x)^i$$

abbreviated ($\lim \int S = \infty$). That is, for any x there exists y such that the duration of S in the intervals with length longer than y is greater than x. A state with infinite limit must be live.

That state S takes v as the limit of its duration over an infinite interval[7] can be specified by

$$\forall \epsilon > 0 \exists y.(\ell > y \Rightarrow |v - \textstyle\int S| < \epsilon)^i$$

abbreviated ($\lim \int S = v$). The above two formulas are actually translations of Cauchy definitions of limits. We can generalize these two definitions to any term τ of DC, and write them as ($\lim \tau = \infty$) and ($\lim \tau = v$).

A live state with limit v can be specified as

$$\forall \epsilon > 0 \exists y.(\ell > y \Rightarrow 0 < (v - \textstyle\int S) < \epsilon)^i$$

abbreviated $\lim_\omega S = v$. Therefore one can measure unbounded liveness by duration limits. For example, for two live states S_1 and S_2 (i.e. $\lceil S_1 \rceil^\omega \wedge \lceil S_2 \rceil^\omega$), the limit ($d$) of their duration difference

$$\lim(\textstyle\int S_1 - \int S_2) = d$$

could be used to define the *distance* of S_1 from S_2 in a *metric* space of live states. Instead of duration difference, one might prefer to use limit (r) of duration ratio to compare liveness of states.

$$\lim(\textstyle\int S_1 - r \cdot \int S_2) = 0$$

Example:
The probabilistic automaton in the Figure has two states S_1 and S_2. We assume that the state transitions take place randomly after each time unit, according to their probabilities (shown in the Figure). The behaviour of the automaton can be specified in DC^i by

$$\lim S_i = \infty \quad (i = 1, 2)$$

and

$$\lim(\textstyle\int S_1 - 1.5 \cdot \int S_2) = 0$$

Namely, each of the two states is live, and has infinite duration. Moreover the liveness ratio between them is 1.5. We believe that extending the Probabilistic Duration Calculus in [13] with infinite intervals can help verify the specifications by showing that the probabilities of the probabilistic automaton satisfying the specifications are equal to 1.

[7] [5] investigates duration limits of states in finite intervals, when the states violate the finite variability in the intervals.

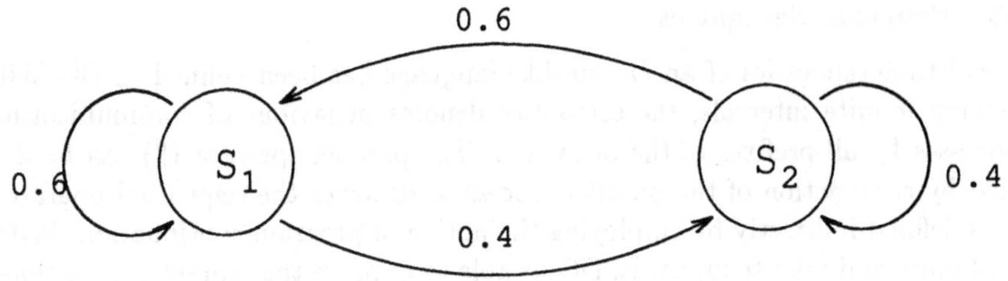

Fig. Probabilistic Automaton

With the inference system of DC^i, one can also establish a calculus of limits of terms, such as

1. $\lim \int S = v \;\Rightarrow\; \lim \int \neg S = \infty$
2. $(\lim \int S_1 = v_1 \wedge \lim \int S_2 = v_2) \;\Rightarrow\; ((\lim \int (S_1 \vee S_2) + \lim \int (S_1 \wedge S_2)) = (v_1 + v_2))$
3. If $\lim \tau_1 = v_1$ and $\lim \tau_2 = v_2$, then
   (a) $\lim(\tau_1 + \tau_2) = (v_1 + v_2)$
   (b) $\lim(\tau_1 - \tau_2) = (v_1 - v_2)$
   (c) $\lim(\tau_1 \cdot \tau_2) = (v_1 \cdot v_2)$

States can be used to model system properties. For any real valued function F, a *point* property of F, such as

$$F \geq v \text{ and } |F - v| < \epsilon,$$

can be regarded as states. Therefore one can specify *divergence* and *convergence* of functions with DC^i. Function F is divergent, designated $\lim F = \infty$, if

$$\forall x \exists y.(\ell > y \;\Rightarrow\; (\ell = y) \frown \lceil F > x \rceil)^i$$

and F is convergent to v, designated $\lim F = v$, if

$$\forall \epsilon > 0 \exists x.(\ell > x \;\Rightarrow\; (\ell = x) \frown \lceil |F - v| < \epsilon \rceil)^i$$

Similarly we can also derive rules for calculating limits of real valued functions. With limits of real valued functions, one can specify properties of continuous variables of control systems, such as system stability [12]. For example, let c stand for the output function of a controlled system, and the *asymptotic* stability of the system can be specified by

$$\exists x.\lceil |c| \leq x \rceil^{*i} \;\wedge\; \lim c = 0$$

That is, c is bounded, and the magnitude of c reaches 0 as time approaches ∞. Let r stand for the input of the system. The *bounded-input bounded-output* stability can be specified by

$$\exists x.\lceil |r| \leq x \rceil^{*i} \;\Rightarrow\; \exists x.\lceil |c| \leq x \rceil^{*i}$$

5.2 Program Semantics

A real time semantics of an Occam-like language has been defined in DC [34]. Lacking infinite intervals, the semantics denotes behaviour of communicating processes by all prefixes of the behaviour. The parallel operator (||) can be defined by conjunction of the parallel processes. However the sequential operator (;) is defined indirectly by employing the notion of program continuation. With both finite and infinite intervals, DC^i is able to improve the semantic definitions given in [34].

Let $c!$ and $c?$ stand for the output and input commands for channel c. With no confusion, we also use $c!$ and $c?$ to designate the states where output to channel c and input from channel c are ready. With DC^i, we can define the semantics of commands $c!$ and $c?$ of a process as

$$\begin{array}{l} [\![c!]\!] \mathrel{\hat{=}} (\lceil c! \wedge \neg c? \rceil^* \frown \lceil c! \wedge c? \rceil)^f \ \vee \ \lceil c! \wedge \neg c? \rceil^{*i} \\ [\![c?]\!] \mathrel{\hat{=}} (\lceil c? \wedge \neg c! \rceil^* \frown \lceil c? \wedge c! \rceil)^f \ \vee \ \lceil c? \wedge \neg c! \rceil^{*i} \end{array}$$

The first formula defines the semantics of command $c!$ by specifying that, when $c!$ is executed, the output partner becomes ready to output (i.e. $\lceil c! \rceil$), and wait synchronization from the input counterpart (i.e. $\lceil c? \rceil$) either forever (i.e. $\lceil c! \wedge \neg c? \rceil^{*i}$), if the communication fails, or for finite time, if the communication succeeds (i.e. $(\lceil c! \wedge \neg c? \rceil^* \frown \lceil c! \wedge c? \rceil)^f$). The second formula defines the semantics of command $c?$ in a symmetric way. In the semantics, we disregard the behaviour of the process on other channels. Please refer to [34] for a complete description.

In order to define the semantics of sequential operator with DC^i, we first introduce a correspondent operator (designated also as ;) in DC^i. By Theorem 7 in Section 4, any DC^i formula $\mathcal{G}$ can be reduced to its prenix normal form

$$D_1^f \ \vee \alpha . D_2^i$$

where α stand for a prenix. Therefore

$$\begin{array}{l} (\mathcal{G} \ \wedge \ fin) \Leftrightarrow D_1^f \\ (\mathcal{G} \ \wedge \ inf) \Leftrightarrow \alpha . D_2^i \end{array}$$

Thus, without loss of generality, the definition of ; can be given as follows. For DC formulas D, C_1 and C_2 , prenix α, and DC^i formulas $\mathcal{G}_1$ and $\mathcal{G}_2$, let

$$\begin{array}{ll} D^f ; (C_1^f \vee \alpha . C_2^i) & \mathrel{\hat{=}} (D \frown C_1)^f \ \vee \ \exists x \alpha . (\ell \geq x \ \Rightarrow \ (D \wedge (\ell = x)) \frown C_2)^i \\ \mathcal{G}_1 ; \mathcal{G}_2 & \mathrel{\hat{=}} (fin \wedge \mathcal{G}_1) ; \mathcal{G}_2 \ \vee \ (inf \wedge \mathcal{G}_1) \end{array}$$

By the definition of ;, a finite behaviour can be sequentially extended by either a finite behaviour or an infinite one. The former is simply defined by $\frown$ (i.e. $(D \frown C_1)^f$), and the latter one by a behaviour, a prefix of which is determined by the extended finite behaviour and the rest part by the extending infinite one (i.e. $\exists x \alpha . (\ell \geq x \ \Rightarrow \ (D \wedge (\ell = x)) \frown C_2)^i$). However, any infinite behaviour $(inf \wedge \mathcal{G}_1)$ cannot be sequentially extended.

A sequential composition of programs can be defined straightforwardly now.

$$[\![S_1; S_2]\!] \;\widehat{=}\; ([\![S_1]\!]; [\![S_2]\!])$$

where S_1 and S_2 stand for two programs.

A semantics of the parallel operator can be defined with DC^i in a way similar to [34]. Let S_1 and S_2 stand for two programs and σ_1 and σ_2 for their *alphabets* (namely, the input and output commands occurring in S_1 and S_2) respectively. For $j = 1, 2$, let

$$\neg\sigma_j \;\widehat{=}\; (\bigwedge_{c!\in\sigma_j} \neg c!) \wedge (\bigwedge_{c?\in\sigma_j} \neg c?)$$

In fact, $\neg\sigma_j$ specifies the *inactive* state of S_j. We use it to define the state of S_j after termination. Therefore

$$(fin \wedge [\![S_1]\!]) \wedge ((fin \wedge [\![S_2]\!]); \lceil\neg\sigma_2\rceil^{*f})$$

specifies the parallel result of finite behaviour of the two programs, where S_2 may terminate before S_1 does. Symmetrically

$$((fin \wedge [\![S_1]\!]); \lceil\neg\sigma_1\rceil^{*f}) \wedge (fin \wedge [\![S_2]\!])$$

specifies the parallel result of finite behaviour of the two programs, where S_1 may terminate before S_2 does. Furthermore

$$(inf \wedge [\![S_1]\!]) \wedge ((fin \wedge [\![S_2]\!]); \lceil\neg\sigma_2\rceil^{*i})$$

specifies the parallel result of infinite behaviour of S_1 and finite behaviour of S_2. Symmetrically we can specify the result of finite behaviour of S_1 and infinite behaviour of S_2 by

$$((fin \wedge [\![S_1]\!]); \lceil\neg\sigma_1\rceil^{*i}) \wedge (inf \wedge [\![S_2]\!])$$

The last specification is for the parallel result of two infinite behaviours of the two programs. That is

$$(inf \wedge [\![S_1]\!]) \wedge (inf \wedge [\![S_2]\!])$$

A disjunction of all previous five formulas defines a semantics of the parallel operator:

$$[\![S_1 \;\|\; S_2]\!]$$

We can therefore derive from the semantics of the parallel operator

$$\vdash_i \; [\![c! \;\|\; c?]\!] \Leftrightarrow \lceil c! \wedge c?\rceil^f$$

With the semantics defined above, program termination can be verified by proving

$$\vdash_i \; [\![S]\!] \Rightarrow fin$$

where S stands for the verified program. For instance, we can prove the termination of $(c! \;\|\; c?)$, since

$$\vdash_i \; \lceil c! \wedge c?\rceil^f \Rightarrow fin$$

6 Discussion

1. [33] and [37] have tried to formalize *integrals*, *mean values* and *function germs* of Boolean valued functions. [5] and DC^i try to formalize a notion of limit. It has been shown that introducing some of continuous mathematics into design calculi can assist the formal technique of programming in specifying and designing computing systems.
2. By introducing limits, DC^i can deal with unbounded liveness and fairness, and can also measure live states by duration limits. One can develop theory of such measurements, and consider interesting applications of it.
3. DC^i introduces infinite intervals by establishing a kind of metalogic of DC (a logic of a metatheory of DC). In the same way, one can introduce infinite intervals into Interval Temporal Logic. We believe that DC^i has paid the least cost to introduce infinite intervals, since, in order to formalize mathematical definition of limits, quantifications is unavoidable.
4. Although the inference system of DC^i seems powerful, we cannot conclude its (relative) completeness in the paper.

References

1. Dang Van Hung and Zhou Chaochen: Probabilistic Duration Calculus for Continuous Time. *UNU/IIST Report No. 25*, 1994.
2. M. Engel, M. Kubica, J. Madey, D.J. Parnas, A.P. Ravn and A.J. van Schouwen: A Formal Approach to Computer Systems Requirements Documentation. In *Proc. the Workshop on Theory of Hybrid Systems*, LNCS 736, R.L. Grossman, A. Nerode, A.P. Ravn and H. Rischel (Editors), pp. 452-474, 1993.
3. M. Engel and H. Rischel: Dagstuhl-Seminar Specification Problem – a Duration Calculus Solution. Personal communication, September 1994.
4. M.R. Hansen: Model-Checking Discrete Duration Calculus. In *Formal Aspects of Computing*. Vol. 6, No. 6A, pp. 826-845,. 1994.
5. M.R. Hansen, P.K. Pandya and Zhou Chaochen: Finite Divergence. In *Theoretical Computer Science*, Vol.138, pp 113-139, 1995.
6. M.R. Hansen and Zhou Chaochen: Semantics and Completeness of Duration Calculus. In *Real-Time: Theory in Practice, REX Workshop*, LNCS 600, J.W. de Bakker, C. Huizing, W.-P. de Roever and G. Rozenberg (Editors), pp. 209-225, 1992.
7. M.R. Hansen, Zhou Chaochen and J. Staunstrup: A Real-Time Duration Semantics for Circuits. In *Proc. of the 1992 ACM/SIGDA Workshop on Timing Issues in the Specification and Synthesis of Digital Systems*, Princeton, March 1992.
8. He Jifeng and J. Bowen: Time Interval Semantics and Implementation of A Real-Time Programming Language. In *Proc. 4th Euromicro Workshop on Real Time Systems*, IEEE Press, June 1992.
9. He Weidong and Zhou Chaochen: A Case Study of Optimization. *UNU/IIST Report No. 34*, December 1994.
10. C.A.R. Hoare: *Communicating Sequential Processes.* Prentice Hall International (UK) Ltd., 1985.

11. Y. Kesten, A. Pnueli, J. Sifakis and S. Yovine: Integration Graphs: A Class of Decidable Hybrid Systems. In *Hybrid Systems*, LNCS 736, R.L. Grossman, A. Nerode, A.P. Ravn and H. Rischel (Editors), pp. 179-208, 1993.
12. B.C. Kuo: *Automatic Control Systems* (sixth edition), Prentice-Hall International Inc., 1991.
13. Liu Zhiming, A.P. Ravn, E.V. Sørensen and Zhou Chaochen: A Probabilistic Duration Calculus. In, *Dependable Computing and Fault-Tolerant Systems Vol. 7: Responsive Computer Systems.* H. Kopetz and Y. Kakuda (Editor), pp. 30-52, Springer Verlag, 1993.
14. Liu Zhiming, A.P. Ravn, E.V. Sørensen and Zhou Chaochen: Towards a Calculus of Systems Dependability. In *Journal of High Integrity System*, Vol. 1, No. 1, Oxford University Press, pp. 49-65, 1994.
15. B. Moszkowski: A Temporal Logic for Multilevel Reasoning about Hardware. In *IEEE Computer*, Vol. 18, No. 2, pp. 10-19, 1985.
16. B. Moszkowski: Some Very Compositional Temporal Properties, In *Programming Concepts, Methods and Calculi (A-56)*, E.-R. Olderog (Editor), Elsevier Science B.V. (North-Holland), pp. 307-326, 1994.
17. B. Moszkowski: Compositional Reasoning about Projected and Infinite Time, *Technical Report EE/0495/M1*, Department of Electrical and Electronic Engineering, University of Newcastle upon Tyne, U.K., 1995.
18. P.H. Pandya: Weak Chop Inverses and Liveness in Duration Calculus. *Technical Report TR-95-1*, Computer Science Group, TIFR, India, 1994.
19. A.P. Ravn and H. Rischel: Requirements Capture for Embedded Real-Time Systems. In *Proc. IMACS-MCTS'91 Symp. Modelling and Control of Technological Systems*, Vol. 2, pp. 147-152, Villeneuve d'Ascq, France, 1991.
20. A.P. Ravn, H. Rischel and K.M. Hansen: Specifying and Verifying Requirements of Real-Time Systems. In *IEEE Trans. Software Eng.*, Vol. 19, No. 1, pp. 41-55, January 1993.
21. R. Rosner and A. Pnueli: A Choppy Logic. In *First Annual IEEE Symposium on Logic In Computer Science*, pp 306-314, IEEE Computer Society Press, June, 1986.
22. J.U. Skakkebæk: Liveness and Fairness in Duration Calculus. In *CONCUR'94: Concurrency Theory*, LNCS 836, B. Jonsson and J. Parrow(Editors), pp. 283–298, 1994.
23. J.U. Skakkebæk, A.P. Ravn, H. Rischel and Zhou Chaochen: Specification of Embedded Real-Time Systems. In *Proc. 4th Euromicro Workshop on Real-Time Systems*, pp. 116-121, IEEE Press, June 1992.
24. J.U. Skakkebæk and N. Shankar: Towards a Duration Calculus Proof Assistant in PVS. In *Formal Techniques in Real-Time and Fault-Tolerant Systems*, LNCS 863, H. Langmaack, W.-P. de Roever and J. Vytopil (Editors), pp. 660-679, Sept. 1994.
25. J.U. Skakkebæk, and P. Sestoft: Checking Validity of Duration Calculus Formulas. *ProCoS II Report ID/DTH JUS 3/1*, January 1993
26. Y. Venema: A Modal Logic for Chopping Intervals. In *Journal of Logic Computation*, Vol. 1, No. 4, pp. 453-476, 1991.
27. H. Weyl: Mathematics and Logic. A Brief Survey Serving as a Preface to a View of "The Philosophy of Bertrand Russell". In *Amer. Math. Monthly*, Vol. 53, pp. 2-13, 1946.
28. B.H. Widjaja, Chen Zongji, He Weidong and Zhou Chaochen: A Cooperative Design for Hybrid Control System. *UNU/IIST Report No.36*, 1995.

29. Yu Huiqun, P.K. Pandya and Sun Yongqiang: A Calculus for Hybrid Sampled Data Systems. In *Formal Techniques in Real-Time and Fault-Tolerant Systems*, LNCS 863, H. Langmaack, W.-P. de Roever and J. Vytopil (Editors), pp. 716-737, Sept. 1994.
30. Yu Xinyao, Wang Ji, Zhou Chaochen and P.K. Pandya: Formal Design of Hybrid Systems. In *Formal Techniques in Real-Time and Fault-Tolerant Systems*, LNCS 863, H. Langmaack, W.-P. de Roever and J. Vytopil (Editors), pp. 738-755, Sept. 1994.
31. Zheng Yuhua and Zhou Chaochen: A Formal Proof of the Deadline Driven Scheduler. In *Formal Techniques in Real-Time and Fault-Tolerant Systems*, LNCS 863, H. Langmaack, W.-P. de Roever and J. Vytopil (Editors), pp. 756-775, Sept. 1994.
32. Zhou Chaochen: Duration Calculi: An Overview. In *the Proceedings of Formal Methods in Programming and Their Applications*, LNCS 735, D. Bjørner, M. Broy and I.V. Pottosin (Editors), pp. 256-266, July 1993.
33. Zhou Chaochen, C.A.R. Hoare and A.P. Ravn: A Calculus of Durations. In *Information Processing Letters*, Vol. 40, No. 5, pp. 269-276, 1991.
34. Zhou Chaochen, M.R. Hansen, A.P. Ravn and H. Rischel: Duration Specifications for Shared Processors. In *Proc. of the Symposium on Formal Techniques in Real-Time and Fault-Tolerant Systems*, LNCS 571, J. Vytopil (Editor), pp. 21-32, January 1992.
35. Zhou Chaochen, M.R. Hansen and P. Sestoft: Decidability and Undecidability Results for Duration Calculus. In *Proc. of STACS '93. 10th Symposium on Theoretical Aspects of Computer Science,* LNCS 665, P. Enjalbert, A. Finkel and K.W. Wagner (Editor), pp. 58-68, Feb. 1993.
36. Zhou Chaochen and Li Xiaoshan: Infinite Duration Calculus. Draft, August 1992.
37. Zhou Chaochen and Li Xiaoshan: A Mean Value Calculus of Durations. In *A Classical Mind (Essays in Honour of C.A.R. Hoare)*, A.W.Roscoe (Editor), Prentice-Hall, pp. 431-451,1994.
38. Zhou Chaochen, A.P. Ravn and M.R. Hansen: An Extended Duration Calculus for Hybrid Real-Time Systems. In *Hybrid Systems*, LNCS 736, R.L. Grossman, A. Nerode, A.P. Ravn and H. Rischel (Editors), pp. 36-59, 1993.
39. Zhou Chaochen, Zhang Jingzhong, Yang Lu and Li Xiaoshan: Linear Duration Invariants. In *Formal Techniques in Real-Time and Fault-Tolerant Systems*, LNCS 863, H. Langmaack, W.-P. de Roever and J. Vytopil (Editors), pp. 86-109, Sept. 1994.

A Delegation-based Object Calculus with Subtyping

Kathleen Fisher* and John C. Mitchell**

Computer Science Department, Stanford University, Stanford, CA 94305
{kfisher,mitchell}@cs.stanford.edu

Abstract. This paper presents an untyped object calculus that reflects the capabilities of so-called delegation-based object-oriented languages. A type inference system allows static detection of errors, such as *message not understood,* while at the same time allowing the type of an inherited method to be specialized to the type of the inheriting object. The main advance over previous work is the provision for subtyping in the presence of delegation primatives. This is achieved by distinguishing a prototype, whose methods may be extended or replaced, from an object, which only responds to messages for which it already has methods. An advantage of this approach is that we have full subtyping without restricting the "run-time" use of inheritance. Type soundness is proved using operational semantics and an analysis of typing derivations.

1 Introduction

There are two basic forms of object-oriented languages, the most common of which is class-based. In these languages, which include C++ [Str86, ES90], Smalltalk [GR83] and Eiffel [Mey92], an object is created by a class, which is a construct that creates objects that share a common representation but which may have different specific data values. In typed class-based languages, it is common for all objects created by a given class to have the same type. An alternate approach, advanced in the language Self [US91, CU89], allows objects to be created directly or by cloning (copying) existing objects. In class-based languages, inheritance is a mechanism for defining one class from another. In delegation-based languages, inheritance is a mechanism for reusing part of one object to define another.

Following our previous work [Mit90, FHM94], the object calculus presented in this paper is delegation-based. Initially, we regarded the delegation-based approach as a technically simpler method for analyzing the same underlying language concepts. This appeared correct for the study of method specialization carried out in [Mit90, FHM94], but in [FM94] we observed that there appeared to be a fundamental trade-off between delegation and subtyping. Specifically,

* Supported in part by an NSF Graduate Fellowship, NSF Grant CCR-9303099, and a Fannie and John Hertz Foundation Fellowship.

** Supported in part by NSF Grant CCR-9303099 and the TRW Foundation.

if an object may be extended with new methods, then it is important to know at compile-time that certain methods have *not* been defined already. This requirement conflicts with the usual motivation for subtyping, which is to allow code to operate uniformly over all objects having some minimum set of required methods. (Similar and related observations appear in [AC94]; see Section 3.1.)

In this paper, we present one way of resolving this conflict. Intuitively, the main idea is to distinguish between objects that may be extended with additional methods (or have existing methods redefined) and those that cannot. This distinction is achieved by giving different uses of objects different types. In other words, an object may be created, and then have new methods added or existing methods redefined. At this point, only trivial subtyping may be used because the type system must keep track of exactly the set of methods associated with the object. However, such an object may be "converted" to a different kind of object, whose methods can no longer be altered. This conversion is done by changing the type of the object to a form which has the expected subtyping properties. In this way, we allow both delegation and subtyping, at the cost of some increase in the complexity of the type system. An alternate way of extending our previous calculus with subtyping, developed in [BL95], is described in Section 6.

The two forms of object type used in this paper highlight the distinction between two interfaces: the inheritance-interface and the client-interface of an object. This distinction is essentially familiar from object-oriented languages and databases, but often not explicitly mentioned in language documentation. If we write a C++ class such as *stack,* then there are really two separate ways of using this class. The first is by calling the constructor of the class to create *stack* objects and then calling their member functions. The second use of the *stack* class is by defining a derived class. (This "reuse" of implementation is traditionally called inheritance.) One way to see that these are very different uses of a class is to notice that they induce very different notions of class equivalence. If we just want the objects constructed by a class to behave in the same way, then we can perform a number of optimizations and program transformations. However, some transformations that preserve the behavior of constructed objects would observably change the behavior of derived classes. A simple example occurs when a class has two mutually recursive member functions, say f and g. If we replace these functions by two equivalent non-recursive functions, we do not change the behavior of constructed objects. However, this transformation may radically alter the behavior of a derived class if both are virtual functions and one is redefined in the derived class. An innovation of our object calculus, when compared with other recent work such as [AC94, Bru93, FHM94, PT94] (summarized and compared in [FM94]), is the type distinction between these two uses of a class.

The distinction between inheritance and client interfaces leads us to distinguish ***prototypes*** and ***objects.*** Intuitively, a prototype is a collection of methods that may be used to implement one or more objects. The operations on prototypes are (i) sending a message (which results in the invocation of a method), (ii) adding a new method (method addition), and (iii) redefining an existing method without changing its type (method override). The type of a prototype is called a *pro type*. For the reasons described above, only trivial subtyping exists between *pro* types.

Intuitively, an object is created by "packaging" a prototype, an operation that does nothing except "seal" the prototype so that no methods can be added or redefined. In our calculus, the "sealing" operation only changes the type from a *pro* type to an object, or *obj type*. Although the only operations allowed on objects from the "outside" are message sends, the methods within an object can override the methods of their host object. This ability to override methods internally permits the existence of methods such as a `setX` method in a point object which replace the point's `x` method with a new one storing a new location. Preventing external method override and method extension by sealing prototypes is sufficient, however, to make interesting subtyping sound for objects. The internal redefinitions do not cause an unsoundness in the type system because they are type-checked with precise knowledge of the *pro* type of the expression to which they were added.

In addition to subtyping, the distinction between prototypes and objects provides some useful insight into the two uses of classes mentioned above. Inheritance on classes is modeled in our system by operations on prototypes. Creating an object from a class is modeled by "sealing" a prototype to an object. The distinction between *pro* and *obj* types clearly illustrates the difference between a derived class's interface to a class and a client program's view of the objects created from that class. In particular, inheritance depends on the presence *and* absence of the methods in a prototype, whereas a client is only concerned with the presence of the methods it uses.

For simplicity, we work with a functional model of objects. Recent studies such as [AC95, Bv93, Pie93] confirm that imperative objects may be treated in essentially the same framework. In addition, our objects are collections of methods; we do not distinguish "member functions" from "member data" since data may be represented by functions that always return the same result. As in [FHM94], our object calculus provides method specialization. Intuitively, method specialization means that the types of methods in classes may be specialized when new classes are created from them via inheritance. Further discussion of method specialization may be found in [FHM94, FM94].

2 Untyped calculus of objects

This calculus is described in full detail in [FHM94]. We briefly summarize it here for presentational completeness.

2.1 The Objects

The untyped object calculus is the result of adding four new object-related syntactic forms to untyped lambda calculus:

$\langle\rangle$	the empty object
$e \Leftarrow m$	send message m to object e
$\langle e_1 \leftarrow\!\!+ m{=}e_2\rangle$	extend object e_1 with new method m having body e_2
$\langle e_1 \leftarrow m{=}e_2\rangle$	replace e_1's method body for m by e_2

We consider $\langle e_1 \leftarrow\!\!+ m{=}e_2\rangle$ meaningful only if e_1 denotes an object that does not have an m method, and $\langle e_1 \leftarrow m{=}e_2\rangle$ meaningful only if e_1 denotes an object that already has an m method. These conditions will be enforced by the type system. If a method is new, then no other method in the object could have referred to it, so it may have any type. On the other hand, if a method is being replaced, then we must be careful not to violate any typing assumptions in other methods that refer to it.

To simplify notation, we write $\langle m_1 = e_1, \ldots, m_k = e_k\rangle$ for $\langle \ldots \langle\langle\rangle \leftarrow\!\!+ m_1 = e_1\rangle \ldots \leftarrow\!\!+ m_k = e_k\rangle$, where $m_1, \ldots, m_k$ are distinct method names. Using this abbreviation, we may code a one-dimensional point object with field x and method move as follows:

$$\begin{aligned} p \stackrel{def}{=} \langle\, & x = \lambda \texttt{self}.\,3, \\ & \texttt{move} = \lambda \texttt{self}.\,\lambda dx.\langle \texttt{self} \leftarrow x = \lambda s.(\texttt{self} \Leftarrow x) + dx\rangle\rangle \end{aligned}$$

We refer to x as a field because it does not use the self variable.

2.2 Message Send

Sending a message m_i to an object containing such a method is modeled by extracting the corresponding method body e_i from the object and applying it to the object itself:

$$\langle m_1 = e_1, \ldots, m_k = e_k\rangle \Leftarrow m_i \quad \stackrel{eval}{\longrightarrow} \quad e_i\ \langle m_1 = e_1, \ldots, m_k = e_k\rangle$$

For example, sending the message move with a displacement of 2 to p, we get:

$$\begin{aligned} p \Leftarrow \texttt{move}\,2 &= (\lambda \texttt{self}.\,\lambda dx.\langle \ldots \rangle)\ p\ 2 \\ &= \langle p \leftarrow x = \lambda s.(p \Leftarrow x) + 2\rangle \\ &= \langle p \leftarrow x = \lambda s.\,3 + 2\rangle \\ &= \langle p \leftarrow x = \lambda \texttt{self}.\,5\rangle \end{aligned}$$

Using a sound rule for object equality,

$$\langle\langle m_1{=}e_1, \ldots, m_k{=}e_k\rangle \leftarrow m_i{=}e_i'\rangle \quad = \quad \langle m_1{=}e_1, \ldots, m_i{=}e_i', \ldots, m_k{=}e_k\rangle$$

we may reach the conclusion

$$\begin{aligned} p \Leftarrow \texttt{move}\,2 = \langle\, & x = \lambda \texttt{self}.\,5, \\ & \texttt{move} = \lambda \texttt{self}.\,\lambda dx.\langle \ldots \rangle\rangle \end{aligned}$$

showing that the result of sending a move message with an integer parameter is an object identical to p, but with an updated x-coordinate.

2.3 Inheritance

The inheritance mechanism of this calculus is very simple, although its typing is somewhat complex. To define colored points from points, we simply need to add a color method to our point object p.

$$\texttt{cp} \stackrel{def}{=} \langle p \leftarrow\!\!+\, \texttt{myColor} = \lambda\texttt{self}.\,\texttt{blue}\rangle$$

If we send the move message to cp with the same parameter, we may calculate the resulting object in exactly the same way as before:

$$\begin{aligned}\texttt{cp} \Leftarrow \texttt{move}\,2 &= (\lambda\texttt{self}.\,\lambda\texttt{dx}.\langle \ldots \rangle)\ \texttt{cp}\ 2\\ &= \langle \texttt{cp} \leftarrow \texttt{x} = \lambda\texttt{s}.(\texttt{cp} \Leftarrow \texttt{x}) + 2\rangle\\ &= \ldots\\ &= \langle \texttt{cp} \leftarrow \texttt{x} = \lambda\texttt{self}.\,5\rangle\end{aligned}$$

with the final conclusion that

$$\begin{aligned}\texttt{cp} \Leftarrow \texttt{move}\,2 = \langle\ & \texttt{x} = \lambda\texttt{self}.\,5,\\ & \texttt{move} = \lambda\texttt{self}.\,\lambda\texttt{dx}.\langle \ldots \rangle,\\ & \texttt{myColor} = \lambda\texttt{self}.\,\texttt{blue}\\ \rangle &\end{aligned}$$

The important feature of this computation is that the color of the resulting colored point is the same as the original one. While move was defined originally for points, the method body performs the correct computation when the method is inherited by an object with additional methods. Hence we have method specialization in the calculus.

2.4 Operational semantics

In defining the operational semantics for our calculus, we must give rules for extracting and applying the appropriate method of an object. A natural way to approach this would be to use a permutation rule

$$\langle\langle \texttt{e}_1 \leftarrow\!\!\circ\ \texttt{n}{=}\texttt{e}_2\rangle \leftarrow\!\!\circ\ \texttt{m}{=}\texttt{e}_3\rangle = \langle\langle \texttt{e}_1 \leftarrow\!\!\circ\ \texttt{m}{=}\texttt{e}_3\rangle \leftarrow\!\!\circ\ \texttt{n}{=}\texttt{e}_2\rangle$$

where m and n are distinct and each occurrence of $\leftarrow\!\!\circ$ may be either $\leftarrow\!\!+$ or $\leftarrow$. This equational rule would let us treat objects as sets of methods, rather than ordered sequences. However, this rule would cause typing complications, since our typing rules only allow us to type object expressions when methods are added in an appropriate order. In particular, if we permute the methods of the object expression

$$\begin{aligned}\langle\langle\langle\rangle \leftarrow\!\!+\, & \texttt{x} = \lambda\texttt{self}.\,3\rangle\\ \leftarrow\!\!+\, & \texttt{x\_succ} = \lambda\texttt{self}.(\texttt{self} \Leftarrow \texttt{x}) + 1\rangle\end{aligned}$$

then the subexpression

$$\langle\langle\rangle \leftarrow\!\!+\, \texttt{x\_succ} = \lambda\texttt{self}.(\texttt{self} \Leftarrow \texttt{x}) + 1\rangle$$

is not well-typed. Therefore, the entire expression cannot be typed.

We circumvent the problem of method order by using a more complicated "standard form" for object expressions, namely,

$$\langle\langle\langle \mathtt{m}_1 = \mathtt{e}_1, \ldots \mathtt{m}_k = \mathtt{e}_k\rangle \leftarrow \mathtt{m}_1 = \mathtt{e}'_1\rangle \ldots \leftarrow \mathtt{m}_k = \mathtt{e}'_k\rangle$$

where each method is defined exactly once, using some arbitrary method body that does not contribute to the observable behavior of the object, and then redefined exactly once by giving the desired method body. Even if the two definitions of a method are the same, this form is useful since it allows us to permute the list of method redefinitions arbitrarily. More formally, in addition to the $\stackrel{eval}{\longrightarrow}$ relation that allows us to evaluate message sends and function applications, the operational semantics includes a subsidiary "bookkeeping" relation $\stackrel{book}{\longrightarrow}$, which allows each object to be transformed into the "standard form" indicated above. The relation $\stackrel{book}{\longrightarrow}$ is the congruence closure of the first four clauses listed in Table 1. These rules also allow the method redefinitions to be permuted arbitrarily. An important property of $\stackrel{book}{\longrightarrow}$, proved in Section 5, is that if $\mathtt{e} \stackrel{book}{\longrightarrow} \mathtt{e}'$, then any type for $\mathtt{e}$ is also derivable for $\mathtt{e}'$. This property would fail if we had the more general permutation rule discussed above.

(*switch ext ov*)	$\langle\langle \mathtt{e}_1 \leftarrow \mathtt{n}=\mathtt{e}_2\rangle \leftarrow\!\!+\, \mathtt{m}=\mathtt{e}_3\rangle$	$\stackrel{book}{\longrightarrow} \langle\langle \mathtt{e}_1 \leftarrow\!\!+\, \mathtt{m}=\mathtt{e}_3\rangle \leftarrow \mathtt{n}=\mathtt{e}_2\rangle$
(*perm ov ov*)	$\langle\langle \mathtt{e}_1 \leftarrow \mathtt{m}=\mathtt{e}_2\rangle \leftarrow \mathtt{n}=\mathtt{e}_3\rangle$	$\stackrel{book}{\longrightarrow} \langle\langle \mathtt{e}_1 \leftarrow \mathtt{n}=\mathtt{e}_3\rangle \leftarrow \mathtt{m}=\mathtt{e}_2\rangle$
(*add ov*)	$\langle \mathtt{e}_1 \leftarrow\!\!+\, \mathtt{m}=\mathtt{e}_3\rangle$	$\stackrel{book}{\longrightarrow} \langle\langle \mathtt{e}_1 \leftarrow\!\!+\, \mathtt{m}=\mathtt{e}_3\rangle \leftarrow \mathtt{m}=\mathtt{e}_3\rangle$
(*cancel ov ov*)	$\langle\langle \mathtt{e}_1 \leftarrow \mathtt{m}=\mathtt{e}_2\rangle \leftarrow \mathtt{m}=\mathtt{e}_3\rangle$	$\stackrel{book}{\longrightarrow} \langle \mathtt{e}_1 \leftarrow \mathtt{m}=\mathtt{e}_3\rangle$
(β)	$(\lambda \mathtt{x}.\, \mathtt{e}_1)\mathtt{e}_2$	$\stackrel{eval}{\longrightarrow} [\mathtt{e}_2/\mathtt{x}]\mathtt{e}_1$
$(\Leftarrow)$	$\langle \mathtt{e}_1 \leftarrow\!\!\circ\, \mathtt{m}=\mathtt{e}_2\rangle \Leftarrow \mathtt{m}$	$\stackrel{eval}{\longrightarrow} \mathtt{e}_2\langle \mathtt{e}_1 \leftarrow\!\!\circ\, \mathtt{m}=\mathtt{e}_2\rangle$ where $\leftarrow\!\!\circ$ may either $\leftarrow\!\!+$ or $\leftarrow$.

Table 1. Bookkeeping and evaluation rules.

3 Static Type System

3.1 Subtyping and delegation

In general, there are two forms of subtyping that are applicable to object types: *width* and *depth* subtyping. If one object type provides all of the methods of another, with the same types, then every object of the first type should be safely substitutable in any context expecting an object with the second type. We call this form of subtyping *width* subtyping. It is also generally possible for one object type to restrict the type of one or more of its methods to a subtype. Again, any object with the more restricted type should be safely substitutable in any context expecting an object with the less restricted type. We call this form of subtyping *depth* subtyping.

As explained in [FM94], pure-delegation based languages, such as that described in [FHM94], do not permit either form of object subtyping. Consider the intuitive definition of a subtype: `A` is a subtype of `B` if we may use an object of type `A` in any context expecting an object of type `B`. If objects of type `A` are to be used as `B`'s, then `A`-objects must have all of the methods of `B`-objects. Because method addition is a legal operation on objects in delegation-based languages, objects with extra methods cannot be used in some contexts where an object with fewer methods may. As an example, a colored point object cannot be used in a context that will add color, but a point object can. For `A` to be a subtype of `B` then, `A`'s must contain exactly the same methods as `B`'s. Hence we do not get width subtyping.

We do not get depth subtyping either, as depth subtyping is unsound in the presence of method override. The following example, given in [AC94] illustrate the problem. Consider the object types `A` and `B`:

$$\mathtt{B} \stackrel{def}{=} \mathbf{prot.}\langle\!|\mathtt{x}:\mathtt{Int},\ \mathtt{y}:\mathtt{Real}|\!\rangle$$
$$\mathtt{A} \stackrel{def}{=} \mathbf{prot.}\langle\!|\mathtt{x}:\mathtt{PosInt},\ \mathtt{y}:\mathtt{Real}|\!\rangle$$

If we allow deep record/object subtyping, $\mathtt{A} <: \mathtt{B}$ since $\mathtt{PosInt} <: \mathtt{Int}$. Now consider an object `a` defined as follows:

$$\mathtt{a} \stackrel{def}{=} \langle \mathtt{x} = 1,\ \mathtt{y} = \mathtt{ln}(\mathtt{self} \Leftarrow \mathtt{x})\rangle$$

We can see that `a` has type A. By the subsumption rule, we may consider `a` to have type B. With this typing, the expression $\langle \mathtt{a} \leftarrow \mathtt{x} = -1\rangle$ is legal. But then sending the message `y` to `a` produces a run-time type error.

In the remainder of this section, we give a type system for the object calculus described in Section 2 that is an extension of the one given in [FHM94]. The major change to the earlier system is the addition of *obj* types, which permit subtyping in both width and depth. In both the earlier system and the one we describe here, expressions with *pro* type (called *class* type in [FHM94]) may have new methods added via method addition ($\leftarrow\!\!+$) and old methods replaced via method override($\leftarrow$). As described above, only trivial subtyping is sound for such types. In the system presented here, we may decide that we are no longer interested in modifying the methods of an expression with *pro* type. At this point, we may "seal" the expression by converting it to a value with an *obj* type. Since $\leftarrow$ and $\leftarrow\!\!+$ are not valid operations on *obj* types, width and depth subtyping are sound for these types. We call expressions given *pro* type *prototypes* because they may either be sent messages or modified to create new kinds of prototypes. Expressions with *obj* type are called *objects* since such expressions may be sent messages or subtyped, but not extended or modified. In the following, we will use the meta-variable *probj* to denote either *obj* or *pro*.

3.2 The Type System

The type defined by the type expression

$$\mathbf{probj}\,\mathtt{t.}\langle\!\langle \mathtt{m}_1 : \tau_1, \ldots, \mathtt{m}_k : \tau_k \rangle\!\rangle$$

is a type $\mathtt{t}$ with the property that when we send the message $\mathtt{m}_i$ for $1 \le i \le k$ to any element $\mathtt{x}$ of this type, the result has type τ_i. A significant aspect of this type is that the bound type variable $\mathtt{t}$ may appear in the types $\tau_1, \ldots, \tau_k$. Thus, when we say $\mathtt{x} \Leftarrow \mathtt{m}_i$ will have type τ_i, we mean type τ_i with any free occurrences of $\mathtt{t}$ in τ_i referring to the type $\mathbf{probj}\,\mathtt{t.}\langle\!\langle \mathtt{m}_1 : \tau_1, \ldots, \mathtt{m}_k : \tau_k \rangle\!\rangle$ itself. Thus, $\mathbf{probj}\,\mathtt{t.}\langle\!\langle \ldots \rangle\!\rangle$ is a special form of recursively-defined type.

The typing rule for message send has the form

$$\frac{\mathtt{e} : \mathbf{probj}\,\mathtt{t.}\langle\!\langle \ldots\, \mathtt{m} : \tau \rangle\!\rangle}{\mathtt{e} \Leftarrow \mathtt{m} : [\mathbf{probj}\,\mathtt{t.}\langle\!\langle \ldots\, \mathtt{m} : \tau \rangle\!\rangle / \mathtt{t}]\tau}$$

where the substitution for $\mathtt{t}$ in τ reflects the recursive nature of object types. This rule may be used in its *pro* form to give the point $\mathtt{p}$ with $\mathtt{x}$ and $\mathtt{move}$ methods type

$$\mathbf{pro}\,\mathtt{t.}\langle\!\langle \mathtt{x} : \mathtt{Int}, \mathtt{move} : \mathtt{Int} \rightarrow \mathtt{t} \rangle\!\rangle$$

since $\mathtt{p} \Leftarrow \mathtt{x}$ returns an integer and $\mathtt{p} \Leftarrow \mathtt{move}\ \mathtt{n}$ has the same type as $\mathtt{p}$.

A subtle but very important aspect of the type system is that when an object is extended with an additional method, the syntactic type of each method does not change. For example, when we extended $\mathtt{p}$ with a color to obtain $\mathtt{cp}$, we obtain an object with type

$$\mathbf{pro}\,\mathtt{t.}\langle\!\langle \mathtt{x} : \mathtt{Int}, \mathtt{move} : \mathtt{Int} \rightarrow \mathtt{t}, \mathtt{myColor} : \mathtt{Color} \rangle\!\rangle$$

The important change to notice here is that although the syntactic type of $\mathtt{move}$ is still $\mathtt{Int} \rightarrow \mathtt{t}$, the meaning of the variable $\mathtt{t}$ has changed. Instead of referring to the type of $\mathtt{p}$ as it did originally, it now refers to the type of $\mathtt{cp}$. This behavior is the method specialization discussed briefly in Section 1. For this kind of reinterpretation of type variables to be sound, the typing rules for object extension and object override must insure that every possible type for a new method will be correct. This check is done through a form of implicit higher-order polymorphism, which is explained more fully in [FHM94].

3.3 Adding subtyping

To insure that subtyping is sound for *obj* expressions, we disable object extension and object override on them, permitting them only to receive messages. Since the methods handling such messages are defined when the expression has a *pro* type, however, it is possible for these methods to redefine themselves or other methods of the host expression. (The polymorphism requirement for method bodies prevents them from adding new methods.) These redefinitions do not cause an unsoundness in the type system, however, because they were type-checked with precise knowledge of the *pro* type of the expression to which they were added. The unsoundness discussed above arises when the type of an expression is promoted via subsumption and then a method is replaced by another method with higher type. This scenario cannot occur in our system because once we seal an expression to a potentially imprecise *obj* type, methods cannot be overridden from the outside, where only the imprecise sealed type is available. Internal methods, type-checked with precise *pro* type information about the expression to which they were added, are type-safe.

Since sealing a prototype $\mathtt{e}$ requires no change to $\mathtt{e}$, we would have liked to be able to use the following rule in conjunction with a subsumption rule to "seal" prototypes:

$$(\textit{seal-unsound}) \qquad \frac{\Gamma \vdash \mathbf{obj}\, t\mathbf{.}R : T}{\Gamma \vdash \mathbf{pro}\, t\mathbf{.}R <: \mathbf{obj}\, t\mathbf{.}R}$$

where the judgement $\Gamma \vdash \mathbf{obj}\, t\mathbf{.}R : T$ denotes that type $\mathbf{obj}\, t\mathbf{.}R$ is well formed. Unfortunately, this rule is unsound when the variable t appears contravariantly (or invariantly) in R. Hence we need a more complicated rule that has the same form as the rule for deriving subtyping judgements on *obj* types. Because of this similarity, we may combine the sealing rule and the *obj* subtyping rule to produce:

$$(<:\ \textit{object}) \qquad \frac{\begin{array}{c} \Gamma \vdash \mathbf{probj}\, s_1\mathbf{.}\langle\!\langle \vec{m} : \vec{\tau},\ \vec{n} : \vec{\sigma} \rangle\!\rangle : T \\ \Gamma \vdash \mathbf{obj}\, s_2\mathbf{.}\langle\!\langle \vec{m} : \vec{\tau}' \rangle\!\rangle : T \\ \Gamma, s_1 <: s_2 \vdash \tau_i <: \tau_i' \\ s_1 \notin FV(\tau_i'),\quad s_2 \notin FV(\tau_i) \end{array}}{\Gamma \vdash \mathbf{probj}\, s_1\mathbf{.}\langle\!\langle \vec{m} : \vec{\tau},\ \vec{n} : \vec{\sigma} \rangle\!\rangle <: \mathbf{obj}\, s_2\mathbf{.}\langle\!\langle \vec{m} : \vec{\tau}' \rangle\!\rangle}$$

The meta-variable *probj* is *pro* when this rule is used to seal prototypes. When used simply for subtyping between object types, *probj* is *obj*. Notice that this rule gives us subtyping in both width and depth for *obj* types. This rule has the somewhat unfortunate consequence that we cannot seal prototypes to *obj* types containing methods that are contravariant (or invariant) in the bound type variable. However, this appears to be a fundamental trade-off and limitation.

The other rules in the type system are either routine or the same as those in [FHM94]. A complete list of the typing rules appears in Appendix A.

3.4 Rows, types, and kinds

More formally, the type expressions are given by:

Types
$$\tau \ ::= \ t \,|\, s \,|\, \tau_1 \to \tau_2 \,|\, \mathbf{pro}\,t.R \,|\, \mathbf{obj}\,t.R$$

Rows
$$R \ ::= \ r \,|\, \langle\!\rangle \,|\, \langle\!R \,|\, m : \tau\!\rangle \,|\, \lambda t.\, R \,|\, R\tau$$

Kinds
$$kind \ ::= \ T \,|\, \kappa$$
$$\kappa \ ::= \ T^n \to [m_1, \ldots, m_k]$$

The row expressions, which are essentially lists of method names and associated types, appear as subexpressions of type expressions. Types and rows are distinguished by kinds. As a notational simplification, we write $[m_1, \ldots, m_k]$ for $T^0 \to [m_1, \ldots, m_k]$. Intuitively, the elements of kind $[m_1, \ldots, m_k]$ are rows that do *not* include method names $m_1, \ldots, m_k$. The reason we must know statically that some method does not appear is to guarantee that methods are not multiply defined. Kinds of the form $T^n \to [m_1, \ldots, m_k]$, for $n \geq 1$, are used to infer a form of higher-order polymorphism of method bodies.

The environments, or contexts, of the system list term, type, and row variables.
$$\Gamma \ ::= \ \epsilon \,|\, \Gamma, x : \tau \,|\, \Gamma, t : T \,|\, \Gamma, r : \kappa \,|\, \Gamma, s_1 <: s_2$$

Note that contexts are ordered lists, not sets.

The judgement forms are standard:

$\Gamma \vdash *$	well-formed context
$\Gamma \vdash e : \tau$	term has type
$\Gamma \vdash \tau : T$	well-formed type
$\Gamma \vdash R : \kappa$	row has kind
$\Gamma \vdash \tau_1 <: \tau_2$	τ_1 subtype of τ_2

4 Examples

To give some intuition for this calculus, we present two examples, neither of which is typable in the system of [FHM94]. Other examples, demonstrating the expressivity of the system, appear in [FHM94]. The first example below illustrates the use of subtyping for code reuse, while the second demonstrates its use for encapsulation.

4.1 Subtyping for code reuse

Consider the function

$$\mathbf{average} \stackrel{def}{=} \lambda \mathtt{p_1}.\, \lambda \mathtt{p_2}.((\mathtt{p_1} \Leftarrow \mathtt{x}) + (\mathtt{p_2} \Leftarrow \mathtt{x}))/2$$

We may give this expression type:

$$\texttt{average} \;:\; \texttt{obj}\,t.\langle\!\langle x : \texttt{int}\rangle\!\rangle \rightarrow \texttt{obj}\,t.\langle\!\langle x : \texttt{int}\rangle\!\rangle \rightarrow \texttt{int}$$

(assuming integer division). Now consider the following two expressions:

$$\begin{aligned} \texttt{p} &\stackrel{def}{=} \langle\, x = \lambda \texttt{self}.3, \\ &\qquad \texttt{move} = \lambda \texttt{self}.\, \lambda dx.\, \langle \texttt{self} \leftarrow x = \lambda s.(\texttt{self} \Leftarrow x) + dx \rangle \rangle \\ \texttt{cp} &\stackrel{def}{=} \langle\, \texttt{p} \leftarrow\!\!+\, c = \lambda \texttt{self}.\texttt{red}\ \rangle \end{aligned}$$

As in the system in [FHM94], we may give p and cp the following types:

$$\begin{aligned} &\texttt{p} : \texttt{pro}\,t.\langle\!\langle x : \texttt{int}, \texttt{move} : \texttt{int} \rightarrow t \rangle\!\rangle \\ &\texttt{cp} : \texttt{pro}\,t.\langle\!\langle x : \texttt{int}, \texttt{move} : \texttt{int} \rightarrow t, c : \texttt{colors} \rangle\!\rangle \end{aligned}$$

In the new system, we may seal these expressions and give them both the following common supertype:

$$\texttt{p}, \texttt{cp} \;:\; \texttt{obj}\,t.\langle\!\langle x : \texttt{int}\rangle\!\rangle$$

Hence we may give the expression **average p cp** type **int**, allowing us to use the function **average** to calculate the average of colored points and points interchangeably.

4.2 Subtyping for encapsulation

Once we change the type of an expression from *pro* type to *obj* type, the methods of that prototype become “read-only”, including those methods that are effectively data fields. Hence if the designers of a prototype wish its users to be able to change the values of some of its methods after it is sealed, they must provide methods to change those values. This restriction provides a mechanism whereby prototype designers can guarantee invariants for the objects created from their prototypes.

Suppose, for example, we were interested in colored points whose color was guaranteed to be either blue or green. Then consider the expression:

$$\begin{aligned} \texttt{cp}_2 \stackrel{def}{=} \langle\, & x = \ldots, \\ & \texttt{move} = \ldots, \\ & c = \lambda \texttt{self}.\texttt{blue}, \\ & \texttt{makeBlue} = \lambda \texttt{self}.\langle \texttt{self} \leftarrow c = \texttt{blue} \rangle, \\ & \texttt{makeGreen} = \lambda \texttt{self}.\langle \texttt{self} \leftarrow c = \texttt{green} \rangle, \\ \rangle & \end{aligned}$$

As in the original system, we may give this expression type:

$$\texttt{cp}_2 : \texttt{pro}\,t.\langle\!\langle x : \texttt{int}, \texttt{move} : \texttt{int} \rightarrow t, c : \texttt{colors}, \texttt{makeBlue} : t, \texttt{makeGreen} : t \rangle\!\rangle$$

With this type, the user may override cp_2's `c` method to give cp_2 any color. However, we may prevent this undesired change by sealing cp_2. We may use the rule ($<:$ *object*) and subsumption to give cp_2 the type:

$$cp_2 : \mathbf{obj}\,t.\langle\!\langle \texttt{x} : \texttt{int},\ \texttt{move} : \texttt{int} \to t,\ \texttt{c} : \texttt{colors},\ \texttt{makeBlue} : t,\ \texttt{makeGreen} : t \rangle\!\rangle$$

With this type, users may no longer set the color of cp_2 by overridding its value for `c`. Now they may only change its color by sending the messages `makeBlue` and `makeGreen`, which guarantees that the object only takes on legal colors.

5 Soundness

We prove the soundness of our type system with respect to the operational semantics given in Section 2.4. We first prove that evaluation preserves type; this property is traditionally called subject reduction. We then show that no typable expression evaluates to *error* using the evaluation rules given in [FHM94]. This fact guarantees that we have no message-not-understood errors for expressions with either *pro* or *obj* types.

The proof begins with two lemmas about substitution for row and type variables. We then prove a normal form lemma that allows us to restrict our attention to derivations of a certain form, simplifying later proofs. Lemmas 4 and 5 give us subject reduction for the bookkeeping rules and β-reduction, respectively. Lemma 6 implies subject reduction for ($\Leftarrow$)-reduction on expressions with *pro* type, while Lemmas 7 and 8 give us subject reduction for ($\Leftarrow$)-reduction on expressions with *obj* type.

The following two substitution lemmas are used in conjunction with row variables to specialize *pro* types to contain additional methods.

Lemma 1. *If* $\Gamma, u_2 : V_2, \Gamma' \vdash U_1 : V_1$ *and* $\Gamma \vdash U_2 : V_2$ *are both derivable, then so is* $\Gamma, [U_2/u_2]\Gamma' \vdash [U_2/u_2]U_1 : V_1$*, where* $U_1 : V_1$ *and* $U_2 : V_2$ *are either both* $\tau : T$ *or both* $R : \kappa$.

Lemma 2. *If* $\Gamma, r : T^n \to [m_1, \ldots, m_k], \Gamma' \vdash e : \tau$ *and* $\Gamma \vdash R : T^n \to [m_1, \ldots, m_k]$ *are both derivable, then so is* $\Gamma, [R/r]\Gamma' \vdash e : [R/r]\tau$.

The type and row equality rules introduce many non-essential judgement derivations, which unnecessarily complicate derivation analysis. We therefore restrict our attention to $\vdash_N$-derivations, which we define as those derivations in which the only appearance of a type or row equality rule is as ($row\beta$) immediately following an occurrence of (*row fn app*). The τnf of a type or row expression is its normal form with respect to ordinary β-reduction. The τnf of a term expression e is just e. Since we are only interested in types and rows in τnf, the following lemma shows we can find a $\vdash_N$-derivation for any judgement of interest.

Lemma 3. *If* $\Gamma \vdash A$ *is derivable, then so is* $\tau nf(\Gamma) \vdash_N \tau nf(A)$.

The proof of this lemma is by induction on the derivation of $\Gamma \vdash A$. Occurrences of equality rules may be eliminated in the $\vdash_N$-derivation because two row or type expressions related via β-reduction must have the same τnf.

From this point on, we will only concern ourselves with contexts and type or row expressions that are in τnf. This limitation is not severe, since any term that has a type has a type in τnf. Future analyses of derivations will consider only $\vdash_N$-derivations, since its restriction on equality rules greatly simplifies the proofs.

The next three lemmas show that the various components of the $\stackrel{eval}{\longrightarrow}$ relation preserve expression types. Lemma 4 is the first of these, showing that the $(\stackrel{book}{\longrightarrow})$ relation has the necessary property.

Lemma 4. *If $\Gamma \vdash \mathtt{e} : \tau$ is derivable, and $\mathtt{e} \stackrel{book}{\longrightarrow} \mathtt{e}'$, then $\Gamma \vdash \mathtt{e}' : \tau$ is derivable.*

The proof of Lemma 4 consists of two parts: the first shows that a derivation from $\Gamma \vdash \mathtt{e} : \tau$ can only depend on the form of τ, not on the form of $\mathtt{e}$. More formally, if $\Gamma \vdash \mathtt{C}[\mathtt{e}] : \tau$ is derived from $\Gamma' \vdash \mathtt{e} : \sigma$ and $\Gamma' \vdash \mathtt{e}' : \sigma$ is also derivable, then so is $\Gamma \vdash \mathtt{C}[\mathtt{e}'] : \tau$. This fact is easily seen by an inspection of the typing rules. The second part shows that if $\Gamma \vdash \mathtt{e} : \tau$ is derivable, and $\mathtt{e} \stackrel{book}{\longrightarrow} \mathtt{e}'$ by $\mathtt{e}$ matching the left-hand side of one of the $(\stackrel{book}{\longrightarrow})$ axioms, then $\Gamma \vdash \mathtt{e}' : \tau$ is also derivable. This fact follows from a case analysis of the four $(\stackrel{book}{\longrightarrow})$ axioms. Row variables and Lemma 2 are essential for the (*switch ext ov*) case.

The fact that (β)-reduction preserves expression types is an immediate consequence of the following lemma:

Lemma 5. *If $\Gamma, \mathtt{x} : \tau_1, \Gamma' \vdash \mathtt{e}_2 : \tau_2$ and $\Gamma \vdash \mathtt{e}_1 : \tau_1$ are both derivable, then so is $\Gamma, \Gamma' \vdash [\mathtt{e}_1/\mathtt{x}]\mathtt{e}_2 : \tau_2$.*

Lemma 5 is proved by induction on the derivation of $\Gamma, \mathtt{x} : \tau_1, \Gamma' \vdash \mathtt{e}_2 : \tau_2$.

The following three lemmas are useful in showing that $(\Leftarrow)$-reduction preserves expression types.

Lemma 6. *If $\Gamma \vdash \langle\langle \mathtt{e} \leftarrow\!\!\!\circ\ \mathtt{m}_1{=}\mathtt{e}_1\rangle \ldots \leftarrow\!\!\!\circ\ \mathtt{m}_k{=}\mathtt{e}_k\rangle : \mathbf{pro}\,t.\langle\!\langle R \,|\, m_1{:}\tau_1, \ldots, m_k{:}\tau_k\rangle\!\rangle$ is derivable, where $m_1, \ldots, m_k$ are distinct and are precisely the method names that occur consecutively to the right of $\mathtt{e}$, then $\Gamma \vdash \mathtt{e}_i : [\mathbf{pro}\,t.\langle\!\langle R \,|\, m_1 : \tau_1, \ldots, m_k : \tau_k\rangle\!\rangle/t](t \rightarrow \tau_i)$. is also derivable.*

The proof of this lemma also has two parts. Part A gives a derivable type for each object obtained by stripping off some number of the $\leftarrow\!\!\!\circ$ operations. The appropriate derivation is obtained by induction on the number of $\leftarrow\!\!\!\circ$ operations removed. Part B gives the required type for each of the $\mathtt{e}_i$'s. Row variables and Lemma 2 are critical to its proof, which is an analysis of the derivation of the type of $\langle\langle \mathtt{e} \leftarrow\!\!\!\circ\ \mathtt{m}_1{=}\mathtt{e}_1\rangle \ldots \leftarrow\!\!\!\circ\ \mathtt{m}_i{=}\mathtt{e}_i\rangle$, given by part A.

The key insight in the soundness of the subtyping of the calculus is that if we derive that an expression $ob \stackrel{def}{=} \langle\langle e \leftarrow\!\!\!\circ_1\ m_1{=}e_1\rangle \ldots \leftarrow\!\!\!\circ_k\ m_k{=}e_k\rangle$ has *obj* type, then a series of lemmas about typing derivations show that we must have derived that *ob* had a related *pro* type. Furthermore, the methods of *ob* must have been typed checked under the assumption that *ob* had that *pro* type. Hence any $\leftarrow$ or $\leftarrow\!\!+$ operations in the methods of *ob* were type checked with full knowledge of *ob*'s most general form and are thus type-safe. Technically, we may state this insight as:

Lemma 7. *If we may derive* $\Gamma \vdash ob : \tau$, *then there exists some type* $\mathbf{pro}\, t.\langle\!\langle R \,|\, \overrightarrow{m} : \overrightarrow{\tau} \rangle\!\rangle$ *such that we may derive in no more steps* $\Gamma \vdash ob : \mathbf{pro}\, t.\langle\!\langle R \,|\, \overrightarrow{m} : \overrightarrow{\tau} \rangle\!\rangle$ *and also* $\Gamma \vdash \mathbf{pro}\, t.\langle\!\langle R \,|\, \overrightarrow{m} : \overrightarrow{\tau} \rangle\!\rangle <: \tau$, *where* $\overrightarrow{m} : \overrightarrow{\tau}$ *is shorthand for* $m_1 : \tau_1, \ldots, m_k : \tau_k$.

To prove subject reduction for the operational semantics, we need to show that each of the basic evaluation rules preserves type. For the ($\Leftarrow$)-reduction rule, this corresponds to showing that if we may derive $\Gamma \vdash \langle \mathsf{e}_1 \leftarrow\!\!\circ\, \mathtt{m} = \mathsf{e}_2 \rangle \Leftarrow \mathtt{m} : \tau$ then we may also derive $\Gamma \vdash \mathsf{e}_2 \langle \mathsf{e}_1 \leftarrow\!\!\circ\, \mathtt{m} = \mathsf{e}_2 \rangle : \tau$. This fact is proved by induction on the derivation of $\Gamma \vdash \langle \mathsf{e}_1 \leftarrow\!\!\circ\, \mathtt{m} = \mathsf{e}_2 \rangle \Leftarrow \mathtt{m} : \tau$. The following lemma is important for the (*object meth app*) case of this proof:

Lemma 8. *If* $\Gamma \vdash \mathbf{probj}\, s_1.\langle\!\langle R_1 \,|\, m : \tau_1 \rangle\!\rangle <: \mathbf{probj}\, s_2.\langle\!\langle R_2 \,|\, m : \tau_2 \rangle\!\rangle$, *is derivable, then* $\Gamma \vdash [\mathbf{probj}\, s_1.\langle\!\langle R_1 \,|\, m : \tau_1 \rangle\!\rangle / s_1]\tau_1 <: [\mathbf{probj}\, s_2.\langle\!\langle R_2 \,|\, m : \tau_2 \rangle\!\rangle / s_2]\tau_2$ *is also derivable.*

Theorem 9 Subject Reduction. *If* $\Gamma \vdash e : \tau$ *is derivable, and* $e \xrightarrow{eval} e'$, *then* $\Gamma \vdash e' : \tau$ *is also derivable.*

The proof is similar in outline to that of Lemma 4; it reduces to showing that each of the basic evaluation steps preserves the type of the expression being reduced. The ($\xrightarrow{book}$) case follows from Lemma 4, the (β) case from Lemma 5 and the ($\Leftarrow$) case from Lemmas 6, 7, and 8.

Theorem 10 Type Soundness. *If the judgement* $\epsilon \vdash e : \tau$ *is derivable, then* $eval(e) \neq error$, *where the function eval is as in [FHM94].*

6 Related work

The authors of [BL95] approach the trade-off between delegation and subtyping in a very different way. Their system, which is an extension of [FHM94], permits a limited form of width subtyping on expressions with (what we call) *pro* type. In particular, they consider $\mathbf{pro}\, t.R$ a subtype of $\mathbf{pro}\, t.R'$ so long as R contains all the methods of R' (with matching types) and none of the "forgotten" methods are referred to in the methods listed in R'. This second condition guarantees that if we forget about the presence of some method m via subsumption and then re-add m with a potentially unrelated type, we cannot violate any typing assumptions the other methods of the object may have made about m. This guarantee is essential to the soundness of this approach to adding subtyping; however, its inability to "forget" methods referred to by "remembered" methods is somewhat unfortunate. In particular, it cannot type examples such as that given in Section 4.2, where desired invariants about an object are maintained by restricting access to certain methods via subsumption.

7 Conclusion

In this paper, we present a type system for an object calculus that extends previous work in [Mit90, FHM94]. The main additions to our previous object calculus are the distinction between prototypes and objects and the addition of subtyping. The distinction between prototypes, which allow method addition and override, and objects, which only receive messages, is essential for subtyping, since previous studies show an incompatibility between object extension, method override, and non-trivial subtyping [FM94]. We have devised a type soundness proof for our calculus of prototypes, objects, and subtyping based on a straightforward operational semantics.

References

[AC94] M. Abadi and L. Cardelli. A theory of primitive objects: untyped and first-order systems. In *Proc. Theor. Aspects of Computer Software*, pages 296–320. Springer-Verlag LNCS 789, 1994.

[AC95] M. Abadi and L. Cardelli. An imperative object calculus: Basic typing and soundness. In *SIPL '95 - Proc. Second ACM SIGPLAN Workshop on State in Programming Languages*. Technical Report UIUCDCS-R-95-1900, Department of Computer Science, University of Illinois at Urbana-Champaign, 1995.

[BL95] V. Bono and L. Liquori. A subtyping for the fisher-honsell-mitchell lambda calculus of objects. Berlin, June 1995. Springer LNCS 933. To appear.

[Bru93] K. Bruce. Safe type checking in a statically-typed object-oriented programming language. In *Proc 20th ACM Symp. Principles of Programming Languages*, pages 285–298, 1993.

[Bv93] K. Bruce and R. van Gent. TOIL: a new type-safe object-oriented imperative language. Manuscript, 1993.

[CU89] C. Chambers and D. Ungar. Customization: Optimizing compiler technology for Self, a dynamically-typed object-oriented programming language. In *SIGPLAN '89 Conf. on Programming Language Design and Implementation*, pages 146–160, 1989.

[ES90] M. Ellis and B. Stroustrop. *The Annotated C^{++} Reference Manual.* Addison-Wesley, 1990.

[FHM94] K. Fisher, F. Honsell, and J.C. Mitchell. A lambda calculus of objects and method specialization. *Nordic J. Computing (*formerly *BIT)*, 1:3–37, 1994. Preliminary version appeared in *Proc. IEEE Symp. on Logic in Computer Science*, 1993, 26–38.

[FM94] K. Fisher and J.C. Mitchell. Notes on typed object-oriented programming. In *Proc. Theoretical Aspects of Computer Software*, pages 844–885. Springer LNCS 789, 1994.

[GR83] A. Goldberg and D. Robson. *Smalltalk-80: The language and its implementation.* Addison Wesley, 1983.

[Mey92] B. Meyer. *Eiffel: The Language.* Prentice-Hall, 1992.

[Mit90] J.C. Mitchell. Toward a typed foundation for method specialization and inheritance. In *Proc. 17th ACM Symp. on Principles of Programming Languages*, pages 109–124, January 1990.

[Pie93] Benjamin C. Pierce. Mutable objects. Draft report; available electronically, June 1993.

[PT94] Benjamin C. Pierce and David N. Turner. Simple type-theoretic foundations for object-oriented programming. *Journal of Functional Programming*, 4(2):207–248, 1994.

[Str86] B. Stroustrop. *The C^{++} Programming Language.* Addison-Wesley, 1986.

[US91] D. Ungar and R.B. Smith. Self: The power of simplicity. *Lisp and Symbolic Computation*, 4(3):187–206, 1991. Preliminary version appeared in *Proc. ACM Symp. on Object-Oriented Programming: Systems, Languages, and Applications*, 1987, 227-241.

A Type System

Types

$$\tau \ ::= \ t \,|\, s \,|\, \tau_1 \rightarrow \tau_2 \,|\, \mathbf{pro}\, t.R \,|\, \mathbf{obj}\, t.R$$

Rows

$$R \ ::= \ r \,|\, \langle\!\langle\rangle\!\rangle \,|\, \langle\!\langle R \,|\, m : \tau \rangle\!\rangle \,|\, \lambda t.\, R \,|\, R\tau$$

Kinds

$$kind \ ::= \ T \,|\, \kappa$$
$$\kappa \ ::= \ T^n \rightarrow [m_1, \ldots, m_k]$$

Contexts:

$$\Gamma \ ::= \ \epsilon \,|\, \Gamma, x : \tau \,|\, \Gamma, t : T \,|\, \Gamma, r : \kappa \,|\, \Gamma, s_1 <: s_2$$

The judgement forms are standard:

$\Gamma \vdash *$	well-formed context
$\Gamma \vdash e : \tau$	term has type
$\Gamma \vdash \tau : T$	well-formed type
$\Gamma \vdash \tau_1 <: \tau_2$	τ_1 subtype of τ_2
$\Gamma \vdash R : \kappa$	row has kind

A.1 Typing rules

General Rules

(*start*)

$$\overline{\epsilon \vdash *}$$

(*projection*)

$$\frac{\Gamma \vdash * \qquad u : v \in \Gamma}{\Gamma \vdash u : v}$$

(*weakening*)

$$\frac{\Gamma \vdash A \qquad \Gamma, \Gamma' \vdash *}{\Gamma, \Gamma' \vdash A}$$

Rules for type expressions

(*type var*)

$$\frac{\Gamma \vdash * \qquad t \notin dom(\Gamma)}{\Gamma, t : T \vdash *}$$

(*type arrow*)

$$\frac{\Gamma \vdash \tau_1 : T \qquad \Gamma \vdash \tau_2 : T}{\Gamma \vdash \tau_1 \rightarrow \tau_2 : T}$$

(*type pro*)

$$\frac{\Gamma, t : T \vdash R : [m_1, \ldots, m_k]}{\Gamma \vdash \mathbf{pro}\, t.R : T}$$

Type and Row Equality

Type or row expressions that differ only in names of bound variables or order of *label* : *type* pairs are considered identical. In other words, we consider α-conversion of type variables bound by λ, *pro*, or *obj* and applications of the principle

$$\langle\!\langle \langle\!\langle R \,|\, n : \tau_1 \rangle\!\rangle \,|\, m : \tau_2 \rangle\!\rangle = \langle\!\langle \langle\!\langle R \,|\, m : \tau_2 \rangle\!\rangle \,|\, n : \tau_1 \rangle\!\rangle$$

within type or row expressions to be conventions of syntax, rather than explicit rules of the system. Additional equations between types and rows arise as a result of β-reduction, written $\rightarrow_\beta$, or β-conversion, written $\leftrightarrow_\beta$.

(*row* β)

$$\frac{\Gamma \vdash R : \kappa, \quad R \rightarrow_\beta R'}{\Gamma \vdash R' : \kappa}$$

(*type* β)

$$\frac{\Gamma \vdash \tau : T, \quad \tau \rightarrow_\beta \tau'}{\Gamma \vdash \tau' : T}$$

(*type eq*) $$\frac{\Gamma \vdash e : \tau, \quad \tau \leftrightarrow_\beta \tau', \quad \Gamma \vdash \tau' : T}{\Gamma \vdash e : \tau'}$$

Rules for rows

(*empty row*) $$\frac{\Gamma \vdash *}{\Gamma \vdash \langle\!\rangle : [m_1, \ldots, m_k]}$$

(*row var*) $$\frac{\Gamma \vdash * \qquad r \notin dom(\Gamma)}{\Gamma, r : T^n \rightarrow [m_1, \ldots, m_k] \vdash *}$$

(*row label*) $$\frac{\Gamma \vdash R : T^n \rightarrow [m_1, \ldots, m_k] \qquad \{n_1, \ldots, n_\ell\} \subseteq \{m_1, \ldots, m_k\}}{\Gamma \vdash R : T^n \rightarrow [n_1, \ldots, n_\ell]}$$

(*row ext*) $$\frac{\Gamma \vdash R : [m, m_1, \ldots, m_k] \qquad \Gamma \vdash \tau : T}{\Gamma \vdash \langle\!\langle R \,|\, m : \tau \rangle\!\rangle : [m_1, \ldots, m_k]}$$

(*row fn abs*) $$\frac{\Gamma, t : T \vdash R : T^n \rightarrow [m_1, \ldots, m_k]}{\Gamma \vdash \lambda t.\, R : T^{n+1} \rightarrow [m_1, \ldots, m_k]}$$

(*row fn app*) $$\frac{\Gamma \vdash R : T^{n+1} \rightarrow [m_1, \ldots, m_k] \qquad \Gamma \vdash \tau : T}{\Gamma \vdash R\tau : T^n \rightarrow [m_1, \ldots, m_k]}$$

Rules for assigning types to terms

(*exp var*) $$\frac{\Gamma \vdash \tau : T \qquad x \notin dom(\Gamma)}{\Gamma, x : \tau \vdash *}$$

(*exp abs*) $$\frac{\Gamma, x : \tau_1 \vdash e : \tau_2}{\Gamma \vdash \lambda x.\, e : \tau_1 \rightarrow \tau_2}$$

(*exp app*) $$\frac{\Gamma \vdash e_1 : \tau_1 \rightarrow \tau_2 \quad \Gamma \vdash e_2 : \tau_1}{\Gamma \vdash e_1\, e_2 : \tau_2}$$

(*empty object*) $$\frac{\Gamma \vdash *}{\Gamma \vdash \langle\rangle : \mathbf{pro}\, t.\langle\!\rangle}$$

(*pro meth app*) $$\frac{\Gamma \vdash e : \mathbf{pro}\,t.\langle\!\langle R \mid m : \tau \rangle\!\rangle}{\Gamma \vdash e \Leftarrow m : [\mathbf{pro}\,t.\langle\!\langle R \mid m : \tau \rangle\!\rangle / t]\tau}$$

(*obj ext*) $$\frac{\begin{array}{l}\Gamma \vdash e_1 : \mathbf{pro}\,t.\langle\!\langle R \mid m_1 : \tau_1, \ldots, m_k : \tau_k \rangle\!\rangle \\ \Gamma, t : T \vdash R : [m_1, \ldots, m_k, n] \\ \Gamma, r : T \to [m_1, \ldots, m_k, n] \vdash \\ \qquad e_2 : [\mathbf{pro}\,t.\langle\!\langle rt \mid \vec{m} : \vec{\tau}, n : \tau \rangle\!\rangle / t](t \to \tau) \qquad r \text{ not in } \tau\end{array}}{\Gamma \vdash \langle e_1 \leftarrow\!\!+\, n = e_2 \rangle : \mathbf{pro}\,t.\langle\!\langle R \mid \vec{m} : \vec{\tau}, n : \tau \rangle\!\rangle}$$

(*obj over*) $$\frac{\begin{array}{l}\Gamma \vdash e_1 : \mathbf{pro}\,t.\langle\!\langle R \mid m_1 : \tau_1, \ldots, m_k : \tau_k \rangle\!\rangle \\ \Gamma, r : T \to [m_1, \ldots, m_k] \vdash \\ \qquad e_2 : [\mathbf{pro}\,t.\langle\!\langle rt \mid \vec{m} : \vec{\tau} \rangle\!\rangle / t](t \to \tau_i)\end{array}}{\Gamma \vdash \langle e_1 \leftarrow m_i = e_2 \rangle : \mathbf{pro}\,t.\langle\!\langle R \mid \vec{m} : \vec{\tau} \rangle\!\rangle}$$

Here $\vec{m} : \vec{\tau}$ is used as an abbreviation for $m_1 : \tau_1, \ldots, m_k : \tau_k$.

A.2 New Rules

(*type object*) $$\frac{\Gamma, t : T \vdash R : [m_1, \ldots, m_k]}{\Gamma \vdash \mathbf{obj}\,t.R : T}$$

(<: *var*) $$\frac{\begin{array}{c}\Gamma \vdash * \\ s_1, s_2 \notin dom(\Gamma) \\ s_1 \neq s_2\end{array}}{\Gamma, s_1 <: s_2 \vdash *}$$

(<: *proj*) $$\frac{\begin{array}{c}\Gamma \vdash * \\ s_1 <: s_2 \in \Gamma\end{array}}{\Gamma \vdash s_1 <: s_2}$$

(<: *var proj*) $$\frac{\begin{array}{c}\Gamma \vdash * \\ s_1 <: s_2 \in \Gamma\end{array}}{\Gamma \vdash s_i : T}$$

(<: *refl*) $$\frac{\Gamma \vdash \tau : T}{\Gamma \vdash \tau <: \tau}$$

$$(<:\ trans)\qquad \frac{\Gamma \vdash \sigma <: \tau \qquad \Gamma \vdash \tau <: \rho}{\Gamma \vdash \sigma <: \rho}$$

$$(<:\ arrow)\qquad \frac{\Gamma \vdash \tau_1 <: \sigma_1 \; \Gamma \vdash \sigma_2 <: \tau_1}{\Gamma \vdash \sigma_1 \rightarrow \sigma_2 <: \tau_1 \rightarrow \tau_2}$$

$$(<:\ object)\qquad \frac{\begin{array}{c}\Gamma \vdash \mathbf{probj}\, s_1 . \langle\!\langle \vec{m} : \vec{\tau},\ \vec{n} : \vec{\sigma} \rangle\!\rangle : T \\ \Gamma \vdash \mathbf{obj}\, s_2 . \langle\!\langle \vec{m} : \vec{\tau}' \rangle\!\rangle : T \\ \Gamma, s_1 <: s_2 \vdash \tau_i <: \tau_i' \\ s_1 \notin FV(\tau_i'),\ \ s_2 \notin FV(\tau_i)\end{array}}{\Gamma \vdash \mathbf{probj}\, s_1 . \langle\!\langle \vec{m} : \vec{\tau},\ \vec{n} : \vec{\sigma} \rangle\!\rangle <: \mathbf{obj}\, s_2 . \langle\!\langle \vec{m} : \vec{\tau}' \rangle\!\rangle}$$

$$(subsumption)\qquad \frac{\Gamma \vdash \sigma <: \tau \qquad \Gamma \vdash e : \sigma}{\Gamma \vdash e : \tau}$$

$$(object\ meth\ app)\qquad \frac{\Gamma \vdash e : \mathbf{obj}\, t . \langle\!\langle R \,|\, m : \tau \rangle\!\rangle}{\Gamma \vdash e \Leftarrow m : [\mathbf{obj}\, t . \langle\!\langle R \,|\, m : \tau \rangle\!\rangle / t]\tau}$$

Model–Checking for Real–Time Systems *

Kim G. Larsen[1] *Paul Pettersson*[2] *Wang Yi*[2]

[1] BRICS*** , Aalborg University, DENMARK
[2] Uppsala University, SWEDEN

Abstract. Efficient automatic model-checking algorithms for real-time systems have been obtained in recent years based on the state-region graph technique of Alur, Courcoubetis and Dill. However, these algorithms are faced with two potential types of explosion arising from parallel composition: explosion in the space of control nodes, and explosion in the region space over clock-variables.
This paper reports on work attacking these explosion problems by developing and combining *compositional* and *symbolic* model-checking techniques. The presented techniques provide the foundation for a new automatic verification tool UPPAAL . Experimental results show that UPPAAL is not only substantially faster than other real-time verification tools but also able to handle much larger systems.

1 Introduction

Within the last decade model–checking has turned out to be a useful technique for verifying temporal properties of finite–state systems. Efficient model–checking algorithms for finite–state systems have been obtained with respect to a number of logics. However, the major problem in applying model–checking even to moderate–size systems is the potential combinatorial explosion of the state space arising from parallel composition. In order to avoid this problem, algorithms have been sought that avoid exhaustive state space exploration, either by *symbolic* representation of the states space using Binary Decision Diagrams [5], by application of *partial order* methods [11, 21] which suppresses unnecessary interleavings of transitions, or by application of *abstractions* and *symmetries* [7, 8, 10].

In the last few years, model–checking has been extended to real–time systems, with time considered to be a dense linear order. A timed extension of finite automata through addition of a finite set of real–valued clock–variables has been put forward [3] (so called timed automata), and the corresponding model–checking problem has been proven decidable for a number of timed logics including timed extensions of CTL (TCTL) [2] and timed μ–calculus (T_μ) [14].

* This work has been supported by the European Communieties under CONCUR2, BRA 7166, NUTEK (Swedish Board for Technical Development) and TFR (Swedish Technical Research Council)

*** Basic Research in Computer Science, Centre of the Danish National Research Foundation.

A state of a timed automaton is of the form (l, u), where l is a control–node and u is a clock–assignment holding the current values of the clock–variables. The crucial observation made by Alur, Courcoubetis and Dill and the foundation for decidability of model–checking is that the (infinite) set of clock–assignments may effectively be partitioned into finitely many *regions* in such a way that clock–assignments within the same region induce states satisfying the same logical properties.

Model–checking of real–time systems based on the region technique suffers two potential types of explosion arising from parallel composition: *Explosion in the region space*, and *Explosion in the space of control–nodes*. We report on attacks on these problems by development and combination of two new verification techniques:

1. A *symbolic* technique reducing the verification problem to that of solving simple constraint systems (on clock–variables), and
2. A *compositional* quotient construction, which allows components of a real–time system to be gradually moved from the system into the specification. The intermediate specifications are kept small using minimization heuristics.

The property-independent nature of regions leads to an extremely fine (and large) partitioning of the set of clock–assignments. Our symbolic technique allows the partitioning to take account of the particular property to be verified and will thus in practice be considerably coarser (and smaller).

For the explosion on control–nodes, recent work by Andersen [4] on (untimed) finite–state systems gives experimental evidence that the quotient technique improves results obtained using Binary Decision Diagrams [5]. The aim of the work reported is to make this new successful compositional model–checking technique applicable to real–time systems. For example, consider the following typical model–checking problem

$$\left(A_1 \mid \ldots \mid A_n\right) \models \varphi$$

where the A_i's are timed automata. We want to verify that the parallel composition of these satisfies the formula φ without having to construct the complete control–node space of $(A_1 \mid \ldots \mid A_n)$. We will avoid this complete construction by removing the components A_i one by one while simultaneously transforming the formula accordingly. Thus, when removing the component A_n we will transform the formula φ into the *quotient* formula $\varphi \,/\, A_n$ such that

$$\left(A_1 \mid \ldots \mid A_n\right) \models \varphi \quad \text{if and only if} \quad \left(A_1 \mid \ldots \mid A_{n-1}\right) \models \varphi \,/\, A_n \qquad (1)$$

Now clearly, if the quotient is not much larger than the original formula we have succeeded in simplifying the problem. Repeated application of quotienting yields

$$\left(A_1 \mid \ldots \mid A_n\right) \models \varphi \quad \text{if and only if} \quad \mathbf{1} \models \varphi \,/\, A_n \,/ A_{n-1} \,/\, \ldots \,/\, A_1 \qquad (2)$$

where $\mathbf{1}$ is the unit with respect to parallel composition. However, these ideas alone are clearly not enough as the explosion may now occur in the size of the final formula instead. The crucial and experimentally "verified" observation by Andersen was that each quotienting should be followed by a minimization of the formula based on a small collection of efficiently implementable strategies. In our setting, Andersen's collection is extended to include strategies for propagating and simplifying timing constraints.

We report on a new symbolic and compositional verification technique developed for the real–time logics $\mathcal{L}_\nu$ [17] and a fragment $\mathcal{L}_s$ designed specifically for expressing safety and bounded liveness properties. Comparatively less expressive than TCTL and T_μ, the fragment $\mathcal{L}_s$ is still sufficiently expressive for practical purposes allowing a number of operators of other logics to be derived. Most importantly, the somewhat restrictive expressive power of $\mathcal{L}_s$ allows for extremely efficient model-checking as demonstrated by our experimental results, which includes a comparison with other existing automatic verification tools for real–time systems (HyTech, Kronos and Epsilon).

For the logics TCTL and T_μ, [14] offers a symbolic verification technique. However, due to the high expressive power of these logics the partitioning employed in [14] is significantly finer (and larger) and implementation–wise more complicated than ours. An initial effort in applying the compositional quotienting technique to real–time systems has been given in [18].

The outline of this paper is as follows: In the next section we give a short presentation of the notions of timed automata and network. In section 3, the logic $\mathcal{L}_\nu$ and its fragment $\mathcal{L}_s$ are presented and their expressive power illustrated. Section 4 reviews region–based model–checking for $\mathcal{L}_\nu$, whereas Section 5 reports on a symbolic verification technique for the fragment $\mathcal{L}_s$ based on constraint solving. Section 6 describes the compositional quotienting technique. Finally, in Section 7 we report on our experimental results, which shows that UPPAAL is not only substantially faster than other real–time verification tools but also able to handle much larger systems.

2 Real–Time Systems

We shall use *timed transition systems* as a basic semantical model for real–time systems. The type of systems we are studying will be a particular class of timed transition systems that are syntactically described by *networks of timed automata* [22, 18].

2.1 Timed Transition Systems

A timed transition system is a labelled transition system with two types of labels: atomic actions and delay actions (i.e. positive reals), representing discrete and continuous changes of real–time systems.

Let *Act* be a finite set of actions ranged over by a, b etc, and $\mathcal{P}$ be a set of atomic propositions ranged over by p, q etc. We use $\mathbf{R}$ to stand for the set of non–negative real numbers, Δ for the set of delay actions $\{\epsilon(d) \mid d \in \mathbf{R}\}$, and L for the union $Act \cup \Delta$.

Definition 1. *A timed transition system over actions Act and atomic propositions $\mathcal{P}$ is a tuple $\mathcal{S} = \langle S, s_0, \longrightarrow, V\rangle$, where S is a set of states, s_0 is the initial state, $\longrightarrow \subseteq S \times L \times S$ is a transition relation, and $V : S \to 2^{\mathcal{P}}$ is a proposition assignment function.* □

Note that the above definition is standard for labelled transition systems except that we introduced a proposition assignment function V, which for each state $s \in S$ assigns a set of atomic propositions $V(s)$ that hold in s.

In order to study compositionality problems we introduce a parallel composition between timed transition systems. Following [16] we suggest a composition parameterized with a synchronization function generalizing a large range of existing notions of parallel compositions. A *synchronization function* f is a partial function $(Act \cup \{0\}) \times (Act \cup \{0\}) \hookrightarrow Act$, where 0 denotes a distinguished no–action symbol [4]. Now, let $\mathcal{S}_i = \langle S_i, s_{i,0}, \longrightarrow_i, V_i\rangle$, $i = 1, 2$, be two timed transition systems and let f be a synchronization function. Then the *parallel composition* $\mathcal{S}_1 \mid_f \mathcal{S}_2$ is the timed transition system $\langle S, s_0, \longrightarrow, V\rangle$, where $s_1 \mid_f s_2 \in S$ whenever $s_1 \in S_1$ and $s_2 \in S_2$, $s_0 = s_{1,0} \mid_f s_{2,0}$, $\longrightarrow$ is inductively defined as follows:

- $s_1 \mid_f s_2 \xrightarrow{c} s_1' \mid_f s_2'$ if $s_1 \xrightarrow{a}_1 s_1'$, $s_2 \xrightarrow{b}_2 s_2'$ and $f(a, b) = c$
- $s_1 \mid_f s_2 \xrightarrow{\epsilon(d)} s_1' \mid_f s_2'$ if $s_1 \xrightarrow{\epsilon(d)}_1 s_1'$ and $s_2 \xrightarrow{\epsilon(d)}_2 s_2'$

and finally, the proposition assignment function V is defined by $V(s_1 \mid_f s_2) = V_1(s_1) \cup V_2(s_2)$.

Note also that the set of states and the transition relation of a timed transition system may be infinite. We shall use networks of timed automata as a finite syntactical representation to describe timed transition systems.

2.2 Networks of Timed Automata

A timed automaton [3] is a standard finite–state automaton extended with a finite collection of real–valued clocks [5]. Conceptually, the clocks may be considered as the system clocks of a concurrent system. They are assumed to proceed at the same rate and measure the amount of time that has been elapsed since they were reset. The clocks values may be tested (compared with natural numbers) and reset (assigned to 0).

Definition 2. *(Clock Constraints) Let C be a set of real–valued clocks ranged over by x, y etc. We use $\mathcal{B}(C)$ to stand for the set of formulas ranged over by g, generated by the following syntax: $g ::= c \mid g \wedge g$, where c is an atomic constraint of the form: $x \sim n$ or $x - y \sim n$ for $x, y \in C$, $\sim \in \{\leq, \geq, =, <, >\}$ and n being a natural number. We shall call $\mathcal{B}(C)$ clock constraints or clock constraint systems over C. Moreover, $\mathcal{B}_M(C)$ denotes the subset of $\mathcal{B}(C)$ with no constant greater than M.* □

[4] We extend the transition relation of a timed transition system such that $s \xrightarrow{0} s'$ iff $s = s'$.

[5] Timed transition systems may alternatively be described using timed process calculi.

We shall use $\mathtt{tt}$ to stand for a constraint like $x \geq 0$ which is always true, and $\mathtt{ff}$ for a constraint $x < 0$ which is always false as clocks can only have non–negative values.

Definition 3. *A timed automaton A over actions Act, atomic propositions $\mathcal{P}$ and clocks C is a tuple $\langle N, l_0, E, I, V\rangle$. N is a finite set of nodes (control–nodes), l_0 is the initial node, and $E \subseteq N \times \mathcal{B}(C) \times Act \times 2^C \times N$ corresponds to the set of edges. In the case, $\langle l, g, a, r, l'\rangle \in E$ we shall write, $l \xrightarrow{g,a,r} l'$ which represents an edge from the node l to the node l' with clock constraint g (also called the enabling condition of the edge), action a to be performed and the set of clocks r to be reset. $I : N \rightarrow \mathcal{B}(C)$ is a function, which for each node assigns a clock constraint (also called the invariant condition of the node), and finally, $V : N \rightarrow 2^{\mathcal{P}}$ is a proposition assignment function which for each node gives a set of atomic propositions true in the node.* □

Note that for each node l, there is an invariant condition $I(l)$ which is a clock constraint. Intuitively, this constraint must be satisfied by the system clocks whenever the system is operating in that particular control–node.

Informally, the system starts at node l_0 with all its clocks initialized to 0. The values of the clocks increase synchronously with time at node l as long as they satisfy the invariant condition $I(l)$. At any time, the automaton can change node by following an edge $l \xrightarrow{g,a,r} l'$ provided the current values of the clocks satisfy the enabling condition g. With this transition the clocks in r get reset to 0.

Example 1. Consider the automata A_m, B_n and $C_{m,n}$ in Figure 1 where m, n, m' and n' are natural numbers used as parameters. The automaton $C_{m,n}$ has four nodes, l_0, l_1, l_2 and l_3, two clocks x and y, and three edges. The edge between l_1 and l_2 has b as action, $\{x, y\}$ as reset set and the enabling condition for the edge is $x > m$. The invariant conditions for nodes l_1 and l_2 are $x \leq m'$ and $y \leq n'$ respectively. □

Now we introduce the notion of a *clock assignment*. Formally, a clock assignment u for C is a function from C to $\mathbf{R}$. We denote by $\mathbf{R}^C$ the set of clock assignments for C. For $u \in \mathbf{R}^C$, $x \in C$ and $d \in \mathbf{R}$, $u + d$ denotes the time assignment which maps each clock x in C to the value $u(x) + d$. For $C' \subseteq C$, $[C' \mapsto 0]u$ denotes the assignment for C which maps each clock in C' to the value 0 and agrees with u over $C \backslash C'$. Whenever $u \in \mathbf{R}^C$, $v \in \mathbf{R}^K$ and C and K are disjoint, we use uv to denote the clock assignment over $C \cup K$ such that $(uv)(x) = u(x)$ if $x \in C$ and $(uv)(x) = v(x)$ if $x \in K$. Given a clock constraint $g \in \mathcal{B}(C)$ and a clock assignment $u \in \mathbf{R}^C$, $g(u)$ is a boolean value describing whether g is satisfied by u or not. When $g(u)$ is true, we shall say that u is a solution og g.

A *state* of an automaton A is a pair (l, u) where l is a node of A and u a clock assignment for C. The initial state of A is (l_0, u_0) where u_0 is the initial clock assignment mapping all clocks in C to 0.

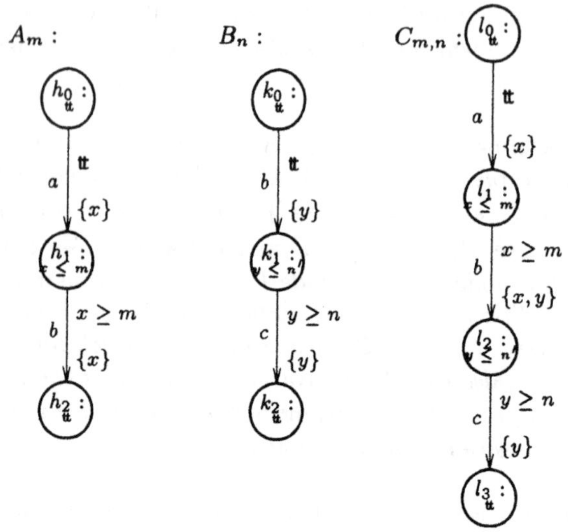

Fig. 1. Three timed automata

The semantics of A is the timed transition system $\mathcal{S}_A = \langle S, \sigma_0, \longrightarrow, V\rangle$, where S is the set of states of A, σ_0 is the initial state (l_0, u_0), $\longrightarrow$ is the transition relation defined as follows:

- $(l, u) \xrightarrow{a} (l', u')$ if there exist r, g such that $l \xrightarrow{g,a,r} l'$, $g(u)$ and $u' = [r \to 0]u$
- $(l, u) \xrightarrow{\epsilon(d)} (l', u')$ if $(l = l')$, $u' = u + d$ and $I(u')$

and V is extended to S simply by $V(l, u) = V(l)$.

Example 2. Reconsider the automaton $C_{m,n}$ of Figure 1. Assume that $d \geq 0$, $m \leq e \leq m'$ and $n \leq f \leq n'$. We have the following typical transition sequence:

$$(l_0, (0,0)) \xrightarrow{\epsilon(d)} (l_0, (d,d)) \xrightarrow{a} (l_1, (0,d)) \xrightarrow{\epsilon(e)} (l_1, (e, d+e)) \xrightarrow{b} (l_2, (0,0)) \xrightarrow{\epsilon(f)}$$
$$(l_2, (f,f)) \xrightarrow{c} (l_3, (f,0))$$

Note that we need to assume that $m \leq e \leq m'$ and $n \leq f \leq n'$ because of the invariant conditions on l_1 and l_2. □

Parallel composition may now be extended to timed automata in the obvious way: for two timed automata A and B and a synchronization function f, the parallel composition $A \,|_f\, B$ denotes the timed transition system $\mathcal{S}_A \,|_f\, \mathcal{S}_B$. Note that the timed transition system $\mathcal{S}_A \,|_f\, \mathcal{S}_B$ can also be represented finitely as a timed automaton. In fact, one may effectively construct the product automaton $A \otimes_f B$ such that its timed transition system $\mathcal{S}_{A \otimes_f B}$ is bisimilar to $\mathcal{S}_A \,|_f\, \mathcal{S}_B$. The nodes of $A \otimes_f B$ is simply the product of A's and B's nodes, the invariant conditions on the nodes of $A \otimes_f B$ are the conjunctions of the conditions on respective A's and B's nodes, the set of clocks is the (disjoint) union of A's and B's clocks, and the edges are based on synchronizable A and B edges with enabling conditions conjuncted and reset–sets unioned.

Example 3. Let f be the synchronization function defined by $f(a, 0) = a$, $f(b, b) = b$ and $f(0, c) = c$. Then the automaton $C_{m,n}$ in Figure 1 is timed bisimilar to the part of $A_m \otimes_f B_n$ which is reachable from (h_0, k_0). □

3 Timed Logics

We first introduce the syntax and semantics of the dense–time logic $\mathcal{L}_\nu$ presented in [17]. For the practical goal of verification of real–time systems, we find that it suffices to consider a certain fragment $\mathcal{L}_s$ especially designed to express safety and bounded liveness properties. Most importantly, as we shall show in subsequent sections, the rectriction to $\mathcal{L}_s$ allows for extremely efficient model–checking algorithms.

3.1 Syntax and Semantics

We first consider a dense–time logic $\mathcal{L}_\nu$ with clocks and recursion. This logic may be seen as a certain fragment [6] of the μ–calculus T_μ presented in [14]. In [17] it has been shown that this logic is sufficiently expressive that for any timed automaton one may construct a single *characteristic* formula uniquely characterizing the automaton up to timed bisimilarity. Also, decidability of a satisfiability [7] problem is demonstrated.

Definition 4. *Let K be a finite set of clocks. We shall call K formula clocks. Let Id be a set of identifiers. The set $\mathcal{L}_\nu$ of formulae over K, Id, Act, and $\mathcal{P}$ is generated by the abstract syntax with φ and ψ ranging over $\mathcal{L}_\nu$:*

$$\varphi ::= c \mid p \mid \varphi \wedge \psi \mid \varphi \vee \psi \mid \exists \varphi \mid \forall \varphi \mid \langle a \rangle \varphi \mid [a] \varphi$$
$$\mid x \text{ in } \varphi \mid x + n \sim y + m \mid Z$$

where c is an atomic clock constraint in the form of $x \sim n$ or $x - y \sim n$ for $x, y \in K$ and natural number n, $p \in \mathcal{P}$ is an atomic predicate, $a \in Act$ is an action, $z \in K$ and $Z \in Id$ is an identifier. □

The meaning of the identifiers is specified by a declaration $\mathcal{D}$ assigning a formula of $\mathcal{L}_\nu$ to each identifier. When $\mathcal{D}$ is understood we write $Z \stackrel{\text{def}}{=} \varphi$ for $\mathcal{D}(Z) = \varphi$.

Given a timed transition system $\mathcal{S} = \langle S, s_0, \longrightarrow, V \rangle$ described by a network of timed automata, we interpret the $\mathcal{L}_\nu$ formulas over an extended state $\langle s, u \rangle$ where $s \in S$ is a state of $\mathcal{S}$, and u is a clock assignment for K. A formula

6 allowing only maximal recursion and using a slightly different notion of model

7 Bounded in the number of clocks and maximal constant allowed in the satisfying automata.

$$\begin{array}{ll}
\langle s,u\rangle \models c & \Rightarrow c(u) \\
\langle s,u\rangle \models p & \Rightarrow p \in V(s) \\
\langle s,u\rangle \models \varphi \vee \psi & \Rightarrow \langle s,u\rangle \models \varphi \text{ or } \langle s,u\rangle \models \psi \\
\langle s,u\rangle \models \varphi \wedge \psi & \Rightarrow \langle s,u\rangle \models \varphi \text{ and } \langle s,u\rangle \models \psi \\
\langle s,u\rangle \models \forall\varphi & \Rightarrow \forall d, s' : s \xrightarrow{\epsilon(d)} s' \Rightarrow \langle s', u+d\rangle \models \varphi \\
\langle s,u\rangle \models [a]\,\varphi & \Rightarrow \forall s' : s \xrightarrow{a} s' \Rightarrow \langle s', u\rangle \models \varphi \\
\langle s,u\rangle \models x \text{ in } \varphi & \Rightarrow \langle s,v'\rangle \models \varphi \text{ where } v' = [\{x\} \to 0]v \\
\langle s,u\rangle \models Z & \Rightarrow \langle s,u\rangle \models \mathcal{D}(Z)
\end{array}$$

Table 1. Definition of satisfiability.

of the form: $x \sim m$ and $x - y \sim n$ is satisfied by an extended state $\langle s,u\rangle$ if the values of x, y in u satisfy the required relationship. Informally, an extended state $\langle s,u\rangle$ satisfies $\forall\varphi$ means that all future states reachable from $\langle s,u\rangle$ by delays will satisfy property φ. Thus $\forall$ denotes universal quantification over delay transitions. Similarly, $\exists$ denotes existential quantification over delay transitions. A state $\langle s,u\rangle$ satisfies $[a]\varphi$ means that all intermediate states reachable from $\langle s,u\rangle$ by an a–transition (performed by s will satisfy property φ. Thus $[a]$ denotes universal quantification over a–transitions. Similarly, $\langle a\rangle$ denotes existential quantification over a–transitions. The formula $(x \text{ in } \varphi)$ initializes the formula clock x to 0; i.e. an extended state satisfies the formula in case the modified state with x being reset to 0 satisfies φ. Finally, an extended state satisfies an identifier Z if it satisfies the corresponding declaration (or definition) $\mathcal{D}(Z)$. Let $\mathcal{D}$ be a declaration. Formally, the satisfaction relation $\models_{\mathcal{D}}$ between extended states and formulas is defined as the largest relation satisfying the implications of Table 1. We have left out the cases for $\exists$ and $\langle a\rangle$ as they are immediate duals.

Any relation satisfying the implications in Table 1 is called a *satisfiability* relation. It follows from standard fixpoint theory [20] that $\models_{\mathcal{D}}$ is the union of all satisfiability relations. For simplicity, we shall omit the subscript $\mathcal{D}$ and write $\models$ instead of $\models_{\mathcal{D}}$ whenever it is understood from the context. We say that $\mathcal{S}$ satisfies a formula φ and write $\mathcal{S} \models \varphi$ when $\langle s_0, v_0\rangle \models \varphi$ where s_0 is the initial state of $\mathcal{S}$ and v_0 is the assignment with $v_0(x) = 0$ for all x. Similarly, we say that a timed automaton A satisfies φ in case $\mathcal{S}_A \models \varphi$. We write $A \models \varphi$ in this case.

Example 4. Consider the following declaration $\mathcal{F}$ of the identifiers X_i and Z_i where i is a natural number.

$$\mathcal{F} = \left\{ X_i \stackrel{\text{def}}{=} [a]\Big(z \text{ in } Z_i\Big) \quad , \quad Z_i \stackrel{\text{def}}{=} (\mathsf{at}(l_3) \vee \Big(z < i \wedge [a]Z_i \wedge [b]Z_i \wedge [c]Z_i \wedge \forall Z_i\Big)\right\}$$

Assume that $\mathsf{at}(l_3)$ is an atomic proposition meaning that the system is operating in control–node l_3. Then, X_i expresses the property that after an a–transition, the system must reach node l_3 within i time units. Now, reconsider the automata A_m, B_n and $C_{m,n}$ of Figure 1 and Examples 1 and 2. Then it may be argued that $C_{m,n} \models X_{m'+n'}$ and (consequently), that $A_m \mid_f B_n \models X_{m'+n'}$. □

$$\mathsf{INV}(\varphi) \equiv X \text{ where } X \stackrel{\text{def}}{=} \varphi \wedge \forall X \wedge [Act]X$$

$$\varphi \ \mathsf{UNTIL}\ \psi \equiv X \text{ where } X \stackrel{\text{def}}{=} \psi \vee \left(\varphi \wedge \forall X \wedge [Act]X\right)$$

$$\varphi \ \mathsf{UNTIL}_{<n}\ \psi \equiv z \text{ in } \left((\varphi \wedge z < n) \ \mathsf{UNTIL}\ \psi\right)$$

$$\psi \ \mathsf{BEFORE}\ n \equiv \mathtt{tt} \ \mathsf{UNTIL}_{<n}\ \psi$$

Table 2. Derived Operators

3.2 Derived Operators

The property Z_i described in Example 3 is an attempt to specify bounded liveness properties: namely that a certain proposition must be satisfied within a given time bound. We shall use the more informative notation at(l_3) BEFORE i to denote Z_i. In the following, we shall present several such intuitive operators that are definable in our logic.

For simplicity, we shall assume that the set of actions Act is a finite set $\{a_1 ... a_m\}$, and use $[Act]\varphi$ to denote the formula $[a_1]\varphi \wedge ... \wedge [a_m]\varphi$. Now, let φ and ψ be a general formulas and n be a natural number. A collection of derived operators are given in Table 2.

The intuitive meanings of these operators are the following: INV(φ) is satisfied by a timed automaton provided φ holds in any reachable state; i.e. φ is an invariant property of the automaton. φ UNTIL ψ is satisfied by a timed automaton provided φ holds until the property ψ becomes true. Due to the maximal fixedpoint semantics this derived operator is the *weak* UNTIL–operator in that there is no guarantee that ψ ever becomes true. The bounded version of the UNTIL–construct φ UNTIL$_{<n}$ ψ is similar to φ UNTIL ψ except that ψ must be true within n time units. A simpler version of this operator is ψ BEFORE n meaning that property ψ must be true within n time units.

3.3 A Logic for Safety and Bounded Liveness Properties

It has been pointed out [13, 22], that the practical goal of verification of real–time systems, is to verify simple safety properties such as deadlock–freeness and mutual exlusion. Similarly, we have found that for practical purposses it (often) suffices to use only a fragment of $\mathcal{L}_\nu$.

Formally, the logic for Safety and Bounded Liveness Properties, $\mathcal{L}_s$, is the fragment of $\mathcal{L}_\nu$ obtained by eliminating the use of the existential quantifiers $\exists$ (over delay transitions) and $\langle a \rangle$ (over a–transitions), and restricting the use of disjunction to formulas of the forms $c \vee \varphi$ (an atomic clock constrain) and $p \vee \varphi$ (an atomic proposition). The logic $\mathcal{L}_s$ is sufficiently expressive that we may specify a number of safety and bounded liveness properties. In particular, restricting ψ to c and p in Table 2 yields (restricted) derived operators expressible in $\mathcal{L}_s$. Consequently the formulas of Example 4 are in $\mathcal{L}_s$. Most importantly, the restriction to $\mathcal{L}_s$ allows for extremely efficient automatic verification.

4 Region–Based Model–Checking

We have presented a model to describe real–time systems, i.e. networks of timed automata, and logics to specify properties of such systems. The next question is how to check whether a given logical formula is satisfied by a given network of automata. This is the so–called model–checking problem. The model-checking problem for $\mathcal{L}_\nu$ consists in deciding if a given timed automaton A satisfies a given specification φ in $\mathcal{L}_\nu$. This problem is decidable using the region technique of Alur, Courcoubetis and Dill [3, 2], which provides an abstract semantics of timed automata in the form of finite labelled transition systems with the truth value of $\mathcal{L}_\nu$ formulas being maintained.

The basic idea is that, given a timed automaton A, two states (l, u_1) and (l, u_2) which are close enough with respect to their clocks values (we will say that u_1 and u_2 are in the same *region*) can perform the same actions, and two extended states $\langle(l, u_1), v_1\rangle$ and $\langle(l, u_2), v_2\rangle$ where $u_1 v_1$ and $u_2 v_2$ are in the same region, satisfy the same $\mathcal{L}_\nu$–formulas. In fact the regions are defined as equivalence classes of a relation $\doteq$ over time assignments [14]. Formally, given C a set of clocks and k an integer, we say $v \doteq u$ if and only if v and u satisfy the same conditions of $\mathcal{B}_k(C)$. $[v]$ denotes the region which contains the time assignment v. $\mathcal{R}_k^C$ denotes the set of all regions for a set C of clocks and the maximal constant k. From a decision point of view it is important to note that $\mathcal{R}_k^C$ is finite.

For a region $\gamma \in \mathcal{R}_k^C$, we can define $b(\gamma)$ as the truth value of $b(v)$ for any v in γ. Conversely given a region γ, we can easily build a formula of $\mathcal{B}(C)$, called $\beta(\gamma)$, such that $\beta(\gamma)(v) = \mathtt{tt}$ iff $v \in \gamma$. Thus, given a region γ', $\beta(\gamma)(\gamma')$ is mapped to the value $\mathtt{tt}$ precisely when $\gamma = \gamma'$. Finally, note that $\beta(\gamma)$ itself can be viewed as a $\mathcal{L}_\nu$ formula.

Given a region $[v]$ in $\mathcal{R}_k^C$ and $C' \subseteq C$ we define the following reset operator: $[C' \to 0][v] = [[C' \to 0]v]$. Moreover, for a region $[v]$, we define the successor region (denoted by $succ([v])$) as the region $[v']$, where:

$$v'(x) = \begin{cases} v(x) + f & \forall x \in C.\, v(x) > k \vee \{v(x)\} \neq 0 \\ v(x) + f/2 & \exists x \in C.\, v(x) \leq k \wedge \{v(x)\} = 0 \end{cases}$$

where $f = min\{1 - \{v(x)\} \mid v(x) \leq k\}$ [8]. Informally the change from γ to $succ(\gamma)$ correspond to the minimal elapse of time which can modify the enabled actions of the current state.

We denote by γ^i the ith successor region of γ (i.e. $\gamma^i = succ^i(\gamma)$). From each region γ, it is possible to reach a region γ' s.t. $succ(\gamma') = \gamma'$, and we denote by i_γ the required number of step s.t. $\gamma^{i_\gamma} = succ(\gamma^{i_\gamma})$.

[8] if this set is empty, then $f = 0$

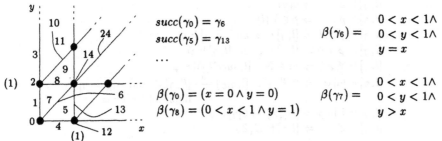

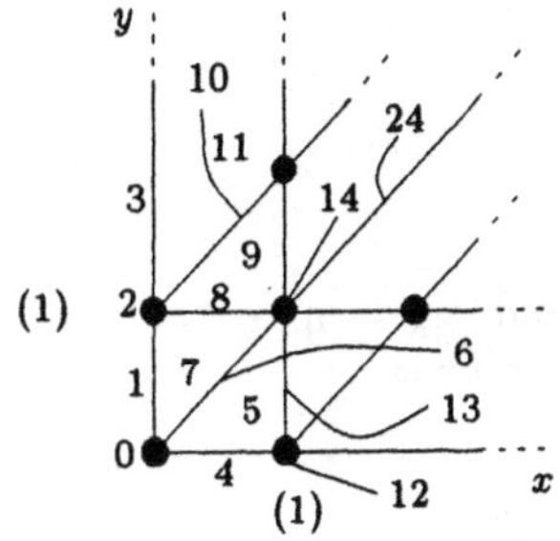

$succ(\gamma_0) = \gamma_6$
$succ(\gamma_5) = \gamma_{13}$
...

$\beta(\gamma_0) = (x = 0 \wedge y = 0)$
$\beta(\gamma_8) = (0 < x < 1 \wedge y = 1)$

$\beta(\gamma_6) = \begin{matrix} 0 < x < 1 \wedge \\ 0 < y < 1 \wedge \\ y = x \end{matrix}$

$\beta(\gamma_7) = \begin{matrix} 0 < x < 1 \wedge \\ 0 < y < 1 \wedge \\ y > x \end{matrix}$

Fig. 2. $\mathcal{R}_k^C$ with $C = \{x, y\}$ and $k = 1$

Example 5. The Figure 2 gives an overview of the set of regions defined by two clocks x and y, and the maximal constant 1. In this case there are 32 different regions. In general successor regions are determined by following 45° lines upwards to the right. □

Let $A = \langle N, l_0, E, I, V \rangle$ be a timed automaton over actions *Act*, atomic propostions $\mathcal{P}$ and clocks C. Let k_A denotes the maximal constant occurring in the enabling condition of the edges E. Then for any $k \geq k_A$ we can now define a region–based semantics of A over region–states $[l, \gamma]$ where $l \in N$ and $\gamma \in \mathcal{R}_k^C$ as follows: for any $[l, \gamma]$ we have $[l, \gamma] \xrightarrow{a} [l', \gamma']$ iff $\exists\ v \in \gamma,\ \ (l, v) \xrightarrow{a} (l', v')$ and $v' \in \gamma'$.

Consider now $\mathcal{L}_\nu$ with respect to formula clock set K and maximal constant k_L (assuming that K and C are disjoint). Then an *extended region–state* is a pair $[l, \gamma]$ where $l \in N$ and $\gamma \in \mathcal{R}_k^{C \cup K}$ with $k = max(k_A, k_L)$. We define now the *region–based semantics* for $\mathcal{L}_\nu$, i.e. the truth value of $\mathcal{L}_\nu$ formulas over extended region–states. Formally, $\vdash_D$ is the largest relation satisfying the implications of Table 3[9]. We have left out the cases for $\exists$ and $\langle a \rangle$ as they are immediate duals. Also, when no confusion can occur we omit the subscript and write $\vdash$ instead of $\vdash_D$. This symbolic interpretation of $\mathcal{L}_\nu$ is closely related to the standard interpretation as stated by the following important result: Let φ be a formula of $\mathcal{L}_\nu$, and let $\langle (l, u), v \rangle$ be an extended state over some timed automaton A, then we have

$$\langle (l, u), v \rangle \models \varphi \quad \text{if and only if} \quad [l, [uv]] \vdash \varphi$$

It follows that the model checking problem for $\mathcal{L}_\nu$ is decidable since it suffices to check the truth value of any given $\mathcal{L}_\nu$ formula φ with respect to the finite transition system corresponding to the extended region–state semantics of A.

[9] $\gamma_{|C}$ (resp. $\gamma_{|K}$) denotes the set of time-assignments in γ restricted to the automata (resp. formula) clocks.

$$
\begin{array}{ll}
[l,\gamma] \vdash c & \Rightarrow c(\gamma) \\
[l,\gamma] \vdash p & \Rightarrow p \in V(l) \\
[l,\gamma] \vdash \varphi \wedge \psi & \Rightarrow [l,\gamma] \vdash \varphi \text{ and } [l,\gamma] \vdash \psi \\
[l,\gamma] \vdash \varphi \vee \psi & \Rightarrow [l,\gamma] \vdash \varphi \text{ or } [l,\gamma] \vdash \psi \\
[l,\gamma] \vdash \mathbf{\forall} \varphi & \Rightarrow \forall i \in \mathbf{N}.\ [l, succ^i(\gamma)] \vdash \varphi \\
[l,\gamma] \vdash [a]\varphi & \Rightarrow \forall [l',\gamma'].\ [l,\gamma_{|C}] \xrightarrow{a} [l',\gamma'_{|C}] \text{ and } \gamma'_{|K} = \gamma_{|K} \\
& \qquad \text{implies } [l',\gamma'] \vdash \varphi \\
[l,\gamma] \vdash x \,\mathsf{in}\, \varphi & \Rightarrow [l,[\{x\} \rightarrow 0]\gamma] \vdash \varphi \\
[l,\gamma] \vdash Z & \Rightarrow [l,\gamma] \vdash D(Z)
\end{array}
$$

Table 3. Definition of region–based satisfiablity.

5 Symbolic Model–Checking

The region–graph technique applied in the previous section allows the state space of a real time system to be partitioned into finitely many regions in such a way that states within the same region satisfy the same properties. However, as the notion of region is essentially property–independent and the number of such regions depends highly on the constants used in the clock constraints of an automaton, the region partitioning is extremely fine (and large). In this section we shall offer a much coarser (and smaller) partitioning of the state space yielding extremely efficient model–checking for the safety logic $\mathcal{L}_s$.

Recall that a semantical state of a network of timed automata is a pair (l, u) where l is a control-node and $u \in \mathbf{R}^C$ is a clock assignment. The model-checking problem is in general to check whether an extended state in the form $\langle (l, u), v\rangle$ satisfy a given formula φ, that is,

$$\langle (l, u), v\rangle \models \varphi$$

Note that u is a clock assignment for the automata clocks and v is a clock assignment for the formula clocks. Now, the problem is that we have too many (in fact, infinitely many) such assignments to check in order to conclude $\langle (l, u), v\rangle \models \varphi$.

In this section, we shall use clock constraints $\mathcal{B}(C \cup K)$ for automata clocks C and formula clocks K, as defined in section 2 to symbolically represent clock assignments. We shall use D to range over $\mathcal{B}(C \cup K)$. For safety formulas $\varphi \in \mathcal{L}_s$ we develop an algorithm to simultaneously check

$$[l, D] \models \varphi$$

which means that for each u and v such that uv is a solution to the constraint system D, we have $\langle (l, u), v\rangle \models \varphi$.

Thus the space $\mathbf{R}^{C \cup K}$ is partitioned in terms of clock constraints. As for a given network and a given formula, we have only finite many such constraints to check, the problem becomes decidable, and in fact as the partitioning takes account of the particular property, the number of partitions is in practice considerably smaller compared with the region-technique.

5.1 Operations on Clock Constraints

To develop the model-checking algorithm, we need a few operations to manipulate clock constraints. Given a clock constraint D, we shall call the set of clock assignments satisfying D, the *solution set* of D.

Definition 5. *Let A and A' be the solution sets of clock constraints $D, D' \in \mathcal{B}(C \cup K)$. We define*

$$\begin{aligned} A^{\uparrow} &= \{w + d \mid w \in A \text{ and } d \in \mathbf{R}\} \\ A^{\downarrow} &= \{w \mid \exists d \in \mathbf{R}:\ w + d \in A\} \\ \{x\}A &= \{[\{x\} \mapsto 0]w \mid w \in A\} \\ A \wedge A' &= \{w \mid w \in A \text{ and } w \in A'\} \end{aligned}$$

□

First, note that $A \wedge A'$ is simply the intersection of the two sets. Consider the set A for the case of two clocks, shown in (a) of Figure 3. The three operations $A^{\uparrow}$, $A^{\downarrow}$ and $\{x\}A$ are illustrated in (b), (c) and (d) respectively of Figure 3. Intuitively, $A^{\uparrow}$ is the largest set of time assignments that will eventually reach A after some delay; whereas $A^{\downarrow}$ is the dual of $A^{\uparrow}$: namely that it is the largest set of time assignments that can be reached by some delay from A. Finally, $\{y\}A$ is the projection of A down to the x-axis. We extend the projection operator to sets of clocks. Let $r = \{x_1 \ldots x_n\}$ be a set of clocks. We define $r(A)$ recursively by $\{\}(A) = A$ and $\{x_1 \ldots x_n\}(A) = \{x_1\}(\{x_2 \ldots x_n\}A)$.

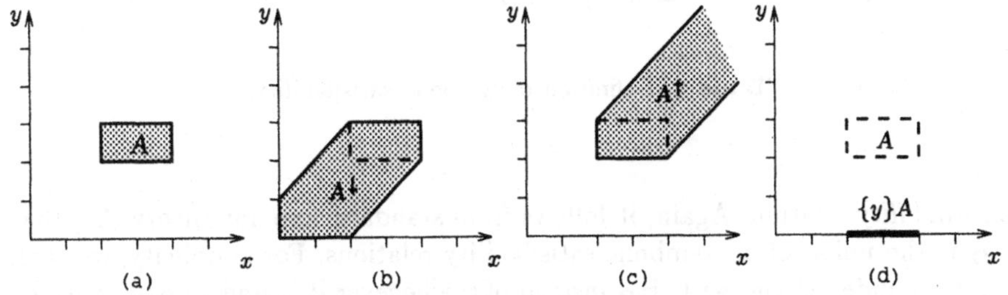

Fig. 3. Operations on Solution Sets

The following Proposition establishes that the class of clock constraints $\mathcal{B}(C \cup K)$ is closed under the four operations defined above.

Proposition 6. *Let $D, D' \in \mathcal{B}(C \cup K)$ with solution sets A and A', and $x \in C \cup K$. Then there exist $D_1, D_2, D_3, D_4 \in \mathcal{B}(C \cup K)$ with solution sets $A^{\uparrow}$, $A^{\downarrow}$, $\{x\}A$ and $A \wedge A'$ respectively.* □

In fact, the resulted constraints D_i's can be effectively constructed from D and D', as shown in section 4.3. In order to save notation, from now on, we shall simply use $D^{\uparrow}$, $D^{\downarrow}$, $\{x\}D$ and $D \wedge D'$ to denote the clock constraints which are guaranteed to exist due to the above proposition. We will also need a few *predicates* over clock constraints for the model-checking procedure. We write $D \subseteq D'$ to mean that the solution set of D is included in the solution set of D' and $D = \emptyset$ to mean that the solution set of D is empty.

5.2 Model–Checking by Constraint Solving

Given a network of timed automaton A over clocks C, we shall interprete formulas of $\mathcal{L}_s$ over clocks K with respect to symbolic states of the form $[l, D]$ where l is a control-node of A and D is a clock constraint of $\mathcal{B}(C \cup K)$. Let $\mathcal{D}$ be a declaration. The symbolic satisfaction relation $\vdash_{\mathcal{D}}$ between symbolic states and formulas of $\mathcal{L}_s$ is defined as the largest relation satisfying the implications in Table 4. We call a relation satisfying the implications in Table 4 a *symbolic*

$$
\begin{array}{ll}
D = \emptyset & \Rightarrow [l, D] \vdash \varphi \\
{[l, D]} \vdash c & \Rightarrow D \subseteq c \\
{[l, D]} \vdash p & \Rightarrow p \in V(s) \\
{[l, D]} \vdash c \vee \varphi & \Rightarrow [l, D] \vdash [l, D \wedge \neg c] \vdash \varphi \\
{[l, D]} \vdash p \vee \varphi & \Rightarrow [l, D] \vdash p \text{ or } [l, D] \vdash \varphi \\
{[l, D]} \vdash \varphi_1 \wedge \varphi_2 & \Rightarrow [l, D] \vdash \varphi_1 \text{ and } [l, D] \vdash \varphi_2 \\
{[l, D]} \vdash [a]\,\varphi & \Rightarrow [l', r(D \wedge g)] \vdash \varphi \text{ whenever } l \xrightarrow{g,a,r} l' \\
{[l, D]} \vdash \forall\varphi & \Rightarrow [l, D] \vdash \varphi \text{ and } [l, (D \wedge I(n))^{\uparrow} \wedge I(l)] \vdash \varphi \\
{[l, D]} \vdash x \text{ in } \varphi & \Rightarrow [l, \{x\}D] \vdash \varphi \\
{[l, D]} \vdash Z & \Rightarrow [l, D] \vdash \mathcal{D}(Z)
\end{array}
$$

Table 4. Definition of symbolic satisfiability.

satisfiability relation. Again, it follows from standard fixpoint theory [20] that $\vdash_{\mathcal{D}}$ is the union of all symbolic satisfiability relations. For simplicity, we shall omit the index $\mathcal{D}$ and write $\vdash_{\mathcal{D}}$ instead of $\vdash$ whenever it is understood from the context.

The following Theorem shows that the symbolic interpretation of $\mathcal{L}_s$ in Table 4 expresses the sufficient and necessary conditions for a timed automata to satisfy a formula φ [10].

---

[10] Note that Theorem cannot be extended to a logic with general disjunction (or existential quantifications): the obvious requirement that $[l, D] \models \varphi_1 \vee \varphi_2$ should imply either $[l, D] \models \varphi_1$ or $[l, D] \models \varphi_2$ will fail to satisfy the Theorem.

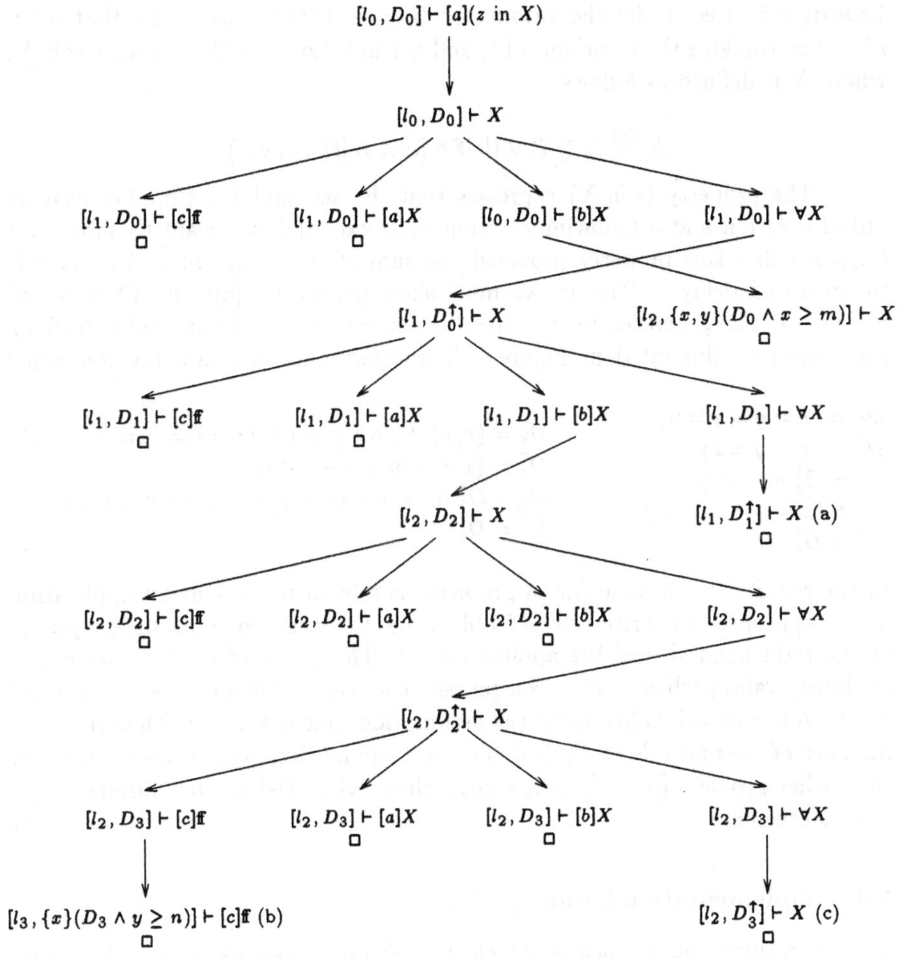

Fig. 4. Rewrite Tree of $[l_0, D_0] \vdash [a](z \text{ in } X)$.

Theorem 7. *Let A be a timed automaton over clock set C and φ a formula of $\mathcal{L}_s$ over K. Then the following holds:*

$$A \models \varphi \quad \textit{if and only if} \quad [l_0, D_0] \vdash \varphi$$

where l_0 is the initial node of A and D_0 is the linear constraint system $\{x = 0 \mid x \in C \cup K\}$. □

Given a symbolic satisfaction *problem* $[l, D] \vdash \varphi$ we may determine its validity by using the implications of Table 4 as rewrite rules. Due to the maximal fixed point property of $\vdash$, rewriting may be terminated successfully in case cycles are encountered. As the rewrite graph of any given problem $[l, D] \vdash \varphi$ can be shown to be finite this yields a decision procedure for model checking.

Example 6. Reconsider the automaton $C_{m,n}$ in Figure 1 assuming that $m' = n' = +\infty$ (making the invariants of l_1 and l_2 true). Consider the property (z in X) where X is defined as follows:

$$X \stackrel{\text{def}}{=} (z \geq i) \vee ([c]\mathrm{ff} \wedge [a]X \wedge [b]X \wedge \forall X)$$

The property (z in X) expresses that the accumulated time between an initial a-action and a following c-action must exceed i. We want to show that $C_{m,n}$ satisfies this property provided the sum of the delays m and n exceeds the required delay i. That is, we must show $[l_0, D_0] \vdash [a](z \text{ in } X)$ provided $n + m \geq i$. The generated rewrite tree (i.e. execution tree of our model checking procedure) is illustrated in Figure 4. The constraints used are the following:

$$\begin{aligned}
D_0 &= \{x = y = z = 0\} \\
D_0^{\uparrow} &= \{x = y = z\} \\
D_1 &= D_0^{\uparrow} \wedge (z < i) \\
&\equiv \{x = y = z, z < i\} \\
D_1^{\uparrow} &= D_0^{\uparrow}
\end{aligned}$$

$$\begin{aligned}
D_2 &= \{x, y\}(D_1 \wedge x \geq m) \equiv \{x = y = 0, m \leq z < i\} \\
D_2^{\uparrow} &= \{x = y, m \leq z - x < i\} \\
D_3 &= D_2^{\uparrow} \wedge z < i = \{x = y, m \leq z - x < i, z < i\} \\
D_3^{\uparrow} &= D_2^{\uparrow}
\end{aligned}$$

In the rewrite tree a node (i.e. a problem) is related to its sons by application of the appropriate rewrite rule of Table 4: i.e. the sons represent the conjuncts of the right-hand side of the applied rule [11]. The leaves of the tree are either obviously valid problems or reoccurrences. The leaf-problem labeled (b) is valid as $(D_3 \wedge y \geq n) = \emptyset$ holds under the assumption that $n + m \geq i$. Thus (b) is an instance of the first rule of Table 4. The problem labeled (a) is a reoccurrence of the earlier problem $[l_1, D_0{}^{\uparrow}]$ as it can be shown that $D_0{}^{\uparrow} = D_1{}^{\uparrow}$. Similarly, (c) is a reoccurrence. □

5.3 Implementation Issues

The operations and predicates on clock constraint systems discussed in Section 5.1 can be efficiently implemented by representing constraint systems as weighted directed graphs. The basic idea is to use a shortest–path algorithm to close a constraint system under entailment so that operations and predicates can be easily computed.

Given a clock constraint system D over a clock set C, we represent D as a weighted directed graph with vertices $C \cup \{0\}$. The graph will have an edge from x to y with weight m provided $x - y \leq m$ is a constraint of D. Similarly, there will be an edge from 0 to x (from x to 0) with weight m whenever $x \leq m$ ($x \geq -m$) is a constraint of D [12].

A clock constraint system D is *closed under entailment* if no constraint of D can be strengthened without reducing the solution set. For closed constraint systems D and D' the inclusion and emptiness predicates are easy to decide:

[11] For problems involving an identifier, the tree reflects *two* successive rule applications starting with the unfolding of the identifier.

[12] In this presentation we have made the simplifying assumption that D does not contain any strict constraints, i.e. constraints of the form $x - y < n$.

$D \subseteq D'$ holds iff for any constraint in D' there is a tighter constraint in D (e.g. whenever $(x - y \leq m) \in D'$ then $(x - y \leq m') \in D$ for some $m' \leq m$); $D = \emptyset$ holds if D contains two contradicting constraints (e.g. $x - y \leq m$ and $x - y \geq n$ where $m < n$). To close a clock constraint system D amounts to solve the shortest-path problem for its graph and can thus be computed in $\mathcal{O}(n^3)$ (which is also the complexity for the inclusion and emptiness predicates), where n is the number of clocks.

Given constraint systems D and D' the operations $D^{\uparrow}$, $D^{\downarrow}$, $\{x\}D$ and $D \wedge D'$ can be computed in $\mathcal{O}(n^2)$. The complexity of the operation $c \wedge D$, where c is an atomic constraint, is $\mathcal{O}(1)$.

6 Compositional Model–Checking

The symbolic model–checking presented in the previous section provides an efficient way to deal with the potential explosion caused by the addition of clocks. However, a potential explosion in the node–space due to parallel composition still remains. In this section we attack this problem by development of a quotient construction, which allows components to be gradually moved from the parallel system into the specification, thus avoiding explicit construction of the global node space. The intermediate specifications are kept small using minimization heuristics. Recent experimental work by Andersen [4] demonstrates that for (untimed) finite–state systems the quotient technique improves results obtained using Binary Decision Diagrams. Also, an initial experimental investigation of the quotient technique to real–time systems in [18] has indicated that these promising results will carry over to the setting of real–time systems. In this section we shall provide a new (and compared with [18] simple) quotient construction and show how to integrate it with the symbolic technique of the previous section.

6.1 Quotient Construction

Given a formula φ of $\mathcal{L}_\nu$, and two timed automata A and B, we aim at constructing a formula (called the *quotient*) $\varphi /_{f} B$ such that

$$A \,|_f\, B \models \varphi \quad \text{if and only if} \quad A \models \varphi /_{f} B \tag{3}$$

The bi–implication indicates that we are moving parts of the parallel system into the formula. Clearly, if the quotient is not much larger than the original formula, we have simplified the task of model–checking, as the (symbolic) semantics of A is significantly smaller than that of $A \,|_f\, B$. More precisely, whenever φ is a formula over K, B is a timed automaton over C and l is a node of B, we define

$$c/_{\!f}\, l = c$$

$$p/_{\!f}\, l = \begin{cases} \mathtt{tt} & ; p \in V(l) \\ p & ; p \notin V(l) \end{cases}$$

$$(\varphi_1 \wedge \varphi_2)/_{\!f}\, l = (\varphi_1/_{\!f}\, l) \wedge (\varphi_2/_{\!f}\, l)$$

$$(\varphi_1 \vee \varphi_2)/_{\!f}\, l = (\varphi_1/_{\!f}\, l) \vee (\varphi_2/_{\!f}\, l)$$

$$(\forall\varphi)/_{\!f}\, l = \forall\Big(I(l) \Rightarrow (\varphi/_{\!f}\, l)\Big)$$

$$(\exists\varphi)/_{\!f}\, l = \exists\Big(I(l) \wedge (\varphi/_{\!f}\, l)\Big)$$

$$(x \,\mathsf{in}\, \varphi)/_{\!f}\, l = x \,\mathsf{in}\, (\varphi/_{\!f}\, l)$$

$$([a]\varphi)/_{\!f}\, l = \bigwedge_{l \xrightarrow{g,c,r} l' \,\wedge\, f(b,c)=a} \Big(g \Rightarrow [b](r \,\mathsf{in}\, \varphi/_{\!f}\, l')\Big)$$

$$X/_{\!f}\, l = X_l \text{ where } X_l \stackrel{\text{def}}{=} \mathcal{D}(X)/_{\!f}\, l$$

Table 5. Definition of Quotient $\varphi/_{\!f}\, l$

the quotient formula $\varphi/_{\!f}\, l$ over $C \cup K$ in Table 5 on the structure of φ [13] [14]. We have left out the case for $\langle a\rangle$ as it is dual to that of $[a]$.

The quotient $\varphi/_{\!f}\, l$ expresses the sufficient and necessary requirement to a timed automaton A in order that the parallel composition $A \,|_f\, B$ with B at node l satisfies φ. In most cases quotienting simply distributes with respect to the formula construction. The quotient construction for $\forall\varphi$ reflects that $A \,|_f\, B$ can only delay provided $I(l)$ is satisfied. The quotient construction for $[a]\varphi$ must quantify over all actions of A which can possibly lead to an a–transition of $A \,|_f\, B$: according to the semantics of parallel composition, b is such an action provided B (at node l) can perform a synchronizable action c (according to some edge $l \xrightarrow{g,c,r} l'$) such that $f(b,c) = a$. The guard as well as the reset set of the involved A-edge $l \xrightarrow{g,c,r} l'$ is reflected in the quotient formula.

Note that the quotient construction for identifiers introduces new identifiers of the form X_l. These new identifiers and their definitions are collected in the (quotient) declaration $\mathcal{D}_B$.

For l_0 the initial node of a timed automaton B, the quotient $\varphi/_{\!f}\, l_0$ ex-

[13] For $g = c_1 \wedge \ldots c_n$ a clock constraint we write $g \Rightarrow \varphi$ as an abbreviation for the formula $\neg c_1 \vee \ldots \vee \neg c_n \vee \varphi$. This is an $\mathcal{L}_s$–formula as atomic constraint are closed under negation.

[14] In the rule for $[a]\varphi$, we assume that all nodes l of a timed automaton are extended with a 0–edge $l \xrightarrow{\mathtt{tt},0,\emptyset} l$.

presses the sufficient and necessary requirement to a timed automaton A in order that the parallel composition $A \|_f B$ satisfies φ. More precisely:

Theorem 8. *Let A and B be two timed automata and let l_0 be the initial node of B. Then*

$$A \|_f B \models_{\mathcal{D}} \varphi \quad \textit{if and only if} \quad A \models_{\mathcal{D}_B} \left(\varphi /_f \, l_0\right)$$

□

Example 7. Reconsider the network, synchronization function and property from Examples 1, 2, 3 and 6. We want to establish that the network $A_m \|_f B_n$ satisfies the following property Y provided $n + m \geq i$:

$$Y \stackrel{\text{def}}{=} [a]\left(z \text{ in } X\right) \qquad X \stackrel{\text{def}}{=} (z \geq i) \vee \left([c]\mathrm{ff} \wedge [a]X \wedge [b]X \wedge \forall X\right)$$

From Theorem 8 it follows that the sufficient and necessary requirement to A_m in order that $A_m \|_f B_n$ satisfies Y is that A_m satisfies $Y /_f \, k_0$. Using the quotient definition from Table 5 we get:

$$\begin{aligned} Y /_f \, k_0 &\stackrel{\text{def}}{=} z \text{ in } (X /_f \, k_0) \\ X /_f \, k_0 &\stackrel{\text{def}}{=} (z \geq i) \vee \left([b](y \text{ in } X /_f \, k_1) \wedge \forall (X /_f \, k_0)\right) \\ X /_f \, k_1 &\stackrel{\text{def}}{=} (z \geq i) \vee \left((y \geq n \Rightarrow [c]\mathrm{ff}) \wedge \forall (X /_f \, k_1)\right) \end{aligned}$$

□

6.2 Minimizations

It is obvious that repeated quotienting leads to an explosion in the formula. The crucial observation made by Andersen in the (untimed) finite–state case is that simple and effective transformations of the formulas in practice may lead to significant reductions.

In presence of real–time we need, in addition to the minimization strategies of Andersen, heuristics for propagating and eliminating constraints on clocks in formulas and declarations. Below we describe the transformations considered:

Reachability: When considering an initial quotient formula $\varphi /_f \, l_0$ not all identifiers in $\mathcal{D}_B$ may be reachable. In UPPAAL an "on-the-fly" technique insures that only the reachable part of $\mathcal{D}_B$ is generated.

Boolean Simplification Formulas may be simplified using the following simple boolean equations and their duals: $\mathrm{ff} \wedge \varphi \equiv \mathrm{ff}$, $\mathrm{tt} \wedge \varphi \equiv \varphi$, $\langle a \rangle \mathrm{ff} \equiv \mathrm{ff}$, $\exists \mathrm{ff} \equiv \mathrm{ff}$, $x \,\text{in}\, \mathrm{ff} \equiv \mathrm{ff}$, $\langle a \rangle \varphi \wedge [a]\mathrm{ff} \equiv \mathrm{ff}$.

Constraint Propagation: Constraints on formula clocks may be propagated using various distribution laws (see Table 6). In some cases, propagation will

$$\begin{aligned}
\emptyset \Rightarrow \varphi &\equiv \mathtt{tt} \\
D \Rightarrow c &\equiv \mathtt{tt} \ \ ; \text{if } D \subseteq c \\
D \Rightarrow ([a]\varphi) &\equiv [a](D \Rightarrow \varphi) \\
D \Rightarrow (\varphi_1 \wedge \varphi_2) &\equiv (D \Rightarrow \varphi_1) \wedge (D \Rightarrow \varphi_2) \\
D \Rightarrow (x \text{ in } \varphi) &\equiv x \text{ in } (\{x\}D \Rightarrow \varphi) \\
D \Rightarrow (p \vee \varphi) &\equiv p \vee (D \Rightarrow \varphi) \\
D \Rightarrow (c \vee \varphi) &\equiv (D \wedge \neg c) \Rightarrow \varphi \\
D \Rightarrow (\mathbf{\forall}\varphi) &\equiv \mathbf{\forall}(D^{\dagger} \Rightarrow \varphi) \ \ ; \text{if } D^{\downarrow} \subseteq D \\
D \Rightarrow X &\equiv D \Rightarrow \mathcal{D}(X)
\end{aligned}$$

Table 6. Constraint Propagation

$$\begin{aligned}
\beta(\gamma) \Rightarrow c &\equiv \begin{cases} \mathtt{tt} & ; c(\gamma) \\ \beta(\gamma) \Rightarrow \mathtt{ff} & ; \text{otherwise} \end{cases} \\
\beta(\gamma) \Rightarrow (\mathbf{\forall}\varphi) &\equiv \mathbf{\forall}\Big(\bigwedge_{i=0\ldots i_\gamma} \beta(\gamma^i) \Rightarrow \varphi \Big) \\
\beta(\gamma) \Rightarrow (\varphi_1 \vee \varphi_2) &\equiv (\beta(\gamma) \Rightarrow \varphi_1) \vee (\beta(\gamma) \Rightarrow \varphi_2)
\end{aligned}$$

Table 7. Region Propagation

lead to trivial clock constraints, which may be simplified to either tt or ff and hence made applicable to Boolean Simplification.

Region Propagation: For constraint identifying single regions, i.e. constraints of the form $\beta(\gamma)$ additional distribution laws are given in Table 7

Constant Propagation: Identifiers with identifier-free definitions (i.e. constants such as tt or ff) may be removed while substituting their definitions in the declaration of all other identifiers.

Trivial Equation Elimination: Equations of the form $X \stackrel{\text{def}}{=} [a]X$ are easily seen to have $X = \mathtt{tt}$ as solution and may thus be removed. More generally, let S be the largest set of identifiers such that whenever $X \in S$ and $X \stackrel{\text{def}}{=} \varphi$ then $\varphi[\mathtt{tt}/S]$ [15] can be simplified to tt. Then all identifiers of S can be removed provided the value tt is propagated to all uses of identifiers from S (as under Constant Propagation). The maximal set S may be efficiently computed using standard fixed point computation algorithms.

[15] $\varphi[\mathtt{tt}/S]$ is the formula obtained by substituting all occurrences of identifiers from S in φ with the formula tt.

Equivalence Reduction: If two identifiers X and Y are semantically equivalent (i.e. are satisfied by the same timed transition systems) we may collapse them into a single identifier and thus obtain reduction. However, semantical equivalence is computationally very hard [16]. To obtain a cost effective strategy we approximate semantical equivalence of identifiers as follows: Let $\mathcal{R}$ be an equivalence relation on identifiers. $\mathcal{R}$ may be extended homomorphically to formulas in the obvious manner: i.e. $(\varphi_1 \wedge \varphi_2)\mathcal{R}(\vartheta_1 \wedge \vartheta_2)$ if $\varphi_1 \mathcal{R} \vartheta_1$ and $\varphi_2 \mathcal{R} \vartheta_2$, $(x \text{ in } \varphi)\mathcal{R}(x \text{ in } \vartheta)$ and $[a]\varphi\mathcal{R}[a]\vartheta$ if $\varphi\mathcal{R}\vartheta$ and so on. Now let $\cong$ be the maximal equivalence relation on identifiers such that whenever $X \cong Y$, $X \stackrel{\text{def}}{=} \varphi$ and $Y \stackrel{\text{def}}{=} \vartheta$ then $\varphi \cong \vartheta$. Then $\cong$ provides the desired cost effective approximation: whenever $X \cong Y$ then X and Y are indeed semantically equivalent. Moreover, $\cong$ may be efficiently computed using standard fixed point computation algorithms.

In the following Examples we apply the above transformation strategies to the quotient formula obtained in Example 7. In particular, the strategies will find the quotient formula to be trivially true in certain cases.

Example 8. Reconsider Example 7 with Y_0, X_0 and X_1 abbreviating $Y/_f\, k_0$, $X/_f\, k_0$ and $X/_f\, k_1$. Now Y_0 is the sufficient and necessary requirement to A_m in order that $A_m \,|_f\, B_n$ satisfies Y. From the definition of satisfiability for timed automata we see that:

$$A_m \models Y_0 \quad \text{if and only if} \quad A_m \models \mathtt{tt} \Rightarrow \Big(y \text{ in } Y_0\Big)$$

This provides an initial basis for constraint propagation. Using the propagation laws from Table 6 we get:

$$\mathtt{tt} \Rightarrow \Big(y \text{ in } Y_0\Big) \;\equiv\; \mathtt{tt} \Rightarrow \Big(\{y,z\} \text{ in } X_0\Big) \;\equiv\; \{y,z\} \text{ in } \Big(D_0 \Rightarrow X_0\Big)$$

where $D_0 = (y = 0 \wedge z = 0)$. This makes the implication $D_0 \Rightarrow X_0$ applicable to constraint propagation as follows:

$$\begin{aligned}(D_0 \Rightarrow X_0) &\equiv D_0 \Rightarrow \Big[(z \geq i) \vee \Big([b](y \text{ in } X_1) \wedge \forall X_0\Big)\Big] \\ &\equiv \Big(D_0 \Rightarrow [b](y \text{ in } X_1)\Big) \wedge \Big(D_0 \Rightarrow \forall X_0\Big) \qquad \text{as } (z < i \wedge D_0) = D_0 \\ &\equiv [b]\Big(y \text{ in } (D_0 \Rightarrow X_1)\Big) \wedge \forall\Big({D_0}^{\uparrow} \Rightarrow X_0\Big)\end{aligned}$$

Continuing constraint propagation yields the equations in Table 8, where $D_1 = (y = 0 \wedge z < i)$. □

Example 9. (Example 8 Continued) Now consider the case when $n \geq i$. That is the delay n of the component B_n exceeds the delay i required as a minimum by the property Y. Thus the component B_n ensures on its own the satisfiability of Y; i.e. for *any* choice of A the system $A \,|_f\, B_n$ will satisfy Y. In

[16] For the full logic T_μ the equivalence problem is undecidable.

$$(D_0 \Rightarrow X_0) \equiv [b]\big(y \,\mathsf{in}\, (D_0 \Rightarrow X_1)\big) \wedge \mathbf{\forall}\big({D_0}^{\dagger} \Rightarrow X_0\big)$$
$$({D_0}^{\dagger} \Rightarrow X_0) \equiv [b]\big(y \,\mathsf{in}\, (D_1 \Rightarrow X_1)\big) \wedge \mathbf{\forall}\big({D_0}^{\dagger} \Rightarrow X_0\big)$$
$$(D_1 \Rightarrow X_1) \equiv \big((D_1 \wedge y \geq n) \Rightarrow [c]\mathbf{ff}\big) \wedge \mathbf{\forall}({D_1}^{\dagger} \Rightarrow X_1)$$
$$(D_0 \Rightarrow X_1) \equiv \big((D_0 \wedge y \geq n) \Rightarrow [c]\mathbf{ff}\big) \wedge \mathbf{\forall}({D_0}^{\dagger} \Rightarrow X_1)$$
$$({D_0}^{\dagger} \Rightarrow X_1) \equiv \big(({D_0}^{\dagger} \wedge z < i \wedge y \geq n) \Rightarrow [c]\mathbf{ff}\big) \wedge \mathbf{\forall}(({D_0}^{\dagger} \wedge z < i)^{\dagger} \Rightarrow X_1)$$
$$({D_1}^{\dagger} \Rightarrow X_1) \equiv \big(({D_1}^{\dagger} \wedge z < i \wedge y \geq n) \Rightarrow [c]\mathbf{ff}\big) \wedge \mathbf{\forall}(({D_1}^{\dagger} \wedge z < i)^{\dagger} \Rightarrow X_1)$$

Table 8. Equations after Constraint Propagation

$$(D_0 \Rightarrow X_0) \equiv [b]\big(y \,\mathsf{in}\, (D_0 \Rightarrow X_1)\big) \wedge \mathbf{\forall}\big({D_0}^{\dagger} \Rightarrow X_0\big)$$
$$({D_0}^{\dagger} \Rightarrow X_0) \equiv [b]\big(y \,\mathsf{in}\, (D_1 \Rightarrow X_1)\big) \wedge \mathbf{\forall}\big({D_0}^{\dagger} \Rightarrow X_0\big)$$
$$(D_1 \Rightarrow X_1) \equiv \big(\mathbf{ff} \Rightarrow [c]\mathbf{ff}\big) \wedge \mathbf{\forall}({D_1}^{\dagger} \Rightarrow X_1)$$
$$(D_0 \Rightarrow X_1) \equiv \big(\mathbf{ff} \Rightarrow [c]\mathbf{ff}\big) \wedge \mathbf{\forall}({D_0}^{\dagger} \Rightarrow X_1)$$
$$({D_0}^{\dagger} \Rightarrow X_1) \equiv \big(\mathbf{ff} \Rightarrow [c]\mathbf{ff}\big) \wedge \mathbf{\forall}({D_0}^{\dagger} \Rightarrow X_1)$$
$$({D_1}^{\dagger} \Rightarrow X_1) \equiv \big(\mathbf{ff} \Rightarrow [c]\mathbf{ff}\big) \wedge \mathbf{\forall}({D_1}^{\dagger} \Rightarrow X_1)$$

Table 9. Equations after Simplification

this particular case (i.e. $n \geq i$) it is easy to see that $({D_i}^{\dagger} \wedge z < i \wedge y \geq n) = \mathbf{ff}$ for $i = 0, 1$ as ${D_i}^{\dagger}$ ensures $z \geq y$. Also for $i = 0, 1$, $(D_i \wedge y \geq n) = \mathbf{ff}$ as $D_i \Rightarrow y = 0$ and we assume $n > 0$. Finally, it is easily seen that $({D_i}^{\dagger} \wedge z < i)^{\dagger} = {D_i}^{\dagger}$ for $i = 0, 1$. Inserting these observations — which all may be efficiently computed — in the equations of Table 8 we get the simplified equations in Table 9. Now, the conjuncts $\mathbf{ff} \Rightarrow [c]\mathbf{ff}$ are obviously equivalent to $\mathbf{tt}$ and will thus be removed by the boolean simplification transformations. Now, using our strategy for Trivial Equation Elimination, it may be found that all the equations in Table 9 are trivial and may consequently be removed (simplified to $\mathbf{tt}$). To see this, simply observe that substituting $\mathbf{tt}$ for $D_i \Rightarrow X_j$ and ${D_i}^{\dagger} \Rightarrow X_j$ on all right-hand sides in Table 9 leads to formulas which clearly can be simplified to $\mathbf{tt}$. Thus, in the case $n \geq i$, our minimization heuristics will yield $\mathbf{tt}$ as the property required of A in order that $A \,|_r\, B_n$ satisfies Y. □

7 Experimental Results

The techniques presented in previous sections have been implemented in our verification tool UPPAAL in C++. We have tested UPPAAL by various examples. We also perform experiments on three existing real–time verification tools: HyTech (Cornell), Kronos (Grenoble), and Epsilon (Aalborg). Though the compositional model–checking technique is still under implementation, our experimental results show that UPPAAL is not only substantially faster than the other tools but also able to handle much larger systems.

In particular, we have used Fisher's mutual exclusion protocol in our experiments on the tools. The reason for choosing this example is that it is well–known and well–studied by researchers in the context of real–time verification. More importantly, the size of the example can be easily scaled up by simply increasing the number of processes in the protocol, thus increasing the number of control–nodes — causing state–space explosion — and the number of clocks — causing region–space explosion.

7.1 Fischer's Mutual Exclusion Protocol

The protocol is to guarantee mutual exclusion in a concurrent system consisting of a number of processes, using clock constraints and a shared variable. We shall model each of the processes as a timed automaton, and the protocol as a network of timed automata.

Assume a concurrent system with n processes $\mathsf{P}_1 ... \mathsf{P}_n$. Each process P_i with i being its identifier, has a clock x_i. We model the shared variable as a timed automaton V over the set of atomic actions: $\mathsf{A}_\mathsf{V} = \{v := i \mid i = 0...n\} \cup \{v = i \mid i = 0...n\}$, and $\mathsf{V} = \langle N, h_0, E, I, V \rangle$ where $N = \{\mathsf{V}_0 ... \mathsf{V}_n\}$, $h_0 = \mathsf{V}_0$, $E = \{\langle \mathsf{V}_i, \mathtt{tt}, v := j, \emptyset, \mathsf{V}_j \rangle \mid i, j = 0...n\} \cup \{\langle \mathsf{V}_i, \mathtt{tt}, v = i, \emptyset, \mathsf{V}_i \rangle \mid i = 0...n\}$, I is defined by $I(\mathsf{V}_i) = \mathtt{tt}$ for all $i \leq n$ and we simply assume V is defined by $V(\mathsf{V}_i) = \emptyset$ for all $i \leq n$.

The automaton for a typical process P_i is shown in Fig 5. We assume that the invariant conditions on nodes are all tt in this particular example. Moreover, we assume that the proposition assignment function is defined in such a way that $\mathsf{at}(l') \in V(l)$ if $l' = l$ and $\neg \mathsf{at}(l') \in V(l)$ if $l' \neq l$ for all nodes l and l'. Note that in the clock constraints $x_i < m_1$ and $x_i > m_2$, we have used two parameters. They can be any natural numbers satisfying the condition $m_1 \leq m_2$. Now, the whole protocol is described as the following network:

$$\mathsf{FISCHER}_n \equiv (\mathsf{P}_1 | \mathsf{P}_2 | ... | \mathsf{P}_n) || \mathsf{V}$$

where $|$ and $||$ are the full interleaving and full syncronization operators, induced by synchronization functions f and g respectively, defined by $f(0, a) = a$, $f(a, 0) = a$, and $g(a, a) = a$.

This is a simplified version of the original protocol and has been studied in e.g. [1, 19], which permits only one process to enter the critical section and never exits it. Recovery actions from failure to enter the critical section are omitted. However, it can be easily extended to model the full version of the protocol.

Intuitively, the protocol behaves as follows: The constraints on the shared variable V ensure that a process must reach B–node before any process reaches C–node; otherwise, it will never move from A–node to B–node. The timing constraints on the clocks ensure that all processes in C–nodes must wait until all

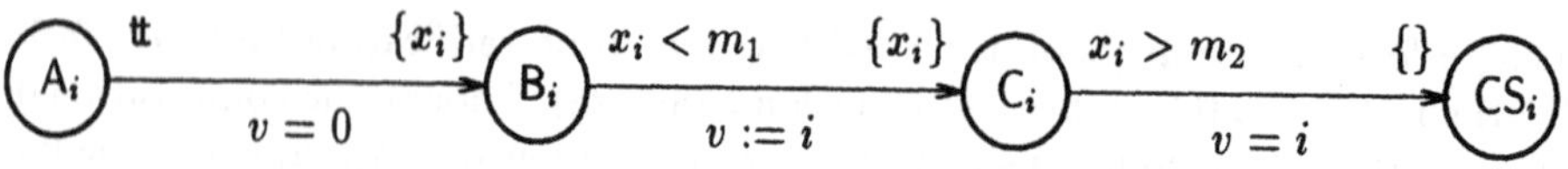

Fig. 5. Fischer's Mutual Exclusion Protocol

processes in B–nodes reach C–nodes. The last process that reaches C–node and sets V to its own identifier gets the right to enter its critical section.

We need to verify that there will never be more than one process in its critical section. An instance of this general requirement can be formalized as an invariant property:

$$\mathsf{MUTEX}_{1,2} \equiv \mathsf{INV}\Big(\neg\mathsf{at}(\mathsf{CS}_1) \vee \neg\mathsf{at}(\mathsf{CS}_2)\Big)$$

So we need to prove the theorem

$$\mathsf{FISCHER}_n \models \mathsf{MUTEX}_{1,2}$$

7.2 Performance Evaluation

Using the current version of our tool UPPAAL , installed on a SparcStation 10 running SunOS 4.1.2 with 64MB of primary memory and 64MB of swap memory, we have verified the mutual exclusion property of Fischers protocol for the cases [17] $n = 2, \ldots, 8$. The time–performance of this experiment can be found in Table 10 and Figure 6. Execution times have been measured in seconds with the standard UNIX program `time`. We have also attempted to verify Fischers protocol using three other existing real–time verification tools: HyTech 0.6 [15], Kronos 1.1c [9], Epsilon 3.0 [6] using the same machine as for the UPPAAL experiment. As illustrated in Table 10 and Figure 6 the experiment showed that UPPAAL is significantly faster than all these tools (50–100 times) and able to deal with much larger systems; all the other tools failed [18] to verify Fischers protocol for more than 4 processes (indicated by $\perp$ in the Table).

The four tools can be devided into two categories: HyTech and Kronos both produce the product of the automata network before the verification is carried out, whereas Epsilon and UPPAAL verifies properties on–the–fly without ever explicitly producing the product automaton. A potential advantage of the first strategy is the reusability of the product automaton. The obvious advantage of the second strategy is that only the necessary part of product automaton needs to be examined saving not only time but also (more importantly) space. For HyTech and Kronos we have measured both the total time as well as the part spent on the actual verification (marked v in Table 10), i.e. not measuring the time for producing the product automaton.

[17] In fact we have verified the case of 9 processes, but on a different machine.

[18] Failure occured either because the verification ran out of memory, never terminated or did not accept the produced product automaton.

	2	3	4	5	6	7	8	9
HyTech	6.0	83.5	⊥					
HyTech^v	3.6	26.4	⊥					
Epsilon	0.8	10.6	242.6	⊥				
Kronos	0.5	4.0	50.5	⊥				
Kronos^v	0.2	3.4	46.9	⊥				
UPPAAL	0.2	0.2	0.7	5.5	18.8	145.0	1107.5	⊥

Table 10. Execution Times (seconds).

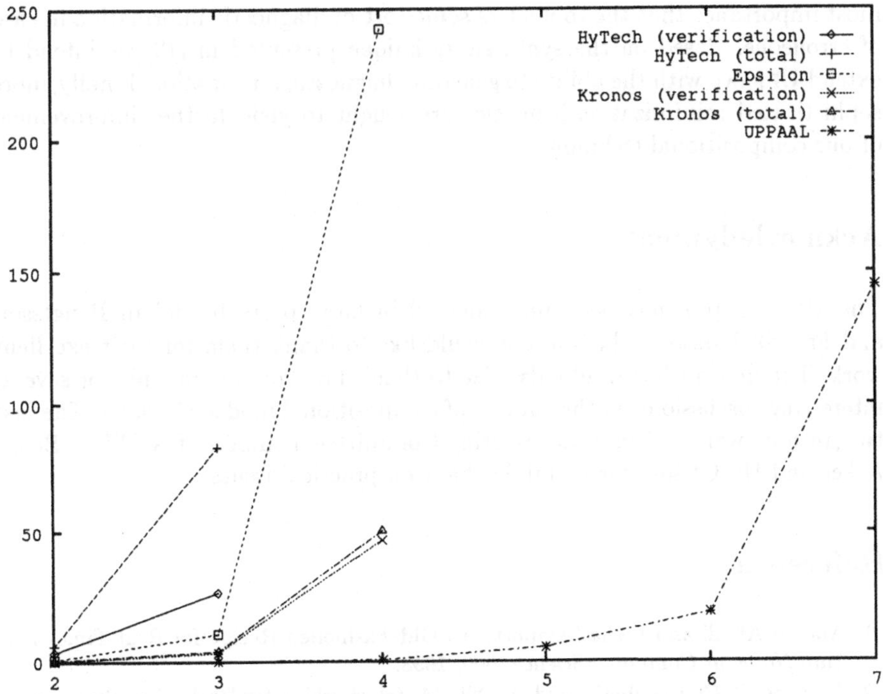

Fig. 6. Execution Times (seconds).

8 Conclusion and Future Work

In developing automatic verification algorithms for real-time systems, we need to deal with two potential types of explosion arising from parallel composition: explosion in the space of control nodes, and explosion in the region space over clock-variables. To attack these explosion problems, we have developed and combined compositional and symbolic model–checking techniques. These techniques have been implemented in a new automatic verification tool UPPAAL . Experimental results show that UPPAAL is not only substantially faster than other real-time verification tools but also able to handle much larger systems.

We should point out that the safety logic we designed in this paper enables the presented techniques to be implemented in a very efficient way. Though the logic is less expressive than the full version of the timed μ–calculus T_μ, it is expressive enough to specify safety properties as well as bounded liveness properties. As future work, we shall study the practical applicability of this logic and UPPAAL by further examples. Our experience shows that the practical limits of UPPAAL is caused by the space–complexity rather than the time–complexity of the model–checking algorithms. Thus, future work includes development of more space–efficient methods for representation and manipulation of clock constraints. For a verification tool to be of practical use in a design process it is of out most importance that the tool offers some sort of diagnostic information in case of erroneous. Based on the synthesis technique presented in [12] we intend to extend UPPAAL with the ability to generate diagnostic information. Finally, more sophisticated minimization heuristics are sought to yield further improvement of our compositional technique.

Acknowledgment

The UPPAAL tool has been implemented in large parts by Johan Bengtsson and Fredrik Larsson. The authors would like to thank them for their excellent work. The first author would also like to thank Francois Laroussinie for several interesting discussions on the subject of compositional model–checking. The last two authors want to thank the Steering Committee members of NUTEK, Bengt Asker and Ulf Olsson, for useful feedback on practical issues.

References

1. Martin Abadi and Leslie Lamport. An Old–Fashioned Recipe for Real Time. *Lecture Notes in Computer Science*, 600, 1993.
2. R. Alur, C. Courcoubetis, and D. Dill. Model–checking for Real–Time Systems. In *Proceedings of Logic in Computer Science*, pages 414–425. IEEE Computer Society Press, 1990.
3. R. Alur and D. Dill. Automata for Modelling Real–Time Systems. *Theoretical Computer Science*, 126(2):183–236, April 1994.
4. H. R. Andersen. Partial Model Checking. *To appear in Proceedings of LICS'95*, 1995.
5. J. R. Burch, E. M. Clarke, K. L. McMillan, D. L. Dill, and L. J. Hwang. Symbolic Model Checking: 10^{20} states and beyond. *Logic in Computer Science*, 1990.
6. K. Cerans, J. C. Godskesen, and K. G. Larsen. Timed modal specifications — theory and tools. *Lecture Notes in Computer Science*, 697, 1993. In Proceedings of CAV'93.
7. E. M. Clarke, T. Filkorn, and S. Jha. Exploiting Symmetry in Temporal Logic Model Checking. *Lecture Notes in Computer Science*, 697, 1993. In Proceedings of CAV'93.

8. E. M. Clarke, O. Grümberg, and D. E. Long. Model Checking and Abstraction. *Principles of Programming Languages*, 1992.
9. C. Daws, A. Olivero, and S. Yovine. Verifying ET–LOTOS programs with KRONOS. *In Proceedings of 7th International Conference on Formal Description Techniques*, 1994.
10. E. A. Emerson and C. S. Jutla. Symmetry and Model Checking. *Lecture Notes in Computer Science*, 697, 1993. In Proceedings of CAV'93.
11. P. Godefroid and P. Wolper. A Partial Approach to Model Checking. *Logic in Computer Science*, 1991.
12. J.C. Godskesen and K.G. Larsen. Synthesizing Distinghuishing Formulae for Real Time Systems — Extended Abstract. *Lecture Notes in Computer Science*, 1995. To occur in Proceedings of MFCS'95. Also BRICS report series RS–94–48.
13. Nicolas Halbwachs. Delay Analysis in Synchronous Programs. *Lecture Notes in Computer Science*, 697, 1993. In Proceedings of CAV'93.
14. T. A. Henzinger, Z. Nicollin, J. Sifakis, and S. Yovine. Symbolic model checking for real-time systems. In *Logic in Computer Science*, 1992.
15. Thomas A. Henzinger and Pei-Hsin Ho. HyTech: The Cornell HYbrid TECHnology Tool. *To appear in the Proceedings of TACAS, Workshop on Tools and Algorithms for the Construction and Analysis of Systems*, 1995.
16. H. Hüttel and K. G. Larsen. The use of static constructs in a modal process logic. *Lecture Notes in Computer Science, Springer Verlag*, 363, 1989.
17. F. Laroussinie, K. G. Larsen, and C. Weise. From Timed Automata to Logic — and Back. *Lecture Notes in Computer Science*, 1995. To occur in Proceedings of MFCS. Also BRICS report series RS–95–2.
18. F. Laroussinie and K.G. Larsen. Compositional Model Checking of Real Time Systems. *Lecture Notes in Computer Science*, 1995. To appear in Proceedings of CONCUR'95. Also BRICS report series RS–95–19.
19. N. Shankar. Verification of REal–Time Systems Using PVS. *Lecture Notes in Computer Science*, 697, 1993. In Proceedings of CAV'93.
20. A. Tarski. A lattice–theoretical fixpoint theorem and its applications. *Pacific Journal of Math.*, 5, 1955.
21. A. Valmari. A Stubborn Attack on State Explosion. *Theoretical Computer Science*, 3, 1990.
22. Wang Yi, Paul Pettersson, and Mats Daniels. Automatic Verification of Real–Time Systems By Constraint–Solving. *In the Proceedings of the 7th International Conference on Formal Description Techniques*, 1994.

On Polynomial Ideals, Their Complexity, and Applications

Ernst W. Mayr

Institut für Informatik
Technische Universität München
D-80290 München, GERMANY
e-mail: MAYR@INFORMATIK.TU-MUENCHEN.DE

Abstract. A polynomial ideal membership problem is a $(w+1)$-tuple $P = (f, g_1, g_2, \ldots, g_w)$ where f and the g_i are multivariate polynomials over some ring, and the problem is to determine whether f is in the ideal generated by the g_i. For polynomials over the integers or rationals, it is known that this problem is exponential space complete. We discuss complexity results known for a number of problems related to polynomial ideals, like the word problem for commutative semigroups, a quantitative version of Hilbert's Nullstellensatz, and the reachability and other problems for (reversible) Petri nets.

1 Introduction

Polynomial rings and their ideals are fundamental in many areas of mathematics, and they also have a surprising number of applications in various areas of computer science, like language generating and term rewriting systems, tiling problems, the complexity of algebraic manifolds, and the complexity of some models for parallel systems. They have also been used in some constrained logic programming software systems, like [1].

The decidability of the membership problem for polynomial ideals over a field or ring can, in a sense, be traced back to ideas in Hilbert's work, and was established in [16], [34], and [33]. The computational complexity of the polynomial ideal membership problem was first discussed in [28] where the special case of the word problem for commutative semigroups was investigated. The bounds derived there imply an exponential space lower bound for the membership problem in polynomial ideals over $\mathbb{Z}$ (the integers) or $\mathbb{Q}$ (the rationals), in fact over arbitrary fields, as well as a doubly exponential lower bound for the time requirements for any Turing machine solving the polynomial ideal membership problem over the rationals. Other, rather special cases of the polynomial ideal membership problem (given by restrictions on the form of the generators) and their complexity have been investigated in [19], and, for the case of special p, in *e.g.* [6], [2], [3], and [15].

In this paper, we give a survey on basic algorithmic problems involving polynomial ideals, on the complexity bound known for these problems and algorithms for them, and on some applications of polynomial ideals in other areas of computer science. It must be stressed, however, that this survey is not intended to be comprehensive and complete, a remark that also applies to the list of references cited at the end.

2 Notation, Fundamental Concepts

2.1 Polynomials and Ideals

Consider the finite set $\{x_1, \ldots, x_n\}$ of indeterminates and let $\mathbb{Q}[x]$ denote the (commutative) ring of polynomials in $x_1, \ldots, x_n$ with rational coefficients. An *ideal* in $\mathbb{Q}[x]$ is any subset I of $\mathbb{Q}[x]$ satisfying

(i) $p, q \in I \;\Rightarrow\; p - q \in I$;
(ii) $p \in I,\, r \in \mathbb{Q}[x] \;\Rightarrow\; rp \in I$.

For polynomials $g_1, \ldots, g_w \in \mathbb{Q}[x]$, let $(g_1, \ldots, g_w) \subseteq \mathbb{Q}[x]$ denote the ideal generated by $\{g_1, \ldots, g_w\}$, *i.e.*,

$$(g_1, \ldots, g_w) = \left\{ \sum_{1 \le i \le w} p_i g_i \,;\, p_i \in \mathbb{Q}[x] \right\}.$$

If $I = (g_1, \ldots, g_w)$, $\{g_1, \ldots, g_w\}$ is called a *basis* of I.

A *monomial* m in $x_1, \ldots, x_n$ is a product of the form

$$m = x_1^{\alpha_1} x_2^{\alpha_2} \cdots x_n^{\alpha_n},$$

with $\alpha = (\alpha_1, \ldots, \alpha_n) \in \mathbb{N}^n$ the *degree vector* of m and $\deg(m) = \sum_{j=1}^{n} \alpha_j$ the *total degree* of m. For succinctness, we also write $m = x^\alpha$.

Each *polynomial* $f(x_1, \ldots, x_n) \in \mathbb{Q}[x]$ is a finite sum

$$f(x_1, \ldots, x_n) = \sum_{1 \le i \le r} c_i \cdot x^{\alpha_i},$$

with $c_i \in \mathbb{Q} - \{0\}$ the coefficient and $\alpha_i \in \mathbb{N}^n$ the degree vector of the ith monomial of f. The product $c_i \cdot x^{\alpha_i}$ is called the ith term of the polynomial f. The total degree of a polynomial is the maximum of the total degrees of its monomials.

Example: Consider $\mathbb{Q}[x_1, x_2]$, the ring of polynomials in x_1 and x_2 with rational coefficients. Then the ideal (x_1^2, x_2) consists of all polynomials $f \in \mathbb{Q}[x_1, x_2]$ such that each monomial of f is divisible by x_1^2 or by x_2.

An *admissible term ordering* in $\mathbb{Q}[x]$ is given by any total order $<$ on $\mathbb{N}^n$ satisfying the following two properties:

1. $\alpha > (0, \ldots, 0)$ for all $\alpha \in \mathbb{N}^n - \{(0, \ldots, 0)\}$;
2. for all $\alpha, \beta, \gamma \in \mathbb{N}^n$,

$$\alpha < \beta \Rightarrow \alpha + \gamma < \beta + \gamma.$$

If $\alpha > \beta$, we say that any term $c \cdot x^\alpha$ is greater in the term ordering than any term $c' \cdot x^\beta$, and, for a polynomial $f(x) = \sum_{i=1}^{r} c_i \cdot x^{\alpha_i}$, we always assume that $\alpha_1 > \alpha_2 > \ldots > \alpha_n$. We call $LM(f) = x^{\alpha_1}$ the *leading monomial* and $LT(f) = c_1 \cdot x^{\alpha_1}$ the *leading term* of f. Since we are dealing with polynomials with coefficients from the field $\mathbb{Q}$, we shall also usually assume that polynomials are normalized, *i.e.*, their leading coefficient c_1 is one. In an abuse of notation, we also write $<$ for the term ordering induced by the order $<$ on the degree vectors.

Example: Let $<$ be the lexicographic ordering on $\mathbb{N}^n$, *i.e.*, if $\alpha, \beta \in \mathbb{N}^n$, $\alpha \neq \beta$, $\alpha = (\alpha_1, \ldots, \alpha_n)$ and $\beta = (\beta_1, \ldots, \beta_n)$ then

$$\alpha < \beta \text{ iff there is an } i \text{ such that for all } j < i \;\; \alpha_j = \beta_j, \text{ and } \alpha_i < \beta_i\,.$$

Then, in the term ordering,

$$x_1 > x_2 > x_3 > 1\,,$$

and the leading term (and the leading monomial) of the polynomial

$$f(x_1, x_2, x_3) = x_1^5 + x_1^2 x_2^4 + x_1^2 x_3^3 + 3x_1 x_2^2 x_3^2 - 1$$

is x_1^5.

Example: Let $<$ be the so-called *graded reverse lexicographic (grevlex)* ordering on $\mathbb{N}^n$, *i.e.*, if $\alpha, \beta \in \mathbb{N}^n$, $\alpha \neq \beta$, $\alpha = (\alpha_1, \ldots, \alpha_n)$ and $\beta = (\beta_1, \ldots, \beta_n)$ then

$$\begin{aligned} \alpha < \beta \text{ iff } & \textstyle\sum_{i=1}^n \alpha_i < \sum_{i=1}^n \beta_i, \text{ or} \\ & \textstyle\sum_{i=1}^n \alpha_i = \sum_{i=1}^n \beta_i, \text{ and there is an } i \text{ such that} \\ & \alpha_j = \beta_j \text{ for all } j > i \text{ and } \alpha_i > \beta_i. \end{aligned}$$

Then, in the term ordering,

$$x_1 > x_2 > x_3 > 1\,,$$

the polynomial in the previous example is written

$$f(x_1, x_2, x_3) = x_1^2 x_2^4 + x_1^5 + 3x_1 x_2^2 x_3^2 + x_1^2 x_3^3 - 1\,,$$

and its leading term is $x_1^2 x_2^4$.

Let I be an ideal in $\mathbb{Q}[x]$, and let some admissible term order $<$ on $\mathbb{Q}[x]$ be given. A finite set $\{g_1, \ldots, g_r\}$ of polynomials from $\mathbb{Q}[x]$ is called a *Standard* or *Gröbner* basis of I (wrt. $<$), if

(i) $\{g_1, \ldots, g_r\}$ is a basis of I;

(ii) $\{LT(g_1), \ldots, LT(g_r)\}$ is a basis of the *leading term ideal* of I, which is the smallest ideal containing the leading terms of all $f \in I$; or, equivalently: if $f \in I$, then

$$LT(f) \in (LT(g_1), \ldots, LT(g_r))\,.$$

Standard and Gröbner bases have been introduced in [17, 18] and [4]. For an excellent exposition of their numerous useful properties, also see [5]. A basis is called *minimal* if it does not strictly contain some other basis of the same ideal. A Gröbner basis is called *reduced* if no term in any one of its polynomials is divisible by the leading monomial of some other polynomial in the basis.

A polynomial $f \in \mathbb{Q}[x]$ is called *homogeneous* (of degree d) if all of its monomials have the same total degree d. Let $f \in \mathbb{Q}[x]$ be some arbitrary polynomial. Then f can uniquely be written as $f = \sum f_i$, where each f_i is homogeneous and $\deg(f_i) \neq \deg(f_j)$ for $i \neq j$. The f_i are called the *homogeneous components* of f. An ideal $I \subseteq \mathbb{Q}[x]$ is called *homogeneous*, if, whenever I contains some polynomial f, it also contains the homogeneous components of f. It can be shown that this is equivalent to the following definition: An ideal $I \subseteq \mathbb{Q}[x]$ is homogeneous if it has a basis consisting of homogeneous polynomials.

2.2 Commutative Semigroups

A *commutative semigroup* $(H, \circ)$ is a set H with a binary operation $\circ$ which is associative and commutative. Usually we shall write ab for $a \circ b$.

A commutative semigroup H is said to be *finitely generated* by a finite subset $S = \{s_1, \ldots, s_n\} \subseteq H$ if

$$H = \{s_1^{\alpha_1} s_2^{\alpha_2} \cdots s_n^{\alpha_n} \,;\, \alpha_i \in \mathbb{N} \text{ for } i = 1, \ldots, n\}.$$

(Note: $s_i^{\alpha_i}$ is short for $\underbrace{s_i \cdots s_i}_{\alpha_i}$.) There is a canonical homomorphism from $\mathbb{N}^n$ to H, mapping $\alpha \in \mathbb{N}^n$ to $s^\alpha \in H$. If this homomorphism actually is a bijection, then H is the free commutative semigroup generated by $\{s_1, \ldots, s_n\}$, which is also denoted by S^*. For a word $m = s_1^{\alpha_1} s_2^{\alpha_2} \cdots s_n^{\alpha_n} \in S^*$, the sum $\alpha_1 + \alpha_2 + \ldots + \alpha_n$ is called the *length* of m.

Note that power a monomial $x^\alpha \in \mathbb{Q}[x]$ can also be looked at as an element of $\{x_1, \ldots, x_k\}^*$.

A *commutative semi-Thue system* over S is given by a finite set $\mathcal{P}$ of productions $l_i \to r_i$, where $l_i, r_i \in S^*$. A word $m' \in S^*$ *is derived in one step* from $m \in S^*$ (written $m \to m'(\mathcal{P})$) applying the production $(l_i \to r_i) \in \mathcal{P}$ iff, for some $\tilde{m} \in S^*$, we have $m = \tilde{m} l_i$ and $m' = \tilde{m} r_i$. The word m *derives* m' iff $m \xrightarrow{*} m'(\mathcal{P})$, where $\xrightarrow{*}$ is the reflexive transitive closure of $\to$. A sequence $(m_0, \ldots, m_r)$ of words $m_i \in S^*$ with $m_i \to m_{i+1}(\mathcal{P})$ for $i = 0, \ldots, r-1$ is called a *derivation* (of length r) of m_r from m_0 in $\mathcal{P}$.

A *commutative Thue system* is a symmetric commutative semi-Thue system $\mathcal{P}$, *i.e.*,

$$(l \to r) \in \mathcal{P} \Rightarrow (r \to l) \in \mathcal{P}.$$

Clearly, commutative Thue systems and commutative semigroups are equivalent concepts.

Derivability in a (commutative) semigroup establishes a congruence $\equiv_{\mathcal{P}}$ on S^* by the rule

$$m \equiv m' \bmod \mathcal{P} \Leftrightarrow_{\text{def}} m \xrightarrow{*} m'(\mathcal{P}).$$

For semigroups, we also use the notation $l \equiv r \bmod \mathcal{P}$ to denote the pair of productions $(l \to r)$ and $(r \to l)$ in $\mathcal{P}$.

If it is understood that $\mathcal{P}$ is a commutative Thue system then the commutativity productions are not explicitly mentioned in $\mathcal{P}$, nor is their application within a derivation in $\mathcal{P}$ counted as a step.

A commutative Thue system $\mathcal{P}$ is also called a *presentation of the quotient semigroup* $S^*/\equiv_{\mathcal{P}}$. For $m \in S^*$, we use $[m]$ to denote the congruence class of $m \bmod \mathcal{P}$.

We remark that commutative semi-Thue systems appear in the literature in two additional equivalent formulations: *vector addition systems* (see next section) and *Petri nets.* Finitely presented commutative semigroups are equivalent to *reversible* vector addition systems or Petri nets. A reader more familiar with Petri nets may want to think of a vector in $\mathbb{N}^k$ as a marking.

2.3 Vector Addition Systems, Petri Nets, and Semilinear Sets

A *vector addition system (VAS)* is a pair (m, V), with $m \in \mathbb{N}^n$ and V a finite set $\{v_1, \ldots, v_r\}$ of vectors in $\mathbb{Z}^n$. The vector m is called the *start vector*, n is the dimension of the VAS, and the v_i are the *transitions*. A VAS is called *reversible*, if, whenever it contains a transition v, it also contains $-v$.

The *reachability set* of a VAS (m, V) is the smallest set $R(m, V)$ satisfying the following two properties:

(i) $m \in R(m, V)$;
(ii) whenever $z \in R(m, V)$, $v \in V$, and $z + v \in \mathbb{N}^n$ then $z + v \in R(m, V)$.

Thus, $R(m, V)$ is the smallest subset of $\mathbb{N}^n$ containing m which is closed under addition of transitions as long as the sum has only nonnegative components.

A *transition sequence* $(v^{(i)})_{1 \le i \le t}$ of transitions $v^{(i)}$ is *applicable* to some vector $m' \in \mathbb{N}^n$ if $m' + \sum_{j=1}^{i} v^{(j)} \in \mathbb{N}^n$ for all $i = 1, \ldots, t$. In this case, the vector $m'' = m' + \sum_{j=1}^{t} v^{(j)}$ is called *reachable from y in (x, V)*, and the transition sequence is called a *derivation* (of length t) of m'' from m'. For this property, we also use the notation $m' \xrightarrow{*} m''$ (V).

Clearly, reversible VASs can be simulated by commutative semigroups (the other direction is also possible, though not directly; the probably simplest way is to replace vector addition systems by *vector replacement systems (VRS)*). For an n-dimensional reversible VAS (m, V) we use a commutative semigroup with n generators $s_1, \ldots, s_n$, and a congruence $l \equiv r$ for every pair $\{v, -v\} \subseteq V$, with

$$l = s_1^{\max\{0,v_1\}} \cdots s_n^{\max\{0,v_n\}}$$
$$r = s_1^{\max\{0,-v_1\}} \cdots s_n^{\max\{0,-v_n\}},$$

where $v = (v_1, \ldots, v_n) \in \mathbb{Z}^n$.

A *linear* subset L of $\mathbb{N}^n$ is a set of the form

$$L = \left\{ a + \sum_{i=1}^{t} n_i b^{(i)};\ n_i \in \mathbb{N} \text{ for } i = 1, \ldots, t \right\}$$

for some vectors $a, b^{(1)}, \ldots, b^{(t)} \in \mathbb{N}^n$.

A *semilinear* set SL is a finite union of linear sets:

$$SL = \bigcup_{j=1}^{k} \left\{ a_j + \sum_{i=1}^{t_j} n_i b_j^{(i)};\ n_i \in \mathbb{N} \text{ for } i = 1, \ldots, t_j \right\}$$

for some vectors $a_j, b_j^{(1)}, \ldots, b_j^{(t_j)} \in \mathbb{N}^n$, $j = 1, \ldots, k$.

A *uniformly semilinear* subset UL of $\mathbb{N}^n$ is a set of the form

$$UL = \bigcup_{j=1}^{k} \left\{ a_j + \sum_{i=1}^{t} n_i b^{(i)};\ n_i \in \mathbb{N} \text{ for } i = 1, \ldots, t \right\}$$

for some vectors $a_j, b^{(1)}, \ldots, b^{(t)} \in \mathbb{N}^n$, $j = 1, \ldots, k$.

We have (see [11]) the following

Theorem 1 *Let $\equiv$ be any congruence relation on $\mathbb{N}^n$. Then the congruence class $[u]$ of any element $u \in \mathbb{N}^n$ with respect to $\equiv$ is a uniformly semilinear set in $\mathbb{N}^n$.*

For a reversible VAS (m, V) this theorem says that the reachability set $R(n, V)$ is a uniformly semilinear set.

Petri nets [31] are a graphical representation of VASs and VRSs, equivalent to VRSs. A marked Petri net P consists of

(i) a finite bipartite multi-digraph (S, T, F), with
  (a) $S = \{s_1, \ldots, s_n\}$ the set of *places*,
  (b) $T = \{t_1, \ldots, t_r\}$ the set of *transitions*, and
  (c) $F : S \times T \cup T \times S \rightarrow \mathbb{N}$ giving the multiplicity of the arcs;
(ii) and an *initial marking* $m \in \mathbb{N}^n$.

A transition $t \in T$ is said to be *enabled* at some marking $m' \in \mathbb{N}^n$ if

$$m'_i \geq F(s_i, t) \text{ for all } i,\ i = 1, \ldots, n\,.$$

If t is enabled at m', it can (but does not have to) fire, producing the new marking m'' with

$$m''_i = m'_i - F(s_i, t) + F(t, s_i) \text{ for all } i,\ i = 1, \ldots, n\,.$$

We also write $m' \xrightarrow{t} m''$. Note that $m'' \in \mathbb{N}^n$.

The *reachability set* $R(P)$ with initial marking m is the smallest subset of $\mathbb{N}^n$ containing m which is closed under $\xrightarrow{t}$ for all $t \in T$.

A Petri net is called *reversible* if, for every transition t, it contains a transition t^{rev} with

$$F(s, t) = F(t^{rev}, s) \text{ and } F(t, s) = F(s, t^{rev}) \text{ for all } s \in S\,.$$

The following figure shows a simple example of a Petri net. The places are depicted by circles, the transitions by bars, the marking of each place by a corresponding number of dots.

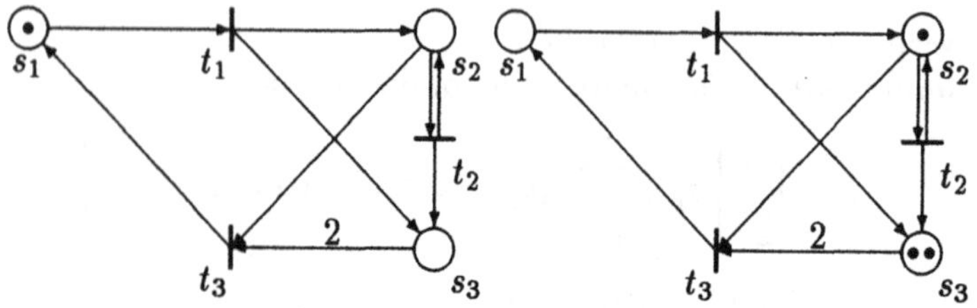

The marking shown on the right is obtained from that on the left by firing t_1 and then t_2. If we make this Petri net reversible by adding transitions t_i^{rev}, for $i = 1, 2, 3$, we obtain a reversible Petri net which is equivalent, as can easily be seen, to a commutative semigroup with generators s_1, s_2, s_3 and the following congruences:

$$s_1 \equiv s_2 s_3$$
$$s_2 \equiv s_2 s_3$$
$$s_2 s_3^2 \equiv s_1$$

3 Basic Problems and Their Complexity

In this section, we are going to consider some of the very basic and fundamental algorithmic problems for the structures we have presented in the previous section. Arguably one of the most central problems for almost all of these structures turns out to be the *uniform word problem for commutative semigroups* which is defined as follows:

Definition 3.1 *Let S be a finite set of generators, and $\mathcal{P}$ a finite set of congruences on S^*. Let $m, m' \in S^*$.*

(i) **Decision Problem:** *Given $S, \mathcal{P}, m$, and m' as input, decide whether*

$$m \equiv m' \bmod \mathcal{P};$$

(ii) **Representation Problem:** *Given $S, \mathcal{P}, m$, and m' as input, decide whether $m \equiv m' \bmod \mathcal{P}$, and if so, find a derivation of m' from m in $\mathcal{P}$.*

Another problem, just as central, is the *polynomial ideal membership problem (PIMP)*. It is

Definition 3.2 *Let $f, g_1, \ldots, g_w$ be polynomials in $\mathbb{Q}[x] = \mathbb{Q}[x_1, \ldots, x_n]$, and let $I = (g_1, \ldots, g_w)$.*

(i) **Decision Problem:** *Given $f, g_1, \ldots, g_w$, decide whether*

$$f \in I;$$

(ii) **Representation Problem:** *Given $f, g_1, \ldots, g_w$, decide whether $f \in I$, and if so, find $p_i \in \mathbb{Q}[x]$ such that*

$$f(x) = \sum_{i=1}^{w} p_i g_i .$$

It is well known (see, e.g., [8]) that the word problem for commutative semigroups can be reduced to PIMP, simply by interpreting each word $m \in S^*$ as a monomial in the indeterminates $s_1, \ldots, s_n$ and observing that

$$m \equiv m' \bmod \mathcal{P} \Longleftrightarrow m' - m \in (r_1 - l_1, \ldots, r_w - l_w) \subseteq \mathbb{Q}[s_1, \ldots, s_n],$$

where $l_i \equiv r_i$, $i = 1, \ldots, w$ are the congruences in $\mathcal{P}$.

In the fundamental paper [16], G. Hermann gave a doubly exponential degree bound for PIMP:

Theorem 2 *Let $f, g_1, \ldots, g_w$ be polynomials $\in \mathbb{Q}[x]$, and let $d = \max\{\deg(g_i); i = 1, \ldots, w\}$. If $f \in (g_1, \ldots, g_w)$, then there exist $p_1, \ldots, p_w \in \mathbb{Q}[x]$ such that*

1. $f = \sum_{i=1}^{w} p_i g_i$; *and*
2. $\deg(p_i) \leq \deg(f) + (wd)^{2^n}$, *for all i, $i = 1, \ldots, w$.*

For improved proofs of this theorem, see [34] and [28].

In [7] and [28] it was shown how to transform this degree bound for PIMP into a space bound for the special case of PIMP, the uniform word problem for commutative semigroups:

Theorem 3 *The uniform word problem for finitely presented commutative semigroups can be decided in exponential space (i.e., space $2^{O(n)}$, with n here the size of the input).*

In [26, 27], this exponential space upper bound was generalized to PIMP:

Theorem 4 *Let P be a polynomial ideal membership problem over $\mathbb{Q}$, and let s be the size of the input for P. Then there is a PRAM algorithm which solves P in parallel time $2^{O(s)}$ using $2^{2^{O(s)}}$ processors.*

Using the Parallel Computation Thesis ([14]) and techniques from [30], one obtains

Theorem 5 *The polynomial ideal membership problem is solvable in sequential space exponential in the size of the problem instance.*

for the decision problem, and also, for the representation problem

Theorem 6 *Let f and $g_1, \ldots, g_w$ be multivariate polynomials over the rationals. If f is an element of the ideal generated by the g_i then a representation*

$$f = \sum_{1 \leq i \leq w} p_i g_i$$

can be found in exponential space.

As is customary, the space bound for the representation problem bounds the work space, not the space on the output tape needed to write down the g_i. This is crucial, since, as we shall see below, their total length can be double exponential in the size of the input. For a detailed proof of these two theorems, see [26].

While the exponential space bound in [26] is based on the classical construction in [16], recently exciting improvements have been obtained for the degree bound for a number of special cases of PIMP. Among them, maybe the most prominent are the following:

Theorem 7 *Let g_i, $i = 1, \ldots, w$, be polynomials in $\mathbb{Q}[x_1, \ldots, x_n]$, let d be the maximal degree of the g_i, and assume that the g_i have no common zero in $\mathbb{C}^n$. Then*

$$1 = \sum_{i=1}^{w} p_i g_i$$

for p_i with $\deg(p_i) \leq \mu n d^{\mu} + \mu d$, *with* $\mu = \min\{n, w\}$.

Using the so-called "Rabinowitsch trick", Brownawell also obtained

Theorem 8 *Let $f, g_i \in \mathbb{Q}[x_1, \ldots, x_n]$ for $i = 1, \ldots, w$, let d and μ be as above, and assume that $f(x) = 0$ for all common zeros x (in $\mathbb{C}^n$) of the g_i. Then there are*

$$e \in \mathbb{N}, \; e \leq (\mu + 1)(n + 2)(d + 1)^{\mu+1},$$
$$p_i \in \mathbb{Q}[x_1, \ldots, x_n], \; \textit{with} \; \deg(p_i) \leq (\mu + 1)(n + 2)(d + 1)^{\mu+2}$$

such that

$$f^e = \sum_{i=1}^{w} p_i g_i .$$

For proofs of these and similar exponential degree bounds, see [2], [3], and [21]. The method of [26] immediately yields

Corollary 8.1 *Whether*

$$1 \in (g_1, \ldots, g_w)$$

can be tested in PSPACE.

Corollary 8.2 *Whether there is an $e \in \mathbb{N}$ such that*

$$g^e \in (g_1, \ldots, g_w)$$

can be tested in PSPACE.

These two corollaries could be termed *quantitative versions* of Hilbert's Nullstellensatz (see, e.g., [36]), one variant of which is

Theorem 9 (Hilbert's Nullstellensatz) *Let k be some algebraically closed field, let $f, g_i \in k[x_1, \ldots, x_n]$, for $i = 1, \ldots, w$, and assume that $f(x) = 0$ for all common zeros x of the g_i. Then (and only then) there is an integer $e \geq 1$ such that*

$$f^e \in (g_1, \ldots, g_w) .$$

There are a few more special cases of PIMP, where we get a PSPACE upper bound. An ideal $I = (g_1, \ldots, g_w) \subseteq \mathbb{Q}[x]$ is called *zero-dimensional* if the common zeros (in $\mathbb{C}^n$) of the g_i are a finite set (for an exact definition of the dimension of an algebraic variety or an ideal we refer the reader to e.g. [9]). For zero-dimensional ideals, an exponential degree upper bound is known for the presentation problem [6]. Such an exponential degree upper bound also holds for complete intersections (the dimension of the algebraic variety defined by the g_i (in $\mathbb{C}^n$) is $n - w$), as shown in [2].

Another "easy" case is when the generators $g_1, \ldots, g_w \in \mathbb{Q}[x]$ are homogeneous. Then the question, whether a general $f \in \mathbb{Q}[x]$ is an element of the ideal $(g_1, \ldots, g_w)$ can be solved by treating each homogeneous component of f separately. Hence, we may assume that f is homogeneous. In this case, $f \in (g_1, \ldots, g_w)$ iff $f(x) = \sum_{i=1}^{w} p_i g_i$ for homogeneous polynomials p_i with $\deg(p_i) = \deg(f) - \deg(g_i)$. Since a homogeneous polynomial in n variables and of degree d can consist of at most $\binom{n+d-1}{n-1}$ distinct monomials, the method of [26] again yields a PSPACE algorithm.

As we have already mentioned, Gröbner bases play an important role in the algorithmic treatment of problems in polynomial ideals. The complexity of algorithms for generating a Gröbner basis from a given set of generators for an ideal has been the subject of intensive study (see e.g. [12] for a rather comprehensive survey). From the numerous complexity result, we would like to mention the following:

Theorem 10 *Let $I = (g_1, \ldots, g_w) \subseteq \mathbb{Q}[x_1, \ldots, x_n]$ be an ideal, let d be the maximal total degree of the g_i, $i = 1, \ldots, w$, and let $<$ be any admissible ordering on $\mathbb{Q}[x]$. Then the reduced Gröbner basis for I consists of polynomials whose total degree is bounded by*

$$2\left(\frac{d^2}{2} + d\right)^{2^{n-1}}.$$

An elegant, elementary proof of this doubly exponential degree bound is given in [10].

Let $g_1, \ldots, g_w \in \mathbb{Q}[x_1, \ldots, x_n]$ be given. A *syzygy* for the g_i is any vector $(p_1, \ldots, p_w) \in (\mathbb{Q}[x])^w$ such that $\sum_{i=1}^{w} p_i g_i = 0$. The set of syzygies forms a (finite dimensional) $\mathbb{Q}[x]$-module [16].

Theorem 11 *Let $g_1, \ldots, g_w \in \mathbb{Q}[x_1, \ldots, x_n]$ be given, and let d be a bound on the total degree of the g_i. Then there is a basis for the module of syzygies whose polynomials have a total degree bounded by*

$$2\left(\frac{d^2}{2} + d\right)^{2^{n-1}}.$$

For a proof, see [16] and [10].

In the remainder of this section, we turn to lower bounds for the algorithmic problems considered so far. The central result here is the lower bound for the uniform word problem for finitely presented commutative semigroups shown in [28]:

Theorem 12 *There is an infinite family of instances $(m^{(i)}, m'^{(i)}, \mathcal{P}^{(i)})$ of the uniform word problem for finitely presented commutative semigroups and a constant $c > 0$ such that each derivation of $m'^{(i)}$ from $m^{(i)}$ in $\mathcal{P}^{(i)}$ contains a word of length $\geq 2^{2^{c \cdot s}}$, where s denotes the input size.*

Using commutative semigroups to simulate counter or Minsky automata [29], this result implies [26]:

Theorem 13 *The uniform word problem for finitely presented commutative semigroups requires exponential space, and, together with the matching upper bound, is therefore exponential space complete.*

Since the word problem for commutative semigroups is a special case of PIMP (the corresponding ideals are also called *binomial ideals*, see [13]), we also obtain an exponential space lower bound (and thus completeness for exponential space) for PIMP. The construction in [28] has been sharpened in [35] (which greatly improves the constant in the exponent from 1/14 to basically 1/2) to yield the following lower bounds:

Theorem 14 *Let n be the number of indeterminates and d the maximal total degree of the generating polynomials in $\mathbb{Q}[x] = \mathbb{Q}[x_1, \ldots, x_n]$. Then there is an infinite family of instances of PIMP, including infinitely many n, such that, for each of these instances, say with generators $g_1, \ldots, g_w$,*

(i) there is a polynomial $f \in \mathbb{Q}[x_1, \ldots, x_n]$ with total degree $\leq d$, such that $f \in (g_1, \ldots, g_w)$ and, whenever

$$f(x) = \sum_{i=1}^{w} p_i g_i ,$$

then the maximal total degree of the p_i is $\geq 2^{2^{n/2-O(\sqrt{n})}}$;

(ii) any syzygy basis for the $g_1, \ldots, g_w$ contains polynomials of degree

$$\geq 2^{2^{n/2-O(\sqrt{n})}} .$$

It is not hard to see that binomial ideals have binomial reduced Gröbner bases, *i.e.*, each polynomial in such a basis is the difference of two terms. Using the relationship of such ideals to (finitely presented) commutative semigroups, we immediately obtain the following lower bounds for Gröbner bases.

Theorem 15 *There are infinitely many $n > 0$ and a $d > 0$ ($d = 5$ suffices) such that for every such n, there is a generating set $g_1, \ldots, g_w$ (with w depending linearly on n), such that each g_i is a difference of two monomials, $\deg(g_i) \leq d$, and there is a constant $c > 0$ (c is roughly $\frac{1}{2}$) such that*

(i) every Gröbner basis for $(g_1, \ldots, g_w)$ contains a polynomial of total degree $\geq 2^{2^{c \cdot n}}$; and

(ii) every Gröbner basis for $(g_1, \ldots, g_w)$ contains at least $2^{2^{c \cdot n}}$ elements.

For a proof, also see [20].

Since we can always homogenize the generators of some ideal in $\mathbb{Q}[x_1, \ldots, x_n]$ introducing an additional indeterminate x_0, these double exponential lower bounds for Gröbner bases also hold for homogeneous ideals.

Finally, we present a PSPACE lower bound for PIMP restricted to homogeneous ideals.

Theorem 16 *The polynomial ideal membership problem, when restricted to homogeneous ideals, requires space $n^{\Omega(1)}$, and hence is PSPACE-complete.*

Proof: We merely sketch a proof here. Let M be any deterministic LBA. Wlog we assume that the tape alphabet of M is $\{0, 1\}$, and that M has a unique accepting and rejecting final configuration. Let m be some input for M of length n. Construct a homogeneous instance of PIMP as follows. Let the set of indeterminates be $\{x_i, y_i, z_i;\ i = 1, \ldots, n\} \cup Q$, where Q is the set of states of the finite control of M. We use x_i and y_i to denote that the contents of the ith cell of M's tape contains a 0 (resp., a 1), and z_i to denote the fact that M's head is positioned over the ith tape cell. Then the initial configuration of M can be represented by a monomial $\tilde{m}$ over these indeterminates, and the unique final accepting configuration by some monomial $\tilde{m}'$. Also, if we allow that each transition of M can also be reversed (*i.e.*, if we turn M from a semi-Thue system into a Thue system), the transition relation of this "symmetric" machine can be represented by a linear (in n) number of polynomials g_j in the above indeterminates, each of which is a difference of two monomials. Each of these polynomials simply expresses the local change that occurs when M, with its head at some position i, executes one step (in forward or backward direction). Also, the polynomial $\tilde{m}' - \tilde{m}$ and the polynomials g_j are homogeneous, the g_j of degree say 4 and $\tilde{m}' - \tilde{m}$ of degree roughly n. Now,

$$M \text{ accepts } m \text{ iff } \tilde{m}' - \tilde{m} \text{ is in the ideal generated by the } g_j\,.$$

As already noted by [32], the fact that we have replaced the semi-Thue system underlying M by a Thue system does not hurt us since M was assumed to be deterministic. □

Using $\tilde{m}' - \tilde{m} + 1$ as an additional generator, we obtain

Corollary 16.1 *Testing whether*

$$1 \in (g_1, \ldots, g_w)$$

is PSPACE-hard.

4 Commutative Semigroups and Petri Nets

As we have seen in the previous section, many lower bounds for problems in polynomial ideals arise from lower bounds for the word problem for commutative semigroups. In this section, we would like to present some upper bounds for commutative semigroup problems which are derived from the double exponential upper bound for the degree of Gröbner bases [10].

An immediate consequence of the Gröbner bases degree bound is

Theorem 17 *Let $\mathcal{P}$ be a finite set of congruences on S^*, $S = \{s_1, \ldots, s_n\}$, let $m \in S^*$ and assume that $b \in [m]$, b minimal wrt the subword ordering (or, equivalently, when interpreted as element of $\mathbb{N}^n$, minimal wrt the standard partial ordering of $\mathbb{N}^n$). Then there is a double exponential upper bound for the length of b.*

With this bound, we can sharpen the upper bound given in [19] for the equivalence problem for commutative semigroups:

Theorem 18 *The equivalence problem for commutative semigroups can be decided in exponential space. It is also exponential space complete.*

As shown in [22], we also get the following exponential space complexity bounds from the results in [28] and [26].

Definition 4.1 *Let $\mathcal{P}$ be a finite set of congruences on S^*, $S = \{s_1, \ldots, s_n\}$, and let $m, m' \in S^*$.*

1. *The* **Boundedness Problem** *is: Given $S, \mathcal{P}$, and m, decide whether $[m]$ is finite.*
2. *The* **Coverability Problem** *is: Given $S, \mathcal{P}, m$, and m', decide whether there is an $m'' \in [m]$ such that m' is a subword of m''.*
3. *The* **Selfcoverability Problem** *is: Given $S, \mathcal{P}$, and m, decide whether there is an $m'' \in [m]$ such that m is a proper subword of m''.*

In [22] we show that, in terms of upper bounds, the boundedness, coverability and selfcoverability problems can all be reduced to instances of PIMP for binomial ideals, and hence are in exponential space. An exponential space lower bound can be obtained by observing that the construction in [28] actually proves the following, slightly stronger statement:

Theorem 19 *There is an infinite family of commutative semigroup word problems $(m, m', \mathcal{P})$ such that for each of them*

(i) *$[m]$ is finite,*
(ii) *m' is not a proper subword of any word in $[m]$, and*
(iii) *any Turing machine requires exponential space on an infinite number of these instances.*

Furthermore, the uniform word problem for finitely generated commutative semigroups with the above restrictions is still complete for exponential space under log-lin reductions.

Using this version, we can reduce exponential space to any of the boundedness, coverability, or selfcoverability problem for commutative semigroups, establishing an exponential space lower bound and thus exponential space completeness for these three problems.

As we have already seen, finitely generated commutative semigroups are equivalent to reversible Petri nets. Therefore, the exponential space lower bound given in [28] improves upon Lipton's original lower bound of $2^{\Omega(\sqrt{n})}$ for the Petri net (or VAS, or VRS) reachability problem [23]. Decidability of this problem has first been established in [24, 25], by means of an algorithm whose complexity is non-primitive recursive, and since then no improvements have been obtained.

From our earlier discussion, it should be clear that the general Petri net reachability problem is equivalent to a version of a (binomial) ideal membership problem where we require that the coefficient polynomials p_i are from the semiring $\mathbb{N}[x]$ (or $\mathbb{Q}^+[x]$, for that matter). However, so far and to the knowledge of this author, this interpretation has not provided any essential insights.

5 Open Problems, Conclusion

In this survey, we have highlighted some of the connections between such different areas as the algebraic theory of multivariate polynomial ideals, elimination theory and complex function theory providing complexity bounds, algebraic geometry, the very fundamental commutative semigroups, and models used in computer science for representing parallel and concurrent processes, like vector addition systems or Petri nets. These interrelationships are quite intriguing since a large number of very basic complexity results for these structures has been obtained using these connections. And this maybe even more so, if one realizes that in several instances, a lower bound has been shown (how else?) using basically string rewriting techniques while matching upper bounds can be established using (sometimes quite elaborate and deep) techniques from analysis or complex function theory.

Another phenomenon that is quite indicative here and possibly typical for other practical areas (and computer algebra and Gröbner bases are being used in practice, even if quite often with some frustration and long waiting hours, as this author can attest to) could be the following: while the worst-case lower bounds for PIMP and Gröbner bases are terrible, seemingly precluding any application in practice, it turns out that much better (more "encouraging") bounds can be derived for the cases that really tend to occur in practical applications, like radical membership or regular intersections. And there are interesting developments to even characterize some *really* applicable cases (bounds better than PSPACE).

While such advances will be necessary in order to apply polynomial ideals in fields like robotics, motion planning, vision, modeling, constrained programming, and others, there also remain a few fundamental questions concerning complexity issues of polynomial ideals and related structures. One is to obtain explicit upper (and possibly better lower) bounds for ideals in $\mathbb{Z}[x]$ (or other nice and effective rings instead of $\mathbb{Z}$). So far, we just have the doubly exponential lower bounds from the word problem for commutative semigroups, and no explicit upper bounds. Another is the complexity of the reachability problem for (general) Petri nets. While this complexity has been characterized for many subclasses of Petri nets, these are all so restricted as of being of little practical value. This means that we should try, on the one hand, to upper bound the complexity of the general Petri net reachability, but also to find characterizations of new subclasses of Petri nets which are of practical relevance and at the same time permit efficient solutions of basic problems like reachability, boundedness, or absence of deadlock. One might object that these goals are contradictory in themselves, since e.g. the reachability problem is already PSPACE-complete for 1-safe Petri nets, but this only says that *different* types of characterizations probably should be investigated, as the example of PIMP seems to indicate in a (slightly?) different area.

References

1. Akira Aiba, Kô Sakai, Yosuke Sato, David J. Hawley, and Ryuzo Hasegawa. Constrained logic programming language CAL. In *Proceedings of the International Conference on Fifth Generation Computer Systems 1988 (Tokyo, Japan, November/December 1988)*, volume 1, pages 263–276. Institute for New Generation Computer Technology, ICOT, 1988.
2. Carlos Berenstein and Alain Yger. Bounds for the degrees in the division problem. *Michigan Math. J.*, 37(1):25–43, 1990.
3. W. Dale Brownawell. Bounds for the degrees in the Nullstellensatz. *Ann. of Math.*, 126:577–591, 1987.
4. Bruno Buchberger. *Ein Algorithmus zum Auffinden der Basiselemente des Restklassenrings nach einem nulldimensionalen Polynomideal.* Ph.d. thesis, Department of Mathematics, University of Innsbruck, 1965.
5. Bruno Buchberger. Gröbner bases: An algorithmic method in polynomial ideal theory. In N.K. Bose, editor, *Multidimensional systems theory*, pages 184–232. D. Reidel Publishing Company, Dordrecht-Boston-London, 1985.
6. Léandro Caniglia, André Galligo, and Joos Heintz. Some new effectivity bounds in computational geometry. In *Proceedings of AAECC-6 (Roma, 1988)*, volume 357 of *LNCS*, pages 131–152, Berlin-Heidelberg-New York-London-Paris-Tokyo-Hong Kong-Barcelona-Budapest, 1988. Springer-Verlag.
7. E. Cardoza, R. Lipton, and A.R. Meyer. Exponential space complete problems for Petri nets and commutative semigroups. In *Proceedings of the 8th Ann. ACM Symposium on Theory of Computing (Hershey, PA)*, pages 50–54, New York, 1976. ACM, ACM Press.
8. Edward W. Cardoza. Computational complexity of the word problem for commutative semigroups. Technical Memorandum TM 67, Project MAC, M.I.T., October 1975.
9. David Cox, John Little, and Donal O'Shea. *Ideals, varieties, and algorithms. An introduction to computational algebraic geometry and commutative algebra.* Springer-Verlag, Berlin-Heidelberg-New York-London-Paris-Tokyo-Hong Kong-Barcelona-Budapest, 1992.
10. Thomas W. Dubé. The structure of polynomial ideals and Gröbner bases. *SIAM J. Comput.*, 19:750–773, 1990.

11. Samuel Eilenberg and M.P. Schützenberger. Rational sets in commutative monoids. *J. Algebra*, 13:173–191, 1969.
12. David Eisenbud and Lorenzo Robbiano, editors. *Computational algebraic geometry and commutative algebra*, volume XXXIV of *Symposia Mathematica*. Cambridge University Press, Cambridge, 1993.
13. David Eisenbud and Bernd Sturmfels. Binomial ideals, June 1994.
14. S. Fortune and J. Wyllie. Parallelism in random access machines. In *Proceedings of the 10th Ann. ACM Symposium on Theory of Computing (San Diego, CA)*, pages 114–118, New York, 1978. ACM, ACM Press.
15. Marc Giusti, Joos Heintz, and Juan Sabia. On the efficiency of effective Nullstellensätze. *Comput. Complexity*, 3:56–95, 1993.
16. Grete Hermann. Die Frage der endlich vielen Schritte in der Theorie der Polynomideale. *Math. Ann.*, 95:736–788, 1926.
17. Heisuke Hironaka. Resolution of singularities of an algebraic variety over a field of characteristic zero: I. *Ann. of Math.*, 79(1):109–203, 1964.
18. Heisuke Hironaka. Resolution of singularities of an algebraic variety over a field of characteristic zero: II. *Ann. of Math.*, 79(2):205–326, 1964.
19. D.T. Huynh. The complexity of the equivalence problem for commutative semigroups and symmetric vector addition systems. In *Proceedings of the 17th Ann. ACM Symposium on Theory of Computing (Providence, RI)*, pages 405–412, New York, 1985. ACM, ACM Press.
20. D.T. Huynh. A superexponential lower bound for Gröbner bases and Church-Rosser commutative Thue systems. *Inf. Control*, 68(1-3):196–206, 1986.
21. J. Kollár. Sharp effective Nullstellensatz. *J. Amer. Math. Soc.*, 1:963–975, 1988.
22. Ulla Koppenhagen and Ernst W. Mayr. The complexity of the boundedness, coverability, and selfcoverability problems for commutative semigroups. Technical Report TUM-I9518, Institut für Informatik, Technische Universität München, May 1995.
23. Richard Lipton. The reachability problem requires exponential space. Research Report 62, Computer Science Dept., Yale University, January 1976.
24. Ernst W. Mayr. An algorithm for the general Petri net reachability problem. In *Proceedings of the 13th Ann. ACM Symposium on Theory of Computing (Milwaukee, WI)*, pages 238–246, New York, 1981. ACM, ACM Press.
25. Ernst W. Mayr. An algorithm for the general Petri net reachability problem. *SIAM J. Comput.*, 13(3):441–460, August 1984.
26. Ernst W. Mayr. Membership in polynomial ideals over $\mathbb{Q}$ is exponential space complete. In B. Monien and R. Cori, editors, *Proceedings of the 6th Annual Symposium on Theoretical Aspects of Computer Science (Paderborn, FRG, February 1989)*, volume LNCS 349, pages 400–406, Berlin-Heidelberg-New York-London-Paris-Tokyo-Hong Kong, 1989. GI, afcet, Springer-Verlag.
27. Ernst W. Mayr. Polynomial Ideals and Applications. *Mitteilungen der Mathematischen Gesellschaft in Hamburg*, XII(4):1207–1215, 1992. Festschrift zum 300jährigen Bestehen der Gesellschaft.
28. Ernst W. Mayr and Albert Meyer. The complexity of the word problems for commutative semigroups and polynomial ideals. *Adv. Math.*, 46(3):305–329, December 1982.
29. Marvin L. Minsky. *Computation: Finite and infinite machines*. Prentice-Hall, Englewood Cliffs, 1967.
30. V. Pan. Complexity of parallel matrix computations. *Theor. Comput. Sci.*, 54(1):65–85, September 1987.

31. C.A. Petri. Kommunikation mit Automaten. Technical Report 2, Institut für Instrumentelle Mathematik, Bonn, 1962.
32. E. Post. Recursive unsolvability of a problem of Thue. *J. Symbolic Logic*, 12:1–11, 1947.
33. Fred Richman. Constructive aspects of Noetherian rings. In *Proceedings of the American Math. Society*, volume 44, pages 436–441, June 1974.
34. A. Seidenberg. Constructions in algebra. *Trans. Am. Math. Soc.*, 197:273–313, 1974.
35. Chee K. Yap. A new lower bound construction for commutative Thue systems, with applications. *J. Symbolic Comput.*, 12:1–28, 1991.
36. Oscar Zariski and Pierre Samuel. *Commutative algebra. Volume I.* Van Nostrand Reinhold Company, New York-Cincinnati-Toronto-London-Melbourne, 1958. The University Series in Higher Mathematics.

From a Concurrent λ-Calculus to the π-Calculus*

Roberto M. Amadio, Lone Leth and Bent Thomsen

CNRS, Sophia-Antipolis ** ECRC, Munich *** ECRC, Munich

Abstract. We explore the (dynamic) semantics of a simply typed λ-calculus enriched with *parallel composition*, *dynamic channel generation*, and *input-output communication primitives*. The calculus, called the $\lambda_{\|}$-calculus, can be regarded as the kernel of concurrent-functional languages such as LCS, CML and Facile, and it can be taken as a basis for the definition of abstract machines, the transformation of programs, and the development of modal specification languages. The main technical contribution of this paper is the proof of adequacy of a compact translation of the $\lambda_{\|}$-calculus into the π-calculus.

1 Introduction

Programming languages that combine functional and concurrent programming, such as LCS [4], CML [11] and Facile [5, 14], are starting to emerge and get applied – some, like Facile, in industrial settings. These languages are conceived for programming of reactive systems and distributed systems. A main motivation for using such languages is that they offer integration of different computational paradigms in a clean and well understood programming model that will allow (formal) reasoning about program behaviour and properties.

Over the past years several formal semantics for such languages have been proposed [3, 6, 13]. Most of these papers focus on defining the (abstract) execution of programs in terms of transition systems. This is clearly important for ensuring correct implementations of the languages. However, it is only the first step towards (formal) reasoning about program behaviour and properties. For the next step it is necessary to have notions of observations and derive equivalences and laws for programs.

As far as we are aware only Facile has been equipped with a notion of observation [6, 13]. However, the semantics and notion of observation proposed for Facile in [6] has several drawbacks. The semantics treats channels as generative constructs. To model this it is necessary to equip the structural operational semantics with a global component keeping track of generated channels. This

* This work is partially supported by ESPRIT BRA 6454 Confer.

** I3S, BP 145, F-06903, Sophia-Antipolis, FRANCE, e-mail: amadio@cma.cma.fr.

*** ECRC, Arabellastrasse 17, D-81925, Munich, GERMANY, e-mail: lone@ecrc.de, bt@ecrc.de.

renders the semantics less abstract and prevents a syntactic treatment. To handle generative channels in the definition of equivalence in [6] pairs of configurations are used and a notion of windows of observations is introduced. This yields a definition of equivalence which is difficult to compare with more "standard" definitions of equivalence, such as bisimulation. Furthermore, the notion of observational equivalence is not known to be a congruence. In fact it was already noted in [6] that it will be difficult to prove that the proposed notion of observational equivalence is a congruence since it does not facilitate a syntax driven proof.

This paper is a first step towards formalising a theory of concurrent functional programming that follows the syntactic traditions of both the λ-calculus and process algebras, such as CCS, π-calculus and Chocs.

To give a more syntactic treatment of the semantics and to crystalise the essential nature of concurrent functional programming we present a simply typed language, say $\lambda_{\|}$, inspired by previous work on the Facile programming language [5, 14, 1] whose three basic ingredients are: (1) A call-by-value λ-calculus extended with the possibility of parallel evaluation of expressions. (2) A notion of channel and primitives to read-write channels in a synchronous way; communications are performed as side effects of expression evaluation. (3) The possibility of dynamically generating new channels during execution. The $\lambda_{\|}$-calculus includes Chocs [12] and π-calculus [9] (at least a simply sorted part of it) as sub-calculi. We use the notion of reduction, commitment and barbed bisimulation [10] as semantic foundation, and to ensure compatibility with the newest trends in process algebra we study an adequate translation into the π-calculus.
The paper is organised as follows. Section 2 introduces the $\lambda_{\|}$-calculus and its semantics. Section 3 describes the translation from $\lambda_{\|}$-calculus to the π-calculus and establishes some basic properties. We also present the fragment of the π-calculus that is used for the translation. Section 4 shows the adequacy of the translation. Finally section 5 indicates some research directions. Due to space limitations all theorems, lemmas and propositions are stated without proof in this version of the paper (see [2][4] for proofs).

2 A concurrent λ-calculus

In this section we present the $\lambda_{\|}$-calculus and define its semantics.

Types are partitioned into values and -one- behaviour type.

$$\sigma ::= o \;\Big|\; (\sigma \to \sigma) \;\Big|\; Ch(\sigma) \;\Big|\; (\sigma \to b) \quad \text{(value type)}$$
$$b \quad \text{(behaviour type)} \qquad \alpha ::= \sigma \;\Big|\; b \quad \text{(value or behaviour type)}$$

An infinite supply of variables $x^\sigma, y^\sigma, \ldots$, labelled with their type, is assumed for any value type σ. We reserve variables $f, g, \ldots$ for functional types $\sigma \to \alpha$. Moreover, an infinite collection of constants $c^\sigma, d^\sigma, \ldots$ is given where σ is either a ground type o or a channel type $Ch(\sigma')$, for some value type σ'. In particular

[4] URL: ftp://ftp.ecrc.de/pub/ECRC_tech_reports/reports/ECRC-95-18.ps.Z

there is a special constant $*^o$. Finally: (i) z, z'... denote variables or constants, (ii) $[e'/x]e$ represents the substitution of e' for x in e, and (iii) $x \in e$ means that x occurs free in e.

$$e ::= c^\sigma \bigm| x^\sigma \bigm| \lambda x^\sigma.e \bigm| ee \bigm| e!e \bigm| e? \bigm| \nu x^{Ch(\sigma)}.e \bigm| \phi \bigm| (e \mid e)$$

The above syntax defines expressions $e, e', \ldots$ inductively generated by the following operators: λ-abstraction $(\lambda x.e)$, application (ee'), parallel composition $(e \mid e')$, restriction $(\nu x.e)$, output $(e!e')$, and input $(e?)$. [5]

Well-typed expressions are defined by the following rules. All expressions are considered up to α-renaming. Parallel composition has to be understood as an associative and commutative operator, with ϕ as identity.

$(asmp)$	$\Rightarrow z^\sigma : \sigma$	(ϕ)	$\Rightarrow \phi : b$
$(\to_I)$	$e : \alpha \Rightarrow \lambda x^\sigma.e : \sigma \to \alpha$	$(\to_E)$	$e : \sigma \to \alpha \quad e' : \sigma \Rightarrow ee' : \alpha$
$(!)$	$e : Ch(\sigma) \quad e' : \sigma \Rightarrow e!e' : o$	$(?)$	$e : Ch(\sigma) \Rightarrow e? : \sigma$
(ν)	$e : \alpha \Rightarrow \nu x^{Ch(\sigma)}.e : \alpha$	$(\mid)$	$e : b \quad e' : b \Rightarrow e \mid e' : b$

Expressions having a σ type are called value expressions and they return a result upon termination. Expressions having type b are called behaviour expressions and they *never* return a result. In particular their semantics is determined only by their interaction capabilities.

Since we are in a call-by-value framework it does not make sense to allow behaviours as arguments of a function. The types' grammar is restricted accordingly in order to avoid these pathological types.

The dynamic semantics of $\lambda_\parallel$ is defined via unlabelled reductions. We describe a rewriting relation (up to structural equivalence) which represents the possible internal computations of a well-typed $\lambda_\parallel$ expression. On top of this relation we build a notion of observation, and notions of barbed bisimulation and equivalence, which is called *barbed bisimulation*, following [10]. Barbed bisimulation can then be refined to a relation called *barbed equivalence*. The latter provides the sought notion of expression equivalence.

A *program* is a closed expression of type behaviour and values are defined as:

$$v ::= c \bigm| x \bigm| \lambda x.e$$

Local evaluation contexts are standard evaluation contexts for call-by-value evaluation[6]:

$$E[\,] ::= [\,] \bigm| E[\,]e \bigm| (\lambda x.e)E[\,] \bigm| E[\,]!e \bigm| z!E[\,] \bigm| E[\,]?$$

[5] It has been shown elsewhere [1] that other operators such as *CML fork* and *Facile spawn* can be adequately represented in this calculus via a *Continuation Passing Style* translation.

[6] For historical reasons ! and ? are written here in infix and postfix notation, respectively. If one writes them in prefix notation then local evaluation contexts are literally call-by-value evaluation contexts.

Local evaluation contexts do not allow evaluation under restriction and parallel composition. In order to complete the description of the reduction relation we need to introduce a notion of *global* evaluation context $C[\,]$:

$$C[\,] ::= [\,] \,\Big|\, (\nu x.C[\,]) \,\Big|\, (C[\,] \mid e)$$

Concerning the restriction operator ν we assume the following structural equivalences:

$$\begin{array}{ll} (\nu_{\mid}) & \nu x.e \mid e' \equiv \nu x.(e \mid e') \quad x \notin e' \\ (\nu_X) & \nu x.\nu y.e \equiv \nu y.\nu x.e \\ (\nu_{E[\,]}) & E[\nu x.e] \equiv \nu x.E[e] \quad x \notin E[\,],\; E[e] : b \end{array}$$

We suppose that the equivalence relation $\equiv$ is closed under global contexts, that is: $e \equiv e'$ implies $C[e] \equiv C[e']$. [7]

Using the notion of local evaluation context two basic reduction rules are defined. The first rule corresponds to local functional evaluation while the second describes inter-process communication. The reduction relation describes the internal computation of a program, therefore it is assumed that $E[\,], E'[\,]$ have type b. The definition of the rewriting relation is extended to all global contexts by the (cxt) rule.

$$\begin{array}{ll} (\beta) & E[(\lambda x.e)v] \to E[[v/x]e] \\ (\tau) & E[z!v] \mid E'[z?] \to E[*] \mid E'[v] \\ (cxt) & \text{if } e \to e' \text{ then } C[e] \to C[e'] \end{array}$$

The derivation tree associated to a one-step reduction of an expression can be seen as having the following structure: (i) at most one application of the (cxt) rule, and (ii) one application of one of the basic reduction rules specified above. We write $e \to_r e'$ if the rule applied in (ii) is $r \in \{\beta, \tau\}$. Observe that by means of structural equivalences it is always possible to display a behaviour expression as: $\nu x_1 ... \nu x_n.(E_1[\Delta_1] \mid ... \mid E_m[\Delta_m])$, where: $n, m \geq 0$, if $m = 0$ then the process can be identified with ϕ, and $\Delta ::= (\lambda x.e)v \,\Big|\, z!v \,\Big|\, z?$.

Convergence for $\lambda_{\|}$-programs is defined as the ability of accepting an input/output communication with the environment, possibly after some internal reduction. A program can commit to make an input or an output operation on a *constant* channel. A relation of immediate commitment $e \downarrow \gamma$ where $\gamma ::= c! \,\Big|\, c?$ is defined as follows:

$$e \downarrow c! \quad \text{if } e \equiv C[E[c!v]] \qquad\qquad e \downarrow c? \quad \text{if } e \equiv C[E[c?]]$$

[7] We could consider a congruent closure but then we would run into marginal technical problems in showing that structurally equivalent $\lambda_{\|}$ terms are translated into equivalent π-processes (the source of the difficulty being the $(\nu_{E[\,]})$ rule). An alternative approach followed in [1] is to transform this equation into a rewriting rule.

Moreover, define $\Rightarrow$ as the reflexive and transitive closure of $\rightarrow$. Then a weak commitment relation $e \Downarrow \gamma$ is defined as: $e \Downarrow \gamma$ if $\exists e'.(e \Rightarrow e' \;\; \wedge \;\; e' \downarrow \gamma)$.

We may now introduce an equivalence on expressions based on a natural extension of Morris style contextual equivalence on λ-terms. A binary relation S between programs is a (weak) barbed simulation if eSf implies: (1) $\forall e'.\exists f'.(e \Rightarrow e'$ *implies* $f \Rightarrow f' \wedge e'Sf')$, and (2) $\forall \gamma.(e \Downarrow \gamma$ *implies* $f \Downarrow \gamma)$. S is a barbed bisimulation if S and S^{-1} are barbed simulations. We write $e \stackrel{\bullet}{\approx} f$ if eSf for some barbed bisimulation S.

Let e, e' be well typed expressions. We define

$$e \approx f \quad if\ any\ P[\,],\ P[e] \stackrel{\bullet}{\approx} P[e']$$

where $P[\,]$ is a one-hole context such that $P[e]$ and $P[e']$ are programs.

Note that by construction $\approx$ is a congruence w.r.t. the calculus operators. Furthermore, it is easy to prove that if $e : \alpha$ then $(\lambda x.e)v \approx [v/x]e$.

3 Translation.

In this section we introduce a translation of the $\lambda_{\|}$-calculus into the π-calculus and we discuss some of the basic properties of the translation. Notably, we produce an optimised translation to which the standard translation reduces by means of administrative reductions.

First we briefly review the fragment of the π-calculus [9] which is sufficient for interpreting the $\lambda_{\|}$-calculus. This fragment is simply sorted and has polyadic channels [8]. Every channel is supposed to be labelled by its *sort*. Sorts are used to constrain the arity of a channel. A channel of sort $Ch(s_1, ..., s_n)$ can carry a tuple $z_1, ..., z_n$, where z_i has sort s_i, for $i = 1, ..., n$.

$$s ::= o \,\big|\, Ch(s_1, ..., s_n) \quad (n \geq 0)$$

Variable and constant channels are denoted with $x, y, ...$ and $c, d, ...$, respectively. As for $\lambda_{\|}$ we assume a constant $*$ of sort o. We feel free to omit writing the ϕ process. As usual, $\mathbf{z}$ stands for $z_1, ..., z_n$ $(n \geq 0)$. π-terms are given by the following grammar:

$$p ::= \phi \,\big|\, \bar{z}\mathbf{z}.p \,\big|\, z(\mathbf{y}).p \,\big|\, \nu x.p \,\big|\, (p \mid p) \,\big|\, !p$$

All expressions are considered up to α-renaming and parallel composition is an associative and commutative operator with ϕ as identity. Concerning the restriction operator ν we assume the standard structural equivalences, i.e. the $(\nu_{|})$ and the (ν_X) from section 2.2. For the sake of simplicity we also assume the following equivalences which concern the replication operator: $!p \equiv p \mid !p$ and $\nu x.!(x(\mathbf{y}).p) \equiv \phi$. We will consider structurally congruent two π-terms which are in the congruent closure of the equations presented above.

The basic reduction rule is:

$$(\tau) \quad \bar{z}\mathbf{z}'.p \mid z(\mathbf{x}).p' \rightarrow p \mid [\mathbf{z}'/\mathbf{x}]p'$$

This rule is extended to certain contexts by the rule:

$$(cxt)\quad p \to p' \;\Rightarrow\; C[p] \to C[p']$$

where: $C[\,] ::= [\,] \,\big|\, C[\,] \mid p \,\big|\, \nu x.C[\,]$. Commitments on constant channels are defined straightforwardly as:

$$p \downarrow c! \quad \text{if } e \equiv C[\overline{c}\mathbf{z}.p] \qquad\qquad e \downarrow c? \quad \text{if } e \equiv C[c(\mathbf{x}).p]$$

A program in the π-calculus is a π-term with no free channels. The notions of barbed bisimulation on programs and barbed equivalence on π-terms are defined along the lines presented for the $\lambda_{\|}$-calculus.

For the purpose of the adequacy proof it is useful to identify certain special reductions which enjoy an interesting confluence property.

- *Administrative reductions* ($\to_{ad}$) fit into the following pattern, where u appears only in the displayed positions. Note that administrative reductions always terminate.

$$(\tau_{ad})\quad C[\nu u.(\overline{u}\mathbf{z} \mid u(\mathbf{x}).p)] \to C[[\mathbf{z}/\mathbf{x}]p]$$

- *β-reductions* ($\to_\beta$) fit into the following pattern, where f may occur in p' only in the pattern $\overline{f}\mathbf{z}'$, for some $\mathbf{z}'$.

$$(\tau_\beta)\quad C[\nu f.(\overline{f}\mathbf{z} \mid !(f(\mathbf{x}).p) \mid p')] \to C[\nu f.([\mathbf{z}/\mathbf{x}]p \mid !(f(\mathbf{x}).p) \mid p')]$$

Proposition 1 Confluence. *Suppose $p \to p_1$ and $p \to_{ad,\beta} p_2$. Then either $p_1 \equiv p_2$ or there is p' such that $p_1 \to_{ad,\beta} p'$ and $p_2 \to p'$.*

A function $\lceil\ \rceil$ from $\lambda_{\|}$ types into π sorts is defined as follows:

$$\lceil o \rceil = o \qquad \lceil Ch(\sigma) \rceil = Ch(\lceil \sigma \rceil)$$

$$\lceil \sigma \to \sigma' \rceil = Ch(\lceil \sigma \rceil, Ch(\lceil \sigma' \rceil)) \qquad \lceil \sigma \to b \rceil = Ch(\lceil \sigma \rceil, Ch())$$

It is possible to statically assign one out of three "colours" to each π-variable involved in the translation. The colours are used to make the functionality of a channel explicit and classify the possible reductions of translated terms. We suppose that in the expression e to be translated all variables of functional type $\sigma \to \alpha$ are represented with $f, g, \ldots$ Channels coming from the original $\lambda_{\|}$ term are represented by $x, y, \ldots$ and channels used for "internal book keeping" in the translation are represented by $u, t, v, \ldots$. Thus we suppose that π-variables are partitioned in three infinite sets: $u, t, w \ldots$; $f, g, \ldots$; and $x, y, \ldots$. Furthermore, we let $r, r', \ldots$, ambiguously denote constants $c, c', \ldots$, variables $x, y, \ldots$, and variables $f, g, \ldots$.

In fig. 1 a function $[\ ]$ from well-typed $\lambda_{\|}$-expressions into well-sorted π-processes is defined. This function is parameterised over a (fresh) channel u. If $e : \sigma$ is a value expression then u has sort $Ch(\lceil \sigma \rceil)$ and it is used to transmit the value (or a pointer to the value) resulting from the evaluation of the expression e. If $e : b$ is a behaviour expression then u is actually of –no use–, we conventionally assign the sort $Ch()$ to the channel u (here we choose to parameterise also the behaviour expressions in order to have a more uniform notation). Each rule using variables r actually stands for *two* rules, one in which r is replaced by a variable $x, y, \ldots$ or a constant and another where it is replaced by a variable $f, g, \ldots$. In the translation only the variables $x, y, \ldots$ can be instantiated by a constant.

$$
\begin{array}{ll}
\lceil r^{\sigma} \rceil u & = \overline{u} r^{\lfloor \sigma \rfloor} \\
\lceil \lambda r^{\sigma}.e \rceil u & = \nu f.(\overline{u} f \mid \lceil f := \lambda r.e \rceil) \\
 & \quad \lceil f := \lambda r.e \rceil = \,!(f(r, w).\lceil e \rceil w) \\
\lceil ee' \rceil u & = \nu t.\nu w.(\lceil e \rceil t \mid t(f).(\lceil e' \rceil w \mid w(r).\overline{f}(r, u))) \\
 & \quad e : \sigma \rightarrow \alpha,\ t : Ch(\lceil \sigma \rightarrow \alpha \rceil),\ w : Ch(\lceil \sigma \rceil) \\
\lceil e!e' \rceil u & = \nu t.\nu w.(\lceil e \rceil t \mid t(x).(\lceil e' \rceil w \mid w(r).\overline{x} r.\overline{u} *)) \\
\lceil e? \rceil u & = \nu t.(\lceil e \rceil t \mid t(x).x(r).\overline{u} r) \\
\lceil \nu x.e \rceil u & = \nu x.\lceil e \rceil u \\
\lceil \phi \rceil u & = \phi \\
\lceil e \mid e' \rceil u & = \lceil e \rceil u \mid \lceil e' \rceil u
\end{array}
$$

Fig. 1. Expression Translation

As expected, reductions at the $\lambda_{\|}$ level are implemented by several reductions at the π level. Roughly the need for a finer description of the computation in the π-calculus relates to two aspects:

- In the π-calculus there is no notion of application. The implicit order of evaluation given by the relative positions of the expressions in the $\lambda_{\|}$-calculus has to be explicitly represented in the π-calculus. In particular the "computation" of the evaluation context is performed by means of certain administrative reductions.
- In the π-calculus it is not possible to transmit functions. Instead a pointer to a function which is stored in the environment by means of the replication operator is transmitted.

Given a process p in π let $\sharp p$ be its normal form w.r.t. administrative reductions on channels coloured $u, t, w, \ldots$.

Definition 2. A binary relation R between programs in $\lambda_{\|}$-calculus and programs in π-calculus is defined as follows, where u is some fresh channel, v_i are λ-abstractions and the substitution is iterated from right to left, as v_j may depend on f_i for $i < j$.

$$[v_1/f_1]\ldots[v_n/f_n]e \; R \; p \quad \textit{if} \quad \sharp p \equiv \sharp \nu f_1 \ldots \nu f_n.(\lceil e \rceil u \mid \lceil f_1 := v_1 \rceil \mid \ldots \mid \lceil f_n := v_n \rceil)$$

In the following $\nu\mathbf{f}$ stands for $\nu f_1 \ldots \nu f_n$; $\lceil \mathbf{f} := \mathbf{v} \rceil$ stands for $\lceil f_1 := v_1 \rceil \mid \ldots \mid \lceil f_n := v_n \rceil$; and $[\mathbf{v}/\mathbf{f}]$ stands for $[v_1/f_1]\ldots[v_n/f_n]$ $(n \geq 0)$.

In order to analyse the structure of $\sharp\nu\mathbf{f}.(\lceil e\rceil u \mid \lceil \mathbf{f} := \mathbf{v}\rceil)$ we define an *optimised* translation of a *generalised* notion of redex and evaluation context.

The optimisation amounts to pre-computing the initial administrative steps of the translation, and the generalisation amounts to the possibility of considering *functional variables as values*. So, for instance, fv is a generalised redex and $fE[\,]$ is a generalised evaluation context.

It is important to note that if $[\mathbf{v}/\mathbf{f}]e \equiv E[\Delta]$ then $e \equiv E'[\Delta']$, where $\Delta', E'[\,]$ are generalised redex and evaluation context, respectively, and $[\mathbf{v}/\mathbf{f}]\Delta' \equiv \Delta$, $[\mathbf{v}/\mathbf{f}]E'[\,] \equiv E[\,]$.

Of course this remark can be extended to the case where: $[\mathbf{v}/\mathbf{f}]e \equiv \nu\mathbf{x}.(E_1[\Delta_1] \mid \ldots \mid E_n[\Delta_n])$.

In the following we assume v is a λ-abstraction. Then Redex translation is defined as follows:

$$
\begin{aligned}
\{(\lambda r'.e)r\}u &= \nu f.(\lceil f := \lambda r'.e\rceil \mid \overline{f}(r,u)) \\
\{(\lambda f'.e)v\}u &= \nu f.\nu f'.(\lceil f := \lambda f'.e\rceil \mid \lceil f' := v\rceil \mid \overline{f}(f',u)) \\
\{fr\}u &= \overline{f}(r,u) \\
\{fv\}u &= \nu f'.(\lceil f' := v\rceil \mid \overline{f}(f',u)) \\
\{z?\}u &= z(r).\overline{u}r \\
\{z!r\}u &= \overline{z}r.\overline{u}* \\
\{z!v\}u &= \nu f'.(\lceil f' := v\rceil \mid \overline{z}f'.\overline{u}*)
\end{aligned}
$$

Context Translation is defined as follows:

$$
\begin{aligned}
\lceil [u']e\rceil u &= \nu w.(u'(f).\lceil e\rceil w \mid w(r).\overline{f}(r,u)) \\
\lceil v[u']\rceil u &= \nu f.(\lceil f := v\rceil \mid u'(r).\overline{f}(r,u)) \\
\lceil f[u']\rceil u &= u'(r).\overline{f}(r,u) \\
\lceil [u']?\rceil u &= u'(x).x(r).\overline{u}r \\
\lceil [u']!e\rceil u &= \nu w.(u'(x).(\lceil e\rceil w \mid w(r).\overline{x}r.\overline{u}*)) \\
\lceil z![u']\rceil u &= u'(r).\overline{z}r.\overline{u}* \\
\lceil E[u']e\rceil u &= \nu t.\nu w.(\lceil E[u']\rceil t \mid t(f).\lceil e\rceil w \mid w(r).\overline{f}(r,u)) \\
\lceil vE[u']\rceil u &= \nu w.\nu f.(\lceil f := v\rceil \mid \lceil E[u']\rceil w \mid w(r).\overline{f}(r,u)) \\
\lceil fE[u']\rceil u &= \nu w.(\lceil E[u']\rceil w \mid w(r).\overline{f}(r,u)) \\
\lceil E[u']?\rceil u &= \nu w.(\lceil E[u']\rceil w \mid w(x).x(r).\overline{u}r) \\
\lceil E[u']!e\rceil u &= \nu t.\nu w.(\lceil E[u']\rceil t \mid t(x).\lceil e\rceil w \mid w(r).\overline{x}r.\overline{u}*) \\
\lceil z!E[u']\rceil u &= \nu t.(\lceil E[u']\rceil t \mid t(r).\overline{z}r.\overline{u}*)
\end{aligned}
$$

This translation is defined for non-trivial contexts only. v is a λ-abstraction.

Lemma 3 Administrative Reductions. *Suppose $E[\,]$ is a non-trivial evaluation context, v_i are λ-abstractions which may depend on f_j for $j < i$. Then*

$$\nu\mathbf{f}.(\lceil E[e]\rceil u \mid \lceil \mathbf{f} := \mathbf{v}\rceil) \to^*_{ad} \nu\mathbf{f}.\nu u'.(\lceil e\rceil u' \mid \lceil E[u']\rceil u \mid \lceil \mathbf{f} := \mathbf{v}\rceil)$$

Note that from the previous lemma one can prove that if $e \equiv e'$ then $\sharp\lceil e\rceil u \equiv \sharp\lceil e'\rceil u$. Furthermore, the previous lemma immediately extends to a general behaviour expression $e \equiv \nu\mathbf{x}.(E_1[\Delta_1] \mid ... \mid E_m[\Delta_m])$ as the translation distributes w.r.t. restriction and parallel composition. In particular, the *administrative normal form* of the behaviour expression e can be characterised as (supposing $E_i[\,]$ is non-trivial, for $i = 1, ..., m$, otherwise just drop the context translation):

$$\sharp\lceil e \rceil u \equiv \nu\mathbf{x}.\nu u_1...u_m.(\{\Delta_1\}u_1 \mid \lceil E_1[u_1] \rceil u \mid ... \mid \{\Delta_1\}u_m \mid \lceil E_m[u_m] \rceil u)$$

4 Adequacy

With the help of the optimised translation described above the following lemma, which relates reductions and commitments modulo the relation R, is derived.

Lemma 4 Relating Reductions and Commitments. *(1) If eRp and $e \to e'$ then $\sharp p \to p'$ and $e'Rp'$.*
(2) Vice versa, if eRp and $\sharp p \to p'$ then $e \to e'$ and $e'Rp'$.
(3) Suppose eRp. Then $e \downarrow \gamma$ iff $\sharp p \downarrow \gamma$.

Theorem 5 Adequacy Translation $\lambda_{\|}$ into π. *(1) Let e, e' be programs in $\lambda_{\|}$. Then $\lceil e \rceil u \stackrel{\bullet}{\approx} \lceil e' \rceil u$ iff $e \stackrel{\bullet}{\approx} e'$. (2) Let e, e' be well-typed expressions in $\lambda_{\|}$. Then $\lceil e \rceil u \stackrel{\bullet}{\approx} \lceil e' \rceil u$ implies $e \stackrel{\bullet}{\approx} e'$.*

5 Conclusion and Directions for Further Work

We have shown that the semantics of high level programming languages integrating functional and concurrent programming, such as Facile and CML, can be put on the same theoretical footing as calculi like the π-calculus and Chocs.

It is an open problem to determine if there is a "natural" fully-abstract translation of the call-by-value λ-calculus into the π-calculus, or in another direction, if there is a "reasonable" extension of the λ-calculus that would make the translation considered here fully-abstract. As far as the latter possibility is concerned, we note that the $\lambda_{\|}$-calculus is an interesting candidate. In particular, we can prove that $\lambda_{\|}$-contexts are *strictly* more discriminating than *pure* call-by-value contexts. For instance, consider the following example (loosely inspired by the *Fickle* function in [7]), where $a!*$ abbreviates $(\lambda w.\phi)a!*$.

$$\begin{array}{ll}
e_1 \equiv \lambda f.f * (f * 1) & : (o \to ((o \to b) \to (o \to b))) \to (o \to b) \\
e_2 \equiv \lambda f.((\lambda h.f * (h1))(f*)) & : (o \to ((o \to b) \to (o \to b))) \to (o \to b) \\
1 \equiv \lambda w.a!* & : o \to b \\
Suc \equiv \lambda w'.\lambda z.\lambda w.(z * \mid a!*) & : o \to ((o \to b) \to (o \to b)) \\
Mul_2 \equiv \lambda w'.\lambda z.\lambda w.(z * \mid z*) & : o \to ((o \to b) \to (o \to b))
\end{array}$$

The expressions e_1 and e_2 are equivalent in the *pure* call-by-value λ-calculus but they can be separated in $\lambda_{\|}$ by the following context:

$$C[\,] \equiv \nu g.([\,](\lambda w.g?w)* \mid g!Suc; g!Mul_2)$$

where: $g!Suc; g!Mul_2 \equiv (\lambda x.\lambda y.\phi)(g!Suc)(g!Mul_2)$.

One may regard our translation into the π-calculus as a way to define the semantics of $\lambda_{\|}$. It would be interesting to determine conditions on $\lambda_{\|}$-programs which assure that their translation falls into a fragment of π for which equivalence is decidable (see [15]). A related issue is to determine a stratification of behaviour expression which guarantees their termination.

References

1. R. Amadio. Translating Core Facile. Technical Report ECRC-94-3, ECRC, Munich, 1994.
2. R. Amadio, L. Leth, and B. Thomsen. From a Concurrent λ-calculus to the π-calculus. Technical Report ECRC-95-18, ECRC, Munich, 1995.
3. D. Berry, R. Milner, and D. N. Turner. A semantics for ML concurrency primitives. In *Proceedings of POPL'92*. ACM, 1992.
4. B. Berthomieu, D. Giralt, and J-P. Gouyon. Lcs users' manual. Technical report 91226, LAAS/CNRS, 1991.
5. A. Giacalone, P. Mishra, and S. Prasad. Facile: A symmetric integration of concurrent and functional programming. *International Journal of Parallel Programming*, 18(2):121–160, 1989.
6. A. Giacalone, P. Mishra, and S. Prasad. Operational and algebraic semantics for facile: A symmetric integration of concurrent and functional programming. In *Proceedings of ICALP 90, LNCS 443*. Springer-Verlag, 1990.
7. R. Milner. Functions as processes. *Journal of Mathematical Structures in Computer Science*, 2(2):119–141, 1992.
8. R. Milner. The polyadic π-calculus: A tutorial. In W. Brauer F.L. Bauer and H. Schwichtenberg, editors, *Logic and Algebra of Specification*. Springer-Verlag, 1993.
9. R. Milner, J. Parrow, and D. Walker. A Calculus of Mobile Process, Parts 1-2. *Information and Computation*, 100(1):1–77, 1992.
10. R. Milner and D. Sangiorgi. Barbed bisimulation. In *SLNCS 623: Proceedings of ICALP 92*. Springer-Verlag, 1992.
11. J. Reppy. Cml: A higher-order concurrent language. In *Proc. ACM-SIGPLAN 91, Conf. on Prog. Lang. Design and Impl.*, 1991.
12. B. Thomsen. Plain chocs. *Acta Informatica*, 30:1–59, 1993. Also appeared as TR 89/4, Imperial College, London.
13. B. Thomsen, L. Leth, and A. Giacalone. Some issues in the semantics of facile distributed programming. In *SLNCS 666: Proceedings of REX School*. Springer-Verlag, 1992. Also appeared as Tech Report ECRC 92-32.
14. B. Thomsen, L. Leth, S. Prasad, T.M. Kuo, A. Kramer, F. Knabe, and A. Giacalone. Facile antigua release programming guide. Technical Report ECRC-93-20, ECRC, Munich, December 1993.
15. Victor, B. and Moller, F. The Mobility Workbench — A Tool for the π-Calculus. In *Proceedings of the 6th International Conference on Computer Aided Verification, CAV'94*, volume 818 of *SLNCS*, pages 428–440. Springer-Verlag, 1994.

Rewriting Regular Inequalities

(extended abstract)

Valentin Antimirov*

CRIN (CNRS) & INRIA-Lorraine,
BP 239, F54506, Vandœuvre-lès-Nancy Cedex, FRANCE
e-mail: Valentin.Antimirov@loria.fr

1 Introduction

In this paper we apply algebraic specification and term-rewriting methods to the *containment problem* in the algebra **Reg**[$\mathcal{A}$] of regular events (languages) on a finite alphabet $\mathcal{A}$ formulated as follows: given two regular expressions r, t (on $\mathcal{A}$), to check if the inequality $r \leq t$ is valid in **Reg**[$\mathcal{A}$].

Standard approaches to the problem are based on translation of the expressions r, t into deterministic finite automata (DFAs); in fact, one can also do with a non-determistic one (NFA) for r. The main peculiarity of our approach is that we avoid such a translation and develop a term-rewriting system (t.r.s.) to reduce $r \leq t$ into a normal form; the latter is *false* whenever $r \leq t$ is not valid in **Reg**[$\mathcal{A}$]. This provides a purely algebraic (symbolic) decision procedure – both for the containment and word problems – that seems to be an interesting contribution *per se*.[2].

Moreover, being extended with some new rewrite rules, our t.r.s. in some cases provides derivations of polynomial size, while any algorithm based on translation of the expressions into DFAs gives rise to an exponential blow-up – we demonstrate this on examples. Of course, the worst-case complexity of our procedure is still exponential – this is not surprising, since the problem is PSPACE-complete [10, 8, 9]. At the end of the paper we provide a more general comparison of our solution with the standard ones by giving an automata-theoretic interpretation of the inference process implemented by our t.r.s.

Some ideas of the present work come from [2]. We also employ a recently introduced technique of *partial derivatives* from [1].

2 Basic Definitions and Notation

We use standard notions and notations of order-sorted algebra [7], term rewriting [5], and finite automata theory [11] without special comments.

---

* On leave from V.M.Glushkov Institute of Cybernetics, Kiev, Ukraine.

[2] It is worth recalling that the ground equational theory of **Reg**[$\mathcal{A}$] is not finitely based [12, 4, 6], so a (non-conditional) t.r.s. – on the signature of **Reg**[$\mathcal{A}$] – solving the word problem in the algebra does not exist. What makes our solution possible is that our t.r.s. works on regular inequalities and conjunctions of those and also involves some auxiliary operations.

Given a set X, we denote its cardinality by $|X|$, its power-set (the set of all subsets of X) by $\mathcal{P}(X)$, and the set of all *finite* subsets of X by $\mathbf{Set}[X]$. Recall that an ***upper semilattice*** is an algebra with a binary operation (called *join*) which satisfies the axioms of associativity, commutativity, and idempotency (*ACI-axioms*, for short). In particular, $\mathbf{Set}[X]$ forms an upper semilattice with the join $\_ \cup \_$ and the least element $\emptyset$. Sometimes we say that an operation has AC- or ACI-properties – this means the operation satisfies corresponding axioms.

Given a finite alphabet $\mathcal{A}$, let $\mathcal{A}^*$ denote the set of all finite words on $\mathcal{A}$ and also the free monoid on $\mathcal{A}$ with concatenation of words $w \cdot u$ as multiplication, and the empty word λ as the neutral element. Let $\mathcal{A}^+ = \mathcal{A} \setminus \{\lambda\}$.

The set $\mathbf{Reg}[\mathcal{A}]$ of regular events (languages) on $\mathcal{A}$ is the least subset of the power-set $\mathcal{P}(\mathcal{A}^*)$ which includes the empty set $\emptyset$, the singletons $\{\lambda\}$ and $\{\alpha\}$ for all $\alpha \in \mathcal{A}$ and is closed under the standard regular operations – concatenation $L_1 \cdot L_2$, union $L_1 \cup L_2$ and iteration (Kleene star) L^*. The set $\mathbf{Reg}[\mathcal{A}]$ together with the operations forms a ***regular algebra***.

Given a regular language L, a ***left quotient of L w.r.t. a word w***, written $w^{-1}L$, is the (regular) language $\{ u \in \mathcal{A}^* \mid w \cdot u \in L \}$. It follows that $w \in L$ is equivalent to $\lambda \in w^{-1}L$ and that $L \subseteq L'$ implies $w^{-1}L \subseteq w^{-1}L'$ for any languages L, L' and word w.

Let $\mathbf{Reg1}[\mathcal{A}]$ be a subset of $\mathbf{Reg}[\mathcal{A}]$ consisting of all the regular languages containing the empty word λ. The complement of this subset w.r.t. $\mathbf{Reg}[\mathcal{A}]$ is denoted by $\mathbf{Reg0}[\mathcal{A}]$. These subsets, together with $\mathcal{A}$, determine an order-sorted structure on $\mathbf{Reg}[\mathcal{A}]$ with a sort $\mathcal{A}$ for the alphabet letters, sorts *Reg0* and *Reg1* for the expressions denoting elements of $\mathbf{Reg0}[\mathcal{A}]$ and $\mathbf{Reg1}[\mathcal{A}]$ respectively, and a sort *Reg* for all regular expressions. The corresponding order-sorted signature *REG* on a given alphabet $\mathcal{A} = \{\alpha_1, \alpha_2, \ldots, \alpha_k\}$ is presented below.

Signature *REG*				
sorts $\mathcal{A}$, *Reg*, *Reg0*, *Reg1*.				
subsorts $\mathcal{A}$ < *Reg0* < *Reg*; *Reg1* < *Reg*.				
constants $\emptyset$: *Reg0*; λ : *Reg1*; $\alpha_1, \alpha_2, \ldots, \alpha_k : \mathcal{A}$.				
operations	$\_+\_$: *Reg Reg* $\to$ *Reg*	$\_\cdot\_$: *Reg Reg* $\to$ *Reg*		
	$\_+\_$: *Reg1 Reg* $\to$ *Reg1*	$\_\cdot\_$: *Reg0 Reg* $\to$ *Reg0*		
	$\_+\_$: *Reg Reg1* $\to$ *Reg1*	$\_\cdot\_$: *Reg Reg0* $\to$ *Reg0*		
	$\_+\_$: *Reg0 Reg0* $\to$ *Reg0*	$\_\cdot\_$: *Reg1 Reg1* $\to$ *Reg1*		
	$\_^*$: *Reg* $\to$ *Reg1*			

Sets of ground terms on the signature *REG* of the sorts *Reg*, *Reg0*, and *Reg1* are defined in the usual way and denoted by $\mathcal{T}_{Reg}$, $\mathcal{T}_{Reg0}$, and $\mathcal{T}_{Reg1}$ correspondingly. Note that $\mathcal{T}_{Reg}$ is a disjoint union of $\mathcal{T}_{Reg0}$ and $\mathcal{T}_{Reg1}$. In what follows we call the elements of $\mathcal{T}_{Reg}$ ***regular terms***. Given a regular term t, we denote by $\|t\|$ the number of all the occurrences of alphabet letters appearing in t and call this an ***alphabetic width*** of t.

Any regular term t denotes a regular language $\mathcal{L}(t)$ and this interpretation is determined by the homomorphism $\mathcal{L}(\_)$ from the term algebra $\mathcal{T}_{Reg}$ to $\mathbf{Reg}[\mathcal{A}]$

(defined in the usual way). This defines a standard interpretation of regular equations and inequalities in the regular algebra: $\mathbf{Reg}[\mathcal{A}] \models r = t$ means $\mathcal{L}(r) = \mathcal{L}(t)$ and $\mathbf{Reg}[\mathcal{A}] \models r \leq t$ means $\mathcal{L}(r) \subseteq \mathcal{L}(t)$. Recall that $\mathbf{Reg}[\mathcal{A}]$ is an upper semilattice w.r.t. the join $\_\cup\_$ and $\_ \leq \_$ is the standard partial ordering associated with this semilattice.

The order-sorted presentation of $\mathbf{Reg}[\mathcal{A}]$ allows us to distinguish syntactically a class of *trivially inconsistent* inequalities of the form $a \leq b$ where $a \in \mathcal{T}_{Reg1}$, $b \in \mathcal{T}_{Reg0}$.

Next we reproduce several definitions and facts from [1].

Let **SReg** be the upper semilattice $\mathbf{Set}[\mathcal{T}_{Reg} \setminus \{\emptyset\}]$ of finite sets of non-zero regular terms. The interpretation $\mathcal{L}$ extends to **SReg** by $\mathcal{L}(R) = \bigcup_{t \in R} \mathcal{L}(t)$.

Given $R \in \mathbf{SReg}$ such that $R = \{t_1, \ldots, t_k\}$ where all t_i are syntactically distinct, we write ΣR to denote a regular term $t_1 + \ldots + t_k$ up to an arbitrary permutation of the summands, i.e., the upper occurences of $\_ + \_$ in this sum are considered modulo ACI-axioms. To complete this, $\Sigma\emptyset$ is defined to be $\emptyset$.[3]

Definition 1. (Linear forms of regular terms) Given a letter $x \in \mathcal{A}$ and a term $t \in \mathcal{T}_{Reg}$, we call the pair $\langle x, t \rangle$ a *monomial*. A *(non-deterministic) linear form* is a finite set of monomials. Let $\mathbf{Lin} = \mathbf{Set}[\mathcal{A} \times \mathcal{T}_{Reg}]$ be the upper semilattice of linear forms. The function $lf(\_) : \mathcal{T}_{Reg} \to \mathbf{Lin}$, returning a linear form of its argument, is defined recursively by the following equations:

$$\begin{aligned} lf(\emptyset) = lf(\lambda) &= \emptyset, & lf(t^*) &= lf(t) \odot t^*, \\ lf(x) &= \{\langle x, \lambda \rangle\}, & lf(r_0 \cdot t) &= lf(r_0) \odot t, \\ lf(r + t) &= lf(r) \cup lf(t), & lf(r_1 \cdot t) &= lf(r_1) \odot t \cup lf(t) \end{aligned}$$

for all $x \in \mathcal{A}$, r, $t \in \mathcal{T}_{Reg}$, $r_0 \in \mathcal{T}_{Reg0}$, $r_1 \in \mathcal{T}_{Reg1}$. These equations involve a binary operation $\_\odot\_ : \mathbf{Lin} \times \mathcal{T}_{Reg} \to \mathbf{Lin}$, which is an extension of concatenation to linear forms defined recursively by the following equations:

$$\begin{aligned} l \odot \emptyset = \emptyset \odot t &= \emptyset, & \{\langle x, \lambda \rangle\} \odot t &= \{\langle x, t \rangle\}, \\ l \odot \lambda &= l, & \{\langle x, p \rangle\} \odot t &= \{\langle x, p \cdot t \rangle\}, \\ \{\langle x, \emptyset \rangle\} \odot t &= \{\langle x, \emptyset \rangle\}, & (l \cup l') \odot t &= (l \odot t) \cup (l' \odot t,) \end{aligned}$$

for all l, $l' \in \mathbf{Lin}$, $t \in \mathcal{T}_{Reg} \setminus \{\emptyset, \lambda\}$, $p \in \mathcal{T}_{Reg} \setminus \{\emptyset, \lambda\}$. □

Definition 2. (Partial derivatives of regular terms). Given $t \in \mathcal{T}_{Reg}$ and $x \in \mathcal{A}$, a regular term p is called a *partial derivative of t w.r.t. x* if $\langle x, p \rangle \in lf(t)$. The equation

$$\partial_x(t) = \{ \, p \in \mathcal{T}_{Reg} \setminus \{\emptyset\} \mid \langle x, p \rangle \in lf(t) \, \}$$

defines a function $\partial\_(\_) : \mathcal{A} \times \mathcal{T}_{Reg} \to \mathbf{SReg}$ which returns a set of (non-zero) partial derivatives of its second argument w.r.t. the first one. The following

---

[3] The idea behind this construction is to take into account the ACI-properties only of *some* occurrences of the operation $\_ + \_$ in regular terms and in this way to reduce the use of AC-matching in the rewrite systems presented below.

equations extend this function allowing words and sets of words as the first argument and sets of regular terms as the second one:

$$\partial_\lambda(t) = \{t\}, \qquad \partial_w(R) = \bigcup_{r\in R} \partial_w(r),$$
$$\partial_{wx}(t) = \partial_x(\partial_w(t)), \qquad \partial_W(t) = \bigcup_{w\in W} \partial_w(t).$$

for all $w \in \mathcal{A}^*$, $W \subseteq \mathcal{A}^*$, $R \subset \mathcal{T}_{Reg}$. An element of the set $\partial_w(t)$ is called a *partial derivative of t w.r.t. w*. A *proper* partial derivative of t is one w.r.t. a non-empty word (i.e., from the set $\partial_{\mathcal{A}^+}(t)$). □

Proposition 3. ([1]) *For any $t \in \mathcal{T}_{Reg}$ and $w \in \mathcal{A}^*$, $\mathcal{L}(\partial_w(t)) = w^{-1}\mathcal{L}(t)$. In particular, any partial derivative $p \in \partial_w(t)$ denotes a subset of $w^{-1}\mathcal{L}(t)$.* □

Theorem 4. ([1]) *For any $t \in \mathcal{T}_{Reg}$, the number of all proper partial derivatives of t is less or equal to the alphabetic width of t, i.e. $|\partial_{\mathcal{A}^+}(t)| \leq \|t\|$.* □

The following is an alternative to the classical definition of *word derivatives* from [3] (cf. also [4, 11]).

Definition 5. (Derivatives) Given a term $t \in \mathcal{T}_{Reg}$ and a word $w \in \mathcal{A}^*$, the regular term $\Sigma\partial_w(t)$ (up to permutation of its summands) is called a *(word) derivative of t w.r.t. w*. Given a set of words $W \subseteq \mathcal{A}^*$, let $\mathcal{D}_W(t)$ be the set $\{ \Sigma\partial_w(t) \mid w \in W \}$ of all the derivatives of t w.r.t. all the words in W. □

It follows from Prop. 3 that $\mathcal{L}(\Sigma\partial_w(t)) = w^{-1}\mathcal{L}(t)$ and from Theorem 4 that the set $\mathcal{D}_{\mathcal{A}^+}(t)$ has at most $2^{\|t\|}$ elements.

3 Containment Calculus for Regular Algebra

As the first step towards our t.r.s, we develop a simple calculus which provides a non-deterministic decision procedure for the containment problem in **Reg**[$\mathcal{A}$]. To formulate the calculus, we first introduce a new notion which will play a central rôle in all our constructions.

Definition 6. (Partial derivatives of regular inequalities) Given a word $w \in \mathcal{A}^*$ and two regular terms a, $b \in \mathcal{T}_{Reg}$, a regular inequality $p \leq q$ is called a *partial derivative of $a \leq b$ w.r.t. w* if $p \in \partial_w(a)$ and $q = \Sigma\partial_w(b)$. Let

$$\partial_w(a \leq b) = \{ p \leq \Sigma\partial_w(b) \mid p \in \partial_w(a) \}$$

be the set of all partial derivatives of $a \leq b$ w.r.t. w. Given a set of inequalities E and a set of words $W \subseteq \mathcal{A}^*$, let $\partial_W(E)$ denote the set $\bigcup_{a\leq b\in E} \bigcup_{w\in W} \partial_w(a \leq b)$ of all the partial derivatives of inequalities in E w.r.t. all the words in W. □

Some basic properties of the partial derivatives are as follows.

Proposition 7. *Given a regular inequality $a \leq b$, the following holds:*

1) $\partial_{wx}(a \leq b) = \partial_x(\partial_w(a \leq b))$ for any $x \in \mathcal{A}$, $w \in \mathcal{A}^*$.

2) If **Reg**[$\mathcal{A}$] $\models a \leq b$, *then* **Reg**[$\mathcal{A}$] $\models \partial_{\mathcal{A}^*}(a \leq b)$. □

Let an *atom* be either a regular inequality, or a boolean constant *true* or *false*. Now we are in a position to formulate the Containment Calculus CC which works on sets of atoms. It consists of the rule (***Disprove***), which infers *false* from a trivially inconsistent inequality, and two rules (***Unfold***) to infer a set of partial derivatives (w.r.t. all alphabet letters) from the other inequalities.

Containment Calculus CC	
$(\textit{Disprove}): a_1 \le b_0 \vdash_{CC} \textit{false}$	**for** $a_1 : \textit{Reg1},\ b_0 : \textit{Reg0}$
$(\textit{Unfold}): \quad a_0 \le b \vdash_{CC} \partial_{\mathcal{A}}(a_0 \le b)$	**for** $a_0 : \textit{Reg0},\ b : \textit{Reg}$
$(\textit{Unfold}): \quad a_1 \le b_1 \vdash_{CC} \partial_{\mathcal{A}}(a_1 \le b_1)$	**for** $a_1 : \textit{Reg1},\ b_1 : \textit{Reg1}$

Given an initial inequality $a \le b$, an inference in the calculus is a sequence

$$S_0 \vdash_{CC} S_1 \vdash_{CC} \dots, \tag{1}$$

of sets of atoms, starting with $S_0 = \{a \le b\}$, such that each set S_{i+1} is an extention of the previous one S_i by consequences produced by one of the inference rules applied to an inequality in S_i. The calculus is sound and complete in the following sense.

Theorem 8. *Given a regular inequality $a \le b$, the following holds:*

1) $a \le b$ is not valid in $\mathbf{Reg}[\mathcal{A}]$ *if and only if a set of atoms containing false is derivable in CC from $a \le b$.*

2) There are at most $\|a\| \cdot 2^{\|b\|}$ different inequalities derivable in CC from $a \le b$. □

It follows that after a finite number of steps, the sequence (1) ends up with a set of atoms S_i which is either inconsistent (contains *false*), or saturated (i.e., $\partial_{\mathcal{A}}(S_i) = S_i$). This implies a non-deterministic decision procedure for (dis)proving regular inequalities. Note that it may take up to $O(|\partial_{\mathcal{A}^+}(a \le b)|)$ inference steps to (dis)prove an inequality $a \le b$. Our next goal is to implement such a procedure as a t.r.s.

4 Proving Regular Inequalities by Rewriting

First we describe an order-sorted t.r.s. CC_{rew} which provides a unique normal form for any regular inequality and in this way makes the procedure presented by CC "more deterministic".

The signature of CC_{rew} is an extension of REG by several new sorts and operations. The sorts with their basic constructors and corresponding axioms and rewrite rules are presented in Table 1. The remaining rewrite rules are placed in Table 2. From here on we use variables declared as follows:

variables $x,\ y : \mathcal{A};\ a,\ b,\ c : \textit{Reg};\ a_0,\ b_0 : \textit{Reg0};\ a_1,\ b_1 : \textit{Reg1}$
$s,\ s',\ s'' : \textit{SReg};\ s_0 : \textit{SReg0};\ s_1 : \textit{SReg1};\ C,\ C',\ C'' : \textit{Conj}$

Table 1. Sorts and basic constructors of $\mathcal{CC}_{rew}$

Module *SREG* over *REG*	
sorts	*SReg*, *SReg0*, *SReg1*.
subsorts	$Reg0 < SReg0 < SReg$, $\quad Reg1 < SReg1 < SReg$, $\quad Reg < SReg$.
operations	$\_\sqcup\_ : SReg\ SReg \to SReg \quad \_\sqcup\_ : SReg\ SReg1 \to SReg1$ $\_\sqcup\_ : SReg1\ SReg \to SReg1 \quad \_\sqcup\_ : SReg0\ SReg0 \to SReg0$
axioms	$s \sqcup s' = s' \sqcup s, \quad (s \sqcup s') \sqcup s'' = s \sqcup (s' \sqcup s'')$
rules	$\emptyset \sqcup s \to s, \quad s \sqcup s \to s$

Module *CONJ* over *SREG*	
sorts	*Bool*, *Ineq*, *Conj*.
subsorts	$Bool < Ineq < Conj$.
const	*true*, *false* : *Bool*
operations	$\_ \leq \_ : Reg\ SReg \to Ineq \qquad \_ \leq_o \_ : Reg\ SReg \to Ineq$ $\_ \wedge \_ : Conj\ Conj \to Conj$
axioms	$C \wedge C' = C' \wedge C, \quad (C \wedge C') \wedge C'' = C \wedge (C' \wedge C'')$
rules	$true \wedge C \to C, \ false \wedge C \to false, \ C \wedge C \to C$

The module *SREG* implements the upper semilattice **SReg** needed to represent right-hand sides of regular inequalities. The join $\_\sqcup\_$ is an "ACI-synonym" of $\_+\_$ (note that we keep treating $\_+\_$ as a free constructor).

The module *CONJ* defines sorts for regular inequalities and conjunctions of those. The constructor $\_ \leq_o \_$ is a synonym of the constructor $\_ \leq \_$; it will be used to mark some inequalities as "old" as explained below.

The module *DERIV* implements the function $der : \mathcal{A}, SReg \to SReg$ to compute derivatives of regular terms w.r.t. alphabet letters. It involves an auxiliary operation[4] $der_2 : \mathcal{A}, Reg, Reg \to SReg$ which satisfies the following invariant: $der_2(x,\ a,\ c) = \Sigma\{\ p{\cdot}c \mid p \in \partial_x(a)\ \}$.

The module *UNFOLD* implements the function $unf : Reg, SReg \to Conj$ to compute the result of unfolding a regular inequality into a conjunction of its partial derivatives. Again, we have to involve an auxiliary operation $unf_2()$.

The upper level of the inference process is implemented in the module *MAIN* through several rewrite rules working on inequalities and conjunctions. The trick here is that the rules (*Unfold*) keep the just unfolded inequality $a \leq s$ in the resulting conjunction, but make it "old". Then, if $a \leq s$ reappears as a result of unfolding of some other inequality, the rules (*Delete*) are used to remove it from a current conjuction. To make the trick work, we have to require that the rules (*Delete*) have a higher priority against the rules (*Unfold*) (i.e., the latter can be applied only when the former can not). This prevents from unfolding the same inequality several times and provides a normalising strategy for $\mathcal{CC}_{rew}$.

[4] We also use conditionals and the equality on $\mathcal{A}$ in two rules, but since the alphabet is finite, the rules can easily be replaced by a finite number of another rewrite rules without these extra operations.

Theorem 9. *CC_{rew} is normalising on ground terms by any rewrite strategy satisfying the priority condition formulated above. The normal form (unique modulo AC-axioms for $\_\sqcup\_$ and $\_\wedge\_$) of a regular inequality $r \leq t$ is either false, if $r \leq t$ is not valid in* **Reg**$[\mathcal{A}]$, *or a conjunction of all partial derivatives of $r \leq t$ otherwise.* □

At this point, we have obtained a term-rewriting solution to the containment problem in **Reg**$[\mathcal{A}]$. It turns out, however, that its efficiency can be substantially improved. Note that the complexity of derivations in CC_{rew} is determined mainly by the number of unfolding steps.[5] Now the idea is to add two group of new rewrite rules aimed at avoiding some redundant unfolding. The first group consists of rules (called *TAU*-rules) eliminating "trivial tautologies", e.g.

$$\emptyset \leq s \rightarrow true, \;\; \lambda \leq s_1 \rightarrow true, \;\; a \leq a \rightarrow true, \;\; a \leq b_1 \cdot a \rightarrow true,$$

and so on. The second (actually, most important) group includes the following *subsumption* rules (*SUB*-rules, for short) eliminating some "trivial logical consequences" of already derived inequalities.

Subsumption rules	
$a \leq s \sqcup s' \wedge a \leq_o s \rightarrow a \leq_o s,$	$a \leq s \sqcup s' \wedge a \leq_o s \wedge C \rightarrow a \leq_o s \wedge C,$
$a \leq s \sqcup s' \wedge a \leq s \rightarrow a \leq s,$	$a \leq s \sqcup s' \wedge a \leq s \wedge C \rightarrow a \leq s \wedge C$

Note that the rules (*Delete*) in CC_{rew} can also be considered as the simplest kind of subsumption. Let us call the extended t.r.s. CC_{opt}.

Despite some of the new rules involve AC-operations in their left-hand sides, one-step application of the rules to a given conjunction can be computed in polynomial time (in the size of this conjunction). The more surprising is the fact that CC_{opt} can in some cases *exponentially* shorten the length of derivations comparing to CC_{rew}. We demonstrate this on the following examples. (We consider regular terms on the alphabet $\mathcal{A} = \{a, b\}$, omit concatenation symbol, and use r^k as a shorthand for k-times concatenation of r, regarding $r^\emptyset$ as λ.)

Example 1. Given a natural constant $n \geq 1$, consider a (valid) inequality $X \leq Y$ where $X = (a^*b)^*a^na^*$ and $Y = (a+b)^*a(a+b)^{n-1}$. It is known that the minimal DFA for Y has 2^n states (see [11, page 30])[6] and one can check that $|\partial_{\mathcal{A}^*}(X \leq Y)|$ is an exponent in n.

However, CC_{opt} rewrites $X \leq Y$ into the following normal form which contains only $n+2$ inequalities and this takes $O(n)$ unfolding and subsumption steps:

$$X \leq_o Y \;\wedge\; (a^*bX \leq_o Y \sqcup Y_1) \;\wedge$$
$$(a^{n-1}a^* \leq_o Y \sqcup Y_1) \;\wedge\; \ldots \;\wedge\; (a^* \leq_o Y \sqcup Y_1 \sqcup \ldots \sqcup Y_n).$$

where Y_k stands for $(a+b)^{n-k}$. □

[5] One unfolding step can be implemented very efficiently using the technique from [1].

[6] Thus, any decision procedure relying upon DFA for Y has to spend an exponential time for constructing it.

Example 2. Given a natural constant $n \geq 2$, consider a (valid) inequality $X \leq Y$ where $X = (b^*a)^*(ba)^{2n-3}b^*$ and $Y = (a+b)^*b(ab^*)^{n-2}((ab^*)^{n-1})^*$.

One can check that Y denotes a language from [13, Theorem 2.1] which is recognised by the minimal DFA with 2^{n-1} states and that $X \leq Y$ has an exponential number of partial derivatives. Nevertheless, $\mathcal{CC}_{opt}$ rewrites $X \leq Y$ by $O(n)$ unfolding and subsumption steps into a normal form containing only $4n-4$ inequalities. □

To provide a more general comparison of the decision procedure represented by our rewrite systems with the procedures based on finite automata, we present the following theorem.

Theorem 10. *Given a regular inequality $a \leq b$, let* $\mathbf{M}$ *be an NFA with the input alphabet $\mathcal{A}$, the set of states $M = \partial_{\mathcal{A}^*}(a \leq b)$, the initial state $\mu_0 = a \leq b$, the transition function $\tau : M \times \mathcal{A} \to \mathbf{Set}[M]$ defined for all $(p \leq q) \in M$, $x \in \mathcal{A}$ by $\tau(p \leq q, x) = \partial_x(p \leq q)$, and the set of final states $F = \{ (p \leq q) \in M \mid p \in \mathcal{T}_{Reg1} \wedge q \in \mathcal{T}_{Reg0} \}$. Then this automaton recognises the language $\mathcal{L}(a) \setminus \mathcal{L}(b)$.*

Now it should be clear that, given an initial inequality $a \leq b$, the rewriting process in $\mathcal{CC}_{rew}$ (and so the inference process of $\mathcal{CC}$) can be interpreted as a gradual "top-down" construction of the non-deterministic automaton $\mathbf{M}$: the rules (*Unfold*) generate new states reachable form a given one and the rule (*Disprove*) checks whether one of the obtained states is final. It can be shown that the automaton $\mathbf{M}$ represents a reachable part of a direct product $M_a \times \neg M_b$ of an NFA M_a for a and a complement of a DFA M_b for b (constructions of M_a and M_b can be found in [1]). Therefore, on the class of valid initial inequalities the complexity of the decision procedure represented by $\mathcal{CC}_{rew}$ is of the same order as the complexity of the algorithm checking emptiness of $M_a \cap \neg M_b$. But the "lazy" nature of the inference procedure gives an advantage on the class of non-valid inequalities: such an inequality can be refuted as soon as we have inferred a "final state" – a partial derivative $p \leq q$ with $p \in \mathcal{T}_{Reg1}$ and $q \in \mathcal{T}_{Reg0}$, and this may happen before we have completed constructing the whole sets of states for M_a and M_b. Thus, on this class of inputs our procedure is certainly better[7] than one that first constructs M_a and $\neg M_b$ and only then checks emptiness of their intersection.

The optimised rewrite system $\mathcal{CC}_{opt}$ provides a decision procedure which has even more advantages: now some valid inequalities can also be proved without constructing the whole sets M_a, M_b. The examples presented above demonstrates that this may be crucial when M_b has a relatively big number of states. But, of course, the subsumption test takes an extra time and may happen to be useless on some inputs. It seems that only experiments with a practical programming implementation of $\mathcal{CC}_{opt}$ may show whether this procedure can compete with the standard ones in *most* cases.

[7] Of course, in the *worst* case one has to generate all partial derivatives in order to find a final state in $M_a \cap \neg M_b$

To conclude, we would like to repeat the point made already in the introduction: the main contribution of the present paper is a new *purely symbolic* method for checking inequalities in $Reg[\mathcal{A}]$ which also provides the concept of *algebraic normal forms* for regular inequalities. The theoretical results and ideas of the present paper may also lead to new practical applications, but this is a topic for another research.

References

1. V. M. Antimirov. Partial derivatives of regular expressions and finite automata constructions. In E. W. Mayr and C. Puech, editors, *12th Annual Symposium on Theoretical Aspects of Computer Science. Proceedings.*, volume 900 of *Lecture Notes in Computer Science*, pages 455–466. Springer-Verlag, 1995.
2. V. M. Antimirov and P. D. Mosses. Rewriting extended regular expressions. *Theoretical Comput. Sci.*, 143:51–72, 1995.
3. J. A. Brzozowski. Derivatives of regular expressions. *J. ACM*, 11:481–494, 1964.
4. J. H. Conway. *Regular Algebra and Finite Machines.* Chapman and Hall, 1971.
5. N. Dershowitz and J.-P. Jouannaud. Rewrite systems. In J. van Leeuwen, A. Meyer, M. Nivat, M. Paterson, and D. Perrin, editors, *Handbook of Theoretical Computer Science*, volume B, chapter 6. Elsevier Science Publishers, Amsterdam; and MIT Press, 1990.
6. Z. Ésik and L. Bernátski. Equational properties of Kleene algebras of relations with conversion. *Theoretical Comput. Sci.*, 137:237–251, 1995.
7. J. A. Goguen and J. Meseguer. Order-sorted algebra I: Equational deduction for multiple inheritance, overloading, exceptions and partial operations. *Theoretical Comput. Sci.*, 105:217–273, 1992.
8. H. B. Hunt III, D. J. Rosenkrantz, and T. G. Szymanski. On the equivalence, containment, and covering problems for the regular and context-free languages. *J. Comput. Syst. Sci.*, 12:222–268, 1976.
9. T. Jiang and B. Ravikumar. Minimal NFA problems are hard. *SIAM J. Comput.*, 22(6):1117–1141, 1993.
10. A. R. Meyer and L. J. Stockmeyer. The equivalence problem for regular expressions with squaring requires exponential space. In *Proceedings of the 13th Ann. IEEE Symp. on Switching and Automata Theory*, pages 125–179. IEEE, 1972.
11. D. Perrin. Finite automata. In J. van Leeuwen, A. Meyer, M. Nivat, M. Paterson, and D. Perrin, editors, *Handbook of Theoretical Computer Science*, volume B, chapter 1. Elsevier Science Publishers, Amsterdam; and MIT Press, 1990.
12. V. N. Redko. On defining relations for the algebra of regular events. *Ukrainian Mat. Z.*, 16:120–126, 1964.
13. S. Yu, Q. Zhuang, and K. Salomaa. The state complexity of some basic operations on regular languages. *Theoretical Comput. Sci.*, 125:315–328, 1994.

Table 2. Further rewrite rules of $\mathcal{CC}_{rew}$

Module *DERIV* over *SREG*	
$\mathrm{der}(x,\ \emptyset)$	$\to \emptyset$
$\mathrm{der}(x,\ \lambda)$	$\to \emptyset$
$\mathrm{der}(x,\ y)$	$\to$ **if** $x = y$ **then** λ **else** $\emptyset$ **fi**
$\mathrm{der}(x,\ a+b)$	$\to \mathrm{der}(x,\ a) \sqcup \mathrm{der}(x,\ b)$
$\mathrm{der}(x,\ a \sqcup s)$	$\to \mathrm{der}(x,\ a) \sqcup \mathrm{der}(x,\ s)$
$\mathrm{der}(x,\ a^*)$	$\to \mathrm{der}_2(x,\ a,\ a^*)$
$\mathrm{der}(x,\ a_0 \cdot b)$	$\to \mathrm{der}_2(x,\ a_0,\ b)$
$\mathrm{der}(x,\ a_1 \cdot b)$	$\to \mathrm{der}_2(x,\ a_1,\ b) \sqcup \mathrm{der}(x,\ b)$
$\mathrm{der}_2(x,\ \emptyset,\ c)$	$\to \emptyset$
$\mathrm{der}_2(x,\ \lambda,\ c)$	$\to \emptyset$
$\mathrm{der}_2(x,\ y,\ c)$	$\to$ **if** $x = y$ **then** c **else** $\emptyset$ **fi**
$\mathrm{der}_2(x,\ a+b,\ c)$	$\to \mathrm{der}_2(x,\ a,\ c) \sqcup \mathrm{der}_2(x,\ b,\ c)$
$\mathrm{der}_2(x,\ a^*,\ c)$	$\to \mathrm{der}_2(x,\ a,\ a^* \cdot c)$
$\mathrm{der}_2(x,\ a_0 \cdot b,\ c)$	$\to \mathrm{der}_2(x,\ a_0,\ b \cdot c)$
$\mathrm{der}_2(x,\ a_1 \cdot b,\ c)$	$\to \mathrm{der}_2(x,\ a_1,\ b \cdot c) \sqcup \mathrm{der}_2(x,\ b,\ c)$
Module *UNFOLD* over *CONJ+DERIV*	
$\mathrm{unf}(\emptyset,\ s)$	$\to \mathrm{true}$
$\mathrm{unf}(\lambda,\ s)$	$\to \mathrm{true}$
$\mathrm{unf}(x,\ s)$	$\to \lambda \leq \mathrm{der}(x,\ s)$
$\mathrm{unf}(a^*,\ s)$	$\to \mathrm{unf}_2(a,\ a^*,\ s)$
$\mathrm{unf}(a+b,\ s)$	$\to \mathrm{unf}(a,\ s) \wedge \mathrm{unf}(b,\ s)$
$\mathrm{unf}(a_0 \cdot b,\ s)$	$\to \mathrm{unf}_2(a_0,\ b,\ s)$
$\mathrm{unf}(a_1 \cdot b,\ s)$	$\to \mathrm{unf}_2(a_1,\ b,\ s) \wedge \mathrm{unf}(b,\ s)$
$\mathrm{unf}_2(\emptyset,\ c,\ s)$	$\to \mathrm{true}$
$\mathrm{unf}_2(\lambda,\ c,\ s)$	$\to \mathrm{true}$
$\mathrm{unf}_2(x,\ c,\ s)$	$\to c \leq \mathrm{der}(x,\ s)$
$\mathrm{unf}_2(a^*,\ c,\ s)$	$\to \mathrm{unf}_2(a,\ a^* \cdot c,\ s)$
$\mathrm{unf}_2(a+b,\ c,\ s)$	$\to \mathrm{unf}_2(a,\ c,\ s) \wedge \mathrm{unf}_2(b,\ c,\ s)$
$\mathrm{unf}_2(a_0 \cdot b,\ c,\ s)$	$\to \mathrm{unf}_2(a_0,\ b \cdot c,\ s)$
$\mathrm{unf}_2(a_1 \cdot b,\ c,\ s)$	$\to \mathrm{unf}_2(a_1,\ b \cdot c,\ s) \wedge \mathrm{unf}_2(b,\ c,\ s)$
Module *MAIN* over *UNFOLD*	
(*Delete*) :	$a \leq s \wedge a \leq_o s \to a \leq_o s$
(*Delete*) :	$a \leq s \wedge a \leq_o s \wedge C \to a \leq_o s \wedge C$
(*Disprove*) :	$a_1 \leq s_0 \to \mathrm{false}$
(*Unfold*) :	$a_0 \leq s \to \mathrm{unf}(a_0,\ s) \wedge a_0 \leq_o s$
(*Unfold*) :	$a_1 \leq s_1 \to \mathrm{unf}(a_1,\ s_1) \wedge a_1 \leq_o s_1$

A simple abstract semantics for equational theories

Gilles Barthe

Faculty of mathematics and informatics
University of Nijmegen, The Netherlands
email: gillesb@cs.kun.nl

Abstract. *We show that a suitable abstraction of the notion of term-algebra, called compositum, can be used to capture in a precise mathematical way the intuition that the category of algebras of most (order-sorted) equational theories is completely characterised by their term-model.*
We also use the relationship between composita and order-sorted equational theories to show that every order-sorted compositum can be canonically embedded into an unsorted one.

1 Introduction

The interplay between syntax and semantics plays a fundamental role in the theory of algebraic specifications. On the one hand, syntax can be used to describe and study syntax-free structures, such as data structures, see [6, 7]. On the other hand, the fundamental nature of syntactic concepts such as unification or term-rewriting is often better understood when presented in an abstract, syntax-free framework (such as monads, algebraic theories or composita), see [5, 10, 12, 14]. In both cases, the switch between the two perspectives (semantic and syntactic) is justified by some adequacy results between the formal system L and the syntax-free framework F under consideration. What is required is that every theory T of the system induces an object $\mathcal{O}(T)$ of the framework. Moreover, we want to

- use L to describe objects of F, so that every object $S \in F$ should be equivalent in some sense to $\mathcal{O}(T)$ for some theory T;
- use F to study some concepts attached to L, so that every theory T should be 'suitably presented' by $\mathcal{O}(T)$.

Here we take the point of view that T is suitably presented by $\mathcal{O}(T)$ if the latter uniquely determines the canonical presentation of T, namely its category of models. For the purpose of this paper, it is convenient to make the notion of adequacy precise in a categorical setting. Let C be a category. A (C-sorted) *formal system* $\mathcal{L}$ is given by a (possibly large) category Theories and a functor Models : $\text{Theories}^{op} \to \mathbb{CAT}/C$, where $\mathbb{CAT}$ is the (large) category of categories and $\mathbb{CAT}/C$ is the comma category over C.

Definition. *A* presentation *of* $\mathcal{L}$ *is a pair* (Pres, P) *where* Pres *is a category and* P : Theories $\to$ Pres *is a functor such that for every two theories* T *and* T', *we* *have* $\mathsf{Models}(T) \cong \mathsf{Models}(T') \Leftrightarrow P(T) \cong P(T')$ *(note that the left isomorphism lives in* $\mathbb{CAT}/C$*).*

Furthermore, we say the presentation is complete *if for every object* 0 *of* Pres, *there exists a theory* T *such that* $P(T) \cong O$.

As usual, $\cong$ is used to denote the existence of an isomorphism betwen two objects in the ambient category.

It is well-known that finitary monads over the category **Set** of sets and algebraic theories provide two complete presentations of the (**Set**-sorted) formal system $\mathcal{E}q$ of equational logic. For order-sorted logics, the situation is less clear. In this paper, we introduce a set-theoretic presentation for an order-sorted equational logic with term-declarations, called QTDL. The presentation uses an order-sorted generalisation of composita[1], a set-theoretic notion which provides an abstract treatment of the notion of term-model ([9]). We show that order-sorted composita present *smooth theories*, a significant class of QTDL theories which includes most of the usual specifications. This result gives (i) a precise mathematical content to the intuition that the categories of algebras of most of order-sorted (and all unsorted) equational theories are uniquely determined by the term-model (ii) a (syntactic) sufficiency criterion to decide for a given theory whether its category of models is uniquely determined by its term-model. This result answers positively to a conjecture of Aczel ([1]) that composita can be used to present equational theories. Yet, unlike the standard presentations of unsorted equational logic, the presentation turns out not to be complete. This will be discussed and remedied elsewhere. Here, we will prove a partial completeness result and use it to show that every Λ-compositum can be embedded into a compositum. The problem was stated (in a different but equivalent form) as an open problem by Williams in [14].

The paper is organised as follows: in section 2, we introduce Quantified Term Declaration Logic and establish its basic properties: soundness, completeness, existence of free algebras. In sections 3 and 4, we introduce Λ-composita and prove that they present smooth theories. Section 5 devoted to showing a partial completeness result for the presentation and to solving the problem of embedding a Λ-compositum into a compositum. We conclude by some remarks on related work. Proofs are omitted; they can be found in [3].

Prerequisites The paper assumes some familiarity with order-sorted logic and order-sorted rewriting (see for example [1, 2, 6, 7, 13]). We also use some very basic category theory, covered for example by [8].

[1] Composita were introduced by A.Nerode in [9] while doing his Ph.D under the supervision of Mac Lane and prior to the apparition of adjunctions, monads or algebraic theories. They have been recently revived and generalised by P.Aczel ([1]). The notion of Λ-compositum considered in this paper is a variant of Aczel's generalised composita.

2 Quantified Term Declaration Logic

Order-sorted logic is by now a well-established field ([6, 7]). The use of term declarations has been studied in [1, 11] and originates from Goguen's seminal paper on order-sorted logic ([4]). Here we introduce a slightly different order-sorted logic with term declarations, called QTDL. It is to the author's knowledge the first order-sorted logic with term declarations using quantified equations as in [7]. As a result, the logic we obtain has initial and free algebras.

2.1 Syntax

QTDL signatures consist of an unsorted signature together with a set of declarations; a declaration can either state that an unsorted term is defined or that an unsorted term is of a given sort.

Definition 1 *An (unsorted)* signature *is of a family of pairwise disjoint sets* $\Phi = (\Phi_n)_{n \in N}$. *Elements of* $\bigcup_{n \in N} \Phi_n$ *are called* function symbols. *If f is a function symbol, its arity is the unique number n such that $f \in \Phi_n$.*

Terms and substitutions are defined as usual. We let $\mathsf{var}(t)$ denote respectively the set of variables of t. A term t is *ground* if $\mathsf{var}(t) = \emptyset$. We now turn to the definition of QTDL signatures. In the sequel, we let Λ be a partial order: $\Lambda^{\downarrow}$ denotes its extension with a top element $\downarrow$. Let $V = \{x_i^\tau | i \in \mathbb{N}, \tau \in \Lambda\}$ be fixed.

Definition 2 *A Λ-signature Σ consists of an unsorted signature Φ and a set $D \subseteq T_\Phi \times \Lambda^{\downarrow}$ of declarations (here T_Φ denotes the set of Φ-terms over V).*

Declarations will be written as $t : \sigma$. Declarations of the form $t : \tau$, where $\tau \in \Lambda$, are *sorting declarations* and state that t has sort τ. Declarations of the form $t : \downarrow$ are *definedness declarations* and state that t is defined, i.e. is a meaningful term of the theory (hence the choice of $\downarrow$ as a top element in $\Lambda^{\downarrow}$). The closure $D^\bullet$ of D is the smallest set containing D and closed under the rules below, where σ, τ and the τ_i's are elements of $\Lambda^{\downarrow}$.

$$\frac{}{x_i^\tau : \tau} \qquad \frac{t : \sigma \quad \sigma \leq \tau}{t : \tau} \qquad \frac{f(t_1, \ldots, t_n) : \downarrow}{t_i : \downarrow} \qquad \frac{t : \sigma \quad s_1 : \tau_1, \ldots, s_n : \tau_n}{t[s_1/x_{i_1}^{\tau_1}, \ldots, s_n/x_{i_n}^{\tau_n}] : \sigma}$$

The set T_Σ of Σ*-terms* is the subset of Φ-terms such that $(t : \downarrow) \in D^\bullet$. Furthermore, we say that a Σ-term t *has sort* τ if $(t : \tau) \in D^\bullet$ $(\tau \in \Lambda^{\downarrow})$.

Definition 3 *A Σ-*presubstitution *is a map $\vartheta : V \to T_\Sigma$ such that for every $i \in \mathbb{N}$ and $\tau \in \Lambda$, ϑx_i^τ is a term of sort τ. Every Σ-presubstitution can be extended into a Σ-*substitution *$\hat{\vartheta}$ by taking*

$$\hat{\vartheta}(x) = \vartheta(x) \qquad \text{if } x \in V$$
$$\hat{\vartheta}(f(t_1, \ldots, t_n)) = f(\hat{\vartheta}t_1, \ldots, \hat{\vartheta}t_n) \qquad \text{if } f(t_1, \ldots, t_n) \in T_\Sigma$$

A Σ*-equation* is a formula of the form $\forall X\ s \doteq t$, where s and t are Σ-terms and $\mathsf{var}(s) \cup \mathsf{var}(t) \subseteq X \subseteq V$. In case $\mathsf{var}(s) \cup \mathsf{var}(t) = X$, we shall simply write $s \doteq t$.

Definition 4 *A* Λ-specification *consists of a pair* $\mathcal{S} = (\Sigma, E)$ *where* Σ *is a* Λ*-signature and* E *is a set of* Σ*-equations.*

The *theorems* of a specification $\mathcal{S} = (\Sigma, E)$ is defined as the smallest set containing E and closed under the rules below.

$$\frac{}{\forall X\ s \doteq s}\ (\mathrm{R}) \qquad \frac{\forall X\ s \doteq t}{\forall X\ t \doteq s}\ (\mathrm{S}) \qquad \frac{\forall X\ s \doteq t \quad \forall X\ t \doteq u}{\forall X\ s \doteq u}\ (\mathrm{T})$$

$$\frac{\forall X\ s_1 \doteq t_1 \quad \ldots \quad \forall X\ s_n \doteq t_n}{\forall X\ f(s_1, \ldots, s_n) \doteq f(t_1, \ldots t_n)}\ (\mathrm{C}) \qquad \frac{\forall X\ s \doteq t}{\forall Y\ \theta s \doteq \theta t}\ (\mathrm{I})$$

where it is assumed in the instantiation (I) rule that θ is a Σ-substitution and $\bigcup\{\mathsf{var}(\theta x) \mid x \in X\} \subseteq Y$. We write $\mathcal{S} \vdash \forall X\ s \doteq t$ whenever $\forall X\ s \doteq t$ is a theorem of $\mathcal{S}$.

2.2 Semantics

Let Λ be a poset. A Λ*-set* consists of a set A and a Λ-indexed family $(A_\tau)_{\tau \in \Lambda}$ of subsets of A such that $A_\tau \subseteq A_\sigma$ whenever $\tau \leq \sigma$. Let $\mathcal{A} = (A, (A_\tau)_{\tau \in \Lambda})$ and $\mathcal{B} = (B, (B_\tau)_{\tau \in \Lambda})$ be two Λ-sets. A map $f : A \to B$ is a map of Λ-sets if for every $\tau \in \Lambda$, $f(A_\tau) \subseteq B_\tau$. Λ-sets form a category $\mathbf{Set}_\Lambda$.

Definition 5 *A* partial algebra *for a* Λ*-signature* Σ *consists of a* Λ*-set* $\mathcal{A}$ *and an interpretation* $f^{\mathcal{A}} : A^n \rightharpoonup A$ *for every function symbol* f *of* Φ.

As usual, an assignment (for an algebra $\mathcal{A}$) is a partial map $\alpha : V \rightharpoonup A$ such that $x_i^\tau \in A_\tau$ for every variable $x_i^\tau \in \mathsf{dom}\ \alpha$. Every assignment α can be extended to a map $[\![.]\!]_\alpha^{\mathcal{A}} : T_\Sigma \rightharpoonup A$ by taking

$$[\![x]\!]_\alpha^{\mathcal{A}} \simeq \alpha x$$
$$[\![f(t_1, \ldots, t_n)]\!]_\alpha^{\mathcal{A}} \simeq f^{\mathcal{A}}([\![t_1]\!]_\alpha^{\mathcal{A}}, \ldots, [\![t_n]\!]_\alpha^{\mathcal{A}})$$

Definition 6 *A* Σ-algebra *is a partial* Σ*-algebra* $\mathcal{A}$ *such that:*
- for every declaration $t : \downarrow$ *(resp.* $t : \tau$*) and assignment* α *such that* $\mathsf{var}(t) \subseteq \mathsf{dom}\ \alpha$, $[\![t]\!]_\alpha^{\mathcal{A}}$ *is defined (resp.* $[\![t]\!]_\alpha^{\mathcal{A}} \in A_\tau$*);*
- for every function symbol f *and* $(a_1, \ldots, a_n) \in \mathsf{dom}\ f^{\mathcal{A}}$*, there exist* Σ*-terms* $t_1, \ldots, t_n$ *and an assignment* α *such that (i)* $[\![t_i]\!]_\alpha^{\mathcal{A}} = a_i$ *for* $i = 1, \ldots, n$ *(ii)* $f(t_1, \ldots, t_n)$ *is a* Σ*-term.*

A Σ-algebra $\mathcal{A}$ *satisfies* an equation $\forall X\ s \doteq t$ (written $\mathcal{A} \models \forall X\ s \doteq t$) if $[\![s]\!]_\alpha^{\mathcal{A}} = [\![t]\!]_\alpha^{\mathcal{A}}$ for all assignments α satisfying $X \subseteq \mathsf{dom}\ \alpha$.

Definition 7 *A* model *of a specification* $\mathcal{S} = (\Sigma, E)$ *(or* $\mathcal{S}$*-algebra for short) is a* Σ*-algebra satisfying all the equations in* E.

Models of a specification form a full subcategory $\mathcal{S}$-Alg of Σ-Alg. There is a forgetful functor $U_{\mathcal{S}} : \mathcal{S}\text{-}\mathsf{Alg} \to \mathbf{Set}_\Lambda$ for every specification $\mathcal{S}$.

Lemma 8 *U_S has a left adjoint F_S.*

In particular, $\mathcal{S}$ has an initial model and a model $T_{(\mathcal{S},X)}$ over X for every $X \subseteq V$. These can be used to prove the soundness/completeness theorem for QTDL. Write $\mathcal{S} \models \forall X\ s \doteq t$ if $\mathcal{A} \models \forall X\ s \doteq t$ for every $\mathcal{S}$-algebra $\mathcal{A}$.

Theorem 9 *For every Λ-specification $\mathcal{S}$, $\mathcal{S} \vdash s \doteq t \Leftrightarrow \mathcal{S} \models s \doteq t$.*

2.3 Smooth specifications

One of the main aims of the paper is to characterise those specifications for which the term-model $T_{(\mathcal{S},V)}$ determines uniquely the category of models. This cannot be true for an arbitrary specification because all term-algebras are used in the proof of the completeness theorem. Hence we must restrict our attention to those specifications for which completeness can be proved from the term-model. This motivates the following definition.

Definition 10 *A specification $\mathcal{S} = (\Sigma, E)$ is* standard *if for every Σ-term t, there exists a Σ-term t' (the* standard form *of t) such that*

(i) $\mathcal{S} \vdash t \doteq t'$,
(ii) for every term u, $\mathcal{S} \vdash u \doteq t \quad \Rightarrow \quad \mathsf{var}(u) \subseteq \mathsf{var}(t')$,
(iii) for every sort τ, $(t : \tau) \in D^\bullet \quad \Rightarrow \quad (t' : \tau) \in D^\bullet$.

A Σ-term is standard *if it is its own standard form.*

The soundness/completeness theorem for a standard specification $\mathcal{S}$ can be proved using the term-model $T_{(\mathcal{S},V)}$ only. Examples of standard specifications include fully inhabited, regular and computable specifications. We recall that a Λ-specification $\mathcal{S} = (\Sigma, E)$ is

- *fully inhabited* if for every sort $\tau \in \Lambda$, there exists a ground term of sort τ;
- *regular* if every equation in E is of the form $s \doteq t$ with $\mathsf{var}\ (s) = \mathsf{var}\ (t)$;
- *computable* if E is confluent and sort-decreasing ([2, 13]).

Hence all the usual specifications are standard. Note that the notion of standard specification is slightly stronger than the notion of context-invariant specification, where a specification $\mathcal{S} = (\Sigma, E)$ is *context-invariant* if for every equation $\forall X\ s \doteq t$, $\mathcal{S} \vdash \forall X\ s \doteq t \Rightarrow \mathcal{S} \vdash s \doteq t$.

One would hope to be able to prove that for standard specifications, the category of models is uniquely determined by the term-model. It is however not so because of the notion of algebra used. The problem is due to overloading. By essence, order-sorted logics with term-declarations only support weak overloading, in the sense that every expression of an algebra has at most one value ([7]). Yet there might be more than one way to assign an expression of an algebra to a term; in our semantics, these expressions must all be equal. This condition on the definition of algebra prevents Λ-composita, introduced in the next section, to present standard theories (see section 4.1). In fact, we must restrict our attention to *smooth* specifications, where a specification $\mathcal{S}$ is smooth if its signature is standard and strictly overloaded:

Definition 11 *A signature Σ is* strictly overloaded *if for every function symbol f of arity n, there exist variables $x_1^{\tau_1}, \ldots, x_n^{\tau_n}$ such that $f(t_1, \ldots, t_n)$ is a Σ-term iff it is an instance of $f(x_1, \ldots, x_n)$. In this case, we say that $(x_1, \ldots, x_n)$ is the* scope *of f.*

Note that strict overloading is a syntactic condition on signatures which ensures that the weak and strong overloading semantics coincide, where in the latter an expression of an algebra is allowed to have more than one value (see [6, 7]).

3 Λ-composita

Λ-composita are an abstraction of the notion of term-model. There is a Λ-set $\mathcal{A}$, whose elements are called objects, a subset X of the carrier of $\mathcal{A}$, the set of parameters, and a monoid of substitutions over $\mathcal{A}$. The idea is that every sort-preserving map from parameters to objects induces a unique substitution. This gives us our main definition. Note that Λ-composita differ from generalised composita ([1]) by having the partial order of sorts as a primitive and the substituvity relation as a derived notion.

Definition 12 *A* Λ-compositum *consists of a triple $\mathbb{A} = (\mathcal{A}, X, S)$ where $\mathcal{A} = (A, (A_\tau)_{\tau \in \Lambda})$ is a Λ-set, $X \subseteq A$ and S is a submonoid of the Λ-maps over $\mathcal{A}$ such that every Λ-map from X (viewed as a Λ-set) to $\mathcal{A}$ has a unique extension in S.*

Λ-composita can be made into a category $\mathbf{Comp}_\Lambda$ by taking as morphisms between two Λ-composita $\mathbb{A} = (\mathcal{A}, X, S)$ and $\mathbb{B} = (\mathcal{B}, Y, S')$ pairs $\Pi = (\pi_o, \pi_s)$, where $\pi_o : A \to B$ (we let $\mathcal{A} = (A, (A_\tau)_{\tau \in \Lambda})$ and $\mathcal{B} = (B, (B_\tau)_{\tau \in \Lambda})$) and $\pi_s : S \to S'$ such that π_o is a Λ-map, $\pi_o(X) \subseteq Y$ and $\pi_o(sa) = (\pi_s s)\ (\pi_o a)$ for every $a \in A$ and $s \in S$.

Moreover, we say that Π is *persistent* if if for every $y \in Y \setminus \{\pi_o x \mid x \in X\}$ and $s \in S$, $(\pi_S s)\ y = y$.

Lemma 13 *Every Λ-specification $\mathcal{S}$ induces a functor $\mathsf{C}_\mathcal{S} : \mathbf{Set}_\Lambda \to \mathbf{Comp}_\Lambda$.*

Proof sketch: for every Λ-set $\mathcal{A} = (A, (A_\tau)_{\tau \in \Lambda})$, the free $\mathcal{S}$-algebra $T_{(\mathcal{S},\mathcal{A})}$ over $\mathcal{A}$ can be turned into a Λ-compositum by taking as set of parameters A and as monoid of substitutions the monoid of endomorphisms of algebras over $T_{(\mathcal{S},\mathcal{A})}$.

4 Λ-composita present smooth Λ-specifications

In this section, we prove that Λ-composita provide a presentation of smooth specifications. By convention, we abbreviate $\mathsf{C}_\mathcal{S}(V)$ by $\mathsf{C}(\mathcal{S})$.

Theorem 14 *Let $\mathcal{S}_1$ and $\mathcal{S}_2$ be two smooth Λ-specifications. Then*

$$\mathsf{C}(\mathcal{S}_1) \cong \mathsf{C}(\mathcal{S}_2) \quad \Leftrightarrow \quad \mathcal{S}_1\text{-}\mathsf{Alg} \cong_s \mathcal{S}_2\text{-}\mathsf{Alg}$$

where $\mathcal{S}_1$-$\mathsf{Alg} \cong_s \mathcal{S}_2$-$\mathsf{Alg}$ is used to denote that $U_{\mathcal{S}_1} \cong U_{\mathcal{S}_2}$ in $\mathbb{CAT}/\mathbf{Set}_\Lambda$.

Proof sketch: We focus on the direct implication. The reverse implication follows from some easy 'categorical nonsense'.

Let $\Pi = (\pi_o, \pi_s) : \mathsf{C}(\mathcal{S}_1) \to \mathsf{C}(\mathcal{S}_2)$ be an isomorphism of Λ-composita. We define a functor $I : \mathcal{S}_1\text{-}\mathsf{Alg} \to \mathcal{S}_2\text{-}\mathsf{Alg}$. The functor is defined by a back-and-forth construction.

Stage 1: *constructing an isomorphism from the class of $\mathcal{S}_1$-algebras to the class of $\mathcal{S}_2$-algebras.*

We first build a partial Σ_2-algebra $\mathcal{B}$ from a $\mathcal{S}_1$-algebra. Let $\mathcal{A}$ be a $\mathcal{S}_1$-algebra with underlying Λ-set $\mathcal{A} = (A, (A_\tau)_{\tau \in \Lambda})$.

The underlying Λ-set $(B, (B_\tau)_{\tau \in \Lambda})$ of $\mathcal{B}$ is defined by $B = A$ and $B_\tau = A_\tau$ for every $\tau \in \Lambda$. For every function symbol f of Φ_2, $f^{\mathcal{B}}$ is defined to be the unique partial map such that:

- its function domain is $B_{\tau_1} \times \cdots \times B_{\tau_n}$ where $(x_{i_1}^{\tau_1}, \ldots, x_{i_n}^{\tau_n})$ is the scope of f;
- given $(a_1, \ldots, a_n) \in \mathrm{dom}\ f^{\mathcal{B}}$, $f^{\mathcal{B}}(a_1, \ldots, a_n) = [\![t]\!]_\alpha^{\mathcal{A}}$ where t is a standard Σ_1-term such that $\pi_o\bar{t} = \overline{f(x_1, \ldots, x_n)}$ and α is the partial assignment defined by $\alpha u = a_i$ if $\pi_s\bar{u}^1 = \bar{x}_i{}^2$ for some $i = 1, \ldots, n$.

One can verify that $\mathcal{B}$ is well-defined. Then, one has to check that $\mathcal{B}$ is a model of $\mathcal{S}_2$. The key result (proved by induction on the length of the terms) is

Lemma 15 *Let α be an assignment. Let t be a Σ_2-term such that* $\mathrm{var}(t) \subseteq \mathrm{dom}\ \alpha$. *Let t' be a standard Σ_1-term such that $\pi_o\bar{t'} = \bar{t}$. Then $[\![t]\!]_\alpha^{\mathcal{B}} = [\![t']\!]_\alpha^{\mathcal{A}}$. Moreover, $[\![t]\!]_\alpha^{\mathcal{B}} \in A_\tau$ whenever t has sort τ.*

Stage 2: *making the construction functorial.*

One must lift the construction to a functor I which commutes with the forgetful functors. Necessarily I must be the identity on maps. One can check that the construction defines a functor and that it is moreover an isomorphism of categories. Note that strict overloading plays a crucial role in here.

4.1 Discussion

Theorem 14 is as it stands in its most general form. Indeed, the theorem becomes false if one of the two assumptions standard or strictly overloaded is dropped. We give two counterexamples. In each case, we have $\mathsf{C}(\mathcal{S}_1) \cong \mathsf{C}(\mathcal{S}_2)$ but $\mathcal{S}_1\text{-}\mathsf{Alg} \not\cong_s \mathcal{S}_2\text{-}\mathsf{Alg}$.

Non-strictly overloaded specifications. Let $\Lambda = \{\tau, \sigma\}$ be a flat poset. Consider the two many-sorted Λ-specifications:

$\mathcal{S}_1$	
Declarations:	$f(x^\tau)\downarrow$
	$f(y^\sigma)\downarrow$
Axioms:	

$\mathcal{S}_2$	
Declarations:	$f(x^\tau)\downarrow$
	$g(y^\sigma)\downarrow$
Axioms:	

$\mathcal{S}_1$ is non-strictly overloaded. Obviously, $\mathsf{C}(\mathcal{S}_1) \cong \mathsf{C}(\mathcal{S}_2)$. Yet $\mathcal{S}_1$-Alg $\not\cong_s$ $\mathcal{S}_2$-Alg: let $\mathcal{A} = (A, (A_\tau)_{\tau \in \Lambda})$ be the Λ-set defined by $A = \{a, b\}$ and $A_\sigma = A_\tau = \{a\}$. We can define 4 models of $\mathcal{S}_2$ over $\mathcal{A}$ but only 2 models of $\mathcal{S}_1$. The limitation to strictly overloaded specifications might appear to be quite serious. However, it can be overcome by considering the so-called lax algebras rather than the algebras of this paper; in [2], the notion of lax algebra is introduced and Theorem 14 is adapted to the case of lax algebras. This is not too surprising because lax algebras are essentially the algebras of [7] and are suitable to model ad hoc polymorphism.

Non-standard specifications. Consider the two single-sorted specifications:

$\mathcal{S}_1$	$\mathcal{S}_2$
Declarations: $c\downarrow$	Declarations: $fx\downarrow$
Axioms:	Axioms: $fx = fy$

$\mathcal{S}_2$ is non-standard. Obviously, $\mathsf{C}(\mathcal{S}_1) \cong \mathsf{C}(\mathcal{S}_2)$. Yet $\mathcal{S}_1$-Alg $\not\cong_s$ $\mathcal{S}_2$-Alg: the initial model of the first one is non-empty, while the initial model of the second is. We can generalise Theorem 14 to non-standard specifications by considering the functors $\mathsf{C}_{\mathcal{S}_i}$ rather than $\mathsf{C}(\mathcal{S}_i)$.

Proposition 16 *Let $\mathcal{S}_1$ and $\mathcal{S}_2$ be two strictly overloaded Λ-specifications. The following are equivalent:*
(i) $\mathsf{C}_{\mathcal{S}_1} \cong \mathsf{C}_{\mathcal{S}_2}$,
(ii) $\mathcal{S}_1$-Alg $\cong_s$ $\mathcal{S}_2$-Alg.

However, Proposition 16 is of a lesser interest than Theorem 14, because the latter captures mathematically the intuition that in most cases, the term-model is enough to determine a specification. Proposition 16, although more widely applicable, is less appealing because it involves all term algebras.

5 Embedding a Λ-compositum into a compositum

5.1 Partial completeness of the presentation

Λ-composita do not provide a complete presentation of Λ-theories because the syntax of QTDL requires that each variable has a least sort. Yet there is a partial completeness result.

Definition 17 *A Λ-compositum $\mathbb{A} = (\mathcal{A}, X, S)$ is:*

- levelled *if every element of X has a least sort, i.e. for every $x \in X$, the set $\{\tau \in \Lambda \mid x \in A_\tau\}$ has a minimum element;*
- finitary *if for every $a \in A$, there exists a finite subset* supp (a) *of X such that for every $s_1, s_2 \in S$, $s_1 \restriction \mathsf{supp}\,(a) = s_2 \restriction \mathsf{supp}\,(a) \quad \Rightarrow \quad s_1 a = s_2 a$.*

Note that every Λ-compositum can be seen as a levelled $\mathcal{P}(\Lambda)$-compositum, where $\mathcal{P}(\Lambda)$ is the powerset of Λ with set inclusion as a partial order.

Theorem 18 *Let $\mathbb{A} = (\mathcal{A}, X, S)$ (where $\mathcal{A} = (A, (A_\tau)_{\tau \in \Lambda})$) be a finitary levelled Λ-compositum. Let $\mathcal{X} = (X, (X \cap A_\tau)_{\tau \in \Lambda})$. Then there exists a strictly overloaded computable Λ-specification $\mathcal{S}$ such that $\mathsf{C}_{\mathcal{S}}(\mathcal{X}) \cong \mathbb{A}$.*

5.2 The embedding

We now apply the partial completeness result to solve the problem of embedding a Λ-compositum into a compositum. In this section, we use an equivalent but more convenient definition of compositum; a compositum is a Λ-compositum $\mathbb{A} = (\mathcal{A}, X, S)$ such that $A_\tau = A$ for all $\tau \in \Lambda$.

Theorem 19 *Every finitary Λ-compositum $\mathbb{A} = (\mathcal{A}, X, S)$ can be embedded into a finitary compositum $\mathbb{A}^\Delta = (\mathcal{A}^\Delta, X, S^\Delta)$; in other words, there exists a morphism of Λ-composita $\Pi^\Delta : \mathbb{A} \to \mathbb{A}^\Delta$ whose first component is an injection.*

Proof sketch: without loss of generality, we can assume that $\mathbb{A}$ is levelled. By Theorem 18, there exists a specification $\mathcal{S}$ such that $\mathsf{C}_\mathcal{S}(\mathcal{X}) \cong \mathbb{A}$. To simplify matters, assume that $\mathsf{C}(\mathcal{S}) = \mathbb{A}$. Let $\mathcal{S}^\bullet$ be the unsorted specification induced by $\mathcal{S}$, $\mathcal{S}^\bullet = (\Phi, E)$. Define $\mathbb{A}^\Delta = \mathsf{C}(\mathcal{S}^\bullet)$. Take as morphism Π^Δ the canonical morphism from $\mathsf{C}(\mathcal{S})$ to $\mathsf{C}(\mathcal{S}^\bullet)$. To prove that Π^Δ is an embedding, it is enough to show that for every Σ-terms s and t, $\mathcal{S}^\bullet \vdash s \doteq t \Rightarrow \mathcal{S} \vdash s \doteq t$. As $\mathcal{S}$ is a computable rewrite system, the right-hand side is equivalent to $s \downarrow_\mathcal{S} t$, where $s \downarrow_\mathcal{S} t$ is used to denote that s and t have a common reduct. As $\mathcal{S}^\bullet$ is confluent, the left-hand side is equivalent to $s \downarrow_{\mathcal{S}^\bullet} t$. So we must prove $s \downarrow_{\mathcal{S}^\bullet} t \Rightarrow s \downarrow_\mathcal{S} t$. This follows from the following fact: if s is a Σ-term, t is a Φ-term and $s \to_{\mathcal{S}^\bullet} t$, then t is a Σ-term and $s \to_\mathcal{S} t$.

Remark: Theorem 19 can be strengthened to prove that the category of finitary composita with persistent morphisms as arrows is a reflective subcategory of the category of finitary Λ-composita with persistent morphisms as arrows.

6 Concluding remarks

We have introduced QTDL, an order-sorted logic with term-declarations and quantified equations, and provided an abstract presentation for it in terms of composita. The main result of the paper, Theorem 14, gives a precise mathematical content to the intuition that in most cases the term-model characterises uniquely (the category of models of) a specification. It turns out that the presentation described in this paper is not complete. The point is that order-sorted logics cannot be extended immediately to lattice-sorted logics, in which variables may not have a minimal sort. In [2], we show that the naive extension is not sound and complete and remedy the problem by looking at a new logic, called ETDL, in which rules for typing and equational resoning are combined.

In this paper, we have restricted our attention to composita. Order-sorted monads provide another (incomplete) presentation of QTDL (see [3]). Again, the presentation can be made complete by looking at ETDL. This will be reported elsewhere.

Acknowledgements

I am grateful to Peter Aczel for introducing me to composita. I would also like to thank Andrezj Tarlecki, who carefully read my thesis and suggested several improvements which I hope are reflected in this paper. Finally, I am indebted to the anonymous referees who made some valuable comments on the first draft of the paper.

References

1. P.Aczel. *Term Declaration Logic and Generalised Composita*, in: Proceedings of the sixth Symposium on Logic and Computer Science, I.E.E.E. Computer Society Press, 1991.
2. G.Barthe. *Term Declaration Logic and Generalised Composita*, Ph.D thesis, University of Manchester, 1993.
3. G.Barthe. *A simple abstract semantics for equational theories*, to appear as a technical report, University of Nijmegen, 1995.
4. J.Goguen. *Order-sorted algebra*, Technical report, UCLA, 1968.
5. J.Goguen. *What is Unification?*, in Hassan Ait-Kaci and Maurice Nivat editors, Resolution of Equations in Algebraic Structures, Vol. 1, pp 217-261, Academic Press, 1989.
6. J.Goguen and R.Diaconescu. *An Oxford survey of Order Sorted Algebra*, Mathematical Structures in Computer Science, Vol. 4, pp 363–392, 1994
7. J.Goguen and J.Meseguer. *Order-Sorted Algebra I: Equational Deduction for Multiple Inheritance, Overloading, Exceptions and Partial Operations*, Theoretical Computer Science, Vol. 105, pp 217-273, 1992.
8. S.Mac Lane. *Categories for the Working Mathematician*, Graduate Texts in Mathematics, Vol. 5, Springer Verlag, 1971.
9. A.Nerode. *Composita, Equations and Freely Generated Algebras*, Transactions of the American Mathematical Society, pp 139-151, 1959.
10. D.Rydeheard and R.Burstall. *Computational Category Theory*, International Series in Computer Science, Prenctice-Hall, 1988.
11. M.Schmidt-Schauß. *Computational Aspects of Order-Sorted Logic with Term Declarations*, Springer Lecture Notes in Artificial Intelligence, Vol. 395, 1989.
12. M.Schmidt-Schauß and J.Siekmann. *Unification Algebras: An Axiomatic Approach to Unification, Equation Solving and Constraint Solving*, Universität Kaiserslautern, SEKI-REPORT SR-88-23, 1988.
13. G.Smolka, W.Nutt, J.Goguen and J.Meseguer. *Order-Sorted Equational Computation*, in Hassan Ait-Kaci and Maurice Nivat editors, Resolution of Equations in Algebraic Structures, Vol. 2, Academic Press, 1989.
14. J.Williams. *Instantiation Theory*, Springer Lecture Notes in Artificial Intelligence, Vol. 518, 1991.

Processes with Multiple Entries and Exits

J.A. Bergstra[1] and Gh. Ştefănescu[2]

[1]University of Amsterdam,Programming Research Group,
Kruislaan 403, 1098 SJ Amsterdam, The Netherlands
and
Utrecht University, Department of Philosophy,
Heidelberglaan 8, 3584 CS Utrecht, The Netherlands

[2]Institute of Mathematics of the Romanian Academy,
P.O. Box 1-764, 70700 Bucharest, Romania

E-mail: janb@fwi.uva.nl -- ghstef@imar.ro

Abstract. This paper is an attempt to integrate the algebra of communicating processes (ACP) and the algebra of flownomials (AF). Basically, this means to combine axiomatized parallel and looping operators. To this end we introduce a model of process graphs with multiple entries and exits. In this model the usual operations of both algebras are defined, e.g. alternative composition (this covers both the sum of ACP and the disjoint sum of AF), sequential composition, feedback, parallel composition, left merge, communication merge, encapsulation, etc. The main results consist of correct and complete axiomatisations of process graphs modulo isomorphism and modulo bisimulation.

Key words & Phrases: process algebra, feedback, flowchart theories.

1 Introduction

This paper is an attempt to integrate the algebra of communicating processes (ACP) and the algebra of flownomials (AF). Basically, this means to combine axiomatized parallel and looping operators.

There are three axiomatized looping operations that may be combined with the existing process algebra (ACP): Kleene's star "$*$" (repetition [Kle56]) used in regular algebra, Elgot's dagger "$\dagger$" (iteration [Elg75]) used in iteration theories, or uparrow "$\uparrow$" (feedback [Ste86]) used in the algebra of flownomials. A study of process algebra with an iteration operation (originary Kleene's star) has already been presented in [BeBP94]. Here we combine process algebra with the feedback operation.

Our goal is to define a process algebra on which all operators of ACP [BeK84] are present as well as feedback, the key iteration construct of flowchart theories. To this end we combine the results of [BaB95], [BeS94] and various other results on flowchart theories [Elg75, Ste87, Ste86, CaS90, BlE93, Ste94]. Like in flowchart schemes [Ste86] feedback and alternative composition " Φ " suffice to express all finite state systems. (" Φ " is a mixture of disjoint sum "$\oplus$", left-composition with converses of functions and right-composition with functions.) In fact alternative composition and feedback allow the construction of normal forms modulo isomorphism of transition systems.

(A similar result on undirected networks is presented in [Par93], an approach that simplifies the earlier algebra of flow graphs of Milner [Mil79].)

Technically we depart from the graph isomorphism model for the operators of ACP that was outlined in [BaB95]. This model is adapted to allow for process graphs with multiple entries and exits. Then the feedback operator is introduced. For technical reasons renaming operators for entries and exits are needed. We notice that entries and exits are just a particular kind of states.

In the process of the design of this model we have taken several decisions which we want to make clear from the very beginning.

– Our transition systems allow for multiple entries and multiple exits. This feature drastically increases the expressive power of the algebra. All finite state processes are represented by closed terms built up from atomic actions and some constants by using two operations only, i.e., alternative composition and feedback. Moreover, almost all process graphs are represented. To be precise: (1) in the case without ϵ (i.e., without empty transitions) all process graphs with no incoming edges into the entries, no outgoing edges from the exits and such that no entry is an exit are represented; (2) if ϵ is allowed, then all process graphs are represented.

– We have decided to make the operations total by providing a default system of working in the case the types do not agree. E.g., seqential composition is defined in the case the outputs of the first process graph do not match the inputs of the second one, as well.

– There are two options regarding the using of port names: (1) to use an arbitrary set of port names and renamings or (2) to order the port names —e.g., to use the first n natural numbers as names for a process with n ports— and an explicit algebra to model the renamings. We have decided to use here the first variant which gives more freedom (i.e., it is more abstract) and perhaps easier to understand. The second version may be more suited for implementations.

– Finally, we had to decide whether we allow or not for ϵ steps. We start here with the model without ϵ steps which seems easier. The loss of expressivity is small in the case of cyclic processes (as we already mention), but important in the case of acyclic ones (not all acyclic processes may be represented using alternative and sequential composition). This ϵ-free case also generates some complications in defining bisimilarity which requires an explicit splitting operator.

In this paper we study the ϵ-free model. The main results are: (i) expressiveness (already mentioned); (ii) a correct and complete axiomatisation of process graphs modulo isomorphism; (iii) a correct and complete axiomatisation of process graphs modulo strong bisimulation.

2 Process graphs modulo isomorphism

2.1 Process graphs

Notations:

– V denotes a set of port names with typical elements p, q, r, s, t, u, v. We assume V is closed to cartesian product. (That is, we assume an injective coding $\bowtie: V \times V \to V$ is given and $A \times B = \{a \bowtie b \mid a \in A,\ b \in B\}$.)

- A denotes a set of atomic actions with typical elements a, b, c, d, e and closed to cartesian product.
- "$\circ$" denotes relational composition; e.g if R, S are binary relations and T a ternary relation, then $R \circ S = \{(x, z) \mid \exists y : (x, y) \in R \text{ and } (y, z) \in S\}$ and $R \circ T \circ S = \{(x, y, z) \mid \exists x', z' : (x, x') \in R, \ (x', y, z') \in T \text{ and } (z, z') \in S\}$.
- Id_A denotes the identity relation on a set A (i.e., $Id_A = \{(x, x) \mid x \in A\}$).
- $\phi|_A$ denotes the restriction of a function ϕ to a subset A of the definition domain.
- $[n] = \{1, \ldots, n\}$
- We write $f \ \cup \ g$ for the union of two functions; if the definition domains are disjoint, the union is a function, too.

Definition 1. (process graphs with multiple entries and multiple exits) A *process graph* P is given by the following data: $P = (\partial_i, E, \ell, \partial_o) : I \xrightarrow{S} O$ where:

- I, S, O are finite subsets of V and $S \cap (I \cup O) = \emptyset$
- E is a finite set
- $\partial_i, \partial_o, \ell$ are functions: $\quad \partial_i : E \to I \cup S, \quad \partial_o : E \to O \cup S, \quad \ell : E \to A$

$G(A)(I, S, O)$ denotes the process graphs with entries in I, exits in O, internal vertices in S and labels in A. The class of process graphs with I and O given and arbitrary S (resp. with arbitrary I, S, O) is denoted by $G(A)(I, O)$ (resp. by $G(A)$). □

The meaning of these data is the following: I specifies the start vertices, S the internal ones and O the end ones; E specifies the edges (transitions); and $\partial_i(e)$, $\partial_o(e)$ and $\ell(e)$ give the source vertex, the target vertex and the label of an edge e, respectively.

Note that $G(A)$ includes all process graphs obeying the following restriction: "a start vertex has no incoming edges, an end vertex has no outgoing edges and a start vertex is not an end vertex".

A process graph $P = (\partial_i, E, l, \partial_o) : I \xrightarrow{S} O$ may also be specified by using the transition relation $T \subseteq (I \cup S) \times A \times (O \cup S)$ consisting of all the transitions $\partial_i(e) \xrightarrow{\ell(e)} \partial_o(e)$, for $e \in E$. Sometimes we write these data as $T : I \xrightarrow{S} O$.

We study two kinds of graph models: (1) with multiplicity (i.e., multiple edges with the same label between two vertices are allowed) and (2) without multiplicity.

Definition 2. Let $P : I \xrightarrow{S} O$ and $P' : I \xrightarrow{S'} O$ be two process graphs and T and T' their transition relations. A bijection $\phi : S \to S'$ is called an *isomorphism* if

$$(Id_I \cup \phi) \circ T' = T \circ (Id_O \cup \phi) \quad \square$$

2.2 Operations on process graphs

The operations will be defined on process graphs modulo isomorphism.

Definition 3. (operations and constants on process graphs)

Constants:

– **Atomic actions a^p_q, for $a \in A$, $p,q \in V$:**

$a^p_q = \langle \partial_i, \{\star\}, \ell, \partial_o \rangle : \{p\} \xrightarrow{\bullet} \{q\}$, where $\partial_i(\star) = p$, $\ell(\star) = a$, $\partial_o(\star) = q$.

– **Empty processes $\emptyset$, $\perp^p$, $\top_p$:**

$$\emptyset = \langle \emptyset,\emptyset,\emptyset,\emptyset \rangle : \emptyset \xrightarrow{\bullet} \emptyset, \quad \perp^p = \langle \emptyset,\emptyset,\emptyset,\emptyset \rangle : \{p\} \xrightarrow{\bullet} \emptyset, \quad \top_q = \langle \emptyset,\emptyset,\emptyset,\emptyset \rangle : \emptyset \xrightarrow{\bullet} \{q\}$$

Operations:

– **Alternative composition** $\oplus$ **:** Let $P = \langle \partial_i, E, \ell, \partial_o \rangle : I \xrightarrow{S} O$ and $P' = \langle \partial'_i, E', \ell', \partial'_o \rangle : I' \xrightarrow{S'} O'$ be given. Assume $S \cap (I' \cup S' \cup O') = \emptyset$, $S' \cap (I \cup S \cup O) = \emptyset$ and $E \cap E' = \emptyset$. Then

$$P \oplus Q = \langle \partial_i \cup \partial'_i,\ E \cup E',\ \ell \cup \ell',\ \partial_o \cup \partial'_o \rangle : I \cup I' \xrightarrow{S \cup S'} O \cup O'$$

Effect: Disjoint union, but the entries (resp. the exits) with the same port name are identified.

– **Feedback** $\uparrow^p_q$**:** Let $P = \langle \partial_i, E, \ell, \partial_o \rangle : I \xrightarrow{S} O$ be given.
  – If $p \notin I$ and $q \notin O$, then $P \uparrow^p_q = P$.
  – Otherwise, take a fresh r (not in $I \cup O \cup S \cup \{p,q\}$) and define

$$\begin{aligned} P \uparrow^p_q \;=\; & \langle \partial_i \circ (Id_{(I-\{p\}) \cup S} \cup t_{\{p\} \cap I \mapsto \{r\}}),\ E,\ \ell,\ \partial_o \circ (Id_{(O-\{q\}) \cup S} \cup t_{\{q\} \cap O \mapsto \{r\}}) \rangle \\ & : I - \{p\} \xrightarrow{S \cup \{r\}} O - \{q\} \end{aligned}$$

where $t_{A \mapsto \{r\}}$ is the unique function from A to $\{r\}$, for a set A with at most one element.

Effect: If there are an entry with label p and an exit with label q, then they are identified as a unique internal vertex r. If q does not appear as an exit and p appears as an entry, then the effect is that we hide enter p that becomes an internal vertex. Similarly if p does not appear as an entry and q appears as an exit. Finally, if p and q do not appear at all, then the result of the feedback is P itself.

– **Renaming port names** $\triangleleft$, $\triangleright$**:** Let $\phi : V \to V$ be a renaming function. Assume $P = \langle \partial_i, E, \ell, \partial_o \rangle : I \xrightarrow{S} O$ is such that $S \cap (\phi(I) \cup \phi(O)) = \emptyset$. Then:

$$\phi \triangleleft P = \langle \partial_i \circ (\phi|_I \cup Id_S),\ E,\ \ell,\ \partial_o \rangle : \phi(I) \xrightarrow{S} O$$

$$P \triangleright \phi = \langle \partial_i,\ E,\ \ell,\ \partial_o \circ (\phi|_O \cup Id_S) \rangle : I \xrightarrow{S} \phi(O)$$

Effect: Rename the entry (resp. exit) ports by means of ϕ and identify the entries (resp. the exits) that get equal renamings.

Notation:

(1) For $I = \{p_1, \ldots, p_m\}$ and $O = \{q_1, \ldots, q_n\}$ we define $\delta^I_O := \perp^{p_1} \oplus \ldots \oplus \perp^{p_m} \oplus \top_{q_1} \oplus \ldots \oplus \top_{q_n}$; $\quad \delta^p_q$ means $\delta^{\{p\}}_{\{q\}}$.

(2) $I(1) := \delta^r_r \uparrow^r_r$, for an $r \in V$ and $I(m) = \oplus^m_{j=1} I(1)$ for an $m \in \mathbb{N}$. Up to an isomorphism $I(m)$ is $\langle \emptyset,\emptyset,\emptyset,\emptyset \rangle : \emptyset \xrightarrow{S} \emptyset$, for an S with m elements.

(3) Let $A, B \subset V$ be such that $A \cap B = \{p_1, \ldots, p_k\}$, $A - B = \{q_1, \ldots, q_m\}$ and $B - A = \{r_1, \ldots, r_n\}$. Then $P \uparrow^A_B$ means $P \uparrow^{p_1}_{p_1} \ldots \uparrow^{p_k}_{p_k} \uparrow^{q_1}_{q'_1} \ldots \uparrow^{q_m}_{q'_m} \uparrow^{r'_1}_{r_1} \ldots \uparrow^{r'_n}_{r_n}$ where $q'_1, \ldots, q'_m, r'_1, \ldots, r'_n$ are fresh variables.

(4) Let $P = \langle \partial_i, E, \ell, \partial_o \rangle : I \xrightarrow{S} O$ and $E' \subseteq E$ be given. We denote by $P|_{E'}$ the restriction $P|_{E'} = \langle \partial_i|_{E'},\ E',\ \ell|_{E'},\ \partial_o|_{E'} \rangle : I \xrightarrow{S} O$.

Operations (continuation):

– **Sequential composition** $\odot$: Let $P : I \to O$ and $P' : I' \to O'$ be given. Take a bijective renaming ϕ such that $\phi(O \cup I')$ contains fresh names only. Then define:

$$P \odot P' = [(P \triangleright \phi) \oplus (\phi \triangleleft P')] \uparrow^{\phi(I')}_{\phi(O)}$$

Effect: Sequential composition. Note that we hide all the exits of P and all the entries of P' that have no correspondent.

– **Parallel composition** $\|$:

$\star$ First, we define $\|$ for processes $P : I \to O$ with $I \cap O = \emptyset$. Let $P = \langle \partial_i, E, \ell, \partial_o \rangle : I \xrightarrow{S} O$ and $P' = \langle \partial'_i, E', \ell', \partial'_o \rangle : I' \xrightarrow{S'} O'$ be two such processes. Define

$$P \parallel P' = \quad \langle \partial^1_i, E^1, \ell^1, \partial^1_o \rangle \quad : I \times I' \xrightarrow{(I \cup S \cup O) \times (I' \cup S' \cup O') - (I \times I' \cup O \times O')} O \times O'$$

where if pr_1 and pr_2 denote the 1st and the 2nd projection, respectively, then

$$\begin{aligned} \partial^1_i &= \partial_i \times \partial'_i \ \cup\ \partial_i \times Id_{I' \cup S' \cup O'} \ \cup\ Id_{I \cup S \cup O} \times \partial'_i, \\ E^1 &= E \times E' \ \cup\ E \times (I' \cup S' \cup O') \ \cup\ (I \cup S \cup O) \times E', \\ \ell^1 &= \ell \times \ell' \ \cup\ pr_1 \circ \ell \ \cup\ pr_2 \circ \ell', \\ \partial^1_o &= \partial_o \times \partial'_o \ \cup\ \partial_o \times Id_{I' \cup S' \cup O'} \ \cup\ Id_{I \cup S \cup O} \times \partial'_o \end{aligned}$$

$\star$ In general, if $P : I \xrightarrow{S} O$ and $P' : I' \xrightarrow{S'} O'$ are two arbitrary graphs, then take two bijective renaming functions ϕ and ψ such that $\phi(O) \cap (I \cup S) = \emptyset$ and $\psi(O') \cap (I' \cup S') = \emptyset$. Define

$$P \parallel P' = [(P \triangleright \phi) \parallel (P' \triangleright \psi)] \triangleright (\phi^{-1} \times \psi^{-1})$$

Effect: Parallel composition, i.e. $P \parallel P'$ performs a transition from P, or a transition from P', or a pair of transitions (one from P and one from P'), if possible.

– **Left merge** $\mathbin{||\!\_}$ and **communication merge** $|$ are defined in a standard way using the definition of $\|$.

– **Encapsulation** ∂_H: Let a subset $H \subseteq A$ and a process $P = \langle \partial_i, E, \ell, \partial_o \rangle : I \xrightarrow{S} O$ be given. Let $E_{A-H} = \{e \in E | \ell(e) \notin A\}$. Then $\partial_H(P) = P|_{A-H}$.

Effect: All the transitions in H are deleted. □

Example 1. An example is given in Figure 1. $P = a^{p_1}_{u''} \oplus a^{u'}_{q_1} \oplus a^{u'}_{v''} \oplus a^{p_2}_{v''} \oplus a^{v'}_{q_2}$ is illustrated in figure (a). The process $Q = P \uparrow^{u'}_{u''} \uparrow^{v'}_{v''}$ is illustrated in figure (c) with an intermediary step shown in (b). Q is particularly interesting since it cannot be represented using atomic actions, constants, alternative and sequential composition, only. □

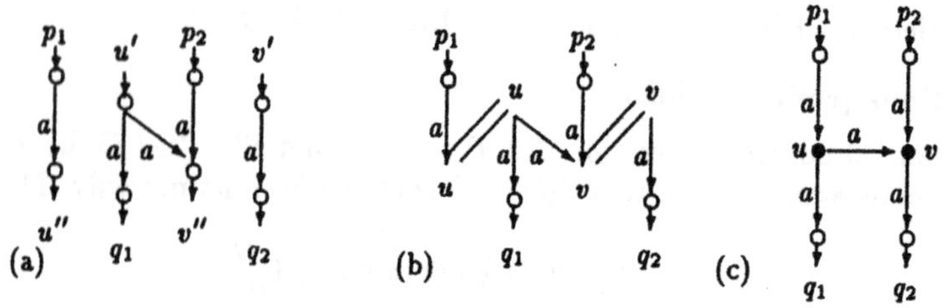

Fig. 1. Process graphs

2.3 Expressiveness

Theorem 4. *1) All graphs in $G(A)$ may be constructed starting with atomic actions or constants $\emptyset$, $\perp$, $\top$ and using alternative composition and feedback only.*

2) There are acyclic processes that cannot be constructed starting with atomic actions and constants and using alternative and sequential composition only.

2.4 Axioms, correctness, completeness

Let $Exp(Act, Const;\ \oplus, \uparrow, \triangleleft, \triangleright)$ be the set of expressions constructed from the atomic actions and constants using alternative composition, feedback and renamings. Define three functions $i, o : Exp \to V$ and $s : Exp \to \mathbb{N}$ giving the sets of port names for the entries and for the exits and the number of the internal vertices, respectively.

We say an expression is in a *prenormal form (pnf)* if it is of the following type

$$(f_1 \oplus \ldots \oplus f_n) \uparrow^{p_1}_{q_1} \ldots \uparrow^{p_k}_{q_k} \quad \text{with } f_i \text{ of type } \perp^p,\ \top_p \text{ or } a^p_q, \text{ for all } i$$

Every expression may be brought to a prenormal form using the axioms in Table 1.

Denote $\perp^I := \oplus_{p \in I} \perp^p$, $\top_O := \oplus_{q \in O} \top_q$. We say an expression E is in a *normal form (nf)* if it is of the following type

$$(\alpha \oplus \beta) \uparrow^{p_1}_{p_1} \ldots \uparrow^{p_k}_{p_k}$$

where

(1) $\alpha = \perp^{I'} \oplus \top_{O'} \oplus I(m)$, for some $I' \subseteq i(E)$, $O' \subseteq o(E)$ and $m \leq s(E)$;

(2) $\beta = \oplus^n_{i=1} a^{r_i}_{s_i}$;

(3) $i(\alpha) \cap i(\beta) = \emptyset$, $o(\alpha) \cap o(\beta) = \emptyset$;

(4) there are no unused feedbacks, i.e, (i) $p_1, \ldots, p_k$ are all distinct and (ii) for each feedback $\uparrow^{p_i}_{p_i}$, $i \in [n]$, there exists either a term $a^{p_i}_q$ or a term $a^r_{p_i}$ in β.

In the case without multiplicity the following condition is added:

(5) β does not contains two equal terms.

A1.	$f \oplus (g \oplus h) = (f \oplus g) \oplus h$	
A2.	$f \oplus g = g \oplus f$	
A3.	$f \oplus \emptyset = f$	
A4.	$f \oplus \bot^p = f$	if $p \in i(f)$
A5.	$f \oplus \top_q = f$	if $q \in o(f)$
A(**)	$f \oplus f = f$	
F1.	$f \uparrow^p_q \uparrow^r_s = f \uparrow^r_s \uparrow^p_q$	if $p \neq r$ and $q \neq s$
F2.	$f \uparrow^p_q = f$	if $p \notin i(f)$ and $q \notin o(f)$
FA1.	$f \oplus (g \uparrow^p_q) = (f \oplus g) \uparrow^p_q$	if $p \notin i(f)$ and $q \notin o(f)$
FA2.	$(f \oplus \bot^p) \uparrow^p_q = f \uparrow^p_q$	if $q \in o(f)$
FA3.	$(f \oplus \top_q) \uparrow^p_q = f \uparrow^p_q$	if $p \in i(f)$

$R1_o$. $a^p_q \triangleright \phi = a^p_{\phi(q)}$	$R1^i$. $\phi \triangleleft a^p_q = a^{\phi(p)}_q$
$R2_o$. $\bot^p \triangleright \phi = \bot^p$	$R2^i$. $\phi \triangleleft \bot^p = \bot^{\phi(p)}$
$R3_o$. $\top_q \triangleright \phi = \top_{\phi(q)}$	$R3^i$. $\phi \triangleleft \top_q = \top_q$
$R4_o$. $\emptyset \triangleright \phi = \emptyset$	$R4^i$. $\phi \triangleleft \emptyset = \emptyset$
RA_o. $(f \oplus g) \triangleright \phi = (f \triangleright \phi) \oplus (g \triangleright \phi)$	RA^i. $\phi \triangleleft (f \oplus g) = (\phi \triangleleft f) \oplus (\phi \triangleleft g)$
$RF1_o$. $(f \uparrow^p_q) \triangleright \phi = (f \triangleright \phi_{\lvert\{q\}\rvert}) \uparrow^p_q$ if $q \notin \phi(o(f) - \{q\})$	$RF1^i$. $\phi \triangleleft (f \uparrow^p_q) = (\phi_{\lvert\{p\}\rvert} \triangleleft f) \uparrow^p_q$ if $p \notin \phi(i(f) - \{p\})$
$RF2_o$. $f \uparrow^p_q = (f \triangleright [s/q]) \uparrow^p_s$ if $s \notin o(f)$	$RF2^i$. $f \uparrow^p_q = ([r/p] \triangleleft f) \uparrow^r_q$ if $r \notin i(f)$

R^i_o. $\phi \triangleleft (f \triangleright \psi) = (\phi \triangleleft f) \triangleright \psi$

Notation: $[q/p](x) =$ "if $x = p$ then q else x";
$\phi_{\lvert A\rvert}(x) =$ "if $x \in A$ then x else $\phi(x)$"

Table 1. Axioms for graph isomorphism

Every expression may be brought to a normal form by using the axioms in Table 1. In addition, the normal form associated to a process graph has a unique empty part α, a unique transition part β and a unique feedback part up to commutations of terms (and identification of equal terms, in the case without multiplicity), permutations of feedbacks and bijective renamings of feedback ports. This may be used to show that two normal forms that represent the same graph may be proved equivalent via the axioms. Hence we get a first axiomatisation result.

Theorem 5. *The axioms in Table 1 (resp. the axioms in Table 1 except for A(**)) are correct and complete for graph isomorphism without (rest. with) multiplicity.*

3 Process graphs modulo bisimulation

3.1 Simulation and bisimulation

Simulation is a standard notion of graph homomorphism that has been used in the study of flow diagram programs (see, e.g. [Ste87, Ste94]). Bisimulation is an equivalence on transition systems introduced by Park [Pa81] in connection with Milner's work on concurrency [Mil80]. They may be defined as follows.

Definition 6. (simulation and bisimulation) Assume two process graphs $P : I \xrightarrow{S} O$ and $P' : I \xrightarrow{S'} O$ are given. Let T and T' be the associated transition relations.
We say P and P' are *similar* via a function $\phi : S \rightarrow S'$ if

$$(Id_I \cup \phi) \circ T' = T \circ (Id_O \cup \phi)$$

We say P and P' are *strongly bisimilar* via a relation $\rho \subseteq S \times S'$ if

$$\begin{aligned}(Id_I \cup \rho) \circ T' &\subseteq T \circ (Id_O \cup \rho)\\ (Id_I \cup \rho^{-1}) \circ T &\subseteq T' \circ (Id_O \cup \rho^{-1})\end{aligned} \quad \square$$

Theorem 7. *[BeS94] Strong bisimulation is the equivalence relation generated by simulation via functions.*

3.2 Axioms, correctness, completeness

Theorem 7 suggests that in order to axiomatize strongly bisimilar processes one has to add to the axioms corresponding to graph isomorphism a new axiom corresponding to simulation via functions, i.e., for a function $\phi : S \rightarrow S'$

$$T \circ (Id_O \ \cup \ \phi) = (Id_O \ \cup \ \phi) \circ T' \quad \Rightarrow \quad T \uparrow^S_S = T' \uparrow^{S'}_{S'}$$

But which is the meaning of the composition here? Clearly, $T \circ (Id_O \ \cup \ \phi)$ may be replaced by $T \triangleright (Id_O \ \cup \ \phi)$. Our model is ϵ-free here, hence we have to define a new operation

$$\psi \triangleright P$$

which has to simulate the effect of the composition on the classes of strongly bisimilar processes.

The intuitive meaning of this operation is to model multiple incoming ϵ-edges to an entry by splitting the entry together with its outgoing transitions in order to have a copy for each incoming edge. This way left composition with functions may be defined on process graphs modulo bisimulation.

Operations (continuation):

– **Splitting input ports** $\triangleright$**:** Let $\phi : V \rightarrow V$ be a function such that $\phi^{-1}(x)$ is finite for all x. Let $P = \langle \partial_i, E, \ell, \partial_o \rangle : I \xrightarrow{S} O$ be such that $S \cap \phi^{-1}(I) = \emptyset$. Suppose $I = \{p_1, \ldots, p_n\}$ and $I_i = \{\phi^{-1}(p_i)\}$, for $i \in [n]$. Let $E_i = \{e \in E | \ \partial_i(e) = p_i\}$, for $i \in [n]$ and $E' = E - \bigcup_{i \in [n]} E_i$. Then

$$\phi \rhd P = \langle \bigcup_{i=1}^{n} pr_2^i \cup \partial_i|_{E'}, \bigcup_{i=1}^{n} E_i \times I_i \cup E', \bigcup_{i=1}^{n} pr_1^i \circ \ell \cup \ell|_{E'},$$
$$\bigcup_{i=1}^{n} pr_1^i \circ \partial_o \cup \partial_o|_{E'} \rangle : \phi^{-1}(I) \xrightarrow{S} O$$

where for $i \in [n]$, pr_1^i and pr_2^i denote the 1st and the 2nd projection of $E_i \times I_i$, respectively.

Effect: In the resulting graph there is a copy of an entry p and of its outgoing edges for each $x \in \phi^{-1}(p)$. In particular, if $\phi^{-1}(p) = \emptyset$ the effect is that p and its outgoing edges are deleted.

Note: This splitting operation may be easier defined by matrices as follows: If T is the matrix associated to P, then the matrix associated to $\phi \rhd P$ is just the matrix composition $(\phi|_{\phi^{-1}(I)} \cup Id_S) \circ T$.

The axioms for this new operation are given in Table 2. Before listing them we make a comment related to axiom SR^i: Let ϕ, $\psi : V \to V$ be two functions such that $\phi^{-1}(x)$ and $\psi^{-1}(x)$ are finite for all x. For $r \in V$ denote by pr_1^r and pr_2^r the 1st and the 2nd projection of $\phi^{-1}(r) \times \psi^{-1}(r)$, respectively. Finally, take $\psi' = \bigcup_{r \in V} pr_1^r$ and $\phi' = \bigcup_{r \in V} pr_2^r$. These functions fulfil $\phi \circ \psi^{-1} = \psi'^{-1} \circ \phi'$.

$S1.$	$\phi \rhd a_q^p = \text{if } \phi^{-1}(p) \neq \emptyset \text{ then } \oplus_{r \in \phi^{-1}(p)} a_q^r \text{ else } \top_q$
$S2.$	$\phi \rhd \bot^p = \text{if } \phi^{-1}(p) \neq \emptyset \text{ then } \oplus_{r \in \phi^{-1}(p)} \bot^r \text{ else } \emptyset$
$S3.$	$\phi \rhd \emptyset = \emptyset$
$S4.$	$\phi \rhd \top_q = \top_q$
$SA.$	$\phi \rhd (f \oplus g) = (\phi \rhd f) \oplus (\phi \rhd g)$
$SF.$	$\phi \rhd (f \uparrow_q^p) = (\phi_{\vert\{p\} \cup \phi^{-1}(p)\vert} \rhd f) \uparrow_q^p$
$SR^i.$	$\phi \rhd (\psi \triangleleft f) = \psi' \triangleleft (\phi' \rhd f)$ where ϕ', ψ' are functions such that $\phi \circ \psi^{-1} = \psi'^{-1} \circ \phi'$
$SR_o.$	$\phi \rhd (f \triangleright \psi) = (\phi \rhd f) \triangleright \psi$

Table 2. Axioms for splitting

Now the axiom corresponding to simulation has a definite phrasing:

$SIM.\ f \triangleright (Id_O \cup \phi) = (Id_I \cup \phi) \rhd f' \quad \Rightarrow \quad f \uparrow_A^A = f' \uparrow_B^B$
for two processes $f : I \cup A \to O \cup A$ and $f' : I' \cup B \to O' \cup B$ and a function $\phi : A \to B$ (A disjoint of I, O; B disjoint of I', O')

Theorem 8. *The axioms of graph isomorphism, those for splitting (in Table 2) and SIM are correct and complete for process graphs modulo bisimulation.*

An axiom scheme analogous to SIM, —sometimes known as a "functoriality axiom"—, was widely used in various axiomatizations related to flowchart schemes (see [CaS90, Ste94], for example). The present SIM axiom is related to the functoriality of functions. A weaker equational instance of the functoriality of functions is the key axiom of iteration theories and was used to get an axiomatisation of synchronization trees, cf. [BlE93].

4 Conclusions

We have presented a model of process graphs with multiple entries and exits. This model allows for a smooth integration of looping and parallel operators. Axiomatisations for such processes under graph isomorphism and strong bisimulation equivalences are given.

An extension of this model is given in [BaBS95]. There more operations and results may be found, including axiomatisation results for parallel operators.

References

[BaB95] J.C.M. Baeten and J.A. Bergstra. Graph isomorphism models for noninterleaving process algebra. In: Proceedings of ACP94 (Eds. A. Ponse, C. Verhoef and S.F.M. van Vlijmen), 299–318. Workshops in Computing, Springer–Verlag, 1995.

[BaBS95] J.C.M. Baeten, J.A. Bergstra and Gh. Stefanescu. Process algebra with feedback. In: Proceedings of the workshop: Three Days of Bisimulation (Eds. A. Ponse, M. de Rijke and Y. Venema). To appear as: CSLI volume, Stanford, 1995.

[BaW90] J. Baeten and W. Weijland. *Process algebra*. Cambridge University Press, 1990.

[BeBP94] J.A. Bergstra, I. Bethke and A. Ponse. Process algebra with iteration and nesting. *The Computer Journal*, 37:243–258, 1994.

[BeK84] J.A. Bergstra and J.W. Klop. Proces algebra for synchronous communication. *Information and Control*, 60:109–137, 1984.

[BeS94] J.A. Bergstra and Gh. Stefănescu. Bisimulation is two-way simulation. *Information Processing Letters*, 52:285–287, 1994.

[BlE93] S.L. Bloom and Z. Esik. *Iteration theories: the equational logic of iterative processes.* EATCS Monographs in Theoretical Compputer Science, Springer Verlag, 1993.

[CaS90] V.E. Căzănescu and Gh. Stefănescu. Towards a new algebraic foundation of flowchart scheme theory. *Fundamenta Informaticae*, 13:171–210, 1990.

[Elg75] C.C. Elgot. Manadic computation and iterative algebraic theories. In *Proceedings Logic Colloquium '73*, pages 175–230, North-Holland, 1975. Studies in Logic and the Foundations of Mathematics, Volume 80.

[Kle56] S.C. Kleene. Representation of events in nerve nets and finite automata. In *Automata Studies*, pages 3-41. Princeton University Press, 1956.

[Mil79] R. Milner. Flowgraphs and flow algebra. *Journal of the Association for Computing Machinery*, 26:794–818, 1979.

[Mil80] R. Milner. *A calculus of communicating systems.* LNCS 92, Springer, 1980.

[Pa81] D. Park. Concurrency and automata on infinite sequences. In *Proceedings 5th GI Conference*, 167–183. LNCS 104, Springer Verlag, 1981.

[Par93] J. Parrow. Structural and behavioural equivalence of networks. *Information and Computation*, 107:58-90, 1993.

[Ste87] Gh. Stefănescu. On flowchart theories. Part I: The deterministic case. *Journal of Computer and System Sciences*, 35:163–191, 1987.

[Ste86] Gh. Stefănescu. Feedback theories (a calculus for isomorphism classes of flowchart schemes). INCREST Preprint No. 24, Bucharest, April 1986. Also in: *Revue Roumaine de Mathematiques Pures et Applique*, 35:73–79, 1990.

[Ste94] Gh. Stefănescu. *Algebra of Flownomials. Part 1: Binary flownomials; Basic Theory.* Report TUM-I9437, Technical University Munich, 1994.

Efficient rewriting in cograph trace monoids

Michael Bertol*
Universität Stuttgart, Institut für Informatik
Breitwiesenstr. 20-22, D-70565 Stuttgart
e-mail: bertol@informatik.uni-stuttgart.de

Abstract. We consider the basic problem of finding irreducible forms w.r.t. a finite noetherian rewriting system over a free partially commutative monoid where the underlying dependence alphabet is a cograph. A linear time algorithm is developed which determines irreducible normal forms w.r.t. finite, length-reducing trace rewriting systems over cograph monoids. This generalizes well-known results for free monoids and commutative monoids and is a significant improvement to the previously known square time algorithm.

1 Introduction

Free partially commutative monoids were introduced in combinatorics by Cartier and Foata [4]. In computer science these monoids are known as trace monoids, cf. Mazurkiewicz [7]. For background material we refer to [1, 8] or [5].

The theory of rewriting over trace monoids combines combinatorial aspects from string rewriting (modulo some commutations) and graph rewriting. The restriction to traces leads to feasible algorithms, but some interesting complexity questions are still open. One of the challenging open problems is to improve the known square time bound for the non-uniform complexity of length-reducing systems. It is well-known that we can compute irreducible descendants over free monoids in linear time [3, Thm. 4.1], if the rewriting system is length-reducing. Obviously, the linear time bound holds for free commutative monoids, too. Our aim is to generalize this bound to a larger class of free partially commutative monoids.

We give a linear time algorithm for determining irreducible forms w.r.t. length-reducing rewriting systems over a natural subclass of finitely generated free partially commutative monoids, the cograph monoids. Cograph monoids are trace monoids where the underlying dependence alphabet is a cograph, i.e. the smallest class of graphs containing singletons and which is closed under finite disjoint union and complementation. From an algebraic viewpoint cograph monoids are the smallest class of monoids containing the finitely generated free and free commutative monoids and which are closed under finite free and finite direct product.

* Partially supported by the ESPRIT Basic Research Action No. 6317 ASMICS II.

These monoids provide a natural model for asynchronous computation. Especially the cograph structure models the execution of "well structured" concurrency, e.g. Dijkstra's construct "`par_begin...par_end`" and the usual sequential composition ";" in a programming language. Since there are many related problems, e.g. (parallel) programming languages that are based on rewriting, it is of interest to know efficient methods for computing irreducible forms.

In the first part of the paper we recall some background about traces, and we establish some facts about cograph monoids and decompositions. Then we develop a linear time algorithm which solves the considered problem.

2 Notations

By Σ we denote a finite *alphabet*; a *dependence alphabet* is a pair (Σ, D), where $D \subseteq \Sigma \times \Sigma$ is a reflexive and symmetric *dependence relation*. The complement $I = \Sigma \times \Sigma \setminus D$ is called the *independence relation* and the quotient monoid $\mathbb{M} = \mathbb{M}(\Sigma, D) = \Sigma^*/\{ab = ba \mid (a,b) \in I\}$ is the *free partially commutative monoid* generated by the dependence alphabet (Σ, D). An element $t \in \mathbb{M}$ is called a trace. We denote by $|t|$ the *length* of a trace t, by $|t|_a$ its *a-length*, for $a \in \Sigma$, and by $\mathrm{alph}(t) = \{a \in \Sigma \mid |t|_a \geq 1\}$ its *alphabet*. It is convenient to extend the independence relation to a relation over $\mathbb{M}$. For $u, v \in \mathbb{M}$ we define $(u,v) \in I$, if $\mathrm{alph}(u) \times \mathrm{alph}(v) \subseteq I$. We denote by $[k]$ the set $\{1, \ldots, k\}$.

A trace may be viewed as a partially ordered multiset (pomset). Let $w = a_1 \ldots a_n \in \Sigma^*$ be a word, $a_i \in \Sigma$, for all $i \in [n]$. Then the pomset of the corresponding trace $t = [w] \in \mathbb{M}$ is given by $(S_t, \sqsubset_t^*)$, where $S_t = \{a_1, \ldots, a_n\}$ is the multiset of letters occurring in w, and $\sqsubset_t^*$ is the partial order induced by $a_i \sqsubset_t a_j$ if both $i < j$ and $(a_i, a_j) \in D$. Interpreted as a directed (acyclic) graph, $(S_t, \sqsubset_t^*)$ is called the dependence graph of t. This provides a nice geometric interpretation.

A *subtrace* s of a trace t is a subgraph of the dependence graph of t such that all directed paths starting and ending in s are entirely contained in s. Every subtrace s of t is a factor, i.e., $t = usv$ (this factorization is not unique, in general). Conversely, if $t = usv$ then the traces u, s and v can be identified as subtraces of t.

Example 1. Let the dependence alphabet (Σ, D) be $(\{a,b,c\}, [a - b - c])$ and the trace $t = [acbacabca] = \{acbaacbac, acbaacbca, \ldots, cabcaabca\}$. In the graph of the trace[2] in the example we find two subgraphs that are subtraces of t representing the trace $l = [acb]$. The occurrences are shaped by polygons. The dotted polygon is a subgraph which is not a factor

$t = [acb][a][acb][ca] \cong$ 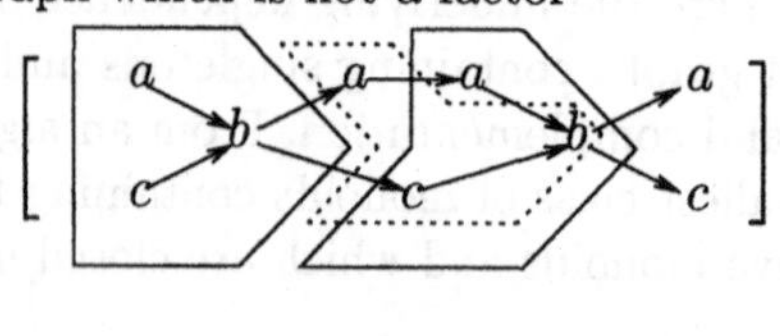

[2] We are drawing (without loss of information for fixed dependence alphabets) the Hasse diagram, only. This is the reduced graph of the partial order.

3 Cograph monoids and decompositions

We restrict our investigations to a subclass of finitely generated free partially commutative monoids, the cograph monoids.

For two disjoint dependence alphabets $(\Sigma, D_\Sigma), (\Delta, D_\Delta)$, i.e., $\Sigma \cap \Delta = \emptyset$, we define the *direct sum* by

$$(\Sigma, D_\Sigma) \oplus (\Delta, D_\Delta) := (\Sigma \cup \Delta, D_\Sigma \cup D_\Delta),$$

and the *complex product* by

$$(\Sigma, D_\Sigma) * (\Delta, D_\Delta) := (\Sigma \cup \Delta, D_\Sigma \cup D_\Delta \cup (\Sigma \times \Delta) \cup (\Delta \times \Sigma)).$$

Definition 1. Let $\mathcal{D}$ be the smallest subclass of dependence alphabets containing all one-letter dependence alphabets $(\{a\}, \{(a,a)\})$ that is closed under taking finite direct sums and complex products. The trace monoids generated by dependence alphabets from $\mathcal{D}$ form the class of *cograph (trace) monoids*, denoted by $\mathcal{M}$.

There is a well-known graph theoretical characterization of cographs, see [10], [11] or [12]. They are precisely undirected (loop-less) graphs $(\Sigma, D - \{(a,a) \mid a \in \Sigma\})$ containing no induced subgraph isomorphic to $L_3 = [a - b - c - d]$.

A *description* for a dependence alphabet (Σ, D) is a well-formed linear term built from the constants $a \in \Sigma$ (representing the one-letter dependence alphabet $(\{a\}, \{(a,a)\})$) and the binary operations $\oplus$ and $*$. We identify descriptions modulo commutativity and associativity of $\oplus$ ($*$, resp.).

For a dependence alphabet (Σ, D) we call the set of connected components its *decomposition set* if the graph (Σ, D) is not connected. If (Σ, D) is connected we call the set of complements of the connected components of the graph $(\Sigma, \Sigma \times \Sigma - D)$ the *decomposition set* of (Σ, D). Note that the decomposition set of $(\Sigma, D) \in \mathcal{D}$ contains disjoint dependence alphabets.

FACT A. Let $T = T_1 \oplus \ldots \oplus T_n$ (or $T = T_i * \ldots * T_n$, resp.), for some $n \in \mathbf{N}$, a description of a dependence alphabet $(\Sigma, D) \in \mathcal{D}$ such that each T_i has no non-trivial $\oplus$-decomposition ($*$, resp.) then the set of dependence alphabets corresponding to the descriptions T_i, $i \in [n]$, is the decomposition set.

FACT B. Let $\mathbf{M}_\Sigma = \mathbf{M}(\Sigma, D_\Sigma)$ and $\mathbf{M}_\Delta = \mathbf{M}(\Delta, D_\Delta)$ be the free partially commutative monoids generated by the disjoint dependence alphabets (Σ, D_Σ) and (Δ, D_Δ). Then

$$\mathbf{M}((\Sigma, D_\Sigma) \oplus (\Delta, D_\Delta)) \cong \mathbf{M}_\Sigma \times \mathbf{M}_\Delta$$
$$\mathbf{M}((\Sigma, D_\Sigma) * (\Delta, D_\Delta)) \cong \mathbf{M}_\Sigma * \mathbf{M}_\Delta.$$

Here $\mathbf{M}_\Sigma * \mathbf{M}_\Delta$ denotes the free product and $\mathbf{M}_\Sigma \times \mathbf{M}_\Delta$ denotes the direct product of monoids $\mathbf{M}_\Sigma$ and $\mathbf{M}_\Delta$.

Let $\mathcal{T}$ be the set of well-formed terms consisting of elements $t \in \mathbf{M}(\Sigma, D)$ and function symbols $*$ and $\times$ (of arbitrary arity). We define the following mapping inductively

$$h_{(\Sigma,D)} : \mathbf{M}(\Sigma, D) \longrightarrow \mathcal{T}$$

If (Σ, D) is not decomposable, i.e. the decomposition set contains only the element (Σ, D) then

$$h_{(\Sigma,D)} : x \longmapsto x$$

is the embedding. Otherwise, (Σ, D) is decomposable with decomposition set $\{(\Sigma_i, D_i) \mid i \in [k]\}$ $(k > 1)$, say.

(i) If the dependence alphabet (Σ, D) is the direct sum $\oplus_{i\in[k]}(\Sigma_i, D_i)$, we define

$$h_{(\Sigma,D)} : \mathbf{M}(\Sigma, D) \cong \prod_{i\in[k]} \mathbf{M}(\Sigma_i, D_i) \longrightarrow \mathcal{T}$$

$$(t_1, \ldots, t_k) \longmapsto \times(h_{(\Sigma_1,D_1)}(t_1), \ldots, h_{(\Sigma_k,D_k)}(t_h))$$

(ii) If (Σ, D) is the complex product $*_{i\in[k]}(\Sigma_i, D_i)$, we define

$$h_{(\Sigma,D)} : \mathbf{M}(\Sigma, D) \cong \mathop{*}_{i\in[k]} \mathbf{M}(\Sigma_i, D_i) \longrightarrow \mathcal{T}$$

$$t_1 \cdot \ldots \cdot t_n \longmapsto *(h_{(\Sigma_{i_1},D_{i_1})}(t_1), \ldots, h_{(\Sigma_{i_n},D_{i_n})}(t_n)),$$

where $t_1 \cdot \ldots \cdot t_n$ denotes the usual free product, i.e., $\{i_1, \ldots, i_n\} \subseteq [k]$ and $i_j \neq i_{j+1}$, for all $j \in [n-1]$, and $t_j \in \mathbf{M}(\Sigma_{i_j}, D_{i_j})$, $j \in [n]$.

Below we write h for $h_{(\Sigma,D)}$.

Remark 2. For cograph dependence alphabets $(\Sigma, D) \in \mathcal{D}$ the images $h(t)$, $t \in \mathbf{M}(\Sigma, D)$, are well-formed terms consisting of words a^n, $n \geq 0$, and the function symbols $*$ and $\times$.

Definition 3. For a (cograph) monoid $\mathbf{M}(\Sigma, D)$ the *trace term* associated with $t \in \mathbf{M}(\Sigma, D)$ is the image $h(t) \in \mathcal{T}$ by the mapping h.

We call non-decomposable dependence alphabets (Σ, D) *nuclear* (i.e. the decomposition set contains one element).

Let $T(\Sigma, D)$ be the h-image $h_{(\Sigma,D)}(\mathbf{M}(\Sigma, D)) \subset \mathcal{T}$. Then the subset $T(\Sigma, D)$ of well-formed terms forms a monoid, where 1 is the empty term and the concatenation $s \circ t$ of two trace terms s and t is defined in the following way.
if $s = \times(s_1, \ldots, s_k)$ and $t = \times(t_1, \ldots, t_k)$ then $s \circ t = \times(s_1 \circ t_1, \ldots, s_n \circ t_n)$
if $s = *(s_1, \ldots, s_m)$ and $t = *(t_1, \ldots, t_n)$ then

$$s \circ t = *(s_1, \ldots s_{m-1}, s_m \circ t_1, t_2, \ldots, t_n), \text{ when } s_m, t_1 \in T(\Sigma_i, D_i), \text{ resp.}$$

$$s \circ t = *(s_1, \ldots, s_m, t_1, \ldots, t_n), \text{ when } s_m \in T(\Sigma_i, D_i) \neq T(\Sigma_j, D_j) \ni t_1$$

Otherwise $s \circ t = st$ (when $s, t \in \mathbf{M}(\Sigma, D)$, and (Σ, D) is nuclear).

Proposition 4. *The mapping* $h : \mathbf{M}(\Sigma, D) \longrightarrow T(\Sigma, D)$ *is injective and a monoid morphism.*

The structure of these objects allows to locate factors of a trace. The following lemma provides a characterization of the factor property.

Lemma 5. *Let* $\{(\Sigma_i, D_i) \mid i \in [k]\}$ *be the decomposition set of the dependence alphabet* (Σ, D), $\mathbf{M}_i = \mathbf{M}(\Sigma_i, D_i)$, *for all* $i \in [k]$, *and let* $t, l \in \mathbf{M}(\Sigma, D)$.

(i) If $t, l \in \prod_{i\in[k]} \mathbf{M}_i$ *(direct product) then* $t = ulv$, *for some* u, v, *iff*

- $\forall i \in [k] : \pi_i(t) = \pi_i(u)\pi_i(l)\pi_i(v)$,

where π_i *is the projection monoid morphism* $\mathbf{M}(\Sigma, D) \longrightarrow \mathbf{M}_i$.

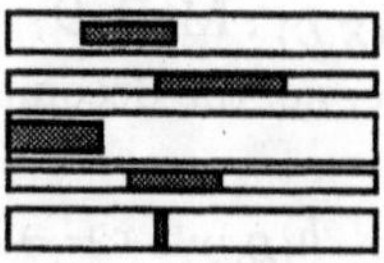

(ii) If $t, l \in *_{i\in[k]}\mathbf{M}_i$ (free product) with $t = t_1 \cdot \ldots \cdot t_n$ and $l = l_1 \cdot \ldots \cdot l_m$ then $t = ulv$, for some u, v, iff either
- $\exists j \in \{0\} \cup [n] \,:\, t_{j+2} = l_2 \wedge \ldots \wedge t_{j+m-1} = l_{m-1}$, and $\exists x, y \,:\, xl_1 = t_{j+1} \wedge l_m y = t_{j+m}$, or
- $\exists i \in [k] \,:\, l \in \mathbf{M}_i$ *with* $t_j = xly$ *for some* $x, y \in \mathbf{M}_i$ *and* $j \in [n]$.

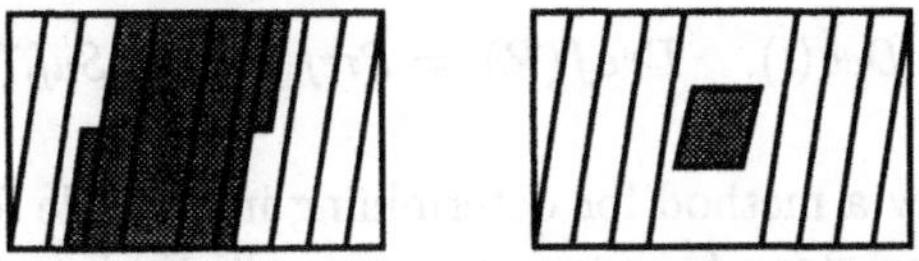

The grey rectangles depict the factor l, the diagonal lines free products, the horizontal lines direct products, and the envelopping square is the shape of the dependence graph of the trace t.

FACT C. Deciding whether two traces t and l are equal, l is a prefix of t, and l is a suffix of t is possible in $\mathcal{O}(\min(|t|, |l|))$ time.

4 Trace rewriting systems

In this section we introduce trace rewriting systems.

Definition 6. A trace-rewriting system is a finite set of rewriting rules $\mathcal{R} \subseteq \mathbf{M}(\Sigma, D) \times \mathbf{M}(\Sigma, D)$.

A trace-rewriting system $\mathcal{R}$ defines a *derivation relation* $\Longrightarrow_{\mathcal{R}}$ by $x \Longrightarrow_{\mathcal{R}} y$ if $x = ulv, y = urv$ for some $u, v \in \mathbf{M}$ and $l \to r \in \mathcal{R}$. By $\stackrel{*}{\Longrightarrow}_{\mathcal{R}}$ we denote the reflexive transitive closure of $\Longrightarrow_{\mathcal{R}}$. A trace-rewriting system $\mathcal{R}$ is called *length-reducing*, if $|l| > |r|$ for all $l \to r \in \mathcal{R}$. The set of *irreducible traces* is given by $\mathrm{Irr}(\mathcal{R}) = \{t \in \mathbf{M} \mid \neg\exists s : t \Longrightarrow_{\mathcal{R}} s\}$.

The following proposition is well-known, see e.g. [6, p.138, Thm. 5.7.1]. The idea is that $\mathrm{Red}(\mathcal{R}) = \mathbf{M} \setminus \mathrm{Irr}(\mathcal{R})$ is a recognizable set. Hence, if $v \in \mathrm{Red}(\mathcal{R})$, then this can be detected by some finite state control and we can also compute a factorization $v = v'lv''$ for some $l \to r \in \mathcal{R}$ via protocols. The rewriting step is then to define $v := v'rv''$. By iterating these two steps we obtain a correct algorithm, but its worst-case behavior is square time. No better time bound is known, in general.

Proposition 7. *Let $\mathcal{R} \subseteq \mathbf{M} \times \mathbf{M}$ be a finite and length-reducing trace rewriting system. Then irreducible forms w.r.t. $\mathcal{R}$ can be computed in square time.*

5 The rewriting algorithm for cographs

In this section we show our main result which is an improvement of Proposition 7. In the following let $\mathcal{R} \subseteq \mathbf{M} \times \mathbf{M}$ be a finite and length-reducing trace rewriting system such that each right hand side is irreducible. By abuse of language we identify the various representations of a trace.

Let us denote the set of prefixes (suffixes resp.) of a trace $t \in \mathbf{M}$ by $\mathit{Pref}(t) := \{\alpha \in \mathbf{M} \mid \alpha\beta = t\}$ ($\mathit{Suff}(t) := \{\beta \in \mathbf{M} \mid \alpha\beta = t\}$, resp.). For a set $S \subseteq \mathbf{M}$ we define $\mathit{Pref}(S) := \bigcup_{s\in S} \mathit{Pref}(s)$ and $\mathit{Suff}(S) := \bigcup_{s\in S} \mathit{Suff}(s)$.

Let $L_{\mathcal{R}}$ denote the set of left hand sides $\{l \mid (l,r) \in \mathcal{R}\}$ of a rewriting system $\mathcal{R} \subseteq \mathbf{M} \times \mathbf{M}$. We denote by $Dec(l) = Dec(h(l))$ the *decomposition of the trace* $l \in L_{\mathcal{R}}$, with $Dec(\times(l_1,\ldots,l_k)) := \bigcup_{i\in[k]} Dec(l_i)$, and $Dec(*(l_1,\ldots,l_n)) := \{l_1 \cdot \ldots \cdot l_n\}$, $n > 1$; $Dec(*(l_1)) := Dec(l_1)$ (is considered as a direct product). Otherwise, for nuclear monoids, $Dec(l) = \{l\}$. Further we define

$$Dec(\mathcal{R}) := \bigcup_{l\in L_{\mathcal{R}}} Dec(l), \quad Pref(\mathcal{R}) := Pref(L_{\mathcal{R}}), \quad Suff(\mathcal{R}) := Suff(L_{\mathcal{R}}).$$

We describe now a method for determining irreducible forms by performing iteratively derivation steps $t = ulv \Longrightarrow_{\mathcal{R}} urv = t'$. Each trace is represented by a tree corresponding to its trace term. Each subterm $h(s)$ corresponding to the trace s (represented by its subtree) is additionally labeled by the sets

$$P(s) = Suff(s) \cap Pref(\mathcal{R}) \quad \text{and} \quad S(s) = Pref(s) \cap Suff(\mathcal{R}).$$

Further, each subterm s is labeled by $L(s)$ where,

(i) if $s = *(s_1,\ldots,s_n)$ then $L(s) = (L_d(s))_{d\in \mathrm{Dec}(\mathcal{R})}$, with $L_d(s)$ is an ordered set of integers $\{i_1,\ldots,i_{n_d}\} \subseteq [n]$ such that $i_j \in L_d(s)$ iff either

$$udv = ud_1 \cdot \ldots \cdot d_k v = s_{i_j} \cdot \ldots \cdot s_{i_j+k-1}, \; k > 1,$$

for some $u, v \in \mathbf{M}$, $j \in [n_d]$, or $i_j \in L_d(s)$ if $d \in L(s_{i_j})$.

(ii) If $s = \times(s_1,\ldots,s_k)$ then the term is labeled by the set of traces $L(s) = \bigcup_{i\in[k]}\{d \in Dec(\mathcal{R}) \mid L_d(s_i) \neq \emptyset\}$.

(iii) If $s = a^n, n > 0$ then $L_d(s) = \emptyset$, for $d \notin a^* \cap Dec(\mathcal{R})$, resp. $L_d(s) = \{j \mid 0 < j \leq n - |d|\}$, for $d \in a^* \cap Dec(\mathcal{R})$.

Applying Fact C iteratively for determining the labels $P(s)$ and $S(s)$ of each subterm s of t we can use this information to determine the L-labeling bottom-up by propagating the information on the path to the root of the tree representing the term. Hence this data structure can be constructed in linear time.

We introduce the following functions and give an implementation of two of them. (Each of the functions maintains the labeling correctly.)

(i) SPLIT($i, *(t_1,\ldots,t_n)$) — split the term $*(t_1,\ldots,t_n)$ into the two terms $*(t_1,\ldots,t_{i-1})$ and $*(t_i,\ldots,t_n)$;

(ii) PROJECT($i, \times(t_1,\ldots,t_k)$) — output the i-th projection t_i of the trace $\times(t_1,\ldots,t_k)$;

(iii) COMPOSE($t_1,\ldots,t_k$) — construct the trace term $\times(t_1,\ldots,t_k)$;

(iv) CONCAT(t_1, t_2) — concatenate two trace terms t_1 and t_2 such that the result represents $t_1 \circ t_2$.

Because of space limitations we only show two of these methods where some trivial cases are omitted.

FUNCTION SPLIT($i, *(t_1,\ldots,t_i,\ldots,t_n)$)
  if $(1 < i < n)$ **then** $\langle t'_1, t'_2\rangle \leftarrow \langle *(t_1,\ldots,t_{i-1}), *(t_i,\cdot\ldots\cdot,t_n)\rangle$
    $P(t'_1) \leftarrow P(t_{i-1}) \cup \{l_1 \cdot \ldots \cdot l_j \in Pref(\mathcal{R}) \mid t_1 \cdot \ldots \cdot t_{i-1} = \alpha l_1 \cdot \ldots \cdot l_j, j > 1\}$
    $P(t'_2) \leftarrow P(t_n) \cup \{l_1 \cdot \ldots \cdot l_j \in Pref(\mathcal{R}) \mid t_i \cdot \ldots \cdot t_n = \alpha l_1 \cdot \ldots \cdot l_j, j > 1\}$
    $S(t'_1) \leftarrow S(t_1) \cup \{l_1 \cdot \ldots \cdot l_j \in Suff(\mathcal{R}) \mid t_1 \cdot \ldots \cdot t_{i-1} = l_1 \cdot \ldots \cdot l_j\beta, j > 1\}$
    $S(t'_2) \leftarrow S(t_i) \cup \{l_1 \cdot \ldots \cdot l_j \in Suff(\mathcal{R}) \mid t_i \cdot \ldots \cdot t_n = l_1 \cdot \ldots \cdot l_j\beta, j > 1\}$

forall $d \in Dec(\mathcal{R})$ **do**
$L_d(t'_1) \leftarrow (L_d(t) \cap [i-1]) - \{i-j \mid d_1 = d'_1 \cdot \ldots \cdot d'_j \in P(t_{i-1}),$
$d_2 \in S(t_i), d_1 d_2 \in Dec(\mathcal{R})\}$
$L_d(t'_2) \leftarrow L_d(t) \cap \{i, \ldots, n\}$
return $(\langle t'_1, t'_2 \rangle$ with labeling S, P, L, resp.)
else if $i = 1$ **then return**$(\langle 1, *(t_1, \cdot \ldots \cdot, t_n)\rangle)$
else if $i = n$ **then return**$(\langle *(t_1, \cdot \ldots \cdot, t_n), 1\rangle)$
otherwise error

The function CONCAT uses the previous functions. Note that each operation maps consistent data structure to consistent structure.

FUNCTION CONCAT(t_1, t_2)
if $t_1 = \times(t'_1, \ldots, t'_k)$ and $t_2 = \times(t''_1, \ldots, t''_k)$
then forall $(1 \le i \le k)$ **do**
$\langle t^1_i, t^2_i \rangle \leftarrow \langle$ PROJECT(i, t_1), PROJECT$(i, t_2)\rangle$
$t'''_i \leftarrow$CONCAT(t^1_i, t^2_i)
return (COMPOSE$(t'''_1, \ldots, t'''_k)$)
if $t_1 = *(t'_1, \ldots, t'_{n_1})$ and $t_2 = *(t''_1, \ldots, t''_{n_2})$ **then**
if $t'_{n_1} \in \mathbb{M}_i \neq \mathbb{M}_j \ni t''_1$, where $\mathbb{M} = *_{i\in[k]}\mathbb{M}_i$
then $s \leftarrow *(t'_1, \ldots, t'_{n_1}, t''_1, \ldots, t''_{n_2})$
else $\langle s'_1, s'_2 \rangle \leftarrow$SPLIT$(n_1, t_1)$ $\langle s''_1, s''_2 \rangle \leftarrow$SPLIT$(2, t_2)$
$s \leftarrow$CONCAT$(s'_1$,CONCAT(CONCAT$(s'_2, s''_1), s''_2))$
let $s = *(s_1, \ldots, s_n)$
$P(s) \leftarrow P(s_n) \cup \{l_1 \cdot \ldots \cdot l_j \in Pref(\mathcal{R}) \mid s_1 \cdot \ldots \cdot s_n = \alpha l_1 \cdot \ldots \cdot l_j, j > 1\}$
$S(s) \leftarrow S(s_1) \cup \{l_1 \cdot \ldots \cdot l_j \in Suff(\mathcal{R}) \mid s_1 \cdot \ldots \cdot s_n = l_1 \cdot \ldots \cdot l_j \beta, j > 1\}$
forall $d \in Dec(\mathcal{R})$ **do**
if there exists $i \in [k]$ with $L_d(s_i) \neq \emptyset$ **then** $L(s) \leftarrow L(s) \cup \{d\}$
$L(s) \leftarrow L(s) \cup \{d_1 d_2 \in Dec(\mathcal{R}) \mid d_1 \in P(t_1), d_2 \in S(t_2)\}$
return s with labeling P, S, L
else ... trivial (base) cases ...

Further each of the four operations has constant time complexity, because the relabeling is (partially) done by inheriting the labeling information and the remaining sets can be calculated locally, hence in constant time (use Fact C). Recall the monoid description and therefore the depth $d(h(t))$ of the term corresponding to the trace t is bounded by the cardinality $|\Sigma|$. Here the depth $d(w) = 1$ for a term $w \in a^*$, and $d(*(t_1, \ldots, t_n)) = d(\times(t_1, \ldots, t_n)) = 1 + \max_{i\in[n]} d(t_i)$.

Suppose we are given three traces $u, v, t \in \mathbb{M}$ with $t = uv$. Then the *right quotient* tv^{-1} is u, and the *left quotient* $u^{-1}t$ is v, respectively.

With the operations presented above we are able to describe a method for calculating a right quotient and left quotient respectively, that is running in linear time and maintains the described data structure. Since we are interested in rewriting of left hand sides it is sufficient considering only quotients tp^{-1} for $p \in Pref(\mathcal{R})$, and $s^{-1}t$ for $s \in Suff(\mathcal{R})$. Again, these procedures run in constant time.

FUNCTION rightQUOTIENT(t, l)
if $l \notin P(t)$ **then error**

else if $t = \times(t_1, \ldots, t_k)$ and $l = \times(l_1, \ldots, l_k)$
  then for all $(1 \leq i \leq k)$ **do**
    $s_i \leftarrow$rightQUOTIENT(PROJECT(i,t),PROJECT(i,l))
   return (COMPOSE$(s_1, \ldots, s_k)$)
else if $t = *(t_1, \ldots, t_n)$ and $l = *(l_1, \ldots, l_m)$
  then if $m = 1$ **then**
    $\langle s_1, s_2 \rangle \leftarrow$SPLIT$(n, t)$
    return CONCAT(s_1,rightQUOTIENT(s_2, l_1))
  else $\langle s, t' \rangle \leftarrow$ SPLIT$(n - m + 1, t)$
   $\langle l', l'' \rangle \leftarrow$ SPLIT$(2, l)$
   return (rightQUOTIENT(s, l'))
else ... trivial (base) cases ...

Note that all functions return on consistent data consistent results, i.e. the interpretation of the resulting term(s) as trace reflects the semantic of the used transformation and the labeling corresponds to the definition of S, P and L.

In the following we may assume that the rewriting system is reduced. In particular, we may assume that all right-hand sides are irreducible. Now we are able to describe an algorithm solving the considered non-uniform normal form problem.

In the method described below, we distinguish between three cases: the base case, when the underlying monoid is nuclear, and the two cases when the monoid is a direct or a free product. In the free product case and the base case we proceed like in Book's algorithm, overlapping factors that occur by concatenation are eliminated. In the direct product case there might arise combinations of factors, i.e., in each projection t_i might occur a factor $d_i \in Dec(\mathcal{R})$, such that $(d_1, \ldots, d_k)$ is a left hand side. Therefore we introduced the labeling L. To reference these factors we use another procedure called PART. PART(t, d) returns for a decomposition $d \in Dec(\mathcal{R})$ a factorization $t = uv$ such that for $d = d_1 \cdot \ldots \cdot d_m$: $u = \tilde{u}d_1$ and $v = d_2 \cdots d_n\tilde{v}$, with $|u|$ minimal.

PROCEDURE PART(t, d)
 if $t = *(t_1, \ldots, t_n)$ and $L_d(t) \neq \emptyset$ **then**
  let $j \leftarrow \min L_d(t)$
  $\langle t', t'' \rangle \leftarrow$SPLIT$(t, j + 1)$
  if $d = *(d_1, \ldots d_m)$ $(m > 1)$ **then return**$\langle t', t'' \rangle$
  if $d = *(d_1)$ **then** $\langle u, v \rangle \leftarrow$SPLIT$(t', j)$)
   $\langle x, y \rangle \leftarrow$PART$(v, d)$
   return($\langle$CONCAT(u, x),CONCAT$(y, t'')\rangle$)
 else if $t = \times(t_1, \ldots, t_k)$ and $d \in L(t)$ **then**
  let $j \in [k]$ with $L_d(t_j) \neq \emptyset$
  forall $(1 \leq i \leq k)$ **do** $t_i \leftarrow$PROJECT(t, i)
  $\langle u, v \rangle \leftarrow$PART$(t_j, d)$
  return($\langle$COMPOSE$(1, \ldots, u, \ldots, 1)$,COMPOSE$(t_1, \ldots, v, \ldots, t_k)\rangle$)
 else if $t, d \in a^*$ and $|t| \geq |d|$ **then return**$(\langle a, a^{|t|-1} \rangle)$
 otherwise error

Using the labeling L extensively this procedure works in constant time and returns consistent data structures (by the same arguments proving the correctness and complexity of e.g. SPLIT).

PROCEDURE CONCATandREDUCE(s, t)
  if $s = *(s_1, \ldots, s_m)$ and $t = *(t_1, \ldots, t_n)$ **then**
    if exists $l = l_1 l_2 \to r \in \mathcal{R}$ with $l_1 \in S(s_m)$ and $l_2 \in P(t_1)$ **then**
      $s' \leftarrow$rightQUOTIENT(s, l_1); $t' \leftarrow$leftQUOTIENT(t, l_2)
      return(CONCATandREDUCE(CONCATandREDUCE(s', r), t'))
    else return(CONCAT(s, t))
  else if $s = \times(s_1, \ldots, s_k)$ and $t = \times(t_1, \ldots, t_k)$ **then**
    $v \leftarrow$CONCAT(s, t)
    for all $(1 \leq i \leq k)$ **do** $u_i \leftarrow 1$; $v_i \leftarrow$PROJECT(i, v)
    while there exists $l = (l_1, \ldots, l_k) \in L_{\mathcal{R}}$ such that $l_j \in L(s) \cup L(t)$,
        for all $j \in [k]$, and $l \to r \in \mathcal{R}$,
        let l be chosen such that x is irreducible in $v = xly$ **do**
      forall $(1 \leq i \leq k)$ **do** $\langle v_i', v_i'' \rangle \leftarrow$PART$(v_i, l_i)$
          let $l_i = l_i^1 l_i^2$ with $l_i^1 \in P(v_i')$ and $l_i^2 \in S(v_i'')$
          $u_i \leftarrow$CONCAT(u_i,rightQUOTIENT(v_i', l_i^1))
          $v_i \leftarrow$leftQUOTIENT(l_i^2, v_i'')
      $u \leftarrow$CONCATandREDUCE(COMPOSE$(u_1, \ldots, u_k), r$)
      $v \leftarrow$COMPOSE$(v_1, \ldots, v_k)$
    end while
    return CONCAT(u, v)
  else if $st = a^n$ and there exists $a^m \to r \in \mathcal{R}$ and $m \leq n$ **then**
    return(CONCATandREDUCE(r, a^{n-m}))
  otherwise return CONCAT(s, t)

Suppose procedure CONCATandREDUCE(s, t) is called on a consistent data structure, i.e. s and t are representations of trace terms with correct labelings, then the procedure has the following invariants.

(i) When leaving the procedure CONCATandREDUCE the data structure is consistent.
(ii) CONCATandREDUCE(s, t) determines a descendant $\hat{t}$ of the trace st, i.e. $st \overset{*}{\Longrightarrow}_{\mathcal{R}} \hat{t}$. Moreover, if s and t are irreducible then $\hat{t}$ is irreducible, too.

Theorem 8. *Let $\mathcal{R} \in \mathbb{M}^2$ be a length-reducing rewriting system over some co-graph monoid $\mathbb{M}$. Further let $t \in \mathbb{M}$ be a trace. Determining an irreducible form of t is solvable in linear time.*

Proof. Without loss of generality $\mathcal{R}$ is normalized, i.e., each right hand side is irreducible and no single letter. Let $t = [a_1 \ldots a_n] \in \mathbb{M}(\Sigma, D)$ be the input trace. We call the routine by "**for** $i = 1$ **to** n **do** $v \leftarrow$CONCATandREDUCE(v, a_i)" with $v = 1$ initially, and return the last value of v. Let $\Psi : \mathbb{M}(\Sigma, D) \longrightarrow \mathbb{N}$ be the linear potential function defined by

$$\Psi(t) = c \cdot |h(t)| + \sum_{t' \text{ is subterm of } h(t)} |P(t')| + |S(t')| + |L(t')|,$$

where $|h(t)|$ is the number of vertices. There exists a constant $c > 0$ (depending on (Σ, D) and $\mathcal{R}$), such that $\Psi(v[a_i \dots a_n])$ is strictly decreasing while performing reduction steps (i.e. doing constant many of the primitive operations for performing one derivation step). Since $\Psi(t) \geq 0$ and $\psi(t) \in \mathcal{O}(|t|)$ we get the result. □

6 Conclusion and further work

We have established a method for solving the non-uniform normal form problem for finite length-reducing trace rewriting systems over cograph monoids working in linear time. We conjecture that it is possible to solve this problem for arbitrary free partially commutative monoids in $\mathcal{O}(n \log n)$ where n denotes the input size, but we doubt that there exists a linear time algorithm, in general.

Acknowledgment: Discussions with Dan Teodosiu motivated me to investigate these structures. I would like to thank Volker Diekert, Anca Muscholl and Zoltán Ésik as well as the anonymous referees for comments and suggestions.

References

1. I.J. Aalbersberg and G. Rozenberg. Theory of traces. *Theoretical Computer Science*, 60:1–82, 1988.
2. A.V. Aho *Algorithms for finding patterns in strings* Handbook of Theoretical Computer Science, Vol. A, J. van Leeuwen, Elsevier 1990
3. R. Book. *Confluent and other types of Thue systems. J. Assoc. Comput. Mach.*, 29:171–182, 1982.
4. P. Cartier and D. Foata. *Problèmes combinatoires de commutation et réarrangements. Lecture Notes in Mathematics 85.* Springer, 1969.
5. V. Diekert and G. Rozenberg. *The Book of Traces. World Scientific*, 1995.
6. V. Diekert. *Combinatorics on Traces.* Number 454 in Lecture Notes in Computer Science. Springer, 1990.
7. A. Mazurkiewicz. Concurrent program schemes and their interpretations. DAIMI Rep. PB 78, Aarhus University, Aarhus, 1977.
8. A. Mazurkiewicz. Trace theory. In W. Brauer et al., editors, *Petri Nets, Applications and Relationship to other Models of Concurrency*, number 255 in Lecture Notes in Computer Science, pages 279–324, Springer, 1987.
9. Vaughan Pratt. Modeling concurrency with partial orders. *International Journal of Parallel Processing*, 15(1):33–71, 1986.
10. Jan Grabowski. On partial languages. *Fundamenta Informatica*, IV(2):427–498, 1981.
11. J. Valdes, R. E. Tarjan, and E. L. Lawler. The recognition of series-parallel digraphs. *SIAM Journal of Computing*, 11(2):298–313, 1981.
12. E.S. Wolk. A note on "The comparabitily graph of a tree". *Amer. Math. Soc.*, 16:17–22, 1965.

Effective Category and Measure in Abstract Complexity Theory

(Extended Abstract)*†

Cristian Calude‡ and Marius Zimand§

Abstract

Strong variants of the Operator Speed-up Theorem, Operator Gap Theorem and Compression Theorem are obtained using an effective version of Baire Category Theorem. It is also shown that all complexity classes of recursive predicates have effective measure zero in the space of recursive predicates and, on the other hand, the class of predicates with almost everywhere complexity above an arbitrary recursive threshold has recursive measure one in the class of recursive predicates.

Keywords: Complexity measure, Operator Speed-up Theorem, Operator Gap Theorem, Compression Theorem, effective Baire classification, effective measure.

1 Introduction

The abstract complexity theory initiated by Blum [2] (see also Bridges [5], Calude [8], Hartmanis and Hopcroft [17], Machtey and Young [23], Seiferas [34]) has revealed fundamental properties of complexity measures. The striking importance of this theory relies in its machine-independent nature. Indeed, the theory is built on just two axioms (Blum axioms) and virtually any conceivable realistic model of computation is bound to satisfy the axioms. Therefore, the major achievements of this theory are primordial facts upon which any theory of complexity on any concrete model is based. The most important results of the abstract complexity theory are of existential type: a) the existence of speedable functions (Blum [2]), b) the existence of computable functions having an

*The work of the first author has been supported by Auckland University Research Grants A18/XXXXX/62090/3414012, A18/XXXXX/62090/F3414022. The second author has been partially supported by grants NSF-CCR-8957604, NSF-INT-9116781/JSPS-ENG-207 and NSF-CCR-9322513 and by the Romanian Department of Education and Science grant 4975-92.

†A full version of this paper will appear in *Theoret. Comput. Sc.* (1996).

‡Computer Science Department, The University of Auckland, Private Bag 92109, Auckland, New Zealand; cristian@cs.auckland.ac.nz.

§Department of Computer Science, University of Rochester, New York, 14627, USA; zimand@cs.rochester.edu.

arbitrary high complexity (Blum [2]), c) the existence of arbitrarily large gaps in the complexity of computable functions (Borodin [4], Trakhtenbrot [36]), d) the existence of a function which defines a complexity class as the union of an r.e. set of complexity classes with bounds satisfying a certain monotonicity condition (Union Theorem, McCreight and Meyer [28]). It is natural to enquiry how typical is the existence of such an object, is it an accident or, in fact, such objects are abounding? The general mathematical practice provides two basic lines for attacking such questions: a topological one using the notion of Baire category and the measure-theoretical one. In this paper, we mainly use the effective Baire category to investigate the size of the class of functions admitting an operator speed-up, the size of the class of functions defining the gap in the Operator Gap Theorem and the size of the class of functions that are almost everywhere hard to compute. This approach has already been followed in abstract complexity theory (Mehlhorn [25], Calude [7, 8, 9], Calude, Istrate, and Zimand [10], Calude and Zimand [12]) and more recently in structural complexity (Lutz [19], Mayordomo [24], Fenner [14], Zimand [39, 40]). Our results may be viewed as strengthenings of the original basic theorems since we show that there are *many* functions verifying them. For example we show that the classes of functions satisfying the Operator Speed-up Theorem and respectively the Operator Gap Theorem are of second category. It is also interesting to note that some topological properties are essential for achieving computational properties. Thus, only the classes of functions s that are meager guarantee that the increase of computational resources from $s(x)$ to $g(x, s(x))$ for any fixed recursive g leads to increased computational power. Also, topological considerations enhance our understanding on the difference between functions that generate complexity gaps and honest functions. Some of the results have implications beyond the topological point of view. For example, we easily derive that given any formal sound system $\mathcal{T}$, there are speedable functions f such that for all machines M computing f, the statement "the function computed by M is speedable" is not provable in $\mathcal{T}$. Similar results are valid for almost everywhere hard functions and for functions generating complexity gaps. In a separate section we also consider the measure-theoretical approach. This is an extremely active topic in structural complexity theory (see the survey of Lutz [21]) and we notice that it can be used in abstract complexity theory too. We classify from the effective measure-theoretical point of view all classes of predicates that were analyzed in the topological setting in the early works of Mehlhorn [25] and Calude [7]. The proofs are not deep and they closely parallel the proofs of similar results in [20] and [24]; our point is that such a line of investigation is feasible and should be pursued in attacking more delicate questions in abstract complexity theory. Table 1 summarizes the currently known results and marks the contribution of this work.

Object	Category	Measure	Where
complexity class	I	0	Category: [25] Measure: here
a.e. complex functions	II	1	Category: here Measure: here
i.o. speedable functions	II	?	[10]
a.e. speedable functions	II	?	[12]
operator speedable functions	II	?	here
functions yielding gaps	II	?	[12]
functions yielding operator gaps	II	?	here
measured set of functions	I	0	Category: [7] Measure: here
r-honest functions	I	-	[25]

Table 1: **The current state of affairs**. Notes: 1. Definitions of objects are provided in the paper. 2. Measure refers to effective measure in the class of recursive predicates. 3. Category refers to effective Baire category. 4. "-" marks incompatibility. 5. "?" indicates an open problem.

2 Notation

We next describe our notations and the definitions related to the category approach. The definitions afferent to the measure-theoretical approach are deferred to Section 5. We shall assume familiarity with, or access to, Bridges [5], Calude [8], Hartmanis and Hopcroft [17], Machtey and Young [23], Seiferas [34], Young [38].

Let $\mathbf{N} = \{0, 1, \ldots\}$ be the set of naturals and let $(\varphi_i)_{i\in\mathbf{N}}$ be an acceptable gödelization of **PR**, the set of unary partial recursive (p.r.) functions from **N** to **N**. Denote by **R** the class of recursive functions and by **RPRED** the class of recursive predicates, i.e. the class of functions in **R** that are $\{0,1\}$-valued. For $\varphi \in \mathbf{PR}$ we put $dom(\varphi) = \{x \in \mathbf{N} \mid \varphi(x) \text{ is defined}\}$. In what follows the term "recursive function" will always refer to a unary recursive function. Let $< \cdot, \cdot >: \mathbf{N}^2 \to \mathbf{N}$ be a fixed pairing function. The set **FR** of p.r. functions whose domain is a finite initial segment of **N** is recursive, *a fortiori* recursively enumerable (r.e.) and we fix an enumeration $(\alpha_i)_{i\in\mathbf{N}}$ of **FR**.

For α in **FR**, we define the length of α by $|\alpha| = 1 + \max\{n \in \mathbf{N} \mid n \in dom(\alpha)\}$. We often consider α in **FR** as being a finite string, where the ith bit of α is $\alpha(i-1)$. We denote the set of such strings by $N^{<\omega}$. If $\alpha, \beta \in \mathbf{N}^{<\omega}$, then $\alpha\beta$ denotes their concatenation. For $f, g \in \mathbf{PR}$, we write $f \sqsubseteq g$ in case $dom(f) \subseteq dom(g)$ and $f(x) = g(x)$, for every x in $dom(f)$. For every $t \in \mathbf{FR}$, put $\mathcal{U}_t \equiv \{f \in \mathbf{PR} \mid t \sqsubseteq f\}$. The family $(\mathcal{U}_t)_{t\in\mathbf{FR}}$ is a system of basic neighborhoods in **PR**; we work with the topology generated by this system (see Calude [7], Mehlhorn [25], Rogers [31], [19]). In the classical framework, a set A in a topological space is *nowhere dense* (or *rare*) if for every open set $\mathcal{O}$ there exists an open subset $\mathcal{O}' \subseteq \mathcal{O}$ such that $\mathcal{O}' \cap A = \emptyset$. A set is *meager* (or of *first Baire category*) if it is a finite or denumerable union of nowhere dense sets, and it is of the *second Baire category* if it is not *meager*. In the effective variant of these notions, there exists a recursive function f which for every basic open set $\mathcal{U}_t$ produces a witness $f(t)$ which indicates the basic open set $\mathcal{U}_{f(t)}$ which is disjoint from the nowhere dense set. This ideas lead to the following definition.

Definition 2.1. *1) A set* $X \subseteq \mathbf{PR}$ *is* **recursively nowhere dense** *if there exists a recursive function* f*, called the* **witness function**, *such that:*

i) $\alpha_n \sqsubseteq \alpha_{f(n)}$, *for all* $n \in \mathbf{N}$,

ii) There exists a natural $j \in \mathbf{N}$ *such that for all natural* $n, |\alpha_n| > j$ *implies* $X \cap \mathcal{U}_{\alpha_{f(n)}} = \emptyset$.

2) A set $X \subseteq \mathbf{PR}$ *is* **recursively meager** *(or* **recursively of first Baire category***) if there exist a sequence of sets* $(X_i)_{i\in\mathbf{N}}$ *and a recursive function* f *such that:*

i) $X = \bigcup_{i\in\mathbf{N}} X_i$,

and for all $i \in \mathbf{N}$:

ii) $\alpha_n \sqsubseteq \alpha_{f(<i,n>)}$, *for all* n,

iii) there exists a natural j *such that for all* $n, |\alpha_n| > j$ *implies* $X_i \cap \mathcal{U}_{\alpha_{f(<i,n>)}} = \emptyset$.

3) A set $X \subseteq \mathbf{PR}$ *is a set of* **recursively second Baire category** *if* X *is not a set of recursively first Baire category.*

For conciseness, we drop most of the times the word **recursively** in the above terminology, as well as the name of the originator of this topological classification, René Baire.

One can easily observe that the extensions $\sqsubseteq$ in the above definition can be taken to be proper ($\sqsubset$) and this will be the case in all our further considerations. The above definition can be stated in terms of the relativized topology of p.r. predicates, i.e. $\{0,1\}$-valued functions, by simply considering that $(\alpha_n)_{n\in\mathbf{N}}$ enumerates **FPRED**, the set of all p.r. predicates having the domain equal to a finite initial segment of $\mathbf{N}$. In this case, the topology is generated by the basic open sets $(\mathcal{U}_t)_{t\in\mathbf{FPRED}}$, where $\mathcal{U}_t = \{f \mid t \subset f, f \text{ is a p.r. predicate}\}$. This abuse of notation will always be clarified by context. See Calude [9] for a general treatment.

A Blum space (see [2]) is a pair $((\varphi_i)_{i\in\mathbf{N}}, (\Phi_i)_{i\in\mathbf{N}})$ where $(\varphi_i)_{i\in\mathbf{N}}$ is an acceptable gödelization of $\mathbf{PR}$ and $(\Phi_i)_{i\in\mathbf{N}}$ is a sequence of p.r. functions (called the *measure complexity* functions) satisfying the following two axioms (called *Blum axioms*): i) $dom(\varphi_i) = dom(\Phi_i)$, for all $i \in \mathbf{N}$, and ii) the ternary predicate $\mathbf{cost}(i,x,y) = 1$, if $\Phi_i(x) \leq y$, and $\mathbf{cost}(i,x,y) = 0$, otherwise, is recursive.

Here, as well as in the rest of the paper, we use the following conventions. If $\Phi \in \mathbf{PR}$ and $x \in \mathbf{N}$ are such that the $\Phi(x)$ is undefined, we write $\Phi(x) = \infty$ and we consider $\infty > y$ for all $y \in \mathbf{N}$. In the sequel we fix a Blum space $\Phi = ((\varphi_i)_{i\in\mathbf{N}}, (\Phi_i)_{i\in\mathbf{N}})$. If g is a recursive function, then the set $C_g^\Phi = \{f \in \mathbf{R} \mid \text{there exists } i \text{ such that } \varphi_i = f, \Phi_i(x) < g(x) \text{ a.e. } x\}$ is called the *complexity class* defined by g. If $P(x,n)$ is a predicate, then we write "$P(x,n)$ a.e. n" in case $P(x,n)$ holds true for all $x \in \mathbf{N}$ and for all but a finite set of $n \in \mathbf{N}$; similarly, "$P(x,n)$ i.o. n" means that $P(x,n)$ holds true for all $x \in \mathbf{N}$ and an infinity of $n \in \mathbf{N}$. An operator $F : \mathbf{PR} \overset{o}{\to} \mathbf{PR}$ is called *effective* if there exists a p.r. function $\psi : \mathbf{N} \overset{o}{\to} \mathbf{N}$ such that for every φ_i in the domain of F, $\psi(i)$ is defined and $F(\varphi_i)(x) = \varphi_{\psi(i)}(x)$ for every x in $\mathbf{N}$ (the notation $\overset{o}{\to}$ is used for partial mappings). The operator is *total* if it is defined on every recursive function and it preserves total recursiveness (i.e. recursive functions are mapped to recursive functions).

3 The Speed-Up Phenomenon

In this section we analyze, from a topological point of view the Operator Speed-up Theorem in its strongest form which involves p.r. predicates. Consequently, in this section we consider the topology generated by $(\mathcal{U}_t)_{t\in\mathbf{FPRED}}$. It is easy to derive all results in this section for the case of functions taking arbitrary integer values. For a total effective operator F, let $SPEED(\Phi, F)$ denote the class of recursive functions having F-speedup almost everywhere. More precisely, $SPEED(\Phi, F) = \{f \in \mathbf{R} \mid \text{for all } \varphi_i = f, \text{ there exists } \varphi_j = f \text{ such that } F(\Phi_j)(x) < \Phi_i(x) \text{ a.e. } x\}$. The main result of this section is stated in the following theorem.

Theorem 3.1. *For every total effective operator F, the set $SPEED(\Phi, F)$ is of second category.*

Proof. Fix a total effective operator F; for compactness, let denote by $SPEED$ the set $SPEED(\Phi, F)$. Assume that $SPEED$ is meager. This means that there exists a decomposition $SPEED = \bigcup_{j\geq 0} SPEED_j$ and a recursive function f such that for every $j \geq 0$, $SPEED_j \cap U_{\alpha_{f(<j,n>)}} = \emptyset$ a.e. n. We construct a recursive function g satisfying for every natural $i \geq 0$ the following two requirements: (1) $R(i)$: $\alpha_{f(<i,m>)} \subset g$ i.o. m, and (2) $Q(i)$: if $\Phi_i(x) < p_i(x)$ i.o. x, then $g \neq \varphi_i$, where (p_i) is a sequence of functions built as in the standard proof of the *Operator Speed-Up Theorem* (see Meyer and Fisher [27], Calude [8]). Conditions $Q(i)$ guarantee that $g \in SPEED$. By the initial assumption, $g \in SPEED_i$, for some i. Condition $R(i)$ implies that $g \in \mathcal{U}_{\alpha_{f(<i,m>)}}$ i.o. m, which contradicts the fact that $SPEED_i \cap \mathcal{U}_{\alpha_{f(<i,m>)}} = \emptyset$ a.e. m. The proof proceeds by defining, in a construction by stages, a family of p.r. functions $(z_s)_{s\geq 0}, z_s : \mathbf{N}^4 \overset{o}{\to} \mathbf{N}$. The function we are interested in will be obtained in two steps: first we take the function z to be the limit (as s goes to ∞) of this construction (we assure that the limit exists); then, in the second step, we fix to convenient values the first three input variables of z (denoted by n, w and v) and finally obtain the desired function g. More precisely, n will be the fixed point deduced from the use of the Recursion Theorem, w will be selected in such a way as to achieve the desired rate of speed-up, and v will be chosen such that α_v will patch the initial segment of the faster program so that it exactly computes the desired function. At stage s we construct the p.r. function z_s, such that $\lambda x.z_s(n, w, v, x)$ *tries* to properly extend $\lambda x.z_{s-1}(n, w, v, x)$. It may be the case that some subcomputations in stage s cannot be performed (more precisely, the fourth condition in what is denoted *Test* below). In this case, the computation loops forever at some point in stage s. In such a situation, naturally, $\lambda x.z_t(n, w, v, x)$ is undefined for all $t \geq s$. However, for the value n_0 obtained through the use of the Recursion Theorem, this will not happen and, consequently, for all w, v and s, $\lambda x.z_s(n_0, w, v, x)$ properly extends $\lambda x.z_{s-1}(n_0, w, v, x)$. The particular feature of the construction is that, for all w and v, the extended part of $\lambda x.z_s(n, w, v, x)$ uses from the previous stages information related to the construction of $\lambda x.z_{s-1}(n, 0, 0, x)$ (and not $\lambda x.z_{s-1}(n, w, v, x)$, as one might suspect). In this way, at all stages s, we extend $\lambda x.z_s(n, w, v, x)$ for all w and v by the same amount. More precisely, we define at each stage s the integer value $Lh_s(n)$ such that for every $w, v \geq 0$, $dom(\lambda x.z_s(n, w, v, x)) = \{0, \ldots, Lh_s(n) - 1\}$. We also use the sets $DIAG_s(n, w, v)$ with the meaning that $i \in DIAG_s(n, w, v)$ if by stage s we have insured that $z_s(n, w, v, x) \neq \varphi_i(x)$ for some $x < Lh_s(n)$. In the computation of $\lambda x.z(n, w, v, x)$ we focus at each stage s on a pair of naturals $ACTIVE_s(n, w, v) = (j, k)$, called the *active pair*, with the intention to fulfill $R(j)$. However, if at some stage we discover an index i with $i << j, k >$ such that $Q(i)$ can be satisfied, we prefer to do it (this is a simple form of the priority method). When $R(j)$ is satisfied, the *next* pair of naturals in a standard ordering of $\mathbf{N}^2$ becomes the new *active pair*.

The construction of $z(n, w, v, x)$:

Stage $s = -1$: Put $z_{-1}(n, w, v, x) \equiv \infty$, for all $n, w, v, x \geq 0$, $Lh_{-1}(n) \equiv 0$, for every $n \geq 0$, $DIAG_{-1}(n, w, v) \equiv \emptyset$, for all $n, w, v \geq 0$, $ACTIVE_{-1}(n, w, v) \equiv (0, 0)$.
Stage $s \geq 0$: Take m such that $\alpha_m = \lambda x.z_{s-1}(n, 0, 0, x)$ and let $(j, k) \equiv ACTIVE_{s-1}(n, 0, 0)$. For x with $0 \leq x < Lh_{s-1}(n)$, define $z_s(n, w, v, x) \equiv z_{s-1}(n, w, v, x)$.

for each x such that $Lh_{s-1}(n) \leq x < |\alpha_{f(<j,m>)}|$ **do**
    if $x \in dom(\alpha_v)$, **then** $z_s(n, w, v, x) \equiv \alpha_v(x)$
    else if there exists i such that:
        1) $i < <j, k>$,
        2) $i \notin DIAG_{s-1}(n, 0, 0)$,
        3) $w \leq i < x$,
        4) $\Phi_i(x) \leq \varphi_n(<i, x>)$
    Conditions 1) - 4) will be further on denoted as the *Test*.
    The *Test* is checked in increasing order of i in the range
    $w, w+1, \ldots, min(x, <j, k>)$. If for some i as above the
    Test cannot be evaluated (because $\varphi_n(<i, x>)$ is not defined)
    then, of course, $z_s(n, w, v, x)$ is not defined and the procedure loops forever.
    then (Satisfy one bit of $Q(i)$):
        choose the least such i,
        $z_s(n, w, v, x) \equiv \max\{1 - \varphi_i(x), 0\}$
        $DIAG_{s-1}(n, w, v) \equiv DIAG_{s-1}(n, w, v) \cup \{i\}$
    else (Satisfy $R(j)$)
        $z_s(n, w, v, x) \equiv \alpha_{f(<j,m>)}(x)$
    end if

end for
if for all x such that $Lh_{s-1}(n) \leq x < |\alpha_{f(<j,m>)}|$,
no i satisfies the *Test* or $x \in dom(\alpha_v)$), **then**
    $ACTIVE_s(n, w, v) \equiv next(ACTIVE_{s-1}(n, w, v))$
else
    $ACTIVE_s(n, w, v) \equiv ACTIVE_{s-1}(n, w, v)$
end if
$DIAG_s(n, w, v) \equiv DIAG_{s-1}(n, w, v)$
$Lh_s(n) \equiv |\alpha_{f(<j,m>)}|$

End of construction

We denote $z(n, w, v, x) \equiv \lim_{s \to \infty} z_s(n, w, v, x)$ (the limit exists by the way the function z_s extends z_{s-1} for each s). Let t be a recursive function such that $\varphi_{t(n,w,v)}(x) = z(n, w, v, x)$. Now we turn to the definition of functions $(p_i)_{i \geq 0}$. First, let ψ be the p.r. function defined by the following clauses:

- $\psi(n, <i, x>) = 0$, if $x \leq i$ or there exists $m \leq i$ with $\Phi_n(<0, m>) \geq x$;

- $\psi(n, < i, x >) = \max\{F(\Phi_{t(n,i+1,v)})(x) | v \leq x\}$, if the first condition fails to hold, but $\Phi_n(< j, y >)$ is defined for all $y \leq x$ and all $i < j \leq x$;
- $\psi(n, < i, x >) = \infty$, in the remaining situations.

By the Recursion Theorem, there exists n_0 such that $\varphi_{n_0}(u) = \psi(n_0, u)$, for all u. Fix such an n_0 and denote $p_i(x) = \varphi_{n_0}(< i, x >)$; the complexity of $p_i(x)$ is then $\Phi_{n_0}(< i, x >)$. One has:

$$p_i(x) = \begin{cases} 0 & \text{if } x \leq i \text{ or there exists } m \leq i \text{ with } \Phi_{n_0}(< 0, m >) \geq x, \\ \max\{F(\Phi_{t(n_0,i+1,v)})(x) | v \leq x\}, & \text{if } p_j(y) \text{ is defined for all } y \leq x \text{ and all } j, i < j \leq x, \\ \infty, & \text{otherwise.} \end{cases} \quad (1)$$

Fact 3.2. *For all natural i, x, $\varphi_{n_0}(< i, x >)$ is defined, i.e. p_i is recursive for all i.*

Fact 3.3. *For all natural v, $p_i(x) \geq F(\Phi_{t(n_0,i+1,v)})(x)$, a.e. x.*

Fact 3.4. *For all natural i, there exists v such that $\varphi_{t(n_0,i,v)} = \varphi_{t(n_0,0,0)}$.*

Fact 3.5. *Define the recursive function g by $g(x) \equiv \varphi_{t(n_0,0,0)}(x)$, $x \in \mathbf{N}$. For every natural j, there exist infinitely many m such that $\alpha_{f(<j,m>)} \sqsubseteq g$, i.e. $R(j)$ is satisfied for every j.*

Fact 3.6. *If $g = \varphi_i$, then $\Phi_i(x) \geq p_i(x)$ a.e. x.*

The main ingredients in the proofs are the classical ideas involved in the whole family of speed-up theorems (Fact 3.2 and Fact 3.3), the priority schema for satisfying $R(i)$ and $Q(i)$ for all natural i (Fact 3.5 and Fact 3.6) and, crucially, our strategy according to which stage s in the construction of $\lambda x.z_s(n, w, v, x))$ uses information related to $\lambda x.z_s(n, 0, 0, x))$ (Fact 3.4).
We finish the proof of the theorem by showing that $g \in SPEED$. Indeed, let $\varphi_i = g$. Then $\Phi_i(x) \geq p_i(x) \geq F(\Phi_{t(n_0,i+1,v)})(x)$ a.e. x, for all $v \in \mathbf{N}$. The first inequality comes from Fact 3.6 and the second one from (1). Take v such that $\varphi_{t(n_0,i+1,v)} = \varphi_{t(n_0,0,0)}$ and $j = t(n_0, i+1, v)$. Then $\varphi_j = g$ and $\Phi_i(x) \geq F(\Phi_j)(x)$ a.e. x. □

It is known that the Speed-Up Theorem is ineffective in many aspects. Blum has shown in [3] that for no speedable function φ_i can one find algorithmically from i the faster program. Related results are due to Meyer and Fischer [27], Helm and Young [18], Schnorr [32], Fulk [16] and Bridges and Calude [6]. The topological analysis of the Speed-Up Theorem easily yields another facet of this phenomenon. Given any sound formal system, it is not possible to detect, with the exception of a tiny meager set of functions, that a function is speedable.

Corollary 3.7. *Let $\mathcal{T}$ be any sound formal system and F a total effective operator. There exists a function $h \in SPEED = SPEED(\Phi, F)$ such that, for each machine M computing h, the sentence* "The function computed by M belongs to $SPEED$" *is not a theorem of $\mathcal{T}$. Moreover the set of such functions h is of the second category.*

Proof. Suppose there exists a formal system$\mathcal{T}$ such that for all $h \in SPEED$ there exists a Turing machine M computing f and such that the sentence "*The function computed by M belongs to $SPEED$*" is a theorem of $\mathcal{T}$. By extracting from the theorems of $\mathcal{T}$ the ones having the above form, we obtain an r.e. sequence of machines M_i such that $\{h \in \mathbf{PR} |\ h \text{ is computed by some } M_i\} = SPEED$. The set $A_i = \{h \in \mathbf{PR} |\ h \text{ is computed by } M_i\}$ is nowhere dense via the witness function f_i which is defined by $f_i(v) = vy$, where $y = \max\{1 - M_i(|v|), 0\}$. It follows that $SPEED$ is meager, which contradicts Theorem 3.1. The second (stronger) assertion follows also by the same reasoning. □

Let g be an increasing recursive function and define the set $HARD(g) = \{f \in \mathbf{R} |$ for every $\varphi_i = f$, $\Phi_i(x) > g(x)$ a.e. $x\}$, of all recursive functions requiring at least $g(x)$ complexity. An well known result, due to Rabin [30] (see also Calude [8]) asserts that $HARD(g)$ is non-empty. Here we can easily derive a stronger result:

Theorem 3.8. *For every increasing recursive function g, $HARD(g)$ is of the second category.*

Proof. Consider the total effective operator F defined by $F(\varphi_i)(x) = g(x)$, for all $i, x \in \mathbf{N}$, and apply Theorem 3.1 to F, as $SPEED(\Phi, F) \subseteq HARD(g)$. □

Note that the analogue of the above theorem for the case of polynomial-time has been established by Mayordomo [24]. Similarly to Corollary 3.7, we obtain the following:

Corollary 3.9. *Let $\mathcal{T}$ be any sound formal system and g an increasing recursive function. There exists a function $h \in HARD(g)$ such that, for each machine M computing h, the sentence* "The function computed by M belongs to $HARD(g)$" *is not a theorem of $\mathcal{T}$. Moreover the set of such functions h is of the second category.*

4 Gap and Compression

A natural problem in computational complexity is to investigate to what extent by allocating more resources we get more computational power. If F is a total effective operator such that $F(f)$ is much bigger than f, are we guaranteed that $C_f^{\Phi} \subset C_{F(f)}^{\Phi}$? The Operator Gap Theorem (see Constable [13], Calude [8] and Young [38]) gives a surprisingly negative result to this question: There are functions f such that $C_f^{\Phi} = C_{F(f)}^{\Phi}$. Our next result shows that there are *many* such functions.

Theorem 4.1. *Let F be a total effective operator such that for all natural i and x, $F(\varphi_i)(x) \geq \varphi_i(x)$. Then the set $GAP(\Phi, F) = \{t \in \mathbf{R} \mid C_t^\Phi = C_{F(t)}^\Phi\}$ is of second category.*

Proof. (Sketch) For each fixed recursive function f, we can construct $t \in GAP(\Phi, F)$ such that for all natural j there are infinitely many n with $\alpha_{f(<j,n>)} \sqsubset t$. This easily implies that $GAP(\Phi, F)$ is not meager. The construction is by stages. At stage s, we define t_s, an initial finite segment of t. Stage s consists of two steps. In the first step, we simply put $t_s = \alpha_{f(<j,m>)}$, where m and j are given by $t_{s-1} = \alpha_m$ and $s =< j, k >$ for some k. By now, we have guaranteed that $\alpha_{f(<j,m>)} \sqsubseteq t_s \sqsubset t$. In the second step, proceeding basically as in Young's proof of the Operator Gap Theorem (see Young [38], Calude [8]) but taking in consideration the effect of the first step, we further extend t_s in such a way that for all $h \in \{0, \ldots, s-1\}$, there is no x in the domain of t_s with $t'(x) \leq \Phi_h(x) \leq F(t')(x)$ for any recursive t' that extends t_s. This second step clearly guarantees that $t \in GAP(\Phi, F)$. □

Corollary 4.2. *Let $\mathcal{T}$ be any sound formal system and F a total effective operator. There exists a function $h \in GAP(\Phi, F)$ such that, for each machine M computing h, the sentence* "The function computed by M belongs to $GAP(\Phi, F)$" *is not a theorem of $\mathcal{T}$. Moreover the set of such functions h is of the second category.*

The Compression Theorem [2] assures us that for *nice* families of functions T, one can indeed get more computational power by raising the resource bound from $t(x)$ to $g(x, t(x))$, for an appropriate g and for any $t \in T$. *Nice* means here a *measured* set, i.e. an r.e. set of partial functions $\gamma_i : \mathbf{N} \stackrel{o}{\to} \mathbf{N}$ for which the ternary predicate $\gamma_i(n) = m$ is recursive. The Compression Theorem states that if $(\gamma_i)_{i \in \mathbf{N}}$ is a measured set, then there exist two recursive function $g : \mathbf{N}^2 \to \mathbf{N}$ and $k : \mathbf{N} \to \mathbf{N}$ such that for each recursive γ_i (1)if $\varphi_j = \varphi_{k(i)}$, then $\Phi_j(x) > \gamma_i(x)$ a.e. x and (ii) $\Phi_{k(i)}(x) \leq g(x, \gamma_i(x))$ a.e. x. We would like to know what properties should a set of functions have in order to possess a Compression Theorem (being a measured set is probably not a necessary condition). It is known from Calude [7] (see also Calude [8]) that every measured set is meager. (In [7], the result is shown in a different topology, the *superset* topology, which is adequate for investigating the size of sets of partial recursive functions. However the proof from [7] can easily be adapted for the Cantor topology.) Our next result shows that the meagerness of a measured set seems to be essential and gives a partial answer to the above question. If we fix the increasing factor g and consider a second category set A of recursive functions, then the compression property does not hold for any f in A.

Proposition 4.3. *Let $g : \mathbf{N}^2 \to \mathbf{N}$ be a recursive function. Then the set $A_g = \{s \in \mathbf{R} \mid$ there existsi with $\Phi_i(x) \leq g(x, s(x))$ a.e. x and for all j with $\varphi_j = \varphi_i$ one has $\Phi_j(x) > s(x)$ a.e. $x\}$ is meager.*

It is not hard to prove that any function in a measured set is r-honest, for an appropriate recursive function r. (We recall that, in a Blum space (φ_i, Φ_i), a recursive function f is r-honest if there is an i such that $\varphi_i = f$ and $\Phi_i(x) \leq r(x, f(x))$ a.e. x; here $r : \mathbf{N}^2 \to \mathbf{N}$ is a recursive function.) Consequently, the functions in the Compression Theorem are all honest. It is also well known that the classical time and space hierarchy theorems for Turing machines (see Hopcroft and Ullman [22]) require time and, respectively, space constructibility of the involved functions, which is a strong variant of honesty. One may suspect that functions defining the gaps in the "anti-hierarchy" Operator Gap theorem are not honest. Indeed, we can prove that many of them are not honest in an extremely strong sense. Fix a *uniform* sequence of Blum spaces $(\varphi^{(i)}, \Phi^{(i)})_{i \in \mathbf{N}}$, in the sense that there exists a recursive function $V : \mathbf{N}^2 \to \mathbf{N}$ such that $V(i, x) = \Phi^{(i)}(x)$, for all i and x.

Proposition 4.4. *Let r, g be recursive functions and $(\varphi^{(i)}, \Phi^{(i)})$ as above. Then the set of functions in $GAP(\Phi, g)$ which are not r-honest in any Blum space $(\varphi^{(i)}, \Phi^{(i)})$ is of second category.*

Proof. By a result of Mehlhorn [25], the class of r-honest functions is meager. It follows that the set $A = \{f \in \mathbf{PR} \mid f \text{ is } r\text{-honest in } (\varphi^{(i)}, \Phi^{(i)}), \text{ for some natural } i\}$ is also meager, since it is the uniform union of a countable set of meager sets. Since, by Theorem 4.1, $GAP(\Phi, g)$ is of second category, we conclude that $GAP(\Phi, g) \setminus A$ is of second category as well. □

5 Effective Measure

To the best of our knowledge, there has been no prior investigation of the size of objects in abstract computational complexity from an effective measure theoretical point of view. It is the aim of this section to illustrate that such a study is quite feasible. Similar results to the ones below have been established for the case of natural complexity measures in [20] and [24].
The recursive and resource-bounded variants of measure theory have been developed by Freidzon [15], Mehlhorn [26], Schnorr [33] and exhaustively by Lutz [20]. It is important to retain that this theory is applicable only to $\{0, 1\}$-valued functions, i.e. to predicates (and this is a major drawback of effective measure when compared to effective category; for example, the size of the class of r-honest functions cannot be analyzed in the measure-theoretical setting). The following definitions come from the latter paper, restricted to the necessities of the current work. For motivations and connections with classical measure theory, we direct the reader to Lutz [20, 21], Allender and Strauss [1], and Mayordomo [24]. Let Σ^* and Σ^∞ be the sets of finite and, respectively, infinite binary strings and let λ denote the empty string. If $w \in \Sigma^*$, then w can be identified with a function in **FPRED** which for all $i < |w|$ maps i into the $(i+1)$th bit of w. The basic neighborhoods $\mathcal{U}_w$ are defined as in Section 1 (in the topology of p.r. predicates).

Definition 5.1. *(i) A* **density function** *is a function $d : \Sigma^* \to [0, \infty)$ satisfying $d(w) \geq \frac{d(w0)+d(w1)}{2}$, for all strings $w \in \Sigma^*$.*

(ii) The **global value** *of a density function d is $d(\lambda)$. The set covered by a density function d is $S[d] = \bigcup_{w\in\Sigma^*, d(w)\geq 1} \mathcal{U}_w$. A density function d covers a set $X \subseteq \Sigma^\infty$ if $X \subseteq S[d]$.*

(iii) An **1-dimensional density system** *is a function $d : \mathbf{N}\times\Sigma^* \to [0,\infty)$ such that for all $i \in \mathbf{N}$, d_i is a density function, where d_i is defined by $d_i(x) = d(i,x)$.*

Definition 5.2. *(i) A set $X \subseteq \Sigma^\infty$ has* **recursive measure zero** *if there exists an 1-dimensional recursive density system d (i.e. d is a recursive real-valued function) such that for all k, d_k covers X with global value $d_k(\lambda) \leq 2^{-k}$.*

(ii) A set $X \subseteq \Sigma^\infty$ has **recursive measure one**, *if the complement of X has recursive measure zero.*

The general idea is to give an effective touch to the standard method in measure theory consisting in the covering of sets by intervals. This is realized by using recursive martingales (called here density functions) betting on the cylinders $\mathcal{U}_w$. The initial investment is the global value. Thus, roughly speaking, a set has recursive measure zero, if there is a recursive winning betting strategy covering the set that starts with an arbitrary small amount of money.
A function $f \in$ **RPRED** will be identified with the infinite binary string $f(0)f(1)\ldots \in \Sigma^\infty$. Lutz [20] has shown that **RPRED** *does not have recursive measure zero.* This important feature makes a recursive measure meaningful for studying the size of classes of functions in **RPRED** with respect to the whole space **RPRED**.

Definition 5.3. *(i) A set $X \subseteq \Sigma^\infty$ has* **recursive measure zero** *in* **RPRED** *if $X \cap$* **RPRED** *has recursive measure zero. (ii) A set $X \subseteq \Sigma^\infty$ has* **recursive measure one** *in* **RPRED** *if the complement of X has recursive measure zero in* **RPRED**.

The following useful result is proved by Lutz [20]: *A recursive union of recursive measure zero sets has recursive measure zero.* Formally, let $d : N^2\times\Sigma^* \to [0,\infty)$ be a recursive function such that for each j, d_j is a density system (where $d_j(i,x) = d(j,i,x)$, for all $i \in \mathbf{N}$ and $x \in \Sigma^*$). If $X, X_1, X_2, \ldots \subseteq \Sigma^\infty$, $X = \bigcup_{i=0}^\infty X_i$ and for all $i \in \mathbf{N}$, d_i is a witness that X_i has recursive measure zero, then X has recursive measure zero. The following result is a recursive measure-theoretical version of Theorem 3.8; surprisingly the proof is much simpler.

Theorem 5.4. *For every recursive function g, $HARD(g) \cap$* **RPRED** *has recursive measure one in* **RPRED**.

Proof. We have to show that the set $HARD^c(g) = \{f \in RPRED \mid$ there exists $\varphi_i = f, \Phi_i(x) \leq g(x)$ i. o. $x\}$ has recursive measure zero. Let $((\varphi_i)_{i\in\mathbf{N}}, (\Phi_i)_{i\in\mathbf{N}})$ be a Blum space of $\{0,1\}$ valued p.r. functions. It is immediate that $HARD^c(g) \subseteq \bigcup H_i$, where

$$H_i = \begin{cases} \{f \in RPRED \mid \varphi_i = f\}, & \text{if there exist infinitely many } x \text{ such that } \Phi_i(x) \leq g(x), \\ \emptyset, & \text{otherwise.} \end{cases}$$

We describe a recursive function $d : \mathbf{N}^2 \times \Sigma^* \to [0, \infty)$ such that d_i witnesses that H_i has effectively measure zero, where $d_i(k, x) = d(i, k, x)$, for all $i \in \mathbf{N}$, and $x \in \Sigma^*$. In fact, we define the "projections" $d_{i,k}$ as follows: $d_{i,k}(\lambda) = 2^{-k}$, and, inductively,

$$d_{i,k}(x0) = \begin{cases} 0, & \text{if } \Phi_i(|x|) \leq g(|x|) \text{ and } \varphi_i(|x|) = 1, \\ 2d_{i,k}(x), & \text{if } \Phi_i(|x|) \leq g(|x|) \text{ and } \varphi_i(|x|) = 0, \\ d_{i,k}(x), & \text{otherwise,} \end{cases}$$

$$d_{i,k}(x1) = \begin{cases} 2d_{i,k}(x), & \text{if } \Phi_i(|x|) \leq g(|x|) \text{ and } \varphi_i(|x|) = 1, \\ 0, & \text{if } \Phi_i(|x|) \leq g(|x|) \text{ and } \varphi_i(|x|) = 0, \\ d_{i,k}(x), & \text{otherwise.} \end{cases}$$

It can readily be checked that $d_{i,k}$ is a density function. Also, by induction on h, one can see that $d_{i,k}(x) \geq 2^{h-k}$ if and only if there exist more than h integers y with $y \leq |x| - 1$ and $\Phi_i(y) < g(y)$. Therefore, it follows that $H_i \subseteq S[d_{i,k}]$, for all $k \in \mathbf{N}$. □

The above proof also implies the effective measure-theoretical analogue of the basic result of Mehlhorn [25], Calude [7] stating that *all complexity classes are meager*. Indeed, if $((\varphi_i)_{i \in \mathbf{N}}, (\Phi_i)_{i \in \mathbf{N}})$ is a Blum space of $\{0, 1\}$-valued p.r. functions, and g is a recursive function, then the complexity class defined by g, C_g^{Φ}, is included in $HARD^c(g)$, the set defined in the above proof and shown to have effective measure zero in **RPRED**. Therefore:

Theorem 5.5. *Let $((\varphi_i)_{i \in \mathbf{N}}, (\Phi_i)_{i \in \mathbf{N}})$ be a Blum space of $\{0, 1\}$-valued p.r. functions and g a recursive function. Then C_g^{Φ} has effective measure zero in* **RPRED**.

We pass to the study of measured set of predicates (see the definition in the previous section). We are interested in measured sets of recursive predicates, i.e. sets of recursive $\{0, 1\}$-valued functions γ_i. Many natural classes of recursive predicates form measured sets, e.g. the class of primitive recursive predicates, every r.e. complexity class of predicates and, in fact, any r.e. set of recursive predicates. It follows from the next theorem that all these classes (and all their subclasses, like all levels in Grzegorczyk's hierarchy) have effective measure zero.

Theorem 5.6. *If $\Gamma = (\gamma_i)_{i \in \mathbf{N}}$ is a measured set of recursive predicates, then Γ has effective measure zero.*

Proof. We decompose $\Gamma = \bigcup \Gamma_i$, where $\Gamma_i = \{\gamma_i\}$. It is immediate to build, for each i, a density system d_i witnessing that Γ_i has recursive measure zero. Namely, $d_{i,k}(\lambda) = 2^{-k}$, and $d_{i,k}(x0) = 2d_{i,k}(x)$, if $\gamma_i(|x|) = 0$, $d_{i,k}(x0) = 0$, otherwise, $d_{i,k}(x1) = 2d_{i,k}(x)$, if $\gamma_i(|x|) = 1$, otherwise. □

References

[1] E. Allender and M. Strauss. Measure on small complexity classes, with applications for BPP, *FOCS'94*, 1994, 807–818.

[2] M. Blum. A machine-independent theory of the complexity of recursive functions, *J. Assoc. Comput. Mach.* 14(2) (1967), 322–336.

[3] M. Blum. On effective procedures for speeding up algorithms, *J. Assoc. Comput. Mach.* 18(2) (1967), 257–265.

[4] A. Borodin. Computational complexity and the existence of complexity gaps, *J. Assoc. Comput. Mach.* 19(1) (1972), 158-174.

[5] D. S. Bridges. *Computability—A Mathematical Sketchbook*, Springer-Verlag, Berlin, 1994.

[6] D. S. Bridges and C. Calude. On recursive bounds for the exceptional values in speed-up, *Theoret. Comput. Sci.* 132 (1994), 387-394.

[7] C. Calude. Topological size of sets of partial recursive functions, *Z. Math. Logik Grundlag. Math.* 28(1982), 455–462.

[8] C. Calude. *Theories of Computational Complexity*, North-Holland, Amsterdam, New York, Oxford, Tokyo, 1988.

[9] C. Calude. Relativized topological size of sets of partial recursive functions, *Theoret. Comput. Sci.* 87 (1991), 347–352.

[10] C. Calude, G. Istrate, and M. Zimand. Recursive Baire classification and speedable functions, *Z. Math. Logik Grundlang. Math.* 3 (1992), 169–178.

[11] C. Calude, H. Jürgensen, and M. Zimand. Is independence an exception?, *Applied Math. Comput.* 66 (1994), 63-76.

[12] C. Calude and M. Zimand. On three theorems in abstract complexity theory: A topological glimpse, *Abstracts of the Second International Colloquium on Semigroups, Formal Languages and Combinatorics on Words*, Kyoto, Japan, 1992, 11-12.

[13] R. L. Constable. The operator gap, *J. Assoc. Comput. Mach.* 19(1) (1972), 175–183.

[14] S. Fenner. Notions of resource-bounded category and genericity, *Proc. 6th Structure in Complexity Theory*, 1991, 347–352.

[15] R. Freidzon. Families of recursive predicates of measure zero, *J. Soviet Math.* 6 (1976), 449–455.

[16] M. A. Fulk. A note on a.e. h-complex functions, *J. Comput. System Sciences* 40 (1990), 444–449.

[17] J. Hartmanis and J. E. Hopcroft. An overview of the theory of computational complexity, *J. Assoc. Comput. Mach.* 18(3) (1971), 444-475.

[18] J. Helm and P. Young. On size vs. efficiency for programs admitting speed-ups, *J. Symbolic Logic* 36 (1971), 21–27.

[19] J. Lutz. Category and measure in complexity theory, *SIAM Journal Computing* 19(6) (1990), 1100–1131.

[20] J. Lutz. Almost everywhere high nonuniform complexity, *J. Comput. System Sciences* 44 (1992), 220–258.

[21] J. Lutz. The quantitative structure of exponential time, *Proceedings of the 8th Structure in Complexity Theory Conference*, 1993, 158–175.

[22] J. E. Hopcroft and J. D. Ullman. *An Introduction to Automata Theory, Languages and Computation*, Addison-Wesley, Reading, Mass., 1979.

[23] M. Machtey and P. Young. *An Introduction to the General Theory of Algorithms*, North-Holland, Amsterdam, 1978.

[24] E. Mayordomo. Almost every set in exponential time is p-bi-immune, *Theoret. Comput. Sci.*, 136(1994), 487–506.

[25] K. Mehlhorn. On the size of sets of computable functions, *Annual IEEE Symp. on Switching and Automata Theory*, Univ. Iowa, 1973, 190–196.

[26] K. Melhorn. The almost all theory of subrecursive degrees is decidable, *Proc. Second ICALP*, Lecture Notes in Computer Science, Springer-Verlag, 1974, 1317–325.

[27] A. R. Meyer and P. C. Fischer. Computational speed-up by effective operators, *J. Symbolic Logic* 37(1) (1972), 55–68.

[28] E. McCreight and A. Meyer. Classes of computable functions defined by bounds on computation: preliminary report, *Conf. Rec. ACM Symp. on Theory of Computing*, 1965, 79–88.

[29] A. R. Meyer and K. Winklman. The fundamental theorem of complexity theory, in J. W. de Bakker and J. van Leeuwen (eds.). *Found. Comput. Sci. III Part 1: Automata, Data Structures, Complexity*, Mathematical Centre Tracts, vol. 108, Amsterdam, 1979, 97–112.

[30] M. Rabin. *Degree of Difficulty of Computing a Function*, Hebrew University, Jerusalem, Technical Report 2 (April 25), 1960.

[31] H. Rogers. *Theory of Recursive Functions and Effective Computability*, McGraw-Hill, New York, 1967.

[32] C. P. Schnorr. Does the computational speed-up concern programming?, *Proc. First Internat. Conf. on Automata, Languages and Programming*, 1972, 589–596.

[33] C. P. Schnorr. Process complexity and effective random tests, *J. Comput. System Sciences* 7 (1973), 376–388.

[34] J. Seiferas. Machine-independent complexity theory, in J. van Leeuwen (ed.). *Handbook of Theoretical Computer Science, vol. A*, Elsevier, 1990, 165–186.

[35] J. Seiferas and A. R. Meyer. Characterization of realizable space complexities, *Annals of Pure and Applied Logic* (to appear).

[36] B. A. Trakhtenbrot. *Complexity of Algorithms and Computations*, Course Notes, Novosibirsk, 1967. (Russian)

[37] P. van Emde Boas. Ten years of speed-up, *Proceedings of the Symposium on Mathematical Foundations of Computer Science*, Lecture Notes in Computer Science #32, Springer-Verlag, Berlin, 1975, 232-237.

[38] P. Young. Easy constructions in complexity theory: gap and speed-up theorems, *Proc. Amer. Math. Soc.* 37 (1973), 555–563.

[39] M. Zimand. If not empty, $NP \setminus P$ is topologically large, *Theoret. Comput. Sci.* 119 (1993), 293–310.

[40] M. Zimand. On the topological size of $p-m$-complete degrees, *Theoret. Comput. Sci.* (to appear)

About Planar Cayley Graphs*

Thomas Chaboud†

1 Introduction

Cayley graphs are graphical representations of groups. Their sole use, for many years, has been in group theory; when it helps to understand a particular group better, one draws a Cayley graph of its.

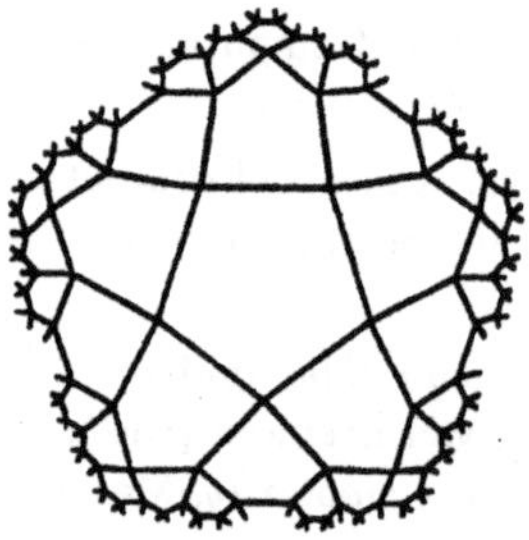

Figure 1: this is not a Cayley graph.

Our approach here is radically different. Many research fields use infinite strongly regular graphs as models. If such a network happens to be a Cayley graph, one immediately has the whole of group theory as a toolbox.

An interconnection network benefits admitting a group labeling on several points: the associated presentation provides a global description of the network, and group properties make studying its communications all the easier; furthermore, one then has an unambiguous local description of each unit's neighbourhood.

Exploration of this field has only recently begun; [4] is an example, although restricted to some finite networks. Cellular Automata, on the other hand, are often studied in various infinite grids. Our approach finds a full justification in Z. Ròka's Ph.D. thesis [9], that uses our results straightforwardly.

Some aspects of finitely presented groups, along with their Cayley graphs, have proven to be powerful tools towards resolving tiling problems, again recently. These studies, [3], [7]... in particular, are what arouse our interest for the subject.

In view of this, one cannot help wanting to know whether a given graph one may use is a Cayley graph or not. This seemingly natural question has only been asked,

*This work was partially supported by the Esprit Basic Research Action "Algebraic and Syntactic Methods In Computer Science" and by the PRC "Mathématique et Informatique".

†LIP, ENS-Lyon, 69364 Lyon Cx 07, France, Email: tchaboud@lip.ens-lyon.fr

and answered, in a few particular cases. For instance, it belongs to folklore that the Petersen graph is a small vertex-transitive graph that is not a Cayley graph.

Even the restriction to planar graphs we will investigate in that paper was left aside, though one may remark it is not entirely trivial: some very simple regular vertex-transitive planar graphs, as that shown in Fig. 1, happen not to be Cayley graphs.

As we use constructive methods to determine whether a graph is a Cayley or not, we are also able to enumerate all group labelings for a given graph. Furthermore, the graphs studied are infinite. We use finite descriptions of these graphs, of course, but among them, some actually correspond with a graph, and some do not. As we will see in the following, these descriptions can be taken as input for an algorithm to enumerate the associated groups. If there is at least a group presentation output, it establishes existence of a graph fitting the description.

In other words, we definitely rather use groups for a better understanding of graphs, than the reverse.

2 Preliminary definitions

Although we will often refer to group presentations in the following, we only state here our particular choices in their use. As for the most basic definitions, they are clearly presented in [8], for instance.

Throughout this paper, a *presentation of a group G* means an expression

$$\langle \{a_i\}_{i\in I} \mid \{R_j\}_{j\in J} \cup \{R_j\}_{j\in J'} \rangle,$$

where $\{a_i\}_{i\in I}$ is a set of *distinct semi-group generators* for G. The relators R_j are chosen in such a way that the $\{R_j\}_{j\in J}$ pair the formal inverses among the $\{a_i\}_{i\in I}$; that is, $|R_j| = 2$ for any j in J, and $|R_j| > 2$ for any j in J'. We will sometimes lighten the notation, using $a, b, c, \ldots$ as generators, thus avoiding subscripts.

The *Cayley graph* of a group G given by such a presentation is the graph $\Gamma(G) = \Gamma(V, E)$ such that there is a bijection $x \mapsto \tilde{x}$, from V to G, and the arc (x,y) is in E, with label a_i, iff $\tilde{x}a_i = \tilde{y}$.

We will say that two presentations are *equivalent*, if there is an isomorphism from one's group to the other's that is a simple renaming of the generators; otherwise, they will be said *different*. In the latter case, we will say, somewhat improperly, that the two groups presented also are *different*.

Let σ denote the permutation of S_d such that $a_{\sigma(i)} = a_i^{-1}$. The permutation σ is an involution, and its fixed points are the i's such that a_i is a generator of order 2. If an arc has label a_i, we may denote the label of its reverse arc either by a_i^{-1}, or by $a_{\sigma(i)}$. Let p_d denote the permutation defined by $p_d(i) = i[d] + 1$; p_d is simply the cyclic order on $\{1, \ldots, d\}$; we often will forget the subscript d.

We use words u, v, w, u_1,... on the alphabet $\{a_i\}_{i\in I} \cup \{a_i^{-1}\}_{i\in I}$, to denote either group elements or paths from a Cayley graph. The identity in G is noted e. A cycle in a graph is said *simple* if it passes only once through each of its vertices. We will call a graph *locally finite* if it has a planar imbedding, on the Euclidean or hyperbolic plane, such that any bounded region of the surface contains only finitely many vertices.

3 Outline of the paper

The general problem we address here is that of deciding, for a planar graph Γ, whether there is a group that has Γ as a Cayley graph, plus, in the affirmative, to enumerate the groups that do. If Γ is finite, simple (brute force) algorithms do the job, so our attention will rather go to infinite planar graphs.

We intend to apply *labeling schemes*, derived from those presented in [1]. This description of a Cayley graph requires that the generators appear in a fixed order around the vertices; this implies that there are finitely many types of faces in the graph, and that they too are ordered around the vertices. This *uniformity*, along with the existence of a Euclidean or hyperbolic imbedding without singularities (*normality*), defines the *pseudo-Archimedean* graphs.

We then address a natural question, namely, among the planar Cayley graphs, which are pseudo-Archimedean? Our partial answer is that all normal planar Cayley graphs are uniform, hence pseudo-Archimedean (Th. 2).

A formal description of the labeling schemes ensues. It is proved that all group presentations one can get from that method label pseudo-Archimedean graphs. Furthermore, we show that any of these presentations corresponds with a uniform tiling of either the Euclidean or the hyperbolic plane by regular convex polygons. As very few such tilings were previously known in the hyperbolic case, one has yet another valuable contribution from group theory to graphs.

Labeling schemes are much more simple than their formalism might indicate: they also have a convenient graphical representation, which is the subject of the next section.

4 Main problem

We suppose to be given an *undirected* planar graph Γ; one can then see any edge in Γ as two opposite arcs. Does there exist a labeling for these arcs such that, with the meaning stated above, Γ is the Cayley graph of a presentation of a group?

We will often prefer arc colours instead of labels, and almost always choose to represent only one from each pair of opposite arcs. If some of the generators have order 2, we will also leave unoriented the edges they label. Our choices are easily understandable; we provide an illustration of them anyway (Fig. 2), for much of the discussion below relies on them.

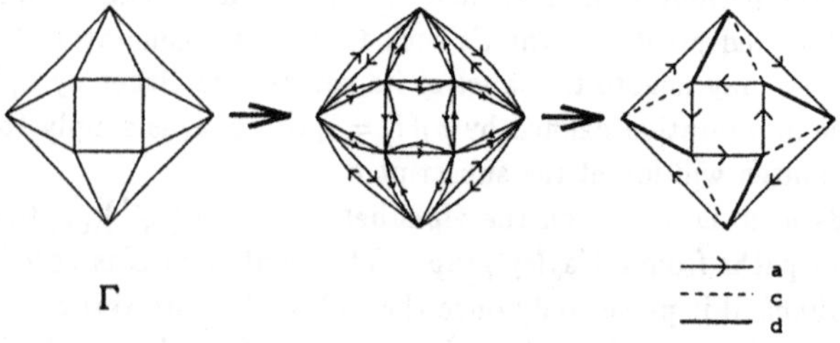

Figure 2: Γ is the Cayley graph of $\langle a, b, c, d \mid ab, c^2, d^2, a^4, bcd \rangle$.

It is obvious that only connected, vertex-transitive, loop- and 2-cycle-free graphs are to be considered, for our purpose. Our preceding result was about planar Cayley graphs with regular dual.

Theorem 1 *let $\Gamma[k,d]$ be a locally finite planar graph, regular with degree d, the dual of which is regular with degree k. $\Gamma[k,d]$ is a Cayley graph iff k has a prime factor less than or equal to d.*

The proof of Theorem 1, to be found in [1], consists in combinatorial computations on group elements viewed as words, for the necessary conditions part. Its sufficient conditions part is mainly based upon the following fact: if such a $\Gamma[k,d]$ is a Cayley graph, then the group generators appear in order around the vertices, up to reversal of orientation. It permitted us to define and apply simple labeling schemes; we will present an extended version of them, that takes care of a larger class of planar graphs, which we will call *normal.*

5 Normal and pseudo-Archimedean graphs

Many of the following definitions are about tilings, for any planar imbedding of a graph can be seen as a planar tiling, with the faces of the former acting as the tiles of the latter. Moreover, some of them are inspired from the formalism for tilings introduced by [5]. We will thus say that a graph has a certain property P if it can be imbedded in the Euclidean or hyperbolic plane so that the tiling induced has property P.

A tiling of the Euclidean or hyperbolic plane is *normal* if:

- N1: every tile is a topological disk;
- N2: the intersection of every two tiles is a connected set;
- N3: the tiles are uniformly bounded.

For a normal graph, N3 means that there exist two positive numbers, r and R, such that any face in the imbedding contains some disk of radius r and lies in a disk of radius R.

In a planar edge-to-edge tiling by polygons, the *type* of a vertex is the n-uple $(k_1, k_2, \ldots, k_n)$ of the number of edges of the tiles that surround it, in cyclic order. If m successive k_i's are equal, we may use the shorthand k_i^m in the type vector; for instance, $(3^3, 4^2)$ will stand for $(3,3,3,4,4)$.

As stated before, we leave the finite graphs aside; hence, from now on, we will assume that every type vector $(k_1, k_2, \ldots, k_n)$ mentioned is such that

- $n \geq 3$;
- $\forall i \in \{1, \ldots, n\}, k_i \geq 3$;
- $\sum_{i=1}^{i=n}((k_i - 2)/k_i) \geq 2$.

This last condition ensures that if the tiling exists, then it is Euclidean if the equality holds, hyperbolic otherwise. All the realizable tilings such that $\sum_{i=1}^{i=n}((k_i - 2)/k_i) < 2$ are finite tilings of the sphere. The methods presented in that paper also take care of those graphs, but we will stick to these three conditions for the sake of simplicity.

An *Archimedean tiling* is a tiling of the Euclidean plane by regular polygons in which all vertices have the same type. We call a tiling *uniform* if all its vertices have the same type. A *pseudo-Archimedean* graph, or *pAg*, is a normal and uniform graph.

Note that if a pAg is Euclidean, then it is Archimedean; this is easily deduced from a result about the duals of such graphs, to be found in [5], p.176. In other words, if a graph is normal and uniform in the Euclidean plane, it can be drawn with its faces as convex (and even regular) polygons.

An essential result is that if a planar Cayley graph is normal, then it is also uniform, and hence a pAg. We only sketch the proof.

Remark 1 *if a word u labels a cycle somewhere in the Cayley graph, then any path labeled by u is a cycle. More generally, two paths labeled by the same word in the Cayley graph have exactly the same self-intersections.*

Remark 2 *if Γ is a normal Cayley graph, then the intersection of any two of its faces is either empty, or a vertex, or again an edge.*

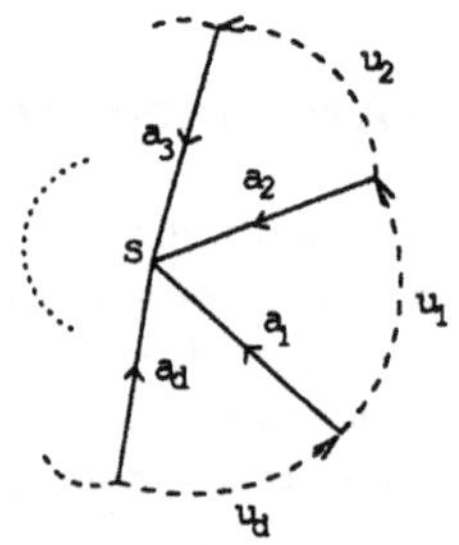

Figure 3: the faces neighbouring s.

Lemma 1 *in a normal planar Cayley graph, the generators appear in cyclic order around the vertices, up to reversal of orientation.*

From here on, we always suppose that the generators' indices are chosen in such a way that they are ordered from a_1 to a_d, counterclockwise, around a given vertex.

By Lemma 1, the same goes around any vertex in Γ, up to reversal of orientation; there only are clockwise and counterclockwise oriented vertices.

Lemma 2 *in a normal planar Cayley graph, if an edge labeled a_1 links two differently oriented vertices, then the same holds for every edge with that same label.*

This tells that for an arc to conserve or invert the orientation between its extremities solely depends on its label, in a normal Cayley graph associated with a given presentation.

This enables one to distinguish, in a group presentation generating a normal Cayley graph, between two types of generators, those that conserve the orientation, and those that invert it. We will thus call a generator a *conserver* or an *inverter*.

Theorem 2 *any planar normal Cayley graph is a pseudo-Archimedean graph.*

This fact considerably simplifies the checking of whether a given normal graph Γ is a Cayley graph; provided that the graph description allows it, to verify if Γ is pseudo-Archimedean will tell if it deserves further consideration.

One should note that, in general, a pAg vertex type vector does not characterize it. However, any further information about the graph makes the decision of whether it is a Cayley graph all the easier. For instance, if no cyclic permutation of $(k_1, \ldots, k_d)$ is a palindrome, one can tell which vertex is clockwise and which is counterclockwise oriented 'at sight'.

Anyway, our algorithm based upon labeling schemes only takes the type vector as an input, and enumerates all the corresponding group presentations. There only are a finite number of them, for any given vector, and a group presentation generates a single Cayley graph. Hence, even if there are many pAg's attached to the vector considered, one can then try and find among these presentations which is or are associated with the graph one wishes.

6 Labeling schemes

Suppose that we are given the number d of semi-group generators, a permutation $\sigma \in S_d$ pairing the indices of the formal inverses, and a mapping $\tau : \{1, \ldots, d\} \rightarrow \{-1, 1\}$ such that $\tau(i) = \tau(\sigma(i))$, for any $i \in \{1, \ldots, d\}$. τ is to be interpreted so that $\tau(i) = -1$ if a_i is an inverter, $\tau(i) = 1$ if a_i is a conserver.

Without loss of generality, one can consider a counterclockwise oriented vertex s. For instance, what does the word labeling that face F whose border contains sa_1^{-1}, s, and sa_2^{-1} look like? Read it clockwise, starting from sa_1^{-1}: its first two arcs, also considered clockwise, are a_1 and $a_{\sigma(p(1))}$, that is, $a_{\sigma(2)}$ (Fig. 4).

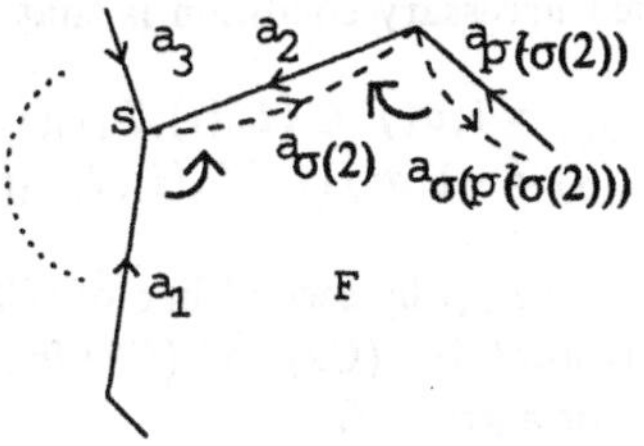

Figure 4: the labels around F, assuming $\tau(2) = -1$.

The next arc's label depends on the type of a_2; its label's index is $i = \sigma(p^{\tau(2)}(\sigma(2))$. Knowing the type of a_i, one can compute the next arc label index, namely $\sigma(p^{\tau(2)\times\tau(i)}(i))$, and so on until the face closes up. Of course, at that time, the last index computed must be 1, and the last subscript upon p must be 1: every cycle contains an even number of inverters.

More formally, let $F_+(i)$ (resp. $F_-(i)$) be the right-hand (resp. the left-hand) face adjacent to an a_i pointing towards a counterclockwise-oriented vertex; $|F_+(i)|$ and $|F_-(i)|$ respectively denote their number of sides. If one considers an a_i pointing towards a clockwise-oriented vertex, $F_+(i)$ is at the left hand and $F_-(i)$ at the right.

Define the following sequence for i in $\{1,\ldots,d\}$, depending only on d, σ and τ:

$$\begin{cases} T_1(i) = 1; \\ L_1(i) = \sigma(p(i)); \\ T_{n+1}(i) = T_n(i) \times \tau(L_n(i)), n \geq 1; \\ L_{n+1}(i) = \sigma(p^{T_{n+1}(i)}(L_n(i))), n \geq 1. \end{cases}$$

In words, the $L_n(i)$ are the label indices of the successive arcs bordering $F_+(i)$, read clockwise. The same iteration, initialized with

$$\begin{cases} T_1(i) = -1; \\ L'_1(i) = \sigma(p^{-1}(i)). \end{cases}$$

provides the label indices of the successive arcs bordering $F_-(i)$, read counterclockwise.

A necessary condition for a pAg with type vector $(k_1,\ldots,k_d)$ to be a Cayley graph is the existence of σ and τ satisfying

$$\forall i \in \{1,\ldots,d\}, \begin{cases} L_{k_i}(i) = i; \\ T_{k_i}(i) = 1. \end{cases} \tag{C1}$$

This condition expresses that every $F_+(i)$ closes up properly. However, it is not sufficient; an additional constraint on $(k_1,\ldots,k_d)$ is that

$$\forall i \in \{1,\ldots,d\}, \forall j \in \{L_n(i)/T_{n+1}(i) = 1, n \geq 1\}, k_i = k_j, \tag{C2}$$

meaning that if there is an a_j pointing a counterclockwise vertex on the border of $F_+(i)$, then $F_+(j) = F_+(i)$.

If a_i is a conserver, $|F_+(i)| = |F_-(\sigma(i))|$ and $|F_-(i)| = |F_+(\sigma(i))|$, of course, hence no direct relation between k_i and $k_{\sigma(i)}$. On the other hand, if a_i is an inverter, then $|F_+(i)| = |F_+(\sigma(i))|$, and a last necessary condition is that

$$\forall i \in \{1,\ldots,d\}, \begin{cases} \forall j \in \{j/\tau(j) = 1, \sigma(j) \in \{L_n(i), T_{n+1}(i) = -1, n \geq 1\}\} \\ \forall j \in \{j/\tau(j) = -1, \sigma(j) \in \{L_n(i), T_{n+1}(i) = 1, n \geq 1\}\} \end{cases}, k_i = k_j. \tag{C3}$$

Note that if $\tau(i) = -1$, then $k_i = k_{\sigma(i)}$ by second line of (C3).

Now, if a couple (σ, τ) satisfies (C1), (C2) and (C3) for a given vector $(k_1,\ldots,k_d)$, one gets a finite presentation for a group G:

$$\langle\{a_1,\ldots,a_d\} \mid \{a_i a_{\sigma(i)}, 1 \leq i \leq d\} \cup \{R_i = (L_1(i)L_2(i)\ldots L_{k_i}(i)), 1 \leq i \leq d\}\rangle.$$

This presentation is redundant: first, its relators set contains both $a_i a_{\sigma(i)}$ and $a_{\sigma(i)} a_i$, if $\sigma(i) \neq i$.

Second, for any R_i, the R_j such that $j \in \{L_n(i)/T_{n+1}(i) = 1, n \geq 1\}$ or $j \in \{j/\tau(j) = 1, \sigma(j) \in \{L_n(i), T_{n+1}(i) = -1, n \geq 1\}\}$ or $j \in \{j/\tau(j) = -1, \sigma(j) \in \{L_n(i), T_{n+1}(i) = 1, n \geq 1\}\}$ all are powers of the same factor or of a conjugate, by definition of L_n, and forced to have the same length by (C2) and (C3). A presentation of G needs only one of them. Once these simplifications done, it becomes apparent that the Cayley graph generated by the presentation obtained is planar, and is a pAg of type vector $(k_1,\ldots,k_d)$.

Let Γ be the Cayley graph associated to such a presentation of a group G.

Γ is planar, from (C2) and (C3), that ensure that every edge belongs to the borders of exactly two faces; Γ is uniform, as coherent with the type vector, from (C1).

A proof that Γ is normal is in [2], as we lack the space here; actually, the result is much stronger: depending on the type vector $T = (k_1, \ldots, k_d)$, Γ can be imbedded so as to tile either the Euclidean or the hyperbolic plane by regular convex polygons.

7 Graphical representation of labeling schemes

The L_n sequences, along with their constraints, might seem a slightly contorted formalism; fortunately, one can easily adapt the graphical representation of the labeling schemes we introduced in [1]. That will also help in discussing which group presentation, obtained from trying to match a given type vector, is different from which other.

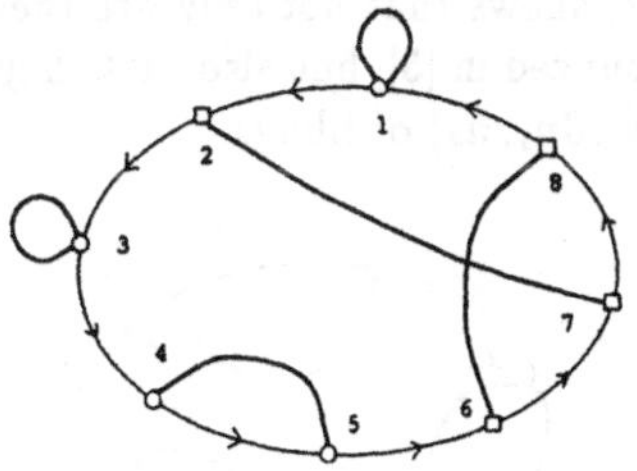

Figure 5: graphical representation of a labeling scheme.

Represent permutation p by a circuit of d vertices numbered from 1 to d, that will stand for the label indices; link every vertex i to its $\sigma(i)$ by an edge. Eventually, replace every vertex by either a spot (for conservers) or a square (for inverters), making sure that no σ edge links a spot to a square (Fig. 5).

Reading the L_n cycles is thus rendered very easy. Starting from any vertex, follow a p arc, then a σ edge. If the arrival vertex is a spot, do the same; if it is a square, follow a p arc backwards from there, then a σ edge. Keep on going backwards on p arcs until the last σ edge arrives at a square; then, start following p arcs onwards again... so on until you are back to your starting vertex in your starting direction. Successive labels of the cycle described are the ends of the σ edges passed through.

On Fig. 5, starting from vertex 2, for instance, one obtains the following word on $\{a_1, \ldots, a_8\}^*$: $a_3a_5a_8a_2$; when following that cycle on the representation, vertices a_3, a_5 and a_2 are passed in the positive direction. That implies that every Cayley graph associated to one of the group presentations given by that scheme is a $(k_1, \ldots, k_8)$-pAg such that $k_2 = k_3 = k_5 = k_7 = 4 \times n$. $k_7 = k_2$ is from constraint (C3).

Applying the same process to the other vertices of that scheme, one finds that it provides group labelings for pAg's with type vectors $(3n_1, 4n_2, 4n_2, n_3, 4n_2, 3n_1, 4n_2, 3n_1)$, assuming $n_1 \geq 1$, $n_2 \geq 1$, and $n_3 \geq 3$, so that $k_i \geq 3$ for any i.

Once having tried to match a few type vectors, there is no doubt one will find handling these graphical schemes very easy.

For instance, Coxeter groups, Cayley graphs of which have any type $(k_1, \ldots, k_d)$ if every k_i is even, are given by the family of schemes suggested by Fig. 6.

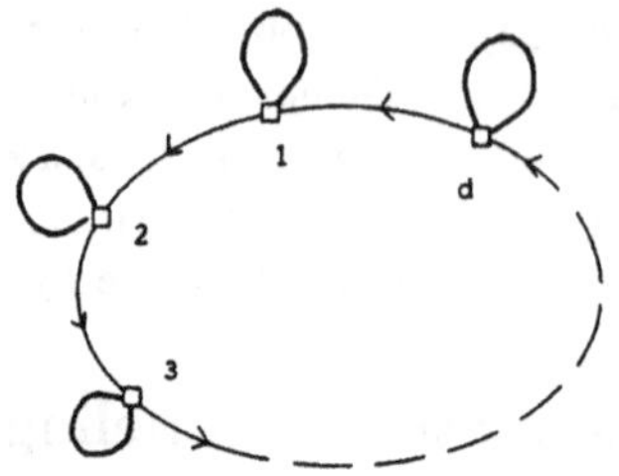

Figure 6: schemes associated to Coxeter groups.

The associated type vectors are $(2n_1, 2n_2, \ldots, 2n_d)$; as the previous section says, this simple drawing proves existence for the corresponding Euclidean and hyperbolic tilings.

Our last example, Fig. 7, shows that not only are there homogeneous tilings with type $(3, n, 3, n, 3, n)$, as announced in [5], but also that they may be viewed as instances of the family $(3n_1, n_2, 3n_1, n_3, 3n_1, n_4)$ of tilings.

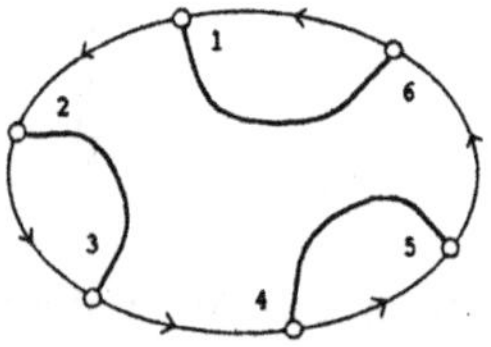

Figure 7: a scheme fitting vector $(3, n, 3, n, 3, n)$, and many others.

It should be obvious that this scheme was chosen only for it generalizes a previous result; any (random) drawing of a representation of a scheme provides infinitely many different and new graphs.

If one finds several group presentations for the same type vector, which are different, in the sense we stated above? This is the point we now turn to.

From here on, a *scheme* will refer either to a triple (p_d, σ, τ), or to its graphical representation. For any d, a given scheme is isomorphic to at most $2d - 1$ others. Two schemes are said *different* if they are not isomorphic. For any scheme S, one can compute its cycle length vector $(x_{i_1}n_{i_1}, \ldots, x_{i_d}n_{i_d})$, where the subscript i_j indicates that generator j belongs to cycle i_j. The x_{i_j} are nonnegative integer constants, and the n_{i_j} formal variables. Examples of that notation were given above as cycle length vector for schemes of figures 5, 6 and 7.

We will say that S *fits* a vector $T = (k_1, \ldots, k_n)$ if there exists an integer assignment to $(n_{i_1}, \ldots, n_{i_d})$ such that $(x_{i_1} \times n_{i_1}, \ldots, x_{i_d} \times n_{i_d}) = T$. A scheme S is *compatible* with a vector T if at least one scheme isomorphic to S fits T.

Suppose that two different schemes are compatible with a given type vector. Of course, they will generate different group presentations, in the sense stated above.

However, the same labeling scheme S might be compatible with a type vector T in more than one way. If several schemes isomorphic to S fit T, each through a single assignment to its cycle length variables, all the presentations obtained are equivalent.

This does not hold any longer if one scheme isomorphic to S fits T through more than one variable assignment. The presentations are equivalent only if these several assignments correspond with symmetries in S. Otherwise, they even might be associated not isomorphic Cayley graphs, still having same type vector, of course.

It is rather easy to derive an algorithm producing all group presentations corresponding to a given type vector, by a direct encoding of labeling schemes. Its complexity is exponential on the length of the vector. Our implementation of that algorithm produced all the possible group labelings for the Archimedean graphs ([2]).

8 Conclusion

Regarding non-normal Cayley graphs, we chose to let them aside from that paper. Though we obtained a few results about these, we could not reach the same complete formalism as for normal ones. Besides, they are rather pathological, in the sense that they do not admit imbeddings without singularities, such as accumulation points. For that reason, one may suppose they will not be as efficient models as normal Cayley graphs, and hence be of lesser interest for the community of computer scientists.

References

[1] Chaboud, T., and Kenyon, C., *Planar Cayley Graphs with Regular Dual*, accepted in Int. J. of Alg. and Comp., 1995.

[2] Chaboud, T., *About Planar Cayley Graphs*, preprint, 1995.

[3] Conway, J.H., and Lagarias, J.C., *Tiling with Polyominoes and Combinatorial Group Theory*, J. of Combinatorial Theory, A 53 (1990), pp. 183-208.

[4] Fiduccia, C., et Zito, J., *Commutative Cayley Graphs as Interconnection Network Architecture*, 7th SIAM Conf. on Disc. Math. (1994)

[5] Grünbaum, B., and Shephard, G.C., *Tilings and Patterns*, W.H. Freeman and Co., New York (1987).

[6] Grünbaum, B., and Shephard, G.C., Incidence symbols and their applications, *Proc. Symp. Pure Math.* 34 (1979).

[7] Lagarias, J.C., et Romano, A Polyomino Tiling of Thurston and its Configurational Entropy, *Journal of Combinatorial Theory*, A 63, pp. 338-358 (1993).

[8] Lyndon, R.C. and Schupp, P.E., *Combinatorial Group Theory*, Springer-Verlag, Berlin (1977).

[9] Róka, Zs., *Automates Cellulaires sur Graphes de Cayley*, PhD thesis, École Normale Supérieure de Lyon, 180-94, 1994.

[10] Thurston, W.P., Conway's Tiling Groups, *Amer. Math. Monthly*, Oct 1990, pp. 757-773.

On Condorcet and Median Points of Simple Rectilinear Polygons

(Extended Abstract)

Victor D. Chepoi * and Feodor F. Dragan **

Department of Mathematics & Cybernetics,
Moldova State University, A. Mateevici str., 60,
Chişinău 277009, Moldova

Abstract Let P be a simple rectilinear polygon with N vertices, endowed with rectilinear metric, and let the location of n users in P be given. There are a number of procedures to locate a facility for a given family of users. If a voting procedure is used, the chosen point x should satisfy the following property: no other point y of the polygon P is closer to an absolute majority of users. Such a point is called a *Condorcet point.* If a planning procedure is used, such as minimization of the average distance to the users, the optimal solution is called a *median point.*
We prove that Condorcet and median points of a simple rectilinear polygon coincide and present an $O(N + n\log N)$ algorithm for computing these sets. If all users are located on vertices of a polygon P, then the running time of the algorithm becomes $O(N + n)$.

Key words: computational geometry; Condorcet point; median point; rectilinear polygon; rectilinear distance.

1 Introduction

Let P be a simple rectilinear polygon in the plane R^2 (i.e., a simple polygon having all edges axis–parallel) with N edges. A *rectilinear path* is a polygonal chain consisting of axis–parallel segments lying inside P. The length of a rectilinear path in the L_1–metric equals the sum of the length of its constituent segments. In other words, the length of a rectilinear path in the L_1–metric is equal to its Euclidean length. For any two points u and v in P, the *rectilinear distance* between u and v, denoted by $d(u, v)$, is defined as the length of the minimum length rectilinear path connecting u and v. The *interval* $I(u, v)$ between two points u, v consists of all points z between u and v, that is

$$I(u,v) = \{z \in P : d(u,v) = d(u,z) + d(z,v)\}.$$

Consider the problem of locating a single facility on a simple rectilinear polygon P on which a given finite number of users are located. Two users may be located at

* Research supported by the Alexander von Humboldt Stiftung
** Research supported by DAAD
e-mail addresses: chepoi@university.moldova.su dragan@university.moldova.su

the same point. Let $\pi(x)$ be the total number of all users located at a point x. The demand is thus described by a *weight function* π from P to the set of non-negative integers. The polygon P is partitioned into three sets with respect to any pair x, y of points:

$$[x \succ y] = \{z \in P : d(x,z) < d(y,z)\},$$
$$[y \succ x] = \{z \in P : d(y,z) < d(x,z)\},$$
$$[x \sim y] = \{z \in P : d(x,z) = d(y,z)\}.$$

If a voting procedure is used to solve the facility location problem, the chosen point x should satisfy the following property:

> No other point y of the polygon P is closer to an absolute majority of users, i.e. $\pi[y \succ x] \leq \pi(P)/2$ for all points $y \in P$,

where for a subset $S \subseteq P$, $\pi(S) = \sum_{p \in S} \pi(p)$. Such a point x is called a *Condorcet point*; see BANDELT [1], HANSEN and THISSE [16], HANSEN et al. [17], LABBÉ [19], WENDELL and MCKELVEY [25], WENDELL and THORSON [26] and HANSEN and LABBÉ [15]. Denote by $Cond(\pi, P)$ the set of all Condorcet points of the polygon P. The weighted distance sum of a point x with respect to π is given by

$$D(x, \pi) = \sum_{i=1}^{n} \pi(p_i) d(x, p_i),$$

where $p_1, p_2, \dots, p_n$ are the points of P where the users are located. If a planning procedure is used, such as minimization of the function $D(x, \pi)$, the optimal solution is a *median* (or a *Weber point*); see [2, 3, 15, 16, 17]. Let $Med(\pi, P)$ be the set of all median points of polygon P with respect to the weight function π.

In some papers the comparison of these two decision making procedures was studied. In particular, in the rectilinear plane [26] and for tree networks [16] Condorcet points and median points coincide. In [1] those networks are characterized on which Condorcet points and median points always coincide. Moreover, in this paper BANDELT presented a complete characterization of those networks on which no Condorcet paradox occurs, i.e. for each distribution of users there exists at least one Condorcet point. In [15] a polynomial algorithm for determining the set of Condorcet points of a network was given.

In [9] we presented an $O(N + nlogN)$ time algorithm for finding a median point of a simple rectilinear polygon P with N vertices. If all n users are located in the vertices of P then the running time becomes $O(n + N)$. In this paper we develop an algorithm of the same complexity for determining the whole set of Condorcet points of a simple rectilinear polygon. We show that in such polygons Condorcet points and median points coincide. The proof is based on some geometric properties of simple rectilinear polygons, in particular on the fact that they are median spaces. One can consider this problem as a kind of constrained facility location problem when we want, e.g., to describe the set of optimal locations of the facility (resulting from a voting procedure) on an urban region or of a some service on a poligonal building floor. Then the rectilinear metric is often a reasonable approximation of travel behavior.

2 Properties of Simple Rectilinear Polygons

Recall that the metric space (X, d) is a *median space* if every triple of points $u, v, w \in X$ admits a unique "median" point $z = m(u, v, w)$, such that

$$d(u, v) = d(u, z) + d(z, v),$$

$$d(u, w) = d(u, z) + d(z, w),$$

$$d(v, w) = d(v, z) + d(z, w).$$

The median spaces represent a common generalization of different mathematical structures such as median semilattices and median algebras [5], median graphs (including trees and hypercubes) [20], median networks [2] and linear spaces with L_1-metric. For classical results on median spaces the reader is referred to [5],[24].

A set M of a metric space (X, d) is *convex* if for any points $x, y \in M$ and $z \in X$ the equality $d(x, z) + d(z, y) = d(x, y)$ implies that $z \in M$. For a subset $S \subset X$ by $conv(S)$ we denote the convex hull of S, i.e. the intersection of all convex sets containing S. A subset H of X is a *half-space* provided both H and $X \setminus H$ are convex. Recall also that the subset M is called *gated* [12], provided every point $x \in X$ admits a *gate* in M, i.e. a point $x_M \in M$ such that $x_M \in I(x, y)$ for all $y \in M$. Any gated subset of a metric space is convex [12]. The converse holds for median spaces:

Lemma 1. *Any convex compact subset of a median space is gated.*

For a proof of this result see [24].

If a metric space X is a union of two gated subspaces X_1 and X_2 with a nonempty intersection, then X is the *gated amalgam* of X_1 and X_2 along $X_1 \cap X_2$. It is known that the gated amalgam of two median spaces is a median space too [24]. Using this result we obtain the next property of rectilinear polygons (for a direct proof see [9]):

Lemma 2. *A simple rectilinear polygon P equipped with L_1-metric is a median space.*

An axis-parallel segment c is called a *cut segment* of a polygon P if it connects two edges of P and lies entirely inside P. Note that any edge or any cut segment of a polygon P is a gated subset of P. Moreover, if P' and P'' are the subpolygons of P defined by the cut c then P is the gated amalgam of P' and P''.

Lemma 3. *If M is a compact convex subset of a simple rectilinear polygon P and $x \in P \setminus M$ then there exists a cut c of P which separates x and M, i.e. $M \cap c = \emptyset$, $x \notin c$, and M and x belong to different subpolygons defined by c.*

Proof of Lemma 3 omitted.

The next property is a particular instance of a result of [4] about the Caratheodory number of gated amalgams of convexity structures; see also [24, p.208]. A direct proof is given in [21].

Lemma 4. *For any finite set S of a simple rectilinear polygon P*

$$conv(S) = \bigcup_{u,v \in S} I(u,v).$$

A subset M of a median space (X,d) is *median stable* [24] provided $m(x,y,z) \in M$ for any triple $x, y, z \in M$. Let S be a finite subset of points of a simple rectilinear polygon P. Consider all horizontal and all vertical cuts which pass through the points of S or the vertices of P. These cuts together with the edges of P generate a *rectilinear grid.* Denote by V the vertices (intersection points) of this grid and by G its graph. Recall that a graph G is *median* [20] if, with respect to the standard graph distance, G is a median space.

Lemma 5 [8]. *V is a median stable subset of P. In particular, G is a median graph.*

3 Results

In this section we investigate the structural properties of sets of Condorcet points and median points of a simple rectilinear polygon P. We use some results for median problem in median graphs and discrete median spaces, established in [2, 3, 22]. Let V be the vertices of the grid of P generated by the set of users and let G be (median) graph of this grid. Denote by $Med(\pi, G)$ the set of median vertices of the graph G.

Lemma 6 [3, 22]. *$Med(\pi, G)$ is convex in G. Moreover, $Med(\pi, G)$ is an interval of G.*

For any cut c and subpolygons P' and P'' defined by this cut we have

$$\pi(P') + \pi(P'') = \pi(P) + \pi(c).$$

Lemma 7 [3, 22]. *If $\pi(P') > \pi(P'')$ then $Med(\pi, P) \subset P'$, otherwise if $\pi(P') = \pi(P'')$ then $Med(\pi, P) \cap c \neq \emptyset$.*

The converse is true for the set $Med(\pi, G)$.

Lemma 8 [22]. *If $Med(\pi, G)$ belongs to the half-space H of G then $\pi(H) > \pi(V)/2 = \pi(P)/2$.*

Lemma 9. *Let x and y be points on a cut c of P. If x and y belong to a common rectangle of the grid then*

$$D(x,\pi) - D(y,\pi) = d(x,y)(\pi[y \succ x] - \pi[x \succ y]).$$

Proof of Lemma 9 omitted.

Lemma 10. $Med(\pi, P) = conv(Med(\pi, G))$.

Proof. By Lemma 6 the set $Med(\pi, G)$ is convex in G. Applying Lemma 4 we deduce that $conv(Med(\pi, G))$ in P coincides with the union of all rectangles (including degenerated ones) of the grid whose all corners belong to $Med(\pi, G)$. Let $R = conv(a, b, c, d)$ be such a rectangle and let x be an arbitrary point of R. First suppose that x belongs to the boundary of R, say $x \in [a, b]$. By Lemma 9 we have

$$D(x, \pi) - D(a, \pi) = d(x, a)(\pi[a \succ x] - \pi[x \succ a]).$$

Observe that $[a \succ x] = [a \succ b]$ and $[x \succ a] = [b \succ a]$. Applying this fact and Lemma 9 to the vertices a and b we get

$$D(a, \pi) - D(b, \pi) = d(a, b)(\pi[b \succ a] - \pi[a \succ b]).$$

Since $D(a, \pi) = D(b, \pi)$ in both G and P [22] we conclude that $\pi[b \succ a] = \pi[a \succ b])$. Therefore $D(x, \pi) = D(a, \pi)$.

Next assume that x is an interior point of R. Let x' and x'' be the boundary points of R which lie on a common horizontal cut with x. Then as we already proved

$$D(x', \pi) = D(a, \pi) = D(b, \pi) = D(x'', \pi).$$

Applying Lemma 9 to the points x and x' we obtain that $D(x, \pi) = D(x', \pi)$. Thus $D(\,\cdot\,, \pi)$ is constant on the set $conv(Med(\pi, G))$.

In order to prove the required equality it is sufficient to establish that $z \notin Med(\pi, P)$ for an arbitrary point $z \notin conv(Med(\pi, G))$. By Lemma 3 there is a cut c which separates the sets $\{z\}$ and $conv(Med(\pi, G))$, i.e.

$$z \in P' \setminus c \ , \ \ conv(Med(\pi, G)) \subseteq P'',$$

where P' and P'' are the subpolygons defined by c. Then $H' = P' \cap V$ and $H'' = P'' \cap V$ represent complementary half-spaces of the graph G. Since $Med(\pi, G) \subset H''$ by Lemma 8 we conclude that $\pi(P'') = \pi(H'') > \pi(P)/2$. Let z^* be gate for z in the subpolygon P''. A straightforward verification shows that

$$D(z^*, \pi) - D(z, \pi) \leq \pi(P') - \pi(P'') < 0,$$

and thus $z \notin Med(\pi, P)$. □

Theorem 1. $Cond(\pi, P) = Med(\pi, P)$.

Proof. First we prove that $Med(\pi, P) \subseteq Cond(\pi, P)$. Assume the contrary, i.e. for some median point x there exists a point y such that $\pi[y \succ x] > \pi(P)/2$. If $x \in conv([y \succ x])$ then by Lemma 4 we have $x \in I(z', z'')$ for two points $z', z'' \in [y \succ x]$. Since $d(y, z') < d(x, z')$ and $d(y, z'') < d(x, z'')$ we obtain a contradiction with the choice of z' and z''. So, assume that $x \notin conv([y \succ x])$. Let x^* be the gate for x in the gated set $conv([y \succ x])$. In the interval $I(x, x^*)$ pick a close neighbor z of x, such that x and z belong to a common rectangle of the grid and to a common cut of P. By Lemma 9

$$D(z, \pi) - D(x, \pi) = d(z, x)(\pi[x \succ z] - \pi[z \succ x]).$$

Since $[z \succ x] \supseteq [x^* \succ x] \supseteq [y \succ x]$ and $\pi[y \succ x] > \pi(P)/2$ we get $D(z,\pi) < D(x,\pi)$, in contradiction with the assumption that $x \in Med(\pi, P)$. Therefore, any median point of P is a Condorcet point.

Conversely, assume that some Condorcet point x is not a median point. As $Med(\pi, P)$ is convex by Lemma 3 there exists a cut c which separates the set $Med(\pi, P)$ and the point x. Let $x \in P' \setminus c$, $Med(\pi, P) \subseteq P''$, where P' and P'' are subpolygons defined by c. Since $Med(\pi, G) \subseteq P'' \cap V$ by Lemma 8 necessarily $\pi(P'') = \pi(P'' \cap V) > \pi(P)/2$. But then for the gate x^* of x in P'' we have $P'' \subset [x^* \succ x]$ and thus $\pi[x^* \succ x] > \pi(P)/2$, a contradiction. □

4 Algorithm

Using the results of the previous section, below an algorithm for computing the set of Condorcet points of a simple rectilinear polygon P is given. The algorithm is based on the Chazelle algorithm for computing all vertex–edge visible pairs [7] and on the Goldman algorithm for finding the median set of a tree [14]. By first algorithm we obtain a decomposition of a polygon P into $O(N)$ rectangles, using only horizontal cuts. The dual graph of this decomposition is a tree $T(P)$: vertices of this tree are the rectangles and two vertices are adjacent in $T(P)$ iff the corresponding rectangles in the decomposition are bounded by a common cut. Denote by $R(v)$ the rectangle which corresponds to a vertex v of $T(P)$. Assign to each vertex of $T(P)$ the weight of their rectangle. (The weight of a rectangle R is the sum of weights of its points(users) minus one half of total weight of interior points of P which belong to the horizontal sides of R.) In order to compute these weights first we have to compute which rectangles of the decomposition of P contain each of the users. Using one of the optimal point location methods ([13],[18]) this can be done in time $O(nlogN)$ with a structure that uses $O(N)$ storage. (Here N is the number of vertices of polygon P, while n is the number of users). Observe that the induced subdivision is monotone, and, hence, the point location structure can be built in linear time. Therefore the weights of vertices of a tree $T(P)$ can be defined in total time $O(nlogN + N)$. When all users are located only on vertices of P then this assignment takes $O(N+n)$ time.

Now using the Goldman algorithm [14] we compute the set $Med(\pi, T(P))$ of median vertices of the tree $T(P)$. It is well known that $Med(\pi, T(P))$ induces a path of $T(P)$, see for example [23]. We claim that $Med(\pi, P) \subset \bigcup\{R(v) : v \in Med(\pi, T(P))\}$ and moreover $Med(\pi, P) \cap R(v) \neq \emptyset$ for every median vertex v of $T(P)$. Indeed, by the majority rule for trees $v \in Med(\pi, T(P))$ if and only if $\pi(T_v) \geq \pi(T_{v'})$ for any neighbor v' of v [14]. (By T_v and $T_{v'}$ we denote the subtrees obtained by deleting the edge (v, v')).

In the polygon P the rectangles $R(v)$ and $R(v')$ are separated by a common horizontal cut c. Let P_v and $P_{v'}$ be the subpolygons defined by c and let $R(v) \subset P_v$ and $R(v') \subset P_{v'}$. All rectangles that correspond to vertices from T_v lie in the subpolygon P_v. If $v \in Med(\pi, T(P))$ then $\pi(P_v) - \pi(c)/2 = \pi(T_v) \geq \pi(T_{v'}) = \pi(P_{v'}) - \pi(c)/2$ for all neighbors of v.

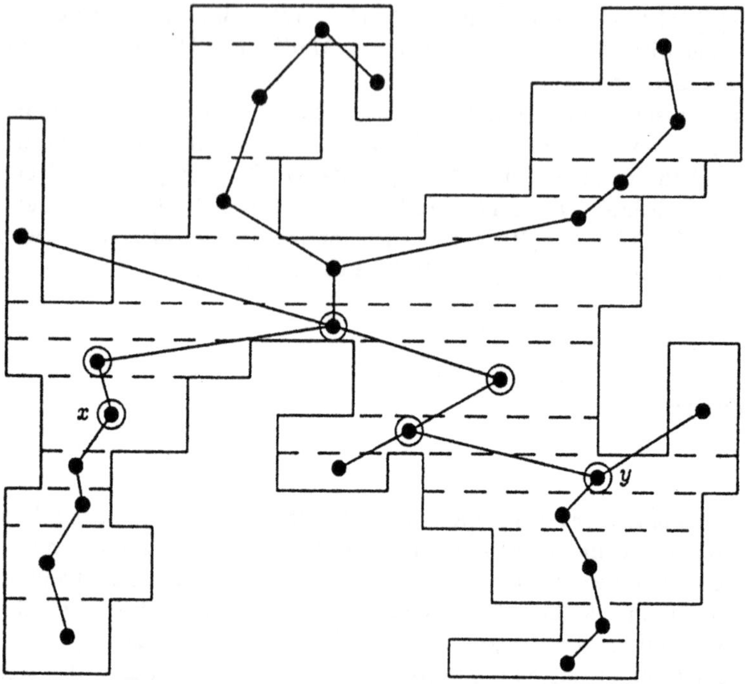

Fig.1. The tree $T(P)$ associated to the partition of P and median vertices of $T(P)$

Since $R(v)$ coincides with the intersection of the subpolygons of the type P_v by Lemma 7 we conclude that $Med(\pi, P) \cap R(v) \neq \emptyset$. Conversely, if $v \notin Med(\pi, T(P))$ then $\pi(P_v) - \pi(c)/2 = \pi(T_v) < \pi(T_{v'}) = \pi(P_{v'}) - \pi(c)/2$ for some vertex v' adjacent to v. By Lemma 7 we obtain that $Med(\pi, P) \cap R(v) = \emptyset$.

Denote by x and y the end-vertices of the path induced by $Med(\pi, T(P))$. Next we concentrate on finding of the median points of rectangles $R(x)$ and $R(y)$. We use for this purpose the method developed in [9]. Suppose that $R(x)$ is bounded by the horizontal cuts c' and c'' of the decomposition of P. Let P' and P'' be the subpolygons of P defined by c' and c'' and disjoint with rectangle $R(x)$, i.e. $P = P' \cup R(x) \cup P''$. For any user z_i let g_i be the gate for z_i in the rectangle $R(x)$. Evidently $g_i \in c'$ if $z_i \in P'$, $g_i \in c''$ if $z_i \in P''$ and $g_i = z_i$ if $z_i \in R(x)$. In order to find these gates, we define the maximal histograms H' and H'' inside P' and P'' with c' and c'' as their bases, respectively. (A *histogram* is a rectilinear polygon that has one distinguished edge, its *base*, whose length is equal to the sum of the lengths of the other edges that are parallel to it; see for example [11].) The vertical edges of these histograms divide the polygons P' and P'' into subpolygons, called *pockets*. Consider for example the pockets from P'. Note that all points from the same pocket have one and the same gate. This is a point of a cut c' which has the same x-coordinate with the cut that separates the pocket and the histogram H'. Hence it is sufficient to find the location of users into the pockets. This can be done by using the partition of P' and P'' into rectangles by vertical vertex-edge visible pairs. Again it is necessary to use the Chazelle algorithm [7] and the optimal point location methods [13],[18]. For any user $z_i \in P' \cup P''$ assign the weight $\pi(z_i)$ to its gate g_i, the weights of users from $R(x)$ remain unchanged. As a result we obtain a median problem in the rectangle $R(x)$. Note that any solution of this problem belongs to $Med(\pi, P) \cap R(x)$. To see this observe that for any two points $z', z'' \in R(x)$ it holds

$$D(z',\pi) - D(z'',\pi) = \sum_{i=1}^{n} \pi(z_i)(d(z',z_i) - d(z'',z_i)) = \sum_{i=1}^{n} \pi(z_i)(d(z',g_i) - d(z'',g_i)).$$

The new median problem on $R(x)$ may be solved by decomposing into two one–dimensional median problems and applying to each of them a modification of the selection algorithm from [6]; see also [10]. Let $M(x) = Med(\pi, P) \cap R(x)$. In a similar way we find the set $M(y) = Med(\pi, P) \cap R(y)$. Both $M(x)$ and $M(y)$ are rectangles (possible degenerate) whose corners are vertices of the grid introduced in Section 2. Denote the corners of $M(x)$ by a_1, a_2, a_3, a_4 and the corners of $M(y)$ by b_1, b_2, b_3, b_4. Suppose that the segments $[a_3, a_4]$ and $[b_3, b_4]$ belong to the sides of $R(x)$ and $R(y)$ which are counterparts of horizontal cuts $c_x = c'$ and c_y, separating $R(x)$ and $R(y)$. By Lemmas 6 and 10 the set $Med(\pi, P)$ coincides with the interval $I(v', v'')$ between two vertices v', v'' of the grid. As cuts c_x and c_y separate the rectangles $R(x)$ and $R(y)$ from the rest of the set $Med(\pi, P)$ we deduce that $v' \in \{a_1, a_2\}$ while $v'' \in \{b_1, b_2\}$. Therefore $Med(\pi, P) = \bigcup_{i,j \in \{1,2\}} I(a_i, b_j)$, i.e. it is enough to compute this union of intervals. Moreover, it is sufficient to find its intersection with each horizontal cut which separates the rectangles $R(x)$ and $R(y)$. Indeed, let $R(v)$ be a rectangle for $v \in Med(\pi, T(P))$. Assume that $R(v)$ is bounded by the horizontal cuts c_1 and c_2 and let

$$Med(\pi, P) \cap c_1 = I' \quad and \quad Med(\pi, P) \cap c_2 = I''.$$

Then $Med(\pi, P) \cap R(v)$ is a rectangle, whose corners can be computed in constant time by finding the intersection of segments I' and I'' with the respective horizontal sides of the rectangle $R(v)$.

Let $x = v_0, v_1, \ldots, v_{k-1}, v_k = y$ be vertices of $Med(\pi, T(P))$ and let $c_1, \ldots, c_k$ be the horizontal cuts of P which correspond to edges of the path induced by $Med(\pi, T(P))$. First we find the gates

$$g^1(a_1), g^1(a_2), g^1(b_1), g^1(b_2), \ldots, g^k(a_1), g^k(a_2), g^k(b_1), g^k(b_2)$$

for points a_1, a_2, b_1, b_2 in the cuts $c_1, \ldots, c_k$, respectively. In order to do this we use the next evident remark: $g^{i+1}(a_1)$ and $g^{i+1}(a_2)$ are the gates for $g^i(a_1)$ and $g^i(a_2)$ in the cut c_{i+1}, while $g^i(b_1)$ and $g^i(b_2)$ are the gates for $g^{i+1}(b_1)$ and $g^{i+1}(b_2)$ in c_i. This follows from the fact that if c cuts P into subpolygons P' and P'' and $x \in P'$ then the gates for x in c and P'' coincide; see [9, Lemma 3].

Let $J_i = [p'_i, p''_i]$ be the smallest segment of the cut c_i containing the points $g^i(a_1), g^i(a_2), g^i(b_1)$ and $g^i(b_2)$. We claim that $[p'_i, p''_i] = Med(\pi, P) \cap c_i$. Assume for example that $Med(\pi, P) = I(a_1, b_1)$. Then necessarily $a_2, b_2 \in I(a_1, b_1)$. The points p'_i, p''_i being the gates for some of the points a_1, a_2, b_1 or b_2 necessarily belong to the interval $I(a_1, b_1)$. By the convexity of the set $I(a_1, b_1)$ we conclude that $[p'_i, p''_i] \subset I(a_1, b_1)$. Next consider a point p outside the segment $[p'_i, p''_i]$. Then

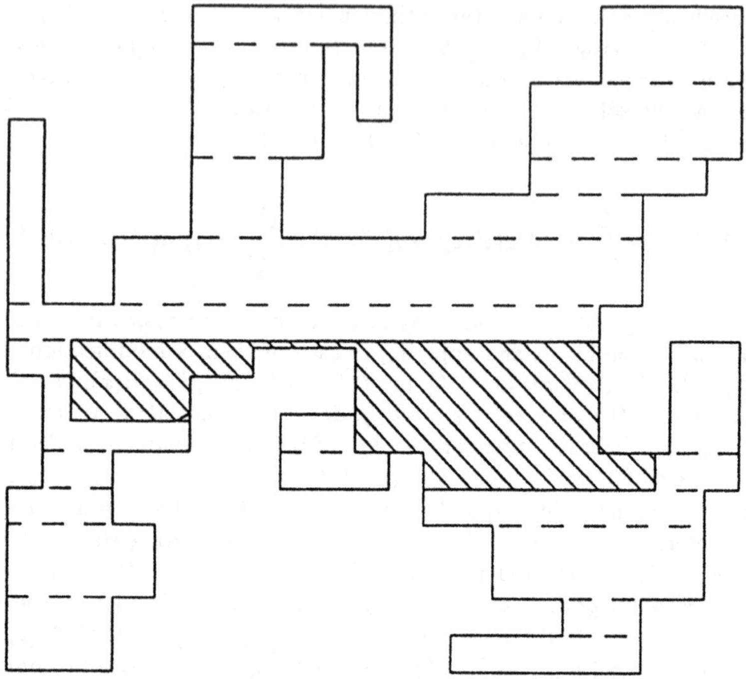

Fig.2. The median and Condorcet subpolygon of P.

$$d(a_1,p) = d(a_1, g^i(a_1)) + d(g^i(a_1), p),$$
$$d(b_1,p) = d(b_1, g^i(b_1)) + d(g^i(b_1), p).$$

Since $d(g^i(a_1),p) + d(g^i(b_1),p) > d(g^i(a_1), g^i(b_1))$ we conclude that $d(a_1,p) + d(b_1,p) > d(a_1,b_1)$, i.e. $p \notin I(a_1,b_1)$. Thus $Med(\pi,P) \cap c_i = [p_i', p_i'']$. The computing of gates for points a_1, a_2, b_1 and b_2 in cuts $c_1, \ldots, c_k$ takes $O(N)$ time. The same number of operations is necessary to compute the segments $J_i = Med(\pi,P) \cap c_i$ and the sets $Med(\pi,P) \cap R(v_i)$, $i = 1, \ldots, k$. Thus we obtain the set $Med(\pi,P)$ as a union of at most N rectangles.

Summarizing the results of this section and taking into account that $Cond(\pi,P) = Med(\pi,P)$ the next result is obtained.

Theorem 2. *The sets of Condorcet and median points of a simple rectilinear polygon P can be found in time $O(nlogN + N)$. If all users are located on vertices of P then the time becomes $O(N + n)$.*

References

1. H.-J. BANDELT. Networks with Condorcet solutions. *European J. Operational Research*, 20(1985), 314–326.

2. H.-J. BANDELT. Single facility location on median networks (submitted).
3. H.-J. BANDELT and J.-P. BARTHELEMY. Medians in median graphs. *Discrete Appl. Math.*, 8(1984), 131–142.
4. H.-J. BANDELT, V.D. CHEPOI and M. VAN DE VEL. Pasch-Peano spaces and graphs. *Preprint* (1993).
5. H.-J. BANDELT and J. HEDLÍKOVÁ. Median algebras. *Discrete Math.*, 45(1983), 1–30.
6. M. BLUM, R.W. FLOYD, V.R. PRATT, R.L. RIVEST and R.E TARJAN. Time bounds for selection. *J. Comput. System Sci.*, 7(1972), 448–461.
7. B. CHAZELLE. Triangulating a simple polygon in linear time. *Discrete Comput. Geom.*, 6(1991), 485–524
8. V.D. CHEPOI. A multifacility location problem on median spaces. *Discrete Appl. Math.* (to appear).
9. V.D. CHEPOI and F.F. DRAGAN. Computing a median point of a simple rectilinear polygon. *Inform. Process. Lett.* 49(1994), 281–285.
10. T.H. CORMEN, C.E. LEISERSON and R.L. RIVEST. Introduction to Algorithms. *MIT Press Cambridge*, MA/Mc Graw-Hill, New-York, 1990.
11. M. DE BERG. On rectilinear link distance. *Computational Geometry: Theory and Applications*, 1(1991), 13–34
12. A. DRESS and R. SCHARLAU. Gated sets in metric spaces. *Aequationes Math.*, 34(1987), 112–120
13. H. EDELSBRUNNER, L.J. GUIBAS and J. STOLFI. Optimal point location in a monotone subdivision. *SIAM J. Comput.*, 15(1985), 317–340
14. A.J. GOLDMAN. Optimal center location in simple networks. *Transportation Sci.*, 5(1971), 212–221
15. P. HANSEN and M. LABBÉ. Algorithms for voting and competitive location on a network. *Transportation Sci.*, 22(1988), 278–288.
16. P. HANSEN and J.-F. THISSE. Outcomes of voting and planning: Condorcet, Weber and Rawls locations. *J. Public Econ.*, 16(1981), 1–15.
17. P. HANSEN, J.-F. THISSE and R.E. WENDELL. Equilibrium analysis for voting and competitive location problems. in: *Discrete Location Theory, R.L. Francis and P.B. Mirchandani (eds.)*, John Wiley & Sons.
18. D.G. KIRKPATRICK. Optimal search in planar subdivisions, *SIAM J.Comput.*, 12(1983), 28–35
19. M. LABBÉ. Outcomes of voting and planning in single facility location problems. *European J. Operational Research*, 20(1985), 299–313.
20. H.M. MULDER. The Interval Function of a Graph. *Math. Centre Tracts (Amsterdam)*, 132(1980).
21. S. SCHUIERER. Helly-type theorem for staircase visibility (submitted).
22. P.S. SOLTAN and V.D. CHEPOI. Solution of the Weber problem for discrete median metric spaces (in Russian). *Trudy Tbilisskogo Math. Inst.*, 85(1987), 53–76
23. B.C. TANSEL, R.L. FRANCIS and T.J. LOWE. Location on networks. Parts 1,2. *Management Sci.*, 29(1983), 482–511
24. M. VAN DE VEL. Theory of Convex Structures. *Elsevier Science Publications (Amsterdam)*, 1993.
25. R.E. WENDELL and R.D. MCKELVEY. New perspectives in competitive location theory. *European J. Operational Research*, 6(1981), 174–182.
26. R.E. WENDELL and S.J. THORSON. Some generalizations of social decisions under majority rules. *Econometrica*, 42(1974), 893–912.

Fast Algorithms for Maintaining Shortest Paths in Outerplanar and Planar Digraphs *

Hristo N. Djidjev[1], Grammati E. Pantziou[2] and Christos D. Zaroliagis[3]

[1] Computer Science Dept, Rice University, P.O. Box 1892, Houston, TX 77251, USA
[2] Computer Science Dept, University of Central Florida, Orlando FL 82816, USA
[3] Max-Planck Institut für Informatik, Im Stadtwald, 66123 Saarbrücken, Germany

Abstract. We present algorithms for maintaining shortest path information in dynamic outerplanar digraphs with sublogarithmic query time. By choosing appropriate parameters we achieve continuous trade-offs between the preprocessing, query, and update times. Our data structure is based on a recursive separator decomposition of the graph and it encodes the shortest paths between the members of a properly chosen subset of vertices. We apply this result to construct improved shortest path algorithms for dynamic planar digraphs.

1 Introduction

The design and analysis of algorithms for dynamic graph problems is one of the most active area of current algorithmic research. For solving a dynamic graph problem one has to design an efficient data structure that not only allows fast answering to a series of queries, but that can also be easily updated after a modification of the input data. Let G be an n-vertex digraph with real valued edge costs but no negative cycles. The *length* of a path p in G is the sum of the costs of all edges of p and the *distance* between two vertices v and w of G is the minimum length of a path between v and w. The path of minimum length between v and w is called a *shortest path* between v and w. Finding shortest path information in graphs is an important and intensively studied problem with many applications. The dynamic version of the problem has also been studied recently [3, 4] and is stated as follows: Given G (as above), build a data structure that will enable fast on-line shortest path or distance queries. In case of edge cost modification of G, update the data structure in an appropriately short time. We will refer to the above, as the *dynamic shortest path* (DSP) problem. The DSP problem has a lot of applications, including dynamic maintenance of a maximum *st*-flow in a network [7], computing a feasible flow between multiple sources and sinks as well as finding a perfect matching in bipartite planar graphs [8].

In this paper, we investigate the DSP problem for particular classes of digraphs, namely those of outerplanar and planar digraphs. An efficient solution

* This work was partially supported by the EU ESPRIT BRA No. 7141 (ALCOM II), by the EU Cooperative Action IC-1000 (ALTEC) and by the NSF grant No. CCR-9409191. Email: hristo@cs.rice.edu, pantziou@cs.ucf.edu, zaro@mpi-sb.mpg.de.

to the DSP problem for outerplanar digraphs has been given in [3] where a data structure is constructed in $O(n)$ time and space and then any distance query is answered in $O(\log n)$ time. The data structure is updated after an edge cost modification in $O(\log n)$ time. (A different approach leading to the same performance characteristics is claimed in [1].) On the other hand, if constant query time is required one can not do much better than the naive approach, i.e. an $O(n^2)$ time and space preprocessing of the input digraph (using e.g. the algorithm in [6]), such that a distance query is answered in $O(1)$ time and a shortest path one in $O(L)$ time, where L is the number of edges of the path. Updating this data structure after an edge cost modification takes $O(n^2)$ time which is equivalent to recomputing the data structure from scratch.

Hence, the interesting question arising is: Can we do better than these two extremes? In particular, can we build a dynamic data structure for the DSP problem on outerplanar digraphs in $o(n^2)$ time and space, such that a query can be answered in $o(\log n)$ time and also the data structure can be updated in appropriately short (say sublinear) time after an edge cost modification?

In this paper, we give an affirmative answer to the above question. More precisely, we present two families of algorithms with $o(\log n)$ query time that achieve an interesting trade-off between preprocessing, update, and query bounds, depending on the choice of a particular parameter ε, $0 < \varepsilon < (1/2)$. The results are stated in Theorems 7 and 9. For the case of constant ε, our results as well as their comparison with previous work are summarized on Table 1. (We would like to mention here that the preprocessing bound has been improved very recently [2] through a completely different method. However, the algorithms presented here are much simpler compared with the ones in [2].)

	[3]	Naive approach	This paper	This paper
Dynamic	Yes	No	Yes	Yes
Preprocessing Time & Space	$O(n)$	$O(n^2)$	$O(n \log\log n)$	$O(n^{1+\varepsilon})$
Single-Pair Dist. Query	$O(\log n)$	$O(1)$	$O(\log\log n)$	$O(1)$
Single-Pair SP Query	$O(L + \log n)$	$O(L)$	$O(L + \log\log n)$	$O(L)$
Update Time	$O(\log n)$	$O(n^2)$	$O(n^{2\varepsilon})$	$O(n^{2\varepsilon})$

TABLE 1: Comparison of results for outerplanar digraphs in the case where ε is an arbitrary constant $0 < \varepsilon < (1/2)$.

Our approach is actually a (non-trivial) generalization of the method given in [3] and is based on: (i) a multilevel decomposition strategy based on graph separators; and (ii) on a sparsification of the input digraph where we keep shortest path information between properly chosen $\Theta(n)$ pairs of vertices.

As a main application of our algorithm we give faster algorithms for shortest path problems in planar digraphs. In the final section of this paper we list also other extensions and generalizations of our results.

2 Preliminaries and Data Structures

Let $G = (V(G), E(G))$ be a connected n-vertex digraph with real edge costs but no negative cycles. A *separation pair* is a pair (x, y) of vertices whose removal divides G into two disjoint subgraphs G_1 and G_2. We add the vertices x, y and the edges $\langle x, y\rangle$ and $\langle y, x\rangle$ to both G_1 and G_2. Let $0 < \alpha < 1$ be a constant. An *α-separator* S of G is a pair of sets $(V(S), D(S))$ where $D(S)$ is a set of separation pairs and $V(S)$ is the set of the vertices of $D(S)$ such that the removal of $V(S)$ leaves no connected component of more than αn vertices. We will call the separation vertices (pairs) of S that belong to any such resulting component H and separate it from the rest of the graph separation vertices (pairs) *attached to* H. It is well known that if G is outerplanar then there exists a 2/3-separator of G which is a single separation pair.

In the sequel, we assume w.l.o.g. that G_o is a biconnected n-vertex outerplanar digraph. Note that if G_o is not biconnected we can add an appropriate number of additional edges of very large costs in order to convert it into a biconnected outerplanar digraph (see e.g. [6]).

An l-decomposition of G_o, where l is an arbitrary positive integer, is a decomposition of G_o into $O(l)$ subgraphs such that (i) each subgraph is of size $O(n/l)$, and (ii) the number of separation pairs attached to each subgraph is $O(1)$. We say that the $O(n/l)$ separation pairs whose removal divides G_o into $O(l)$ subgraphs are *associated with* G_o.

Let λ be an arbitrary function over positive integers whose values are positive integers. Suppose that we compute a $\lambda(n)$-decomposition of G_o and then recursively find a $\lambda(n_i)$-decomposition of each resulting connected component G_i of n_i vertices until we get subgraphs of constant size. We associate a tree with the above decomposition as follows: at each level of recursion, the node associated with G_i is parent of the roots of the trees corresponding to the components of the $\lambda(n_i)$-decomposition of G_i. We call the resulting tree a *λ-decomposition tree* of G_o. Note that each subgraph G_i has $O(1)$ separation pairs attached to it and $O(n_i/\lambda(n_i))$ separation pairs associated with it. We describe a recursive algorithm that constructs a λ-decomposition tree $DT(G_o)$ of G_o that will be used in the construction of a suitable data structure for maintaining shortest path information in G_o. At each level of the recursion, the algorithm recursively decomposes an n_i-vertex graph G_i into $O(\lambda(n_i))$ subgraphs and builds the corresponding part of the decomposition tree.

Let in the algorithm below $\hat{G}$ denote an $\hat{n}$-vertex subgraph of G_o (initially $\hat{G} := G_o$).

ALGORITHM Decomp_Tree($\hat{G}, \lambda, DT(\hat{G})$)
BEGIN

1. For any connected component K of $\hat{G}$ with $|V(K)| > \hat{n}/\lambda(\hat{n})$ do Steps 1.1 and 1.2.

1.1. Denote by S the set of separation pairs (2/3-separators) in $\hat{G}$ found during all previous iterations and denote by n_{sep} the number of separation pairs of S attached to K. Check which of the following cases applies.

1.1.1. If $n_{sep} \leq 3$, then let $p = \{p_1, p_2\}$ be a separation pair of K that divides K into two subgraphs K_1 and K_2 with no more than $2\hat{n}/3$ vertices each.

1.1.2. Otherwise ($n_{sep} > 3$), let $p = \{p_1, p_2\}$ be a separation pair that separates K into subgraphs K_1 and K_2 each containing no more than 2/3 of the number of separation pairs attached to K.

1.2. Add p to S.

2. Find a λ-decomposition tree of each component of $\hat{G}$ by running this algorithm recursively.

3. Create a separator tree $DT(\hat{G})$ rooted at a new node associated with all separation pairs p found in Step 1.1 and whose children are the roots of the λ-decomposition trees of the components of $\hat{G}$.

END.

Following [3], we can implement each recursive step of the algorithm in $O(\hat{n})$ time and space. It is easy to see that by choosing $\lambda(n) = n^{\varepsilon}$, for any $0 < \varepsilon < (1/2)$, the depth of the $DT(G_o)$ is $O((1/\varepsilon) \log\log n)$. Hence, we have:

Lemma 1. *Let $\lambda(n) = n^{\varepsilon}$, where $0 < \varepsilon < (1/2)$ is an arbitrary number. Algorithm Decomp_Tree($G_o, \lambda, DT(G_o)$) constructs a λ-decomposition tree of G_o in $O((1/\varepsilon) n \log\log n)$ time and $O((1/\varepsilon) n \log\log n)$ space.*

Given an outerplanar digraph G_o and a set M of vertices of G_o, *compressing G_o with respect to M* means constructing a new outerplanar digraph of $O(|M|)$ size that contains M and such that the distance between any pair of vertices of M in the resulting graph is the same as the distance between the same vertices in G_o [6].

Definition 2. Let G_o be an n-vertex outerplanar digraph and let $\{p_1, p_2\}$ be a separator pair of G_o that divides G_o into connected components one of which is G. Let S be a set of $\lambda(n)$ separation pairs that divides G into $O(\lambda(n))$ subgraphs K. Let A_K be the set of separator pairs attached to K, $|A_K| = O(1)$. Construct a digraph $SR(G)$ as follows: remove S from G, compress each resulting subgraph K with respect to $(V(S) \cup \{p_1, p_2\}) \cap V(K)$, and join the resulting subgraphs at vertices $V(A_K)$. We call $SR(G)$ the *sparse representative* of G.

Remark: Let e be an edge with both of its endpoints in A_K. It is clear that e can be shared by at most two subgraphs K. In the above definition, when subgraphs are joined at the vertices of $V(A_K)$, we keep as the cost of e the smallest of the (possibly) two different costs that e may have in the two subgraphs.

3 The Dynamic Shortest Path Algorithm for Outerplanar Digraphs

3.1 The Preprocessing Phase

The preprocessing algorithm constructs the λ-decomposition tree $DT(G_o)$. Each node of $DT(G_o)$ is associated with a subgraph G of G_o along with the set of separation pairs associate with it (as they are determined by the decomposition procedure), and also contains a pointer to the sparse representative $SR(G)$ of G. The sparse representative $SR(G)$ is computed for all graphs G of $DT(G_o)$. According to Definition 2, $SR(G)$ consists of the union of the compressed versions of G_i with respect to the separation pairs attached to G plus the associated separation pairs dividing G into the subgraphs G_i, where G_i's are the children of G in $DT(G_o)$. Therefore the size of $SR(G)$ is proportional to the number of separation pairs attached to and associated with G. Note that for each leaf of $DT(G_o)$ we have that $SR(G) \equiv G$, since in this case G is of $O(1)$ size. Moreover, for all graphs G of $DT(G_o)$, the preprocessing algorithm also computes all pairs shortest path (APSP) information between the separation pairs associated with G. Thus, in the query phase, we can answer distance queries in $SR(G)$ in constant time and shortest path queries in time proportional to the number of edges of the path.

In the following, let $\lambda(n) = n^{\varepsilon}$, where $0 < \varepsilon < (1/2)$ is any arbitrary number.

ALGORITHM Pre_1(G_o)
BEGIN
1. Construct a λ-decomposition tree $DT(G_o)$.
2. Compute the sparse representative $SR(G_o)$ of G_o as follows.
   for each child G of G_o in $DT(G_o)$ **do**
(a) **if** G is a leaf of $DT(G_o)$ **then** $SR(G) = G$
    else find $SR(G)$ by running Step 2 recursively on G.
(b) Construct the sparse representative of G_o as described in Definition 2 by using the sparse representatives of the children of G_o.
(c) Run an APSP algorithm on $SR(G_o)$ storing the shortest path information among the $O(\lambda(n))$ separation vertices in a table.
3. Generate a table with entries $[v, G_l]$, where G_l is the leaf subgraph of $DT(G_o)$ containing v. Construct a similar table for the edges of G_o.
4. Preprocess $DT(G_o)$ (using e.g. the algorithm in [9]) such that lowest common ancestor queries can be answered in $O(1)$ time.
END.

From the discussion preceding the algorithm and Lemma 1, we have:

Lemma 3. *Let $\lambda(n) = n^{\varepsilon}$, where $0 < \varepsilon < (1/2)$ is any arbitrary number. Algorithm Pre_1(G_o) takes $O((1/\varepsilon)n \log\log n)$ time and space.*

3.2 Answering a Query

The query algorithm computes the distance between any two vertices v and z of G_o and proceeds as follows. First, use $DT(G_o)$ to find a subgraph G of G_o such that there exists at least one separation pair associated with G that separates v from z. Let $P_{12} = (p_1, p_2)$ be a separation pair associated with G such that v and p_1, p_2 belong to the same child subgraph of G and also P_{12} separates v from z.

Let $P_{34} = (p_3, p_4)$ be another separation pair associated with G which separates v from z and moreover, z and p_3, p_4 belong to the same child subgraph of G. (Note that P_{12} and P_{34} may coincide.) Let $d(v, z)$ denote the distance between v and z. Then clearly,

$$d(v,z) = \min\{\min\{d(v,p_1) + d(p_1,p_3) + d(p_3,z), d(v,p_1) + d(p_1,p_4) + d(p_4,z)\},$$
$$\min\{d(v,p_2) + d(p_2,p_3) + d(p_3,z), d(v,p_2) + d(p_2,p_4) + d(p_4,z)\}\}. \quad (1)$$

Hence, for answering the query it suffices to compute the distances $d(v, p_1)$, $d(p_3, z)$, $d(v, p_2)$, $d(p_4, z)$ and $D(P_{12}, P_{34})$, where $D(P_{12}, P_{34})$ denotes the set of all four distances from a vertex in P_{12} to a vertex in P_{34}. In order to do this we will need the shortest path information stored in the tables of the sparse representatives.

Now we discuss how one can use the information the sparse representatives provide. Let $s = (s_1, s_2)$ be any separation pair *attached to G*. The distance from s_1 to s_2 in $SR(G)$ is, by the preprocessing algorithm, equal to the distance between s_1 and s_2 in G. Note that, in general, the distance from s_1 to s_2 in G might be different from the distance between these vertices in G_o. Before we present the query algorithm, we give a way to determine the distances in G_o between the vertices of certain separation pairs that are used in the computation of the distance between v and z. Let G_v be the subgraph associated with the leaf node of $DT(G_o)$ that contains v. Let $D(G_v)$ be the set of all distances in G_o between the vertices of the separation pairs attached to ancestors of G_v (including G_v itself) in $DT(G_o)$. Then $D(G_v)$ can be found by the following algorithm.

ALGORITHM Attached_Pairs(G_v)
BEGIN

1. Let G' be the parent of G_v in $DT(G_o)$. If $G' = G_o$ then $D(G') := \emptyset$; otherwise compute recursively $D(G')$ by this algorithm.

2. For each separation pair (s'_1, s'_2) attached to G', find $d(s'_1, s'_2)$ and $d(s'_2, s'_1)$ in G_o by using the tables of $SR(G')$ and the information in $D(G')$. Set $D(G_v) := D(G') \cup \{d(s'_1, s'_2), d(s'_2, s'_1)\}$.

END.

Algorithm Attached_Pairs can be used to compute the distances in G_o between the vertices of all separation pairs attached to G_v or to any ancestor of G_v in $DT(G_o)$, so that one can ignore the rest of G_o when computing distances in G_v or in one of its ancestors. (As a consequence, we can also compute the correct distances in $D(P_{12}, P_{34})$ in $O(1)$ time, using the tables of the sparse representative of G.) It is not hard to see that the running time of the above algorithm is $O((1/\varepsilon) \log\log n)$ (i.e. proportional to the depth of $DT(G_o)$).

Next we describe the query algorithm. Let v' be a vertex that belongs to the same subgraph G_v of G_o that is a leaf of $DT(G_o)$ and that contains v. Let $p(v)$ be the pair of vertices v, v'. Similarly define a pair of vertices $p(z)$ that contains z and a vertex z' which belongs to the leaf G_z of $DT(G_o)$ containing z. Then (1) shows that $D(p(v), p(z))$ can be found in constant time, given $D(p(v), P_{12})$,

$D(P_{12}, P_{34})$ and $D(P_{34}, p(z))$. The following recursive algorithm is based on the above fact.

ALGORITHM Query_1(G_o, v, z)
BEGIN

1. Find the subgraphs G_v and G_z as defined above.
2. Run the Algorithm Attached_Pairs on G_v and on G_z .
3. Find pairs of vertices $p(v)$ and $p(z)$ as defined above.
4. Find a subgraph G of G_o such that there exist separation pairs P_{12}, P_{34} associated with G (as defined above) that separate $p(v)$ and $p(z)$ in G. Use the information found at step 2 and the tables of $SR(G)$ to compute the correct distances in $D(P_{12}, P_{34})$.
5. Find $D(p(v), P_{12})$ as follows:

5.1. Let G' be the child of G in $DT(G_o)$ that contains $p(v)$ and P_{12}. If G' is a leaf of $DT(G_o)$, then determine $D(p(v), P_{12})$ directly in constant time.

5.2. If G' is not a leaf then find the child G'' of G' that contains $p(v)$. For each separation pair p' attached to G'' do the following. Compute $D(p(v), p')$ by executing Step 5 recursively with $P_{12} := p'$, and then find $D(p(v), P_{12})$ using (1). (Note that $D(p', P_{12})$ can be taken from the tables of $SR(G')$.) Keep as $D(p(v), P_{12})$ the minimum of the computed distances.

6. Find $D(P_{34}, p(z))$ as in Step 5.
7. Use $D(p(v), P_{12})$, $D(P_{12}, P_{34})$, $D(P_{34}, p(z))$ and (1) to find $D(p(v), p(z))$.

END.

Steps 1, 3 and 4 of algorithm Query_1 take $O(1)$ time by the preprocessing of G_o. Step 2 takes $O((1/\varepsilon) \log \log n)$ time (as discussed above). It is not difficult to see that each recursive execution of Step 5 takes $O(1)$ time and the depth of the recursion is bounded by the depth of $DT(G_o)$. Thus, we have:

Lemma 4. *Algorithm Query_1(G_o, v, z) finds the distance between any two vertices v and z of an n-vertex outerplanar digraph G_o in $O((1/\varepsilon) \log \log n)$ time.*

Algorithm Query_1 can be modified in order to answer path queries. The additional work (compared with the case of distances) involves uncompressing the shortest paths corresponding to edges of the sparse representatives of the graphs from $DT(G_o)$. Uncompressing an edge from a graph $SR(G)$ involves a traversal of a subtree of $DT(G_o)$, where at each step an edge is replaced by $|G|^\varepsilon$ new edges each possibly corresponding to a compressed path. Obviously this subtree will have no more than L leaves, where L is the number of the edges of the output path. Then the traversal time can not exceed the number of the vertices of a binary tree with L leaves in which each internal node has exactly 2 children. Any such tree has $2L - 1$ vertices. Thus the next claim follows.

Lemma 5. *The shortest path between any two vertices v and z of an n-vertex outerplanar digraph G_o can be found in $O(L + (1/\varepsilon) \log \log n)$ time, where L is the number of edges of the path.*

3.3 Updating the Data Structures

In the sequel, we will show how we can update our data structures for answering shortest path and distance queries in outerplanar digraphs, in the case where an edge cost is modified. The algorithm for updating the cost of an edge e in an n-vertex outerplanar digraph G_o is based on the following idea: the edge will belong to at most $O((1/\varepsilon)\log\log n)$ subgraphs of G_o, as they are determined by Algorithm Pre_1. Therefore, it suffices to update (in a bottom-up fashion) the sparse representatives, as well as their tables, of those subgraphs that are on the path from the subgraph G_l containing e (where G_l is a leaf of $DT(G_o)$) to the root of $DT(G_o)$. Let $parent(G)$ denote the parent of a node G in $DT(G_o)$, and $\hat{G}$ denote any sibling of a node G in a $DT(G_o)$ such that G and $\hat{G}$ have a common separation pair attached to them. (Note also that an edge e can belong to at most one other sibling of G_l.) The algorithm for the update operation is the following.

ALGORITHM Update_1($G_o, e, w(e)$)
BEGIN
1. Find a leaf G of $DT(G_o)$ for which $e \in E(G)$.
2. Update the cost of e in G with the new cost $w(e)$.
3. If e belongs also to some $\hat{G}$ then update the cost of e in $\hat{G}$.
4. **While** $G \neq G_o$ **do**
    (a) Update $SR(parent(G))$ by using the new versions of $SR(G)$ and $SR(\hat{G})$ and then by running an APSP algorithm on it.
    (b) $G := parent(G)$.

END.

The first three steps of the above algorithm require $O(1)$ time. Let $U(n)$ be the maximum time required by Step 4. Then it is clear that after updating recursively the child subgraph of G_o containing edge e, we need $O(\lambda(n))$ time to recompute $SR(G_o)$ plus $O(\lambda^2(n))$ time to recompute the APSP tables of $SR(G_o)$. Hence, $U(n) \leq U(n/\lambda(n)) + O(\lambda^2(n))$. Letting $\lambda(n) = n^\varepsilon$, we have:

Lemma 6. *Let $\lambda(n) = n^\varepsilon$, where $0 < \varepsilon < (1/2)$ is an arbitrary number. Algorithm Update_1 updates after an edge cost modification the data structures created by the preprocessing algorithm in $O(f(\varepsilon)n^{2\varepsilon})$ time, where $f(\varepsilon) = 1$, if ε is a constant (independent of n), or $f(\varepsilon) = (1/\varepsilon)\lceil\log\log n\rceil$, if ε depends on n.*

Summarizing all the results in Section 3, we get:

Theorem 7. *Given an n-vertex outerplanar digraph G_o with real-valued edge costs but no negative cycles and an arbitrary number $0 < \varepsilon < (1/2)$, there exists an algorithm for maintaining all pairs shortest paths information in G_o under any edge cost modification, with the following performance characteristics: (i) preprocessing time and space $O((1/\varepsilon)n\log\log n)$; (ii) single-pair distance query time $O((1/\varepsilon)\log\log n)$; (iii) single-pair shortest path query time*

$O(L + (1/\varepsilon)\log\log n)$ (where L is the number of edges of the path); (iv) update time (after an edge cost modification) $O(f(\varepsilon)n^{2\varepsilon})$, where $f(\varepsilon) = 1$, if ε is independent of n, or $f(\varepsilon) = (1/\varepsilon)\lceil\log\log n\rceil$ otherwise.

4 Improving More on the Query Time

In this section we shall describe how the algorithms presented in the previous section can be modified such that a distance query is answered in $O(1)$ time. In the following, let $\lambda(n) = n^{\varepsilon}$, for some arbitrary number $0 < \varepsilon < (1/2)$.

We change the first step of the preprocessing algorithm as follows. Instead of dividing each child subgraph H of G_o into $\lambda(n/\lambda(n))$ subgraphs, we can divide it into $\lambda(n)$ subgraphs. This will reduce the depth of $DT(G_o)$ to $O(1/\varepsilon)$. However, notice that now all the descendant subgraphs of G_o which are leaves of $DT(G_o)$ are of size $O(n^{\varepsilon})$ and there are $O(n^{1-\varepsilon})$ of them. We shall run on these subgraphs an APSP algorithm. Call the new preprocessing algorithm Pre_2(G_o). Hence, we have the following:

Lemma 8. *Algorithm Pre_2(G_o) runs in $O((1/\varepsilon)n + n^{1+\varepsilon})$ time and uses $O((1/\varepsilon)n + n^{1+\varepsilon})$ space.*

A query is answered in the same way as before. But since now the depth of $DT(G_o)$ is $O(1/\varepsilon)$, we can answer a query in this time.

The data structures can be updated (after an edge cost modification) using the same approach as in Section 3.3. This means that we need $O(1/\varepsilon)$ iterations and for each $SR(G)$ of a descendant subgraph G we have to run an APSP algorithm for updating the shortest path information among the separation pairs associated with it. Also we have to run the APSP algorithm to the leaf subgraph of G_o containing the edge whose cost has been modified. This will give us a total of $O((1/\varepsilon)n^{2\varepsilon})$ time for updating our data structures.

The above discussion leads to the following.

Theorem 9. *Given an n-vertex outerplanar digraph G_o with real-valued edge costs but no negative cycles and an arbitrary number $0 < \varepsilon < (1/2)$, there exists an algorithm for maintaining all pairs shortest paths information in G_o under any edge cost modification with the following performance characteristics: (i) preprocessing time and space $O((1/\varepsilon)n + n^{1+\varepsilon})$; (ii) single-pair distance query time $O(1/\varepsilon)$; (iii) single-pair shortest path query time $O(L+1/\varepsilon)$ (where L is the number of edges of the path); (iv) update time (after an edge cost modification) $O((1/\varepsilon)n^{2\varepsilon})$.*

5 Extensions of our Results

The algorithms for the DSP problem for outerplanar digraphs we described in this paper can be used for constructing faster algorithms for planar digraphs. The approach used is the same as the one in [3] and is (partially) based on the hammock decomposition technique introduced by Frederickson in [5, 6]. This technique allows the reduction of the shortest paths problems on planar digraphs with nice topology to similar problems on outerplanar digraphs. Our results for planar digraphs can be obtained by incorporating the results of Theorems 7 and 9 into the algorithms of [3]. (We omit details due to space limitations and the interested reader is referred to [3].)

We mention also the following extensions and generalizations of our results: (i) We can handle efficiently edge deletions. (Note that deletion of an edge e is equivalent to assigning a very large cost to e so that no shortest path will use e.) (ii) Our algorithms can detect a negative cycle (in a way similar to that described in [3]), either if it exists in the initial digraph, or if it is created after an edge cost modification. (iii) Using the ideas of [5], our results can be used to design improved algorithms for the DSP problem on digraphs with small genus. (iv) Although our algorithms do not directly support edge insertion, they are fast enough so that even if the preprocessing algorithm is run from scratch after any edge insertion, they still provide better performance compared with the naive approach. Moreover, our algorithms can support a special kind of edge insertion, called *edge re-insertion*. That is, we can insert any edge that has previously been deleted within the resource bounds of the update operation.

References

1. H. Bondlaender, "Dynamic Algorithms for Graphs with Treewidth 2", *Proc. 19th WG'93*, LNCS 790, pp.112-124, Springer-Verlag, 1994.
2. S. Chaudhuri and C. Zaroliagis, "Shortest Path Queries in Digraphs of Small Treewidth", *Proc. 22nd ICALP*, LNCS, Springer-Verlag, 1995, to appear.
3. H. Djidjev, G. Pantziou and C. Zaroliagis, "On-line and Dynamic Algorithms for Shortest Path Problems", *Proc. 12th STACS*, LNCS 900, pp.193-204, Springer-Verlag, 1995.
4. E. Feuerstein and A.M. Spaccamela, "Dynamic Algorithms for Shortest Paths in Planar Graphs", *Theor. Computer Science*, 116 (1993), pp.359-371.
5. G.N. Frederickson, "Using Cellular Graph Embeddings in Solving All Pairs Shortest Path Problems", *Proc. 30th Annual IEEE Symp. on FOCS*, 1989.
6. G.N. Frederickson, "Planar Graph Decomposition and All Pairs Shortest Paths", *J. ACM*, Vol.38, No.1, January 1991, pp.162-204.
7. R. Hassin, "Maximum flow in (s,t)-planar networks", *Inform. Proc. Lett.*, 13(1981), p.107.
8. G. Miller and J. Naor, "Flows in planar graphs with multiple sources and sinks", *Proc. 30th IEEE Symp. on FOCS*, 1989, pp.112-117.
9. B. Schieber and U. Vishkin, "On Finding Lowest Common Ancestors: Simplification and Parallelization", *SIAM J. Computing*, 17(6), pp.1253-1262, 1988.

r–Domination Problems on Homogeneously Orderable Graphs*

(extended abstract)

Feodor F. Dragan[1] and Falk Nicolai[2]

[1] Department of Mathematics and Cybernetics, Moldova State University
A. Mateevici str. 60, Chişinău 277009, Moldova
[2] Gerhard-Mercator-Universität –GH– Duisburg, FB Mathematik, FG Informatik
D 47048 Duisburg, Germany

Abstract. In this paper we consider r–dominating cliques in homogeneously orderable graphs (a common generalization of dually chordal and distance–hereditary graphs) and their relation to strict r–packing sets. We give a simple criterion for the existence of r–dominating clique and show that the cardinality of a maximum strict r–packing set equals the cardinality of a minimum r–dominating clique provided the last parameter exists and is not two. Finally we present two efficient algorithms. The first one decides whether a given homogeneously orderable graph has a r–dominating clique and, if so, computes both a minimum r–dominating clique and a maximum strict r–packing set of the graph. The second one computes a minimum connected r–dominating set in this graph.

1 Introduction

In a graph $G = (V, E)$ a subset $D \subseteq V$ is a *dominating set* iff each vertex $v \in V \setminus D$ has at least one neighbour in D. Often certain constraints for dominating sets are required : the dominating set must be connected (*connected dominating set*), complete (*dominating clique*), independent (*independent dominating set*) and so on.

Since V itself is a dominating set of G every graph has a dominating set, but computing a minimum one (i.e. a dominating set of smallest size) is in general $\mathbb{NP}$–complete. For special graph classes the situation is sometimes much better. There are a lot of papers concerned with finding minimum dominating sets in special graphs — for a bibliography of domination c.f. [10], for a compact survey of special graph classes we refer to [1].

The dominating clique problem is of a somewhat different nature since not every graph has a dominating clique. Indeed, there are two problems — decide whether a given graph possesses a dominating clique and, if so, then compute a minimum one. For the well–known class of weakly chordal graphs (i.e. those graphs which does not contain an induced cycle or its complement of length larger than four) the decision problem is $\mathbb{NP}$–complete (c.f. [5]). In chordal graphs (the weakly chordal graphs which does not contain an induced 4–cycle) the decision problem is easy but

* First author supported by DAAD. Second author supported by DFG. E-mail addresses : dragan@university.moldova.su, nicolai@marvin.informatik.uni-duisburg.de

it is NP–complete to compute a minimum dominating clique (c.f. [11]). In contrast, computing minimum dominating cliques is polynomial in strongly chordal graphs (c.f. [11]), an important subclass of chordal graphs.

In this paper we investigate the more general problem of r–domination. Given a graph G and a vertex function $r : V(G) \rightarrow \mathbb{N}$ (i.e. a nonnegative integer is assigned to each vertex) a set $D \subseteq V$ *r-dominates* G (is a *r-dominating set*) iff for each vertex v of $V \setminus D$ there is a vertex x in D such that $d_G(v,x) \leq r(v)$ where d_G is as usual the distance metric on G. Obviously, with $r(v) = 1$ for all $v \in V$ the classical domination problem is a special case of the r–domination problem. Again, certain constraints for a r–dominating set are considered yielding the problems r–dominating clique, connected r–dominating set and so on.

Recall that the connected r–dominating set problem is a generalization of the Steiner tree problem. Indeed, given a Steiner set T we assign to each vertex $t \in T$ the value $r(t) := 0$ (for all other vertices v define $r(v) := |V(G)|$) and then compute a minimum connected r–dominating set which is a Steiner tree.

Note that the r–dominating clique problem is a generalization of the central vertex (a vertex with minimal eccentricity) problem (c.f. [7]).

In [8] the existence criterion for chordal graphs given in [11] is generalized in terms of r–dominating cliques and is proved to be valid (in this generalized form) for Helly graphs and chordal graphs. Again the computation of a minimum r–dominating clique is NP–complete for Helly graphs. For dually chordal graphs — a subclass of Helly graphs containing all strongly chordal graphs — a linear time algorithm is presented (see [3]).

In distance–hereditary graphs both the decision and the minimality problem can be solved in linear time as was shown in [7].

In this paper we consider r–dominating cliques in homogeneously orderable graphs and their relation to strict r–packing sets. Homogeneously orderable graphs were introduced in [4] as a common generalization of dually chordal and distance–hereditary graphs.

We prove that a homogeneously orderable graph G possesses a r–dominating clique if and only if for any pair of vertices x, y of G $d(x,y) \leq r(x) + r(y) + 1$ holds where $r : V \rightarrow \mathbb{N}$ is a given vertex function. Again this result is a generalization of the one for dually chordal (cf. [8]) and distance–hereditary graphs (cf. [7]).

Furthermore we show that for homogeneously orderable graphs with r–dominating cliques the cardinality of a maximum strict r–packing set equals the cardinality of a minimum r–dominating clique provided the last parameter is not two.

Finally we present two efficient algorithms with quadratic running time. The first one decides whether a given homogeneously orderable graph has a r–dominating clique and, if so, computes both a minimum r–dominating clique and a maximum strict r–packing set of the graph. The second one computes a minimum connected r–dominating set in homogeneously orderable graphs. For dually chordal graphs we refer to [3, 6], for distance–hereditary graphs see [7, 2].

Note that in [4] we already presented a quadratic time algorithm for the Steiner tree problem on homogeneously orderable graphs.

2 Preliminaries

Throughout this paper all graphs $G = (V, E)$ are finite, undirected, simple (i.e. loop-free and without multiple edges) and connected.

The *(open) neighbourhood* of a vertex v of G is $N(v) := \{u \in V : uv \in E\}$.

A nonempty set $H \subseteq V$ is *homogeneous* in $G = (V, E)$ iff any vertex $w \in V \setminus H$ is adjacent to either all or none of the vertices from H.

A homogeneous set H is *proper* iff $|H| < |V|$. Trivially for each $v \in V$ the singleton $\{v\}$ is a proper homogeneous set. Note also that for a subset $U \subset V$ if a set $H \subseteq U$ is homogeneous in G then it is homogeneous also in the induced subgraph $G(U)$ but not vice versa.

The *distance* $d_G(u, v)$ of vertices u, v is the minimal length of any path connecting these vertices. If no confusion can arise we will omit the index G. For a set $S \subseteq V$ and a vertex $v \in V$ we define the distance of v to S as $d(v, S) := \min\{d(v, x) : x \in S\}$.

The *k-th neighbourhood* $N^k(v)$ of a vertex v of G is the set of all vertices of distance k to v, i.e. $N^k(v) := \{u \in V : d_G(u, v) = k\}$.

The *disk* of radius k centered at v is the set of all vertices of distance at most k to v, i.e. $D(v, k) := \{u \in V : d_G(u, v) \leq k\} = \bigcup_{i=0}^{k} N^i(v)$. For a set $U \subseteq V$ we define $D(U, k) := \bigcup_{u \in U} D(u, k)$.

Let $e(v)$ denote the *eccentricity* of vertex $v \in V$, i.e. $e(v) := \max\{d(v, u) : u \in V\}$. Then, the *radius* $rad(G)$ of G is the minimum over all eccentricities $e(v)$, $v \in V$, whereas the *diameter* $diam(G)$ of G is the maximum over all eccentricities $e(v)$ for v in V. The *k-th power* G^k of G is the graph with the same vertex set as G where two vertices are adjacent iff their distance is at most k in G.

In the sequel a subset U of V is a *k-set* iff U induces a clique in the power G^k, i.e. for any pair x, y of vertices of U we have $d_G(x, y) \leq k$.

Let U_1, U_2 be disjoint subsets of V. If every vertex of U_1 is adjacent to every vertex of U_2 then U_1 and U_2 form a *join*, denoted by $U_1 \bowtie U_2$. A set $U \subseteq V$ is *join-splitted* iff U can be partitioned into two nonempty sets U_1, U_2 such that $U = U_1 \bowtie U_2$.

Next we recall the definition of homogeneously orderable graphs as given in [4] : A vertex v of $G = (V, E)$ with $|V| > 1$ is *h-extremal* iff there is a proper subset $H \subset D(v, 2)$ which is homogeneous in G and for which $D(v, 2) \subseteq D(H, 1)$ holds. A sequence $\sigma = (v_1, \ldots, v_n)$ is a *h-extremal ordering* iff for any $i = 1, \ldots, n-1$ the vertex v_i is h-extremal in $G_i := G(\{v_i, \ldots, v_n\})$. A graph G is *homogeneously orderable* iff G has a h-extremal ordering.

In [4] we proved that a graph is homogeneously orderable iff the hypergraph of the maximal join-splitted sets is a dual hypertree. Moreover,

Theorem 2.1 *A graph G is homogeneously orderable if and only if the square G^2 of G is chordal and each maximal 2-set of G is join-splitted.*

Furthermore we characterized hereditary homogeneously orderable graphs (i.e. those graphs for which every induced subgraph is homogeneously orderable too) as the house-hole-domino-sun-free graphs (HHDS-free graphs).

The local structure of homogeneously orderable graphs implies a simple cubic time recognition algorithm for these graphs which produces a h–extremal ordering too.

As shown in [4] both dually chordal and distance–hereditary graphs are homogeneously orderable graphs.

The following two lemmata are important for the sequel.

Lemma 2.2 ([4]) *If v is a h–extremal vertex with $e(v) \geq 2$ then there is a homogeneous set $H \subseteq N(v)$ dominating $D(v, 2)$.*

Lemma 2.3 ([4]) *If v is a h–extremal vertex of a graph G with $e(v) \geq 2$ then $G \setminus \{v\}$ is an isometric subgraph of G.*

Hereby a (connected) induced subgraph F of G is *isometric* iff the distances within F are the same as in G, i.e. for any pair of vertices x, y of F we have $d_F(x, y) = d_G(x, y)$.

Finally we recall the concept of r–domination. For a vertex function $r : V \to \mathbb{N}$ (note that we assume zero to be a natural number) a set $D \subseteq V$ *r–dominates* G iff for each vertex $v \in V \setminus D$ there is a vertex $x \in D$ such that $d(v, x) \leq r(v)$ holds. A r–dominating set D is *minimal* iff for any vertex $x \in D$ the set $D \setminus \{x\}$ does not r–dominate G. A minimal r–dominating set D is *minimum* iff D has the smallest cardinality among all minimal r–dominating sets of G. Analogously one can define *connected r–dominating sets* and *r–dominating cliques.*

If C is a minimal r–dominating clique of a graph G then the minimality of C implies that for every vertex c of C there must be a vertex x_c in G such that $d(x_c, c) \leq r(x_c)$ and $d(x_c, c') > r(x_c)$ for all $c' \in C \setminus \{c\}$, i.e. x_c is r–dominated only by c. Such a vertex x_c we call *private neighbour* of c.

A dual concept is the following : A set $S \subseteq V$ is called *strict r–packing set* iff for all vertices x, y of S the equation $d(x, y) = r(x) + r(y) + 1$ holds. As above we can define maximal and maximum strict r–packing sets.
So we have the following parameters :

- $\pi_r(G)$ — the size of a maximum strict r–packing set of G,
- $\gamma_r(G)$ — the size of a minimum r–dominating set of G,
- $\gamma_{r,con}(G)$ — the size of a minimum connected r–dominating set of G and
- $\gamma_{r,cl}(G)$ — the size of a minimum r–dominating clique of G or ∞.

Note that for arbitrary graphs G we obviously have $\pi_r(G) \leq \gamma_r(G) \leq \gamma_{r,con}(G) \leq \gamma_{r,cl}(G)$.

3 Theoretical results

For the sequel let G be a graph with vertex function $r : V \to \mathbb{N}$. Define $Z(G)$ to be the set of vertices of G with r–value zero.

The following straightforward lemma handles the case $e(v) \leq 1$ for some vertex v of G. So in the sequel we may assume $e(v) \geq 2$.

Lemma 3.1 *If v is a vertex with $e(v) \leq 1$ and $Z(G)$ is complete then we have :*

1. *If $Z(G) \neq \emptyset$ r–dominates G then $Z(G)$ is both minimum r–dominating clique and maximum strict r–packing set of G.*
2. *If $Z(G) \neq \emptyset$ does not r–dominate G then $Z(G) \cup \{v\}$ is a minimum r–dominating clique and $Z(G) \cup \{u\}$ is a maximum strict r–packing set of G where u is a private neighbour of v.*
3. *If $Z(G) = \emptyset$ then $\{v\}$ is both minimum r–dominating clique and maximum strict r–packing set of G.*

3.1 The existence of r–dominating cliques

Lemma 3.2 *Let G be a homogeneously orderable graph with vertex function $r : V \to \mathbb{N}$, v be a h–extremal vertex such that $e(v) \geq 2$ and let $H \subseteq N(v)$ be a homogeneous set dominating $D(v, 2)$. Furthermore let S be an arbitrary subset of V containing v and fulfilling*

$$(P) \qquad \forall x, y \in S : d_G(x, y) \leq r(x) + r(y) + 1.$$

Define $S' := (S \setminus \{v\}) \cup \{w\}$ where w is either

(H1) *a vertex from $S \cap N(v) \cap Z(G)$ if this intersection is nonempty but $S \cap H \cap Z(G) = \emptyset$, or*
(H2) *a vertex from $S \cap H$ with minimal r–value if $H \cap S \neq \emptyset$, or*
(H3) *a vertex from H with minimal r–value otherwise.*

Then S' fulfills (P) in $G' := G \setminus \{v\}$ with respect to r' where $r'(x) := r(x)$ for all $x \in V \setminus \{w, v\}$ and

$$r'(w) := \begin{cases} 0 & : \ (H1) \text{ or } r(v) = 0, \\ \min\{r(w), r(v) - 1\} & : \ (H2) \text{ and } r(v) \geq 1, \\ r(v) - 1 & : \ (H3) \text{ and } r(v) \geq 1. \end{cases}$$

Proof omitted.

Theorem 3.3 *Let G be a homogeneously orderable graph with vertex function $r : V \to \mathbb{N}$ and let S be a subset of V. Then S is r–dominated by some clique C of G if and only if $d_G(x, y) \leq r(x) + r(y) + 1$ for all $x, y \in S$.*

Proof : Obviously, if S is r–dominated by some clique C then the distance requirements are fulfilled. The converse we prove by induction on the size of G. Let v be a h–extremal vertex. We may assume $e(v) \geq 2$ by Lemma 3.1. Hence we can choose a homogeneous set $H \subseteq N(v)$ dominating $D(v, 2)$. Let S be an arbitrary subset of V which fulfills the distance requirements. If $v \notin S$ then we are done by the induction hypothesis and Lemma 2.3. So let $v \in S$, $G' := G \setminus \{v\}$ and $S' := (S \setminus \{v\}) \cup \{w\}$ where $w \in N(v)$ is choosen according to the rules $(H1)$–$(H3)$ of Lemma 3.2. By Lemma 3.2 and the induction hypothesis S' is r'–dominated (r' defined as in Lemma 3.2) by some clique C' in G'. If $r(v) \geq 1$ we are done since in all cases the vertex r'–dominating w r–dominates v too. If $r(v) = 0$ we have $r'(w) = 0$ and hence $C' \cap N(v)$ is nonempty. Suppose $C := \{v\} \cup (C' \cap N(v))$ does not r–dominate S in G. Since

C' is a r'–dominating clique for S' in G' there must be vertices $x \in C' \cap N^2(v)$ and $y \in S$ such that $d(x,y) = r(y)$ and $d(y,C) > r(y)$. Note that $r(y) \geq 1$ since $r(x) > 0$ and $x \neq y$. Thus $y \notin N(v)$ since otherwise v r–dominates y.

Case $(H2)$, **and** $(H3)$. Here we have $w \in H$. Since $r'(w) = 0$ implies $w \in C$ we obtain $d(w,y) \geq r(y) + 1$. From $d(v,y) = d(w,y) + 1$ and $r(v) = 0$ we conclude $d(v,y) \geq r(v) + r(y) + 2$, a contradiction.

Case $(H1)$. We immediately conclude $r'(w) = r(w) = 0$ and $w \in N(v) \setminus H$. Thus $d(w,y) = d(v,y) = r(y) + 1$. If $C \cap H$ is nonempty then choose any vertex $h \in C \cap H$ and proceed as in the above case. If $C \cap H = \emptyset$ the above considerations imply that the new clique $C := \{v,h\} \cup (C' \cap N(v))$, with $h \in H$, is a r–dominating clique in G. □

Corollary 3.4 *For homogeneously orderable graphs G we have $2rad(G) \geq diam(G) \geq 2(rad(G) - 1)$.*

3.2 Minimum r–dominating cliques and maximum strict r–packing sets

Here we consider the relationship of the parameters $\pi_r(G)$, $\gamma_r(G)$, $\gamma_{r,con}(G)$ and $\gamma_{r,cl}(G)$ for homogeneously orderable graphs with r–dominating cliques. Proofs of the next two lemmata are omitted.

Lemma 3.5 *Let G be a homogeneously orderable graph with h–extremal vertex v, $e(v) \geq 2$, and with a vertex function $r : V \to \mathbb{N}$ such that $r(v) \geq 1$. Moreover, assume that G is not r–dominated by a single vertex but by some minimum clique containing v. Then there is a r–dominating clique of G of the same size which does not contain v.*

In the sequel we will often apply Theorem 3.3 and Lemma 3.2. In all of these cases we will have $S := V(G)$. Thus rule $(H3)$ for the choice of w will never be used.

Lemma 3.6 *Let G be a homogeneously orderable graph which is r–dominated by some clique but not by a single vertex. Let v be any h–extremal vertex of G with $e(v) \geq 2$ and $r(v) \geq 1$. Furthermore let $H \subseteq N(v)$ be a homogeneous set dominating $D(v,2)$. Define w, G' and r' as in Lemma 3.2 with $S := V(G)$. Then any r'–dominating clique C' in G' is a r–dominating clique in G, and if C is a minimum r–dominating clique in G then there exists a r'–dominating clique C' in G' of the same size, i.e. $\gamma_{r,cl}(G) = \gamma_{r',cl}(G')$.*

Theorem 3.7 *If a homogeneously orderable graph G possesses a r–dominating clique and $\gamma_{r,cl}(G) \neq 2$ then $\pi_r(G) = \gamma_{r,cl}(G)$.*

Proof : Since for $\gamma_{r,cl}(G) = 1$ there is nothing to show let $\gamma_{r,cl}(G) \geq 3$. Let v be a h–extremal vertex of G. If $e(v) = 1$ then we are done by Lemma 3.1. So assume $e(v) \geq 2$ and let $H \subseteq N(v)$ be a homogeneous set dominating $D(v,2)$. First we consider the case $r(v) = 0$. If $Z(G)$ r–dominates G then it is both a minimum r–dominating clique and a maximum strict r–packing set of G. Otherwise we show that $C := Z(G) \cup \{h\}$ with $h \in H$ is a r–dominating clique in G. Assume $Z(G) \cap H \neq \emptyset$

and let h' be a vertex from this intersection. We prove that $Z(G)$ r–dominates G. Suppose for the contrary that there is a vertex x with $d(x, Z(G)) > r(x)$. Obviously $x \notin D(v, 1)$ implying $d(v, x) = d(h', x) + 1$. Consequently $d(x, v) > r(x) + r(v) + 1$, a contradiction to Theorem 3.3. Thus $Z(G) \cap H$ is empty and C is complete. By similar arguments (replace h' by h) C r–dominates G. Since $Z(G)$ does not r–dominate G the clique C is minimum and there is a private neighbour x of h. Thus $Z(G) \cup \{x\}$ is a maximum strict r–packing set of G. This settles the case $r(v) = 0$.

To prove the assertion for $r(v) \geq 1$ we proceed by induction on the size of G. Define w, G' and r' as in Lemma 3.2 with $S := V$ (thus case $(H3)$ cannot arrise). From Lemma 3.6 we have $\gamma_{r',cl}(G') = \gamma_{r,cl}(G)$. By the induction hypothesis we have $\pi_{r'}(G') = \gamma_{r',cl}(G')$. Since $\pi_r(G) \leq \gamma_{r,cl}(G)$ it remains to show that $\pi_{r'}(G') \leq \pi_r(G)$.

By Lemma 2.3 we have only to consider the case $r'(w) = r(v) - 1 < r(w)$ and w belongs to a maximum strict r'–packing set P' of G'. Since we only changed the radius of w we have $d(w, y) \neq r(w) + r(y) + 1$ for all vertices $y \in P' \setminus \{w\}$.

Suppose there is a vertex $y \in N(v) \cap P' \setminus \{w\}$. From $d(w, y) = r(v) + r(y) \leq 2$, $r(v) \geq 1$ and the choice of w in case $(H2)$ of Lemma 3.2 we conclude $d(w, y) = 2$, $y \in H$, $r(y) = r(w) = r(v) = 1$. Assume there is a vertex x of $P' \setminus \{w, y\}$. Then we get $d(x, y) = 2 + r(x)$ and $d(w, x) = 1 + r(x)$. Since H is homogeneous and w, y are in H we conclude $x \in H$. But then $r(x) = 0$ which is impossible in case $(H2)$. Thus $\pi_{r'}(G') = 2$ contradicting $\pi_{r'}(G') = \gamma_{r',cl}(G') = \gamma_{r,cl}(G) \geq 3$.

Therefore for all $y \in P' \setminus \{w\}$ we have $d(v, y) \geq 2$ and $d(v, y) = d(w, y) + 1 = r(v) + r(y) + 1$. So the set $P := (P' \setminus \{w\}) \cup \{v\}$ is a strict r–packing set of G and hence $\pi_{r'}(G') \leq \pi_r(G)$. □

Corollary 3.8 *Let G be a homogeneously orderable graph possessing a r–dominating clique.*

1. *If* $\gamma_{r,cl}(G) \neq 2$ *or* $\pi_r(G) > 1$ *then* $\pi_r(G) = \gamma_r(G) = \gamma_{r,con}(G) = \gamma_{r,cl}(G)$.
2. $\gamma_r(G) = \gamma_{r,con}(G) = \gamma_{r,cl}(G)$.

Corollary 3.9 *In a homogeneously orderable graph G any set of pairwise intersecting disks has either a nonempty common intersection or there is an edge such that for each of these disks at least one vertex of the edge belongs to the disk.*

Proof omitted.

4 The algorithms

At first we present an efficient algorithm for computing the distance matrix of certain graphs in quadratic time.

Theorem 4.1 *For any graph G with given ordering* $(v_1, \ldots, v_n)$ *of its vertex set such that for each* $i = 1, \ldots, n-1$ *holds (with* $G_1 := G$*)*

(1) $G_{i+1} := G(\{v_{i+1}, \ldots, v_n\})$ *is an isometric subgraph of* G_i *and*
(2) for a given vertex $w_i \in N(v_i) \cap V(G_i)$ *and for any vertex* x *in* $N^j(v_i) \cap V(G_i)$, $j = 2, 3, \ldots$ *there is a path of length* j *joining* x *and* v_i *and containing* w_i

the distance matrix of G can be computed in quadratic time $O(n^2)$.

Proof omitted.

Note that by Lemma 2.3 any homogeneously orderable graph fulfills the presumptions of Theorem 4.1.

4.1 r–dominating cliques and strict r–packing sets

By Theorem 4.1 we can compute the distance matrix for a given homogeneously orderable graph in quadratic time. Thus by Theorem 3.3 we can decide in the same time whether the given graph possesses a r–dominating clique. Moreover, we can check the graph for a r–dominating vertex in quadratic time too. So assume for the sequel that a given homogeneously orderable graph G is not r–dominated by some vertex but by some clique.

The following algorithm both computes a r–dominating clique of minimum size and a maximum strict r–packing set of G.

The algorithm works in three rounds. In the first round it steps through a given h–extremal ordering and manipulates r by using the rules of Lemma 3.2 and the arguments of Lemma 3.6 until it reaches a vertex v with $r(v) = 0$ or $e(v) = 1$. In the second one we choose a minimum r–dominating clique C and a maximum strict r–packing set P of the current graph according to Lemma 3.1 and to the proof of Theorem 3.7. By Lemma 3.6 the clique C is a minimum r–dominating clique in G too. If $|C| = 2$ then $\pi_r(G) \leq 2$, and we can compute a maximum strict r–packing set P of the initial graph in quadratic time only using the distance matrix.

Otherwise, to find a maximum strict r–packing set in G, the algorithm in the third round goes backwards through the sequence and updates the parameter P according to the arguments of the proof of Theorem 3.7.

The three different cases arrising in our algorithm are the following :

Case 1. $e(v) \geq 2$ and $r(v) \geq 1$.
We define w, G' and r' according to the rules $(H1)$–$(H3)$ of Lemma 3.2. By Lemma 3.6 we have $\gamma_{r,cl}(G) = \gamma_{r',cl}(G')$, and each minimum r'–dominating clique of G' is a minimum r–dominating clique of G.

Case 2. $e(v) = 1$.
In this case we terminate round one, compute the parameters (C, P) of the current graph defined according to Lemma 3.1, and start round three. This can be done in time $O(n^2)$ by stepping through the distance matrix .

Case 3. $e(v) \geq 2$ and $r(v) = 0$.
In this case we terminate round one. Since any r–dominating clique is contained in $D(v, 1)$ either $C = P = Z(G)$ or $C = \{h\} \cup Z(G)$ and $P = Z(G) \cup \{x\}$ where $h \in H$ and x is a private neighbour of h. By the proof of Theorem 3.7 C is a minimum r–dominating clique and P is a maximum strict r–packing set of the current graph. This step can be easily performed in quadratic time by stepping through the distance matrix. Now start round three.

Theorem 4.2 *In homogeneously orderable graphs it can be decided in time $O(n^2)$ whether the given graph is r–dominated by some clique, provided a h–extremal ordering is given. Moreover, if the graph has a r–dominating clique then a minimum one and a maximum strict r–packing set can be computed in the same time.* □

4.2 Connected r–dominating sets

For the sequel let v be a h–extremal vertex with $e(v) \geq 2$ and let $H \subseteq N(v)$ be a homogeneous set dominating $D(v,2)$. Define $A := N(v) \setminus H$, $r_H := \min\{r(h) : h \in H\}$ and $r_A := \min\{r(a) : a \in A\}$ (if $A = \emptyset$ we put $r_A := \infty$). Moreover suppose that v does not r–dominate G. Define $G' := G \setminus \{v\}$ and let S' be a minimum connected r'–dominating set in G' (r' will be defined in the following cases). In what follows we describe how we can obtain a minimum connected r–dominating set S in G from S'. The correctness proof of this algorithm is presented in the full version of this paper.

Case 1. $r(v) > 0$ and $\min\{r_H, r_A\} = 0$.
We define $r'(x) := r(x)$ for all vertices $x \in V(G')$. Then $S := S'$ is a minimum connected r–dominating set in G.

Case 2. $r(v) > r_H$ and $\min\{r_H, r_A\} > 0$.
Again we define $r'(x) := r(x)$ for all vertices $x \in V(G')$. The connected r–dominating set $S := S'$ is minimum in G.

Case 3. $1 < r(v) \leq r_H$ and $r_A > 0$.
Choose a vertex h from H such that $r(h) = r_H$, define $r'(h) := r(v) - 1$ and $r'(x) := r(x)$ for all remaining vertices. The connected r–dominating set $S := S'$ is minimum in G.

Case 4. $r(v) = 1$ and $\min\{r_H, r_A\} > 0$.
We distinguish between two subcases.

Case 4.1. There is a vertex x in $N^i(v)$, $i \geq 2$, such that $d(x,v) \geq r(x) + 2$.
We define $r'(h) := 0$ for an arbitrary vertex h of H and $r'(x) := r(x)$ for all $x \in V(G') \setminus \{h\}$. Then $S := S'$ is a minimum connected r–dominating set in G.

Case 4.2. For all vertices $x \in V \setminus D(v,1)$ we have $d(x,v) \leq r(x) + 1$.
If H is r–dominated by some vertex $h \in H$ then $S := \{h\}$. Otherwise either there is a vertex $a \in A$ which r–dominates G or S contains at least two vertices. But then we may choose $S := \{v, h\}$ where h is an arbitrary vertex from H.

Case 5. $r(v) = 0$.
If there is a vertex $w \in N(v)$ with $r(w) = 0$ then we do not change the r–values. Otherwise define $r'(w) := 0$ for an arbitrary vertex w of H and $r'(x) := r(x)$ for all other vertices. In both cases add the edges between each vertex of $N(v) \setminus \{w\}$ and w in G' (as was shown in [4] G' has the same h–extremal ordering as G). Moreover, the distance matrix of G' can be obtained in linear time from the distance matrix of G. Now it is easy to see that $S := S' \cup \{v\}$ is a minimum connected r–dominating set in G.

Finally consider the case $e(v) \leq 1$. If there are vertices with r–value zero, i.e. $Z(G) \neq \emptyset$, then $Z(G)$ or $Z(G) \cup \{v\}$ is a minimum connected r–dominating set of G. Otherwise v r–dominates G.

Theorem 4.3 *In homogeneously orderable graphs a minimum connected r–dominating set can be computed in time $O(n^2)$ provided a h–extremal ordering is given.*

Proof omitted.

The following table presents some algorithmic results for the considered graph classes and problems. Hereby n is the number of vertices and m is the number of edges of a graph. We will write $O(m)$ instead of $O(n+m)$ since all graphs are connected in this paper.

Class	r–dom. clique		minimum con.
	decision	minimum	r–dom. set
trees	$O(n)$ folk		$O(n)$ folk
chordal graphs	$O(nm)$ [11, 8]	NP [10]	NP [10]
distance–hereditary graphs	$O(m)$ [7]		$O(m)$ [2]
dually chordal graphs	$O(m)$ [3, 8]		$O(n^2)$ [3, 6]
weakly chordal graphs	NP [5]		NP [10]
homogeneously orderable graphs	$O(n^2)$ here		$O(n^2)$ here

Acknowledgement The authors would like to thank their colleague Andreas Brandstädt for useful discussions.

References

1. A. BRANDSTÄDT, Special graph classes - a survey, *Technical Report* Universität Duisburg SM–DU–199, 1991.
2. A. BRANDSTÄDT and F.F. DRAGAN, A Linear–Time Algorithm for Connected r–Domination and Steiner Tree on Distance–Hereditary Graphs, *Technical Report* Universität Duisburg SM–DU–261, 1994.
3. A. BRANDSTÄDT, V.D. CHEPOI and F.F. DRAGAN, The algorithmic use of hypertree structure and maximum neighbourhood orderings, *International Workshop "Graph-Theoretic Concepts in Computer Science"* 1994, to appear.
4. A. BRANDSTÄDT, F.F. DRAGAN and F. NICOLAI, Homogeneously orderable graphs, *Technical Report* Universität Duisburg SM–DU–271, 1994.
5. A. BRANDSTÄDT and D. KRATSCH, Domination problems on permutation and other graphs, *Theoretical Computer Science* 54 (1987), 181–198.
6. F.F. DRAGAN, HT–graphs: centers, connected r—domination and Steiner trees, *Computer Science Journal of Moldova* 1 (1993), No. 2, 64–83.
7. F.F. DRAGAN, Dominating cliques in distance–hereditary graphs, *Proc. of the 4th SWAT*, Aarhus, Denmark, Springer LNCS 824 (1994), 370–381.
8. F.F. DRAGAN and A. BRANDSTÄDT, r–dominating cliques in Helly graphs and chordal graphs, *Proc. of the 11th STACS*, Caen, France, Springer LNCS 775 (1994), 735 – 746.
9. M.C. GOLUMBIC, Algorithmic Graph Theory and Perfect Graphs, *Academic Press, New York* 1980.
10. S.C. HEDETNIEMI and R. LASKAR (eds.), Topics on domination, *Annals of Discr. Math.* 48 (1991), North-Holland.
11. D. KRATSCH, P. DAMASCHKE and A. LUBIW, Dominating cliques in chordal graphs, *Discr. Math.* 128 (1994), 269–275.

Growing Patterns in 1D Cellular Automata*

Bruno Durand and Jacques Mazoyer

Laboratoire de l'Informatique du Parallélisme, ENS-Lyon, CNRS,
46 Allée d'Italie, 69364 Lyon Cedex 07, France

Abstract. We study limit evolutions of cellular automata (CA) both theoretically and experimentally. We show that either all orbits enter the limit set after less than a finitely bounded number of states, or almost all orbits never enter the limit set. We link this result with a classification of CAs according to their limit behavior due to Čulik *et al.*: in the first case, the considered CA belongs to class 1 while in the second case, it belongs to class 2.
By experiments, we try to measure the convergence speed of orbits to their accumulation points. We compute the maximum number of nested growing segments in a space-time diagram representing a finite portion of an orbit. We observe that, in the average case, this criterion depends only on the CA and not on the configuration. We also observe two kinds of CA *wrt.* our criterion which correspond to the intuitive notions of chaos and regularity. We do further experiments to explain the fact that the proportion of chaotic diagrams grows with the number of states of the considered CA.

1 Introduction

The goal of our paper is to analyze the limit behavior of a cellular automaton (CA for short) iterated on a configuration. A configuration is a bi-infinite string of cells, such that to each cell is associated a state which belongs to a finite set. The sequence of all configurations obtained by iterating a CA on a given configuration is called its orbit. The set of configurations that can be obtained after an arbitrarily large number of iterations is called the *limit set* Ω of the considered CA. An interesting classification of CAs based on their limit sets has been presented by Čulik *et al.* in [2]. A CA belongs to class 1 iff its limit set is obtained after a finite number of steps.

Limit sets are also very important to study CA's dynamics but do not always correspond to our physical intuition because they contain some ephemeral configurations having preimages of arbitrary orders. We introduce another notion of limit set called the *accessible set* (Acc) containing exactly all accumulation points of orbits. We prove that Acc $\subset \Omega$. A fundamental difference between Ω and Acc is that the limit set Ω can defined with "shift invariant"configurations, *i.e.* on bi-infinite words (classes of shift-translatable configurations) rather than on configurations in $S^{\mathbf{Z}}$. Accessible sets (Acc) are shift invariant, but are not defined for bi-infinite words

* This work was partially supported by the Esprit Basic Research Action "Algebraic and Syntactical Methods In Computer Science" and by the PRC "Mathématique et Informatique".

and the exemple that follows Theorem 8 shows that there is no natural way to do it.

If we endow the set of configurations with a measure, then we get the following result: "If $\mathcal{A}$ belongs to class one, then for any configuration c its orbit enters Ω. If $\mathcal{A}$ belongs to class two, then for almost all configurations c, the orbit of c and Ω are disjointed."

Observing space-time diagrams obtained by the iteration of a CA on a configuration, we wondered at which speed these configurations were converging to Acc. We explain in Section 2.1 the notion of convergence in the set of configurations: a sequence converges if there exists in this sequence a family of nested growing segments centered on the same cell. The limit configuration is the union of these segments. We wondered also if one can observe these families of growing patterns in a reasonably small space-time diagram.

Thus we made some experiments in order to measure this speed for several different CAs and then for "random" CAs (Section 3). We observed that this speed is an intrinsic property of the chosen CA and does not statistically depend on the configuration. Two classes of CAs appeared, those with high convergence speed which produce obviously simplified space-time diagrams, and those with low convergence speed the diagram of which appeared as almost chaotic. But when the number of states increases, we observe that more and more diagrams are chaotic. We study this evolution in Section 3.2 and prove that this fact can be partially explained by higher transients but we conjecture that the proportion of chaotic CAs increases with the number of states.

2 Definitions

Cellular automata are formally defined as triplets (S, N, f). $S = \{s_1, s_2, \ldots, s_k\}$ is a finite set called the set of *states*. The *neighborhood* N is a v-tuple of distinct vectors of $\mathbb{Z}$. For us, $N = (x_1, \ldots, x_v)$: the x_i's are the relative positions of the neighbor cells with respect to a given center cell. The states of these neighbors are used to compute the new state of the center cell. The *local function* of the cellular automaton $f : S^v \mapsto S$ gives the local transition rule.

A *configuration* is an application from $\mathbb{Z}$ to S. The set of configurations is $S^{\mathbb{Z}}$ on which the *global function* G of the cellular automaton is defined via f:

$$\forall c \in S^{\mathbb{Z}},\ \ \forall i \in \mathbb{Z},\ \ G(c)(i) = f(c(i + x_1), \ldots, c(i + x_v)).$$

Note that cellular automata are characterized by S, N and f, but, even if two cellular automata are syntactically different, they may compute the same global function G. We shall say that a CA is bijective (or injective, surjective) when G is of this kind. Well-known results prove that injectivity is equivalent to bijectivity for CAs (see for instance [5, 6, 4, 1]).

2.1 Topology

Most results of this sections are already known (Theorem 6 excepted) but we present them so that the reader can better understand the motivations of our statistical analysis (section 3).

Let us endow S with the discrete topology for which all subsets are open. The set of all configurations being a countable product of sets S, we endow $S^{\mathbb{Z}}$ with the *product topology*: an open subset of $S^{\mathbb{Z}}$ is a union of finite intersections of sets of the form $\mathcal{O}_{i,a} = \{c \in S^{\mathbb{Z}},\ c(i) = a\}$.

Very often, studies concerning cellular automata use intensively the notion of pattern that we define below. In a topological approach, this notion is very natural since patterns correspond to basic open sets.

More precisely, if we define a *pattern* as a mapping from a finite subset of $\mathbb{Z}$ to S, we can define a *basic open set* associated to a pattern as the set of all configurations equal to the pattern on its domain: $\mathcal{O}_p = \left\{c \in S^{\mathbb{Z}},\ c\left|_{\mathrm{domain}(p)} = p\right.\right\}$. Note that the $\mathcal{O}_p$'s (and the $\mathcal{O}_{i,a}$'s which are special $\mathcal{O}_p$'s) are both open and closed: their complements are equal to finite unions of $\mathcal{O}_q$'s where $\mathrm{domain}(p) = \mathrm{domain}(q)$ and $p \neq q$. Any open set $\mathcal{U}$ can be written as a union of basic open sets: $\mathcal{U} = \bigcup_{p \in \Pi} \mathcal{O}_p$.

Proposition 1. *$S^{\mathbb{Z}}$ is a compact metric space.*

Proof. As S endowed with the discrete topology is compact, the countable product $S^{\mathbb{Z}}$ of compact sets is compact too. This result follows from Tychonoff's theorem but does not require the axiom of choice. A distance on $S^{\mathbb{Z}}$ can be

$$d(c, c') = 2^{-m} \text{ where } m = \min\{d(i, 0),\ c(i) \neq c'(i)\}.$$

The topology induced by this distance is the same as the product topology. Note that this distance is not shift invariant and gives more importance to cells near 0.

The following theorem is very basic and states a fundamental link between continuous functions and CA.

Theorem 2 Richardson 1972 [7]. *A function $f : S^{\mathbb{Z}} \mapsto S^{\mathbb{Z}}$ which commutes with shifts is a* continuous *function if and only if it is the global function of a cellular automaton.*

A shift of vector v transforms a configuration c into a configuration c' if and only if $\forall i \in \mathbb{Z},\ c'(i) = c(i + v)$. The original proof of this theorem is very long and complicated. It can be drastically simplified by using the compactness of $S^{\mathbb{Z}}$.

In the remainder of the section, we use results of Karel Čulik *et al.* [2] to classify cellular automata according to their long term evolution. We use the notion of *limit set* which consists of the set of configurations that can be obtained after an arbitrarily large number of iterations. If this set can be obtained after a finite number of steps, then the cellular automaton is said to belong to class one, otherwise it belongs to class two. This classification has been proved undecidable by Jarkko Kari in [3]. We also endow the set of all configurations with a "natural" measure and prove that configuration orbits almost never (resp. always) enter the limit set for cellular automata belonging to class two (resp. belonging to class one). We present another notion of limit set that we call *accessible set* which consists of the set of all possible accumulation points of all orbits. We prove that this notion is more restrictive than the previous one but describes better CA's dynamics.

2.2 Limit sets

Let G be the global function computed by a cellular automaton. We define its i-th iterate $G^{\circ i}$ as follows:

$$G^{\circ 0} = \mathrm{Id}_{S^{\mathbb{Z}}} \; ; \quad \forall i \in \mathbb{N},\; G^{\circ i+1} = G \circ G^{\circ i} = G^{\circ i} \circ G.$$

If the cellular automaton $\mathcal{A}$ computes G, then a cellular automaton which computes $G^{\circ i}$ (see Theorem 2 for its existence) is called the i-th iterate of $\mathcal{A}$: $\mathcal{A}^{\circ i}$. For all non negative integer i, we define Ω_i, a subset of $S^{\mathbb{Z}}$, by $\Omega_i = G^{\circ i}(S^{\mathbb{Z}})$. The limit set Ω is the intersection of all Ω_i's: $\Omega = \bigcap_{i \in \mathbb{N}} \Omega_i$.

First observe that $\forall i \in \mathbb{N},\ \Omega_{i+1} \subset \Omega_i$. As $S^{\mathbb{Z}}$ is a compact set and G is a continuous function, each Ω_i is a compact set.

One can prove that Ω is a non-empty compact set and that Ω is included in its preimage. In other terms, $\forall c \in \Omega,\ \exists d \in \Omega,\ G(d) = c$. In the following, we call $\mathcal{A}$'s limit set $\Omega^{\mathcal{A}}$. Limit sets are invariant under iteration: $\Omega^{\mathcal{A}} = \Omega^{\mathcal{A}^{\circ j}}$ for all $j > 0$.

2.3 A classification theorem

The classification of cellular automata presented below is very convenient to study the evolution of a cellular automaton. Its interest is reinforced by a basic theorem formulated by Čulik *et al.* [2], given below in Theorem 3. Thus, we classify cellular automata within two classes:

1. $\exists i \in \mathbb{N},\ \Omega = \Omega_i$
2. $\forall i \in \mathbb{N},\ \Omega \neq \Omega_i$ (which is equivalent to $\Omega_i \neq \Omega_{i+1}$).

For instance, all surjective automata are such that $\Omega = S^{\mathbb{Z}}$ and belong to class 1. Another example is the following "projection" automaton:

$$S = \{0, 1, 2\},\quad N = (0_{\mathbb{Z}}),\quad f(0) = 0,\quad f(1) = 1,\quad f(2) = 0.$$

In this case it is clear that $\Omega = \Omega_1$.

But ordinarily, cellular automata belong to class 2. For example consider

$$S = \{0, 1\},\quad N = \{-1, 0, 1\},\quad f(1, 1, 1) = 1,\quad f(.,.,.) = 0 \text{ otherwise}.$$

This automaton belongs to class 2 because the configuration $({}^{\omega}0.1.0^{2n}.1.0^{\omega})_{n \in \mathbb{N}}$ belongs to Ω_n but not to Ω_{n+1}.

The following theorem is due to Čulik *et al.* [2]. We give its proof below because we shall use it farther on.

Theorem 3* Čulik et al. *89. *For each cellular automaton $\mathcal{A}$, $\mathcal{A}$ belongs to class 2 if and only if there exists a countable intersection D of dense open sets such that*

$$\Omega \cap \Big(\bigcup_{i \in \mathbb{N}} G^i(D)\Big) = \emptyset.$$

Proof. Assume that $\mathcal{A}$ is in class 2. We define D as follows:

$$D_i = S^{\mathbb{Z}} - G^{-i}(\Omega)$$

$$D = \bigcap_{i>0} D_i = \bigcap_{i>0} (S^{\mathbb{Z}} - G^{-i}(\Omega)) = \bigcap_{i>0} \left(S^{\mathbb{Z}} - \{c \in S^{\mathbb{Z}},\ G^{\circ i}(c) \in \Omega\}\right).$$

The D_i's are dense and non-empty because they are translation invariant open sets and if a D_i were empty then $\mathcal{A}$ would belong to class 1. Hence we can apply Baire's theorem which states that D is dense because it is a countable intersection of dense open sets. By construction,

$$\left(\bigcup_{i>0} G^{\circ i}(D)\right) \bigcap \Omega = \emptyset.$$

If $\mathcal{A}$ belongs to class 1, then Ω is open hence D cannot exist.

2.4 A measure on configurations

All the following is devoted to an extension of Čulik's *et al.*'s Theorem 3 in terms of probability measures. Thus, we endow $S^{\mathbb{Z}}$ with a measure with exactly the same method used to endow it with a topology.

Definition 4. Let us put on each state of S the weight $1/k$ where k is the cardinality of S. Consider on S the measure defined by these weights. If we consider the product measure, we get the *natural* measure on $S^{\mathbb{Z}}$.

The measure of a set of configurations is shift invariant, and for instance, the measure of $\mathcal{O}_{i,a} = \{c \in S^{\mathbb{Z}},\ c(i) = a\}$ is exactly $1/k$ and the measure of a set $\mathcal{O}_p$ where p is a pattern is $1/k^r$ where r is the cardinal of p's domain.

Lemma 5. *The measure of a shift invariant non-empty open set is 1.*

We present below our interpretation of Čulik *et al.*'s results in terms of probabilities.

Theorem 6. *Consider a cellular automaton $\mathcal{A}$. If $\mathcal{A}$ belongs to class one, then for all configurations $c \in S^{\mathbb{Z}}$, the orbit of c enters Ω. If $\mathcal{A}$ belongs to class two, then for almost all configurations $c \in S^{\mathbb{Z}}$, the orbit of c and Ω are disjoint.*

Recall that the orbit of c is simply the set $\{c, G(c), G^{\circ 2}(c), \ldots, G^{\circ k}(c), \ldots\}$.

Proof. If $\mathcal{A}$ belongs to class 1, the result is trivial since there exists an integer i such that $\Omega = \Omega_i$ hence $G^{\circ i}(c) \in \Omega$.

If $\mathcal{A}$ belongs to class 2, then we come back to the proof of Theorem 3. In this proof, a dense open set D is built.

$$D = \bigcap_{i>0} D_i = \bigcap_{i>0} (S^{\mathbb{Z}} - G^{-i}(\Omega)) = \bigcap_{i>0} \left(S^{\mathbb{Z}} - \{c \in S^{\mathbb{Z}},\ G^{\circ i}(c) \in \Omega\}\right).$$

Remark now that D is a countable intersection of shift invariant open sets of measure 1 (see lemma 5). According to the Σ-additivity of the measure, the measure of D is 1 and our theorem holds.

The consequences of this result will become clearer after the next section in which we present another notion of limit set.

2.5 Accessible set

In this new notion, we wish to capture only the points that can be approached by iteration of the cellular automaton.

Definition 7. Consider a global function G. For any configuration c, $\mathrm{Acc}(c)$ is the set of accumulation points of the sequence $(c, G(c), G^{\circ 2}(c), \ldots, G^{\circ k}(c), \ldots)$.
More formally, $\mathrm{Acc}(c) = \bigcap_{i \in \mathbb{N}} \overline{\mathrm{Acc}_i(c)}$ where $\mathrm{Acc}_i(c) = \bigcup_{j>i} \{G^{\circ j}(c)\}$.
We define the *accessible set* Acc of all accumulation points by $\mathrm{Acc} = \bigcup_{c \in S^{\mathbb{Z}}} \mathrm{Acc}(c)$.

This new notion is interesting because one can approach any accessible point by iterating the cellular automaton on a configuration while this is not always the case for configurations in the limit set Ω as we shall see below.

Theorem 8. *Acc* $\subset \Omega$.

The reverse inclusion $\mathrm{Acc} \supset \Omega$ is not always true, as showned by the following counter-example: $S = \{0,1\}$, $N = \{-1,0,1\}$, $f(1,1,1) = 1$, $f(.,.,.) = 0$ otherwise.

During evolutions of this cellular automaton, blocks of 1's decrease and disappear, hence accumulation points of any orbits can only be either 1 everywhere or 0 everywhere: $\mathrm{Acc} = \{{}^{\omega}0^{\omega}, {}^{\omega}1^{\omega}\}$. But these configurations are not alone in Ω which is formed by

$${}^{\omega}0^{\omega},\ {}^{\omega}1^{\omega},\ {}^{\omega}01^{\omega},\ {}^{\omega}10^{\omega},\ \text{where the first 1 is on any cell of } \mathbb{Z}$$
$${}^{\omega}01^{n}0^{\omega},\ \text{where the first 1 is on any cell of } \mathbb{Z} \text{ and } n \in \mathbb{N}.$$

With the help of accessible sets, we can give a more precise interpretation of our Theorem 6. If a cellular automaton belongs to class 2, then almost every orbit is of the following kind: it never enters Ω hence its accumulation points are never visited.

This is equivalent to saying that in the orbit, there exists a strictly increasing sequence of growing patterns the union of which is a configuration of Ω. Our result is not affected if we replace the notion of growing patterns by the notion of growing segments. This gave us the idea of counting the number of nested segments in such space-time diagrams as will be done in the next section.

3 Statistical analysis

3.1 Our basic experiment

We consider CAs with neighborhood $(-1,0,1)$. We denote by k their number of states. Thus their transition function is a 3-dimensional table filled with k^3 states. We consider in the following that all states are equiprobable and chosen randomly and independently. There is only a finite number of possibility for these CAs (k^{k^3}).

We choose the size for a domain on which we shall observe evolutions of CAs. More precisely, we choose the size of a segment of cells n and a time of study t. We shall observe a space-time $(n \times t)$-rectangle. We then choose an initial configuration. We wish that all cells in the $(n \times t)$-rectangle behave as if the initial configuration was infinite. Thus we consider an initial segment of size $t + n + t$. To each cell of this segment is given a state chosen randomly and independently. We then iterate the chosen CA on the segment, and extract our $(n \times t)$-rectangle filled with states. This object will be the field for the analysis.

On this rectangle, we compute the maximum number of nested segments, centered on the same cell, and included in the rectangle. More precisely, we consider segments of size $2l + 1$; a $(2l + 1)$-segment S_1 is nested into a $(2l + 3)$-segment S_2 iff when one extract the $2l + 1$ center cells of S_2 one gets the segment S_1. The *depth* of a cell will be the maximum number of nested segments centered on this cell. We compute the maximum of depths of all the n cells of the rectangle. We call this value α. It depends on the number of states k, on the number of cells n, on the time t, on the CA $\mathcal{A}$, and on the configuration c. Thus we shall denote it by $\alpha_{k,n,t}(\mathcal{A}, c)$.

Our fundamental motivation for computing α is the hope that the maximal growing sequence of pattern is the beginning of a sequence that converges to an accumulation point extending the segments. Practically, if one looks to space-time diagrams, one sees that it is always the case, even if it can be false in some rare cases that we never saw in our experiments. The integer α measures the speed of convergence of orbits to their accumulation points. If α is low, then orbits converge slowly and the diagram seems random, else the evolution of the CA seems to simplify the diagram.

For now on, we observe the variations of α. If all its parameters are fixed, the configuration c excepted, then we observe that α may only vary a little, even if in theory it can be very high if c is a constant configuration for instance. We have tested a few CAs chosen randomly and we have always obtained a distribution for α of the form of Fig 1, the maximums being different and depending on $\mathcal{A}$. We also tried to make $\mathcal{A}$ vary with c fixed, and we observed exactly the same phenomena than when both c and $\mathcal{A}$ vary. Thus we decided for most of our study to consider that the random variable of α would be the pair $(\mathcal{A}, c)$ and we observed the distribution of α for some values of k. We also tried to make n and t vary but this does not make any noticeable change in our results. This experimental result should be made in perspective with our Theorem 6.

In order to better understand the distribution measured for α, we decide to compare it with the distribution of α measured in a really random distribution of states over a rectangle $n \times t$: we choose at random a state among k in each cell of a $(n \times t)$-rectangle. It is clear that space-time diagrams obtained by iterations of a CA are never random rectangle because three neighbor cells determine uniquely the center state at the next step. All over a space-time diagram, there is such a correlation. But we have been astonished to see through our results that α would not make a difference between a really random diagram and a space-time diagram that could appear random at first sight. In space-time diagrams, when two segments are nested, it implies that the triangles based on these segments are nested too which is of course not the case in random diagrams. If we had computed α with the number of nested triangles instead of the number of nested segments, we would have obtain the same distribution for space-time diagrams but not for random diagrams. We chose to base our study on segments because it is much faster to compute and in order to be able to determine at first sight what kind of evolution is involved.

Figures 2 and 2 show our results for $k = 3$ to $k = 6$. The first observation is that, when k grows, distributions of α resemble more and more distributions for random diagrams. There are two possible explanations: either the proportion of CAs which simplify diagrams decreases with k, or the simplification does not begin at the first steps of the iteration, but begins after a transient the length of which increases with k. We do further experiments in Section 3.2 to explain more precisely this fact.

We also observe that for some CAs, space-time diagrams lead to the same α as random diagrams: the criterion α do not distinguish "true" random diagrams with these space-time diagrams. But it is interesting that our human eyes do so! If you look at these space-time diagrams, you would say that they are random, or at least that they represent a chaotic evolution of a CA. All these space-time diagrams look chaotic. For other space-time diagrams, after a transient, the evolution is drastically simplified. If the simplification occurs soon enough then all these diagrams have the same high, and maximum α. The fact that there exists such a maximum is also due to the shape of our rectangles: we always chose $t = 3n$. The most important observation is that the simplification is in general very brutal and that, a few CA excepted, α classifies CAs in two classes. For theoretical reasons, one cannot say that these classes correspond with Čulik *et al.*'s classes: a bijective CA is in Čulik *et al.*'s class 1 but its evolution can be regular (identity) or chaotic (many chaotic bijective CAs are known).

3.2 A study on the transient

In Fig. 2, the distribution of α for $k = 6$ is very similar to the distribution for random diagrams. We wonder if the reason is that most CAs with 6 states are chaotic or if the proportion of chaotic CAs does not increase with k but that they simplify diagrams later. To answer this question, we have invented a new experiment. Instead of considering a space-time diagram the first row of which is a random configuration, we base our rectangle on the t_0-th iterate of a random configuration. Practically, we randomly choose a configuration c (larger enough) and compute $\alpha_{k,n,t}(\mathcal{A}, \mathcal{A}^{t_0}(c))$. We obtain the distributions of Fig. 3.

Compared with distributions of Fig. 2, it is clear that something happened: there appear slightly less chaotic diagrams. The difference is significant but not drastic. For $k = 5$ the difference appears for $t_0 = 50$ but not for $k = 6$ where we had to use $t_0 = 100$ to see a small difference. It means clearly that the average length of transients increases with k. More experiments would be needed to determine whether the proportion of chaotic CAs increase with k or not. We think that it is really the case but our experiments do not prove it directly.

4 Conclusion

Computations of our criterion α are rather time consuming and we have been very limited in our experiments. We think that our statistical studies could be continued in order to better understand CA's dynamics. Nevertheless, we could capture some interesting phenomena and link them with theory;

- we observe two classes of CA with respect to our criterion: "simple" GAs and "chaotic" ones.
- evolutions of CAs on a random line are not much influenced by the chosen random line.
- the transient length increases with the number of states. We do not know whether it grows very fast or not. We conjecture that the proportion of simple CAs decreases with the number of states.

Even if our criterion is aimed to measure convergence speeds of orbits to their accumulation points, our classes of CAs are not clearly linked to Čulik *et al.*'s. There exist both simple and chaotic class 1 CAs. To link our experiments to Wolfram's classes (see [9, 8] for instance), we need first to point out a fundamental difference: Wolfram's configurations are periodic (or equivalently are restricted to a ring) while ours are infinite. This may lead to many basic differences because evolutions on rings are *a priori* simpler. Wolfram's classes 1 and 2 correspond obviously to simple CAs while class 4 corresponds to chaotic ones. It is more difficult to characterize class 3 but we think that they are not really chaotic. Furthermore, it seems that our criterion corresponds exactly to the intuitive notion of "chaotic evolutions": if α is low then the space-time diagram and the considered CA are chaotic, but if it is high, then the diagram really appears regular which was rather inexpected.

References

1. S. Amoroso and Y.N. Patt. Decision procedures for surjectivity and injectivity of parallel maps for tesselation structures. *J. Comp. Syst. Sci.*, 6:448–464, 1972.
2. K. Čulik, Y. Pachl, and S. Yu. On the limit sets of cellular automata. *SIAM J. Comput.*, 18:831–842, 1989.
3. J. Kari. Rice's theorem for the limit set of cellular automata. *Theoretical Computer Science*, 127(2):229–254, 1994.
4. A. Maruoka and M. Kimura. Conditions for injectivity of global maps for tessallation automata. *Information and Control*, 32:158–162, 1976.
5. E.F. Moore. Machine models of self-reproduction. *Proc. Symp. Apl. Math.*, 14:13–33, 1962.
6. J. Myhill. The converse to Moore's garden-of-eden theorem. *Proc. Am. Math. Soc.*, 14:685–686, 1963.
7. D. Richardson. Tesselations with local transformations. *Journal of Computer and System Sciences*, 6:373–388, 1972.
8. S. Wolfram. Universality and complexity in cellular automata. In T. Toffoli D. Farmer and S. Wolfram, editors, *Cellular Automata*, pages 1–36. North-Holland, 1983.
9. S. Wolfram. Twenty problems in the theory of cellular automata. *Physica Scripta*, T9:170–183, 1985.

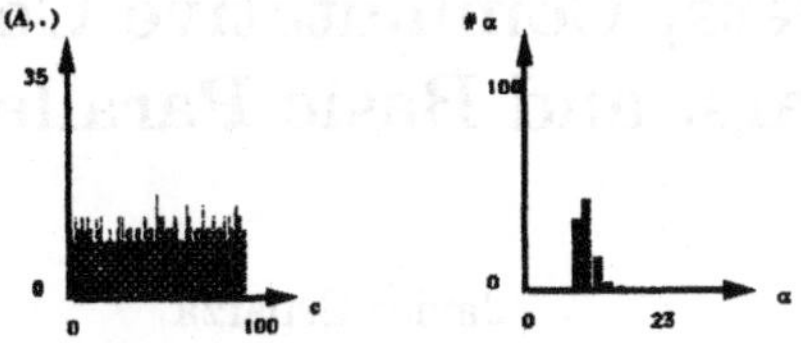

Fig. 1. The configuration varies alone

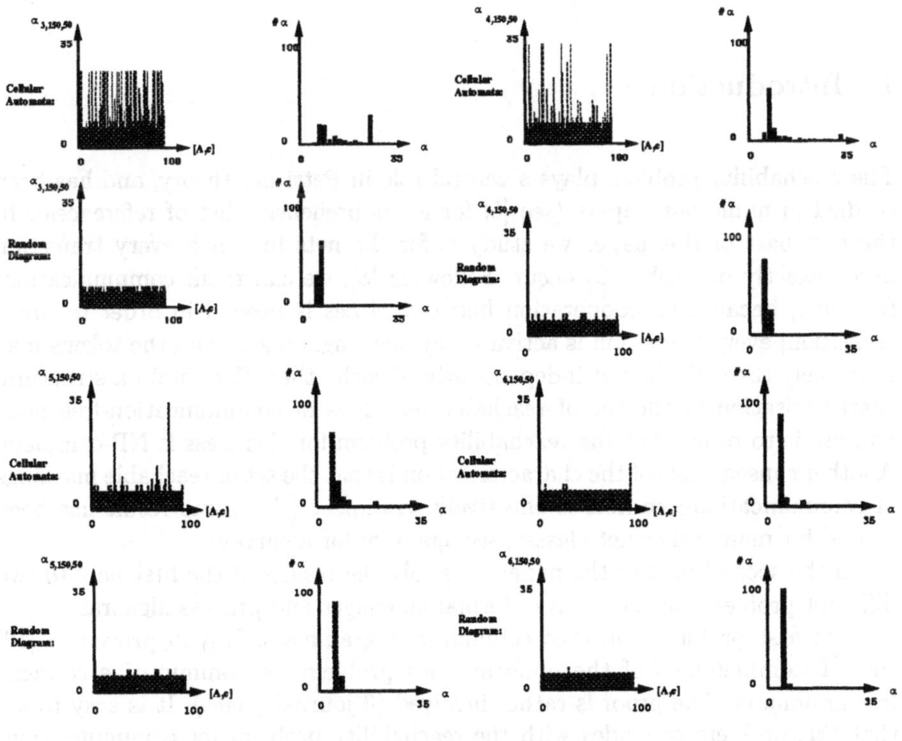

Fig. 2. Distributions of α for $k = 3$, $k = 4$, $k = 5$ and $k = 6$

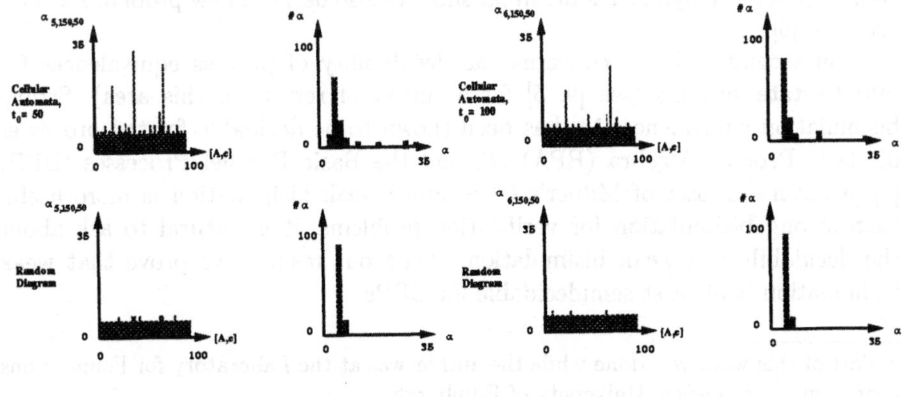

Fig. 3. When rectangles are based on $\mathcal{A}^{t_0}(c)$

Petri Nets, Commutative Context-Free Grammars, and Basic Parallel Processes

Javier Esparza
Institut für Informatik
Technische Universität München*

1 Introduction

The reachability problem plays a central rôle in Petri net theory, and has been studied in numerous papers (see [5] for a comprehensive list of references). In the first part of this paper we study it for the nets in which every transition needs exactly one token to occur. Following [8], we call them communication-free nets, because no cooperation between places is needed in order to fire a transition; every transition is activated by one single token, and the tokens may flow freely through the net independently of each other. We obtain a structural characterisation of the set of reachable markings of communication-free nets, and use it to prove that the reachability problem for this class is NP-complete. Another consequence of the characterisation is that the set of reachable markings of communication-free nets is effectively semilinear (this same result has been proved for many other net classes, see again [5] for a survey).

In the second part of the paper we apply the results of the first part to two different problems in the areas of formal languages and process algebras.

The first problem concerns commutative grammars. Huynh proved in [12] the NP-completeness of the uniform word problem for commutative context-free grammars. The proof is rather involved (8 journal pages). It is easy to see that this problem coincides with the reachability problem for communication-free nets. Therefore, our results lead immediately to a new and considerably shorter proof of Huynh's result. In passing, we also derive a new proof of Parikh's theorem [6].

The second problem concerns the decidability of process equivalences for infinite-state systems (see [9, 5] for a survey of results in this area). Strong bisimulation equivalence [14] has been shown to be decidable for the processes of Basic Process Algebra (BPA) [2], and the Basic Parallel Processes (BPP) [3], a natural subset of Milner's CCS. Since weak bisimulation is more useful than strong bisimulation for verification problems, it is natural to ask about the decidability of weak bisimulation. Using our results, we prove that weak bisimulation is at least semidecidable for BPPs.

---

* Part of this work was done while the author was at the Laboratory for Foundations of Computer Science, University of Edinburgh.

2 Petri nets and labelled Petri nets

For the purposes of this paper it is convenient to describe Petri nets using some notations on monoids. Given a finite alphabet $V = \{v_1, \dots, v_n\}$, the symbols $V^\star$, $V^\oplus$ denote the free monoid and free commutative monoid generated by V, respectively. Given a word w of $V^\star$ or $V^\oplus$, and an element v of V, $w(v)$ denotes the number of times that v appears in w. A word w of $V^\oplus$ will be represented in two different ways:

- as a multiset of elements of V (for instance, $\{v_1, v_1, v_2\}$ is the word containing two copies of v_1 and one of v_2);
- as the vector $(w(v_1), \dots, w(v_n))$.

The context will indicate which representation is being used at each moment. The *Parikh mapping* $\mathcal{P}: V^\star \rightarrow V^\oplus$ is defined by $\mathcal{P}(w) = (w(v_1), \dots, w(v_n))$. Given $u, v \in V^\oplus$, $u + v$ denotes the concatenation of u and v, which corresponds to addition of multisets or sum of vectors.

A *net* is a triple $N = (S, T, W)$, where S is a set of *places*, T is a set of *transitions*, and $W: (S \times T) \cup (T \times S) \rightarrow I\!N$ is a *weight function*. Graphically, places are represented by circles, and transitions by boxes. If $W(x, y) > 0$, then there is an arc from x to y labeled by $W(x, y)$ (when $W(x, y) = 1$, the arc is not labeled for clarity). We denote by ${}^\bullet x$ the set $\{y \mid W(y, x) > 0\}$ and by $x^\bullet$ the set $\{y \mid W(x, y) > 0\}$. For a set X, ${}^\bullet X$ ($X^\bullet$) denotes the union of ${}^\bullet x$ ($x^\bullet$) for every element x of X. An element of $S^\oplus$ is called a *marking* of N. A marking M is graphically represented by putting $M(s)$ *tokens* (black dots) in each place s. A *Petri net* is a pair (N, M_0), where N is a net and M_0 is a marking of N called the *initial marking*. A marking M of a net $N = (S, T, W)$ *enables* a transition t if $M(s) \geq W(s, t)$ for every place $s \in {}^\bullet t$. If t is enabled at M, then it can *occur*, and its occurrence leads to the marking M', given by $M'(s) = M(s) + W(t, s) - W(s, t)$ for every place s. This is denoted by $M \xrightarrow{t} M'$. Given $\sigma = t_1 t_2 \dots t_n$, $M \xrightarrow{\sigma} M'$ denotes that there exist markings $M_1, M_2, \dots, M_{n-1}$ such that $M \xrightarrow{t_1} M_1 \xrightarrow{t_2} M_2 \dots M_{n-1} \xrightarrow{t_n} M'$ and we say that M' is *reachable* from M. We denote by $M \xrightarrow{\sigma}$ (read "σ is enabled at M") the fact that $M \xrightarrow{\sigma} M'$ for some marking M'. The *reachability problem* for a class of Petri nets consists of deciding, given a Petri net (N, M_0) of the class and a marking M of N, if M is reachable form M_0.

In Section 5 we shall consider labelled Petri nets. A *labelled net* is a fourtuple (S, T, W, l), where (S, T, W) is a net and $l: T \rightarrow Act$ is a labelling function on a set *Act* of actions. Markings, enabledness and ocurrence of transitions are defined as for unlabelled nets. $M \xrightarrow{a} M'$ denotes that $M \xrightarrow{t} M'$ for some transition t such that $l(t) = a$.

3 Communication-free Petri nets

Definition 1. Communication-free Petri Nets
A net $N = (S, T, W)$ is *communication-free* if $|{}^{\bullet}t| = 1$ for every $t \in T$, and $W(s,t) \leq 1$ for every $s \in S$ and every $t \in T$. A Petri net (N, M_0) is communication-free if N is communication-free.

We characterise structurally the set $\{\mathcal{P}(\sigma) \mid M_0 \xrightarrow{\sigma}\}$ of a communication-free Petri net (N, M_0). We need the notions of siphon and subnet generated by a set of transitions.

A subset of places R of a net is a *siphon* if ${}^{\bullet}R \subseteq R^{\bullet}$. A nonempty siphon is said to be *proper*. A siphon is *unmarked* at a marking M if none of its places is marked at M. It follows immediately from the definition of the occurrence rule for Petri nets that if a siphon is unmarked at a marking, then it remains unmarked, i.e., it cannot become marked by the occurrence of transitions.

Given a net $N = (S, T, W)$ ans a subset of transitions U, the *subnet of N generated by U* is $({}^{\bullet}U \cup U^{\bullet}, U, W_U)$, where W_U is the restriction of W to the pairs (x, y) such that x or y is a transition of U. Given $X \in T^{\oplus}$, the net N_X is defined as the subnet of N generated by the set of transitions that appear in X.

We can now state the characterisation of the set $\{\mathcal{P}(\sigma) \mid M_0 \xrightarrow{\sigma}\}$.

Theorem 2.
Let (N, M_0) be a communication-free Petri net, where $N = (S, T, W)$, and let $X \in T^{\oplus}$. There exists a sequence $\sigma \in T^{\star}$ such that $M_0 \xrightarrow{\sigma}$ and $\mathcal{P}(\sigma) = X$ iff

(a) $M_0(s) + \sum_{t \in T} (W(t,s) - W(s,t)) \cdot X(t) \geq 0$ *for every* $s \in S$, *and*

(b) every proper siphon of N_X is marked at M_0.

The proof is omitted for lack of space. It would take about two pages of this volume. We illustrate the result by means of an example. Figure 1 shows a communication-free net. There is no occurrence sequence with Parikh mapping $X = (1, 0, 0, 1)$, because for the place s_5 we have

$$M_0(s_5) + \sum_{t \in T} (W(t, s_5) - W(s_5, t)) \cdot X(t) = -X(t_4) = -1$$

and therefore (a) does not hold. There is no occurrence sequence with Parikh mapping $X = (0, 1, 1, 0)$ either. In this case, (a) holds, but not (b): the net N_X contains the transitions t_2, t_3, and the places s_2, s_3, s_4, s_5, and so the set $\{s_2, s_5\}$ is a proper siphon of N_X unmarked at M_0. Finally, the reader may check that $(1, 1, 0, 1)$ satisfies both (a) and (b), and in fact $t_1\, t_2\, t_4$ is an occurrence sequence having this Parikh mapping.

We easily derive the following two results:

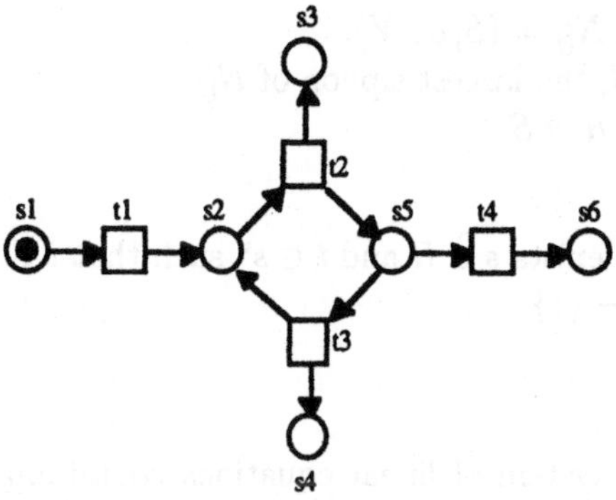

Fig. 1 Illustration of Theorem 2

Theorem 3.
The reachability problem for communication-free Petri nets is NP-complete.

Proof: NP-hardness follows from a straightforward reduction from the satisfiability problem for boolean formulae in conjunctive normal form. Given such a formula F, we construct a communication-free Petri net (N, M_0) and a marking M such that F is satisfiable iff M is reachable from M_0. Figure 2 shows the communication-free Petri net corresponding to the formula $C_1 \wedge C_2$, where

$$C_1 = x_1 \vee \overline{x_2} \vee x_3 \quad C_2 = \overline{x_1} \vee \overline{x_2} \vee \overline{x_3}$$

To prove membership in NP, use the following nondeterministic algorithm,

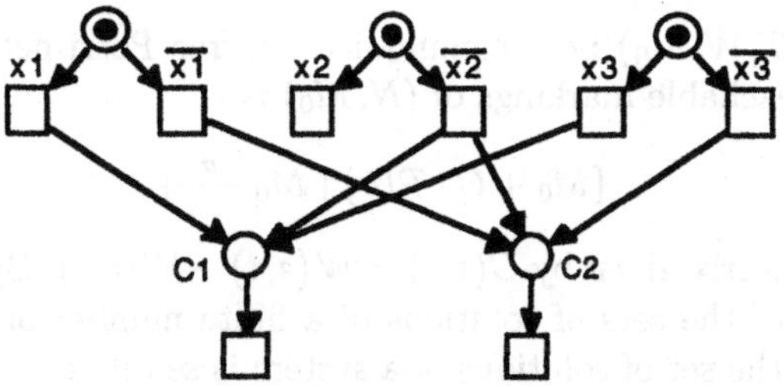

Fig. 2 Communication-free Petri net for the formula of the text. The formula is satisfiable iff the marking $\{C_1, C_2\}$ is reachable

whose correctness follows immediately from Theorem 2. Given a communication-free Petri net (N, M_0) and a marking M, guess a set U of transitions of N. Check in polynomial time whether every proper siphon of N_U is marked at M_0. For that, let N'_U be the net obtained from N_U by removing all places marked by M_0, and use the following greedy algorithm, due to Starke [15], to compute the largest siphon R of N'_U, (clearly, every proper siphon of N_U is marked at M_0 iff $R = \emptyset$):

Input: The net $N'_U = (S, U, W)$.
Output: $R \subseteq S$, the largest siphon of N'_U.
Initialization: $R = S$.

begin
  while there exists $s \in R$ and $t \in s^\bullet$ such that $t \notin {}^\bullet R$ **do**
    $R := R - \{s\}$
  endwhile
end

Now, construct the system of linear equations containing for each place s the equation

$$M(s) = M_0(s) + \sum_{t \in T} (W(t,s) - W(s,t)) \cdot X(t)$$

Guess in polynomial time the smallest solution $X \geq 0$ of the system which satisfies $X(t) = 0$ for every $t \notin U$ (the smallest solution has polynomial size in the input due to the result of von zur Gathen and Sieveking [11]). ■

A subset $\mathcal{X}$ of a free commutative monoid $V^{\oplus}$ is *linear* if there exist elements $v, v_1, \ldots, v_n \in V^{\oplus}$ such that

$$\mathcal{X} = \{v + a_1 v_1 + \ldots + a_n v_n \mid a_1, \ldots, a_n \geq 0\}$$

$\mathcal{X}$ is *semilinear* if it is a finite union of linear sets. We have:

Theorem 4.
The set of reachable markings of a communication-free Petri net is effectively semilinear.

Proof: Let (S, T, W, M_0) be a communication-free Petri net. It is easy to see that the set of reachable markings of (N, M_0) is

$$\{M_0 + C \cdot \mathcal{P}(\sigma) \mid M_0 \xrightarrow{\sigma}\}$$

where C is the matrix given by $C(s,t) = W(s,t) - W(t,s)$. By Theorem 2, this set is the union of the sets of solutions of a finite number of systems of linear equations. Since the set of solutions of a system is semilinear, the result follows. ■

4 Context-free and commutative context-free grammars

A *context-free grammar* is a 4-tuple $G = (Non, Ter, A, P)$, where Non and Ter are disjoint sets, called the sets of nonterminals and terminals, respectively, A is an element of Non called the *axiom*, and P is a finite subset of $Non \times (Non \cup Ter)^*$, called the set of *productions*. The *language* $L(G)$ of a context-free grammar G is defined as usual. A *commutative context-free grammar* (or ccf-grammar for short) is a 4-tuple $G^c = (Non, Ter, A, P^c)$, where Non, Ter, and A are as above, and $P^c \subseteq Non \times (Non \cup Ter)^{\oplus}$. That is, free monoids are replaced by free commutative monoids. Given two commutative words $\alpha, \beta \in (Non \cup Ter)^{\oplus}$, α *directly generates* β, written $\alpha \longrightarrow \beta$, if $\alpha = \alpha_1 + \gamma$, $\beta = \alpha_1 + \delta$, and $(\gamma, \delta) \in P^c$. α *generates* β if $\alpha \overset{*}{\longrightarrow} \beta$, where $\overset{*}{\longrightarrow}$ denotes the reflexive and transitive closure of $\longrightarrow$.

The following ccf-grammar with $Non = \{A, B\}$ and $Ter = \{a, b, c\}$ generates the language $\{\{a, b^{2n}, c^n\} \mid n \geq 0\}$.

$$A \rightarrow \{a\} \quad A \rightarrow \{b, b, A, B\} \quad B \rightarrow \{c\}$$

We define a mapping which assigns to a ccf-grammar $G^c = (Non, Ter, A, P^c)$ a Petri net (S, T, W, M_0). The Petri net of Figure 3 is the one assigned to the grammar above. The reader can possibly guess the definition of the Petri net from

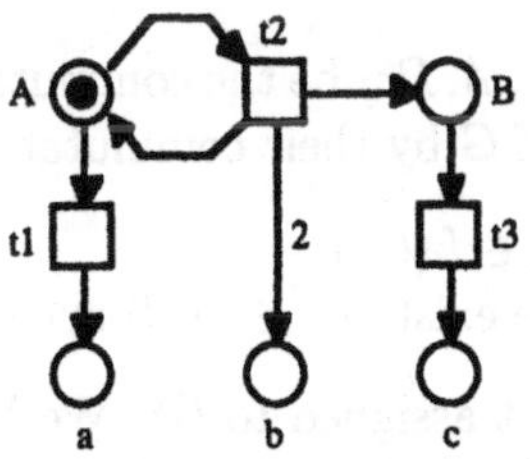

Fig. 3 A Petri net

this example: S is the set $Non \cup Ter$, i.e., there is a place for each terminal and for each nonterminal; T is the set P^c, i.e., there is a transition for each production. Given a place s and a transition $t = (X, w)$, the weight $W(s, t)$ is 1 if $s = X$ and 0 otherwise, whereas the weight $W(t, s)$ is the number of times that s appears in the commutative word w. Finally, M_0 is the marking that puts one token on the axiom A and no tokens on the rest. It follows directly from this description that commutative context-free grammars are assigned communication-free Petri nets.

Every word w of $(Non \cup Ter)^{\oplus}$ is a marking of the net (S, T, W). The application of a production corresponds to the occurrence of a transition. In particular, the relation $\overset{*}{\longrightarrow}$ on words of $(Non \cup Ter)^{\oplus}$ corresponds to the reachability relation on markings.

The uniform word problem for commutative context-free grammars. The *uniform word problem* for commutative grammars is the problem of deciding, given a grammar $G^c = (Non, Ter, A, P^c)$ and a commutative word w of terminals, if $A \overset{*}{\longrightarrow} w$. It follows immediately from our description of the net assigned to G^c that $A \overset{*}{\longrightarrow} w$ iff w is a reachable marking of the Petri net translation of G^c. So the uniform word problem for ccf-grammars can be reduced in linear time to the reachability problem for communication-free Petri nets, and vice versa. Therefore, we have a new proof for the following result of [12]:

Theorem 5. *[12]*
The uniform word problem for commutative context-free grammars is NP-complete. ■

The connection between commutative context-free grammars and Petri nets was pointed out in [12], but not used. The proof of membership in NP of [12] takes 8 journal pages, and is rather involved. Our proof is shorter[2], and it uses only standard techniques of net theory.

Parikh's Theorem. As a second consequence of Theorem 2, we obtain a new proof of Parikh's Theorem.

Theorem 6.
Let $G = (Non, Ter, A, P)$ be a context-free grammar. The set $\{\mathcal{P}(w) \in Ter^{\oplus} \mid w \in L(G)\}$ is semilinear.

Proof: Let $G^c = (Non, Ter, A, P^c)$ be the commutative grammar obtained after replacing the productions of G by their commutative versions. We have:

- if $w \in L(G)$, then $\mathcal{P}(w) \in L(G^c)$;
- if $w \in L(G^c)$, then there exists $w' \in L(G)$ such that $w = \mathcal{P}(w')$.

Let (N, M_0) be the Petri net assigned to G^c. We have $\mathcal{P}(w) \in L(G^c)$ iff $\mathcal{P}(w)$, seen as a marking of N, is reachable from M_0. Therefore, $\mathcal{P}(L(G))$ is the set of reachable markings of (N, M_0) that only put tokens in the places corresponding to terminals. This latter condition can be expressed using linear equations. The result follows now from Theorem 4. ■

We do not claim this proof to be simpler than Parikh's proof, which is rather straightforward, and takes little more than 3 pages in [6]. However, it shows still another connection between Petri nets and formal language theory.

5 Weak bisimilarity in Basic Parallel Processes

We show that communication-free Petri nets are also strongly related to Basic Parallel Processes (BPPs), a subset of CCS.

Basic Parallel Process expressions are generated by the following grammar:

$$\begin{array}{rll} E ::= & \mathbf{0} & \text{(inaction)} \\ \mid & X & \text{(process variable)} \\ \mid & aE & \text{(action prefix)} \\ \mid & E+E & \text{(choice)} \\ \mid & E \parallel E & \text{(merge)} \end{array}$$

where a belongs either to a set of *atomic actions* Act or is the *silent action* τ. A BPP is a family of recursive equations $\mathcal{E} = \{X_i \stackrel{\text{def}}{=} E_i \mid 1 \le i \le n\}$, where the

2 The comparison of the lengths is fair, because both proofs use the result of von zur Gathen and Sieveking.

X_i are distinct and the E_i are BPP expressions at most containing the variables $\{X_1, \ldots, X_n\}$. We further assume that every variable occurrence in the E_i is *guarded*, that is, appears within the scope of an action prefix. The variable X_1 is singled out as the *leading variable* and $X_1 \stackrel{\text{def}}{=} E_1$ is the *leading equation.*

Any BPP determines a labelled transition system, whose transition relations $\xrightarrow{a}$ are the least relations satisfying the following rules:

$$aE \xrightarrow{a} E \qquad \frac{E \xrightarrow{a} E'}{E+F \xrightarrow{a} E'} \quad \frac{E \xrightarrow{a} E'}{E \parallel F \xrightarrow{a} E' \parallel F'}$$

$$\frac{E \xrightarrow{a} E'}{X \xrightarrow{a} E'} (X \stackrel{\text{def}}{=} E) \quad \frac{F \xrightarrow{a} F'}{E+F \xrightarrow{a} F'} \quad \frac{F \xrightarrow{a} F'}{E \parallel F \xrightarrow{a} E \parallel F'}$$

The states of the transition system are the BPP expressions E that satisfy $X_1 \xrightarrow{w} E$ for some string w of actions.

Let $\mathcal{E}$ and $\mathcal{F}$ be two BPPs with disjoint sets of variables. Consider the labelled transition system obtained by putting the transition systems of $\mathcal{E}$ and $\mathcal{F}$ side by side. A binary relation $\mathcal{R}$ between the states of this labelled transition system is a weak bisimulation if whenever $E\mathcal{R}F$ then, for each $a \in Act$,

– if $E \xrightarrow{a} E'$ then $F \xRightarrow{a} F'$ for some F' with $E'\mathcal{R}F'$;
– if $F \xrightarrow{a} F'$ then $E \xRightarrow{a} E'$ for some E' with $F'\mathcal{R}E'$;

and, moreover

– if $E \xrightarrow{\tau} E'$ then $F \xRightarrow{\epsilon} F'$ for some F' with $E'\mathcal{R}F'$;
– if $F \xrightarrow{\tau} F'$ then $E \xRightarrow{\epsilon} E'$ for some E' with $F'\mathcal{R}E'$,

where $\xRightarrow{a} = (\xrightarrow{\tau})^* \xrightarrow{a} (\xrightarrow{\tau})^*$, and $\xRightarrow{\epsilon} = (\xrightarrow{\tau})^*$.

$\mathcal{E}$ and $\mathcal{F}$ are weakly bisimilar if their leading variables are related by some weak bisimulation.

A BPP is in *normal form* if every expression E_i on the right hand side of an equation is of the form $a_1\alpha_1 + \ldots + a_n\alpha_n$, where for each i the expression α_i is a merge of variables. It is proved in [1] that every BBP is weakly (even strongly) bisimilar to a BPP in normal form, which can be effectively (and efficiently) constructed (the proof is very similar to that for Greibach normal form and context-free grammars). Therefore, the problem of deciding strong or weak bisimilarity for BPPs can be reduced to the same problem for BPPs in normal form.

Every BPP in normal form can be translated into a labelled communication-free Petri net. The translation is graphically illustrated by means of an example in Figure 4. The net has a place for each variable X_i. For each subexpression $a_j\alpha_j$ in the defining equation of X_i, a transition is added having the place X_i in its preset, and the variables that appear in α_j in its postset. If a variable appears n times in α_j, then the arc leading to it is given the weight n. The transition is labelled by a_j.

It follows easily from the rules of the operational semantics that every state E of a BPP $\{X_i \stackrel{\text{def}}{=} E_i \mid 1 \leq i \leq n\}$ in normal form can be written as

$$X_1^{i_1} \parallel X_2^{i_2} \parallel \ldots \parallel X_n^{i_n}$$

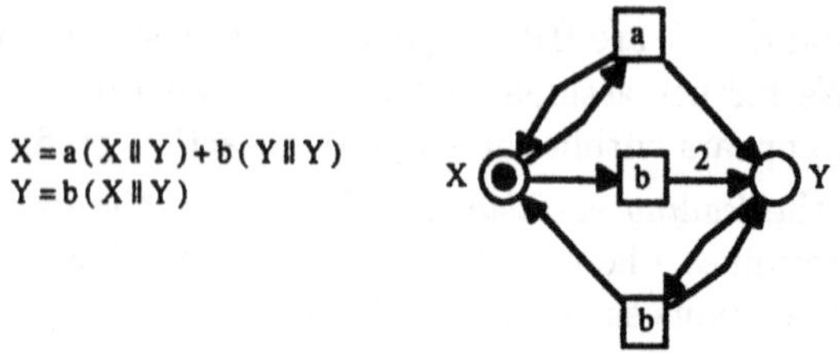

Fig. 4 A BPP and its corresponding Petri net

where $X^i = \underbrace{X \parallel \ldots \parallel X}_{i}$. Such a state corresponds to the marking $X_1^{i_1} X_2^{i_2} \cdots X_n^{i_n}$ of the labelled Petri net assigned to $\mathcal{E}$, and vice versa. Moreover, if M_E and $M_{E'}$ are the markings corresponding to the states E and E', then $E \xrightarrow{a} E'$ iff $M_E \xrightarrow{a} M_{E'}$. So the transition system of a BPP in normal form and the reachability graph of its labelled Petri net are isomorphic. It follows that two BPPs are weakly bisimilar iff their corresponding labelled communication-free Petri nets are weakly bisimilar, where weak bisimilarity for labelled Petri nets is defined as for BPPs, just replacing the states of the labelled transition systems by markings, and the relations $\xrightarrow{a}$ between BPP expressions by the corresponding reachability relations between markings.

5.1 Weak bisimilarity is semidecidable for BPPs

We give a positive test of weak bisimilarity for labelled communication-free Petri nets. We first observe that it suffices to consider instances of the problem in which the two Petri nets have the same underlying nets, i.e., they differ only in the initial markings. To prove it, given two labelled communication-free Petri nets (N_1, M_{01}) and (N_2, M_{02}), we construct another two, (N, M'_{01}) and (N, M'_{02}) as follows. N is the result of putting N_1 and N_2 side by side. M'_{01} coincides with M_{01} on N_1, and puts no tokens on N_2. M'_{02} coincides with M_{02} on N_2, and puts no tokens on N_1. Clearly, (N, M'_{01}) and (N, M'_{02}) are weakly bisimiliar iff (N_1, M_{01}) and (N_2, M_{02}) are weakly bisimilar.

In the sequel we fix a unique labelled communication-free net $N = (S, T, W, l)$, and study the weak bisimilarity of two initial markings M_{01} and M_{02} of N.

It follows immediately from the definitions that the union of two bisimulations is a bisimulation. Therefore, there exists a unique maximal weak bisimulation between the markings of N. Clearly, (N, M_{01}) and (N, M_{02}) are weakly bisimilar iff the pair (M_{01}, M_{02}) belongs to the maximal weak bisimulation. We study the properties of the maximal weak bisimulation. Given a commutative monoid $A^{\oplus}$, an equivalence relation $\mathcal{R} \subseteq A^{\oplus} \times A^{\oplus}$ is a *congruence* iff for every $x, y, z \in A^{\oplus}$ $(x, y) \in \mathcal{R}$ implies $(x + z, y + z) \in R$. We have:

Lemma 7.
The maximal weak bisimulation between markings of a labelled communication-free net is a congruence.

Proof: It is easy to see that the maximal weak bisimulation is an equivalence relation. It remains to prove that if M_1 is weakly bisimilar to M_2, then $M_1 + M$ is weakly bisimilar to $M_2 + M$ for every marking M. Let $\mathcal{R}$ be a weak bisimulation containing (M_1, M_2). The relation $\mathcal{R}' = \{(M_1' + M, M_2' + M) \mid (M_1', M_2') \in \mathcal{R}, M \in S^{\oplus}\}$ contains $(M_1 + M, M_2 + M)$ for every marking M. We show that $\mathcal{R}'$ is a weak bisimulation.

Let $(M_1' + M, M_2' + M)$ be a pair of $\mathcal{R}'$. If $M_1' + M \xrightarrow{a} M''$, then, since a transition needs only one token to occur, we have two possible cases:

(1) $M_1' \xrightarrow{a} M_1''$ and $M'' = M_1'' + M$.
Since $(M_1', M_2') \in \mathcal{R}$, there exists M_2'' such that $M_2' \overset{a}{\Longrightarrow} M_2''$ and $(M_1'', M_2'') \in \mathcal{R}$. Then $M_2' + M \overset{a}{\Longrightarrow} M_2'' + M$ and $(M_1'' + M, M_2'' + M) \in \mathcal{R}'$.

(2) $M \xrightarrow{a} M'$ and $M'' = M_1' + M'$.
Then $M_2' + M \overset{a}{\Longrightarrow} M_2' + M'$ and $(M_1' + M', M_2' + M') \in \mathcal{R}'$.

The other direction and the case $M_1' + M \xrightarrow{\tau} M''$ are proved similarly. ∎

We can now apply (following [13]) the following result of [4],

Theorem 8. *[4]*
Every congruence on a commutative monoid $A^{\oplus}$ is a semilinear subset of $A^{\oplus} \times A^{\oplus}$. ∎

By this theorem, the maximal weak bisimulation between markings of a comunication-free net is a semilinear subset of $S^{\oplus} \times S^{\oplus}$, where S is the set of places of the net.

The set of semilinear relations on a commutative monoid is recursively enumerable. Let $\mathcal{S}_1, \mathcal{S}_2, \mathcal{S}_3 \ldots$ be an effective enumeration of this set. We have to check if some $\mathcal{S}_i$ is a weak bisimulation containing (M_1, M_2).

Whether $(M_1, M_2) \in \mathcal{S}_i$ or not can be decided by solving a set of linear diophantine equations, which can be done in nondeterministic polynomial time.

In order to decide whether $\mathcal{S}_i$ is a weak bisimulation we need some well-known results about semilinear sets and Presburger arithmetic. Presburger arithmetic is the first order theory of addition. More precisely, formulae of Presburger arithmetic are built out of variables, quantifiers, and the symbols $0, \leq, +$. Formulae are interpreted on the natural numbers, and the symbols are interpreted as the number 0, the natural total order on $\mathbb{N}$, and addition.

A subset A of $\mathbb{N}^n$ is *expressible* in Presburger arithmetic if there exists a Presburger formula $\mathbf{A}(x_1, \ldots, x_k)$ with free variables $x_1, \ldots, x_n$ such that for every $(n_1, \ldots, n_k) \in \mathbb{N}^n$, the closed formula $\mathbf{A}(n_1, \ldots, n_k)$ is true if and only if $(n_1, \ldots, n_k) \in A$. We say that $\mathbf{A}(x_1, \ldots, x_k)$ *expresses* A.

Ginsburg and Spanier obtain in [7] the following result:

Theorem 9. *[7]*
A subset $A \subseteq \mathbb{N}^n$ is semilinear iff it is expressible in Presburger arithmetic. Moreover, the transformations between semilinear sets and formulae are effective. ∎

Presburger's classical theorem can be seen as a consequence of this result:

Theorem 10.
It is decidable if a formula of Presburger arithmetic is true. ■

With the help of these theorems and Theorem 2 we can finally prove:

Theorem 11.
Let $N = (S, T, W, l)$ be a labelled communication-free net, and let $\mathcal{R} \subseteq S^{\oplus} \times S^{\oplus}$ be a semilinear relation. It is decidable if $\mathcal{R}$ is a weak bisimulation.

Proof: The following relations on the markings of N are effectively semilinear:

(1) $\{(M, M') \mid M \xrightarrow{a} M'\}$.
The definition of $M \xrightarrow{a} M'$ can be easily encoded in Presburger arithmetic.
(2) $\{(M, M') \mid M \overset{a}{\Longrightarrow} M'\}$.
$M \overset{a}{\Longrightarrow} M'$ iff there exists $X \in T^{\oplus}$ such that conditions (a) and (b) of Theorem 2 hold, and moreover
(c) for every $t \in T$ such that $l(t) \notin \{a, \tau\}$, $X(t) = 0$, and
(d) $\sum_{t \in l^{-1}(a)} X(t) = 1$.

The sets satisfying one of the conditions (a)–(d) are all effectively semilinear. By Theorem 9, its intersection is also effectively semilinear.

By Theorem 9 there exist formulae of Presburger arithmetic expressing the relations (1) and (2) above and the relation $\mathcal{R}$. Using them, it is easy to encode the definition of weak bisimulation as another formula of Presburger arithmetic. The result follows from Theorem 10. ■

The algorithm to decide if a semilinear relation is a weak bisimulation has a high complexity. This is not very surprising, because the algorithms given so far to decide strong bisimulations for BPPs are non-elementary [3]. There exist so far no lower bounds for this problem. However, Hirshfeld, Jerrum and Moller have recently shown in [10] that deciding strong bisimulation is polynomial for the class of *normed* BPPs, in which a process may always terminate.

6 Conclusions

We have solved the reachability problem for communication-free Petri nets using very well-known techniques of net theory. We have then shown that this solution has several applications to context-free and commutative context-free grammars, and Basic Parallel Processes. More precisely, and in order of increasing interest, we have obtained a new proof of Parikh's theorem, a simpler proof of the NP-completeness of the uniform word problem for commutative context-free grammars, and the first positive result about the decidability of weak bisimulation for Basic Parallel Processes.

Acknowledgments

Many thanks to Soren Christensen and Yoram Hirshfeld for helpful discussions. Special thanks are also due to Richard Mayr and an anonymous referee for patiently correcting many typos in an earlier version, and for very useful suggestions.

References

1. S. Christensen. Decidability and Decomposition in Process Algebras. Ph.D. Thesis, University of Edinburgh, CST-105-93, (1993).
2. S. Christensen, H. Hüttel, and C. Stirling. Bisimulation equivalence is decidable for all context-free processes. Proceedings of CONCUR '92, LNCS 630, pp. 138–147 (1992).
3. S. Christensen, Y. Hirshfeld and F. Moller. Bisimulation equivalence is decidable for basic parallel processes. CONCUR '93, LNCS 715, 143–157 (1993).
4. S. Eilenberg and M.P. Schützenberger. Rational sets in commutative monoids. Journal of Algebra 13, 173–191 (1969).
5. J. Esparza and M. Nielsen. Decidability Issues for Petri Nets – a Survey. EATCS Bulletin 52 (1994). Also: Journal of Information Processing and Cybernetics.
6. S. Ginsburg. The Mathematical Theory of Context-free Languages. McGraw-Hill (1966).
7. S. Ginsburg and E.H. Spanier. Semigroups, Presburger formulas and languages. Pacific Journal of Mathematics 16, pp. 285–296 (1966).
8. Y. Hirshfeld. Petri Nets and the Equivalence Problem. CSL '93, LNCS 832 pp. 165–174 (1994).
9. Y. Hirshfeld and Faron Moller. Deciding Equivalences in Simple Process Algebras. In: "Modal Logic and Process Algebra: Proceedings of a 3-day Workshop on Bisimulation", CSLI Press (1995).
10. Y. Hirshfeld, M. Jerrum and F. Moller. A polynomial algorithm for deciding bisimulation of normed basic parallel processes. LFCS report 94-288, Edinburgh University (1994). To appear in Journal of Mathematical Structures in Computer Science.
11. J.E. Hopcroft and J.D. Ullman. Introduction to Automata Theory, Languages and Computation. Addison-Wesley (1979).
12. D.T. Huynh. Commutative Grammars: The Complexity of Uniform Word Problems. Information and Control 57, 21–39 (1983).
13. P. Jančar. Decidability Questions for Bisimilarity of Petri Nets and Some Related Problems. STACS '94, LNCS 775, pp. 581–592 (1994). To appear in Theoretical Computer Science.
14. R. Milner. Communication and Concurrency. Prentice-Hall (1989).
15. P.H. Starke. Analyse von Petri-Netz Modellen. Teubner (1990).

Implementation of a UU-Algorithm for Primitive Recursive Tree Functions*

Heinz Faßbender

Universität Ulm, Abt. Theoretische Informatik, D-89069 Ulm, Germany,
e-mail: fassbend@informatik.uni-ulm.de

Abstract. We present the implementation of an efficient universal unification algorithm for the class of equational theories which are induced by primitive recursive tree functions, on an abstract machine and prove its correctness. This machine extends a graph reduction machine by adding mechanisms for the handling of unification and nondeterminism which results from the existence of free variables in the input terms.

1 Introduction

The general problem of unification of two terms t and s in the presence of a set E of equations is called *E-unification problem.* Two terms t and s are *E-unifiable*, if there exists a substitution φ such that $\varphi(t) =_E \varphi(s)$, where $=_E$ is the equational theory induced by E. The E-unification problem can be generalized by considering a class $\mathcal{E}$ of sets of equations. Then, a *universal unification (for short: uu-) algorithm for* $\mathcal{E}$ [13] is an algorithm which takes as input a set E from the class $\mathcal{E}$ and two terms t and s, and which computes a *complete set of E-unifiers of t and s* [7], i.e., a set of E-unifiers that includes at least every minimal E-unifier of t and s.

In this paper, we will concentrate on a uu-algorithm for the class of sets of equations which are induced by primitive recursive (for short: pr) tree functions. We formalize pr tree functions by particular canonical, totally-defined, not strictly sub-unifiable term rewriting systems (cf. [1, 3]), called *modular tree transducers (for short: mt's)* [2].

In an mt, the ranked alphabet of symbols is partitioned into the ranked alphabets F and Δ of function symbols which are partitioned into modules, and constructors, respectively. For every function symbol f of rank $n+1$ and every constructor σ of rank k there exists exactly one rule with left hand side $f(\sigma(x_1,\ldots,x_k),y_1,\ldots,y_n)$; the right hand side is a term over constructors, variables from the left hand side, and recursive function calls which have to fulfil some conditions. For example, the set R_1 of rules of the mt $M_1 = (F_1, \Delta_1, R_1)$, where $F_1 = \{mult^{(2)}, add^{(2)}\}$, $\Delta_1 = \{\gamma^{(1)}, \alpha^{(0)}\}$, and M_1 has two modules, is shown in Figure 1. By M_1 the multiplication and addition for natural numbers are defined in a primitive recursive way, where the constructors α and γ represent zero and the successor, respectively.

* This paper is an extended abstract. The full version is available from the author.

Module 1:	$mult(\alpha, y_1)$	$\rightarrow$	α	(1)
	$mult(\gamma(x_1), y_1)$	$\rightarrow$	$add(mult(x_1, y_1), y_1)$	(2)
Module 2:	$add(\alpha, y_1)$	$\rightarrow$	y_1	(3)
	$add(\gamma(x_1), y_1)$	$\rightarrow$	$\gamma(add(x_1, y_1))$	(4)

Fig. 1. Rules of the modular tree transducer M_1.

The implemented uu-algorithm improves the uu-algorithm in [8] by allowing only leftmost outermost (for short: lo) narrowing steps (as in the uu-algorithm in [1]) and by applying decomposition steps from the unification algorithm in [10] as early as possible. Since decomposition steps force an inconsistency check for constructors and a particular occur check, some non-successful derivations are stopped earlier than in the uu-algorithm in [1]. From this, together with the fact that successful derivations have the same lengths as in the uu-algorithm in [1], it follows that our uu-algorithm is more efficient than the uu-algorithm in [1] which is, in its turn, more efficient than the uu-algorithm in [8].

Our uu-algorithm is implemented on an abstract machine that is an extension of the G-machine [9] which serves for an implementation of the reduction relation for functional programming languages. The extension consists of adding mechanisms for the handling of nondeterminism which results from the existence of free variables, and for implementing decomposition steps. Since we show that the implementation is correct, it can be considered as an operational description of our uu-algorithm which can be implemented straightforwardly on a computer.

This paper is organized in five sections where the second section contains preliminaries. In Section 3 we recall the concept of mt and the uu-algorithm from [2, 3], respectively. Section 4 deals with the implementation of the uu-algorithm on the abstract machine. Finally, Section 5 contains some concluding remarks, comparisons with related work, and it indicates further research topics.

2 Preliminaries

We recall and collect some notations, basic definitions, and terminology which will be used in the rest of the paper.

For $j \in \mathbb{N}$, $[j]$ denotes the set $\{1, \ldots, j\}$; thus $[0] = \emptyset$. As usual for a set A, A^* denotes the set of words over A. The empty word is denoted by Λ. Let $\Rightarrow \subseteq A \times A$, then $\Rightarrow^+$ and $\Rightarrow^*$ denote the transitive closure and the transitive-reflexive closure of $\Rightarrow$, respectively.

A finite indexed set Ω, where every element in Ω is indexed by its unique *rank* $\in \mathbb{N}$, is called a *ranked alphabet*. The subset $\Omega^{(m)}$ of Ω consists of all symbols of rank m. We write an element $\sigma \in \Omega^{(k)}$ as $\sigma^{(k)}$.

In the rest of the paper $\mathcal{V}$ denotes a fixed enumerable set of *variables* which is the union of the sets $X = \{x_1, x_2, \ldots\}$, $Y = \{y_1, y_2, \ldots\}$, and $FV = \{z_1, z_2, \ldots\}$ of *recursion variables, parameter variables,* and *free variables,* respectively.

Let Ω be a ranked alphabet and let S be an arbitrary set. Then the set of *terms over Ω and S*, denoted by $T\langle\Omega\rangle(S)$, is defined by: (i) $S \subseteq T\langle\Omega\rangle(S)$ and (ii) for every $f \in \Omega^{(k)}$, $k \geq 0$, and $t_1, \ldots, t_k \in T\langle\Omega\rangle(S)$: $f(t_1, \ldots, t_k) \in T\langle\Omega\rangle(S)$.

For a term $t \in T\langle\Omega\rangle(\mathcal{V})$, the set of *occurrences of t*, denoted by $O(t)$, is a subset of $\mathbb{N}^*$ which is defined by: (i) If $t \in \mathcal{V} \cup \Omega^{(0)}$, then $O(t) = \{\Lambda\}$ and (ii) if $t = f(t_1, \ldots, t_n)$, where $f \in \Omega^{(n)}$, $n > 0$, and for every $i \in [n] : t_i \in T\langle\Omega\rangle(\mathcal{V})$, then $O(t) = \{\Lambda\} \cup \bigcup_{i\in[n]}\{iu \mid u \in O(t_i)\}$.

For $t \in T\langle\Omega\rangle(\mathcal{V})$ and $u \in O(t)$, t/u denotes the *subterm of t at occurrence u*, and $t[u]$ denotes the *label of t at occurrence u*. We use $\mathcal{V}(t)$ to denote the set of variables occurring in t.

An assignment $\varphi : \mathcal{V} \to T\langle\Omega\rangle(\mathcal{V})$, where the set $\{x \mid \varphi(x) \neq x\}$ is finite, is called a *$(\mathcal{V}, \Omega)$-substitution*. The set $\{x \mid \varphi(x) \neq x\}$ is denoted by $\mathcal{D}(\varphi)$ and it is called the *domain of φ*. If $\mathcal{D}(\varphi) = \emptyset$, then φ is denoted by $\varphi_\emptyset$. We say, that φ is *ground,* if for every $x \in \mathcal{D}(\varphi) : \mathcal{V}(\varphi(x)) = \emptyset$. The set $\bigcup_{x\in\mathcal{D}(\varphi)} \mathcal{V}(\varphi(x))$ is denoted by $\mathcal{I}(\varphi)$. The set of $(\mathcal{V}, \Omega)$-substitutions and the set of ground $(\mathcal{V}, \Omega)$-substitutions are denoted by $Sub(\mathcal{V}, \Omega)$ and $gSub(\mathcal{V}, \Omega)$, respectively. The *composition* of two substitutions φ and ψ is the substitution which is defined by $\psi(\varphi(x))$ for every $x \in \mathcal{V}$. It is denoted by $\varphi \circ \psi$. Let $V \subset \mathcal{V}$. The *restriction of φ to V* is denoted by $\varphi|_V$.

In the rest of the paper, we let E denote a finite set of equations over Ω and $\mathcal{V}$. The *E-equality,* denoted by $=_E$, is the finest congruence relation over $T\langle\Omega\rangle(\mathcal{V})$ containing every pair $(\psi(t), \psi(s))$, where $(t = s) \in E$ and ψ is an arbitrary $(\mathcal{V}, \Omega)$-substitution. Two terms $t, s \in T\langle\Omega\rangle(\mathcal{V})$ are called *E-unifiable,* if there exists a $(\mathcal{V}, \Omega)$-substitution φ such that $\varphi(t) =_E \varphi(s)$. The set $\{\varphi \mid \varphi(t) =_E \varphi(s)\}$ is denoted by $\mathcal{U}_E(t, s)$ (cf. [13]). Let V be a finite subset of $\mathcal{V}$. We define a preorder $\preceq_E (V)$ on $Sub(\mathcal{V}, \Omega)$ by $\varphi \preceq_E \varphi' \ (V)$, if there exists a $(\mathcal{V}, \Omega)$-substitution ψ such that for every $x \in V : \psi(\varphi(x)) =_E \varphi'(x)$ (cf. [13]).

Let Ω be divided into two disjoint sets F and Δ, let $t, s \in T\langle\Omega\rangle(\mathcal{V})$ and let $V = \mathcal{V}(t) \cup \mathcal{V}(s)$. A set S of $(\mathcal{V}, \Delta)$-substitutions is a *ground complete set of (E, Δ)-unifiers of t and s away from V,* if the following three conditions hold: (i) For every $\varphi \in S$: $\mathcal{D}(\varphi) \subseteq V$ and $\mathcal{I}(\varphi) \cap V = \emptyset$. (ii) $S \subseteq \mathcal{U}_E(t, s) \cap Sub(\mathcal{V}, \Delta)$. (iii) For every $\varphi \in \mathcal{U}_E(t, s) \cap gSub(\mathcal{V}, \Delta)$ there is a $\psi \in S$ such that $\psi \preceq_E \varphi \ (V)$ (cf. [1]).

3 Modular Tree Transducer and the UU-Algorithm

In this section we recall the notion of mt and the uu-algorithm from [2, 3], resp..

Definition 3.1 An *mt* is a tuple (F, mod, Δ, R), where F and Δ are ranked alphabets of function symbols and constructors, resp., such that $F^{(0)} = \emptyset$, $mod : F \to \mathbb{N}$ is a mapping, and R is the finite set of rewrite rules; for every $f \in F^{(n+1)}$ and $\sigma \in \Delta^{(k)}$ with $n, k \geq 0$, there is exactly one *(f, σ)-rule* in R, i.e., a rule of the form $f(\sigma(x_1, \ldots, x_k), y_1, \ldots, y_n) \to t$, where $t \in T\langle F \cup \Delta\rangle(\{x_1, \ldots, x_k\} \cup \{y_1, \ldots, y_n\})$, such that for every function symbol $g \in F$ which occurs in t, the following two conditions hold: (i) $mod(g) \geq mod(f)$ and (ii) if $mod(g) = mod(f)$, then the first argument of g in t is in the set $\{x_1, \ldots, x_k\}$.

An example for an mt is presented in Section 1. Since the mapping *mod* has no influence in our further considerations, we always denote the fixed mt M by (F, Δ, R) and E_M denotes the set $\{l = r \mid l \rightarrow r \in R\}$.

Remark that every mt is a ctn-trs [3], because it is a canonical [2], totally defined [2], and not strictly sub-unifiable term rewriting system which follows from [1] and the definition of the rules' left hand sides.

In [3], an efficient uu-algorithm for the class of sets of equations which are induced by ctn-trs's is introduced. Now, we present an informal explanation of this algorithm for the class of equational theories which are induced by mt's. For the formal definition, the reader is referred to [3]. By this algorithm a ground complete set of (E_M, Δ)-unifiers of two terms t and s is computed by derivations of a narrowing relation, called *ulo-narrowing relation* [3]. In derivations of this relation, decomposition steps of the unification algorithm in [10] are applied during the lo-narrowing derivation as early as possible. For realizing decomposition steps, M is extended to the canonical term rewriting system $\widehat{M} = (\widehat{F}, \Delta, \widehat{R})$ by adding a new binary symbol *equ* to F and by adding for every $\sigma \in \Delta$ of rank $k \geq 0$ the *σ-decomposition-rule* to R which has the following form:

$$equ(\sigma(x_1, \ldots, x_k), \sigma(x_{k+1}, \ldots, x_{2k})) \rightarrow \sigma(equ(x_1, x_{k+1}), \ldots, equ(x_k, x_{2k})).$$

For every mt M and two input terms t and s, the uu-algorithm considers every derivation of the ulo-narrowing relation $\overset{u}{\rightsquigarrow}_{\widehat{M}}$ beginning with the derivation form $(equ(t, s), \varphi_\emptyset)$. In particular, derivation forms of $\overset{u}{\rightsquigarrow}_{\widehat{M}}$ are pairs $(t, \varphi) \in T\langle\widehat{F} \cup \Delta\rangle(FV) \times Sub(FV, \Delta)$. We only allow free variables in the derivation forms for preventing conflicts with variables occurring in rewrite rules.

Let (t, φ) be the current derivation form. In the following derivation step by $\overset{u}{\rightsquigarrow}_{\widehat{M}}$, the leftmost occurrence of the symbol *equ* in t, denoted by $impO(t)$, is considered. Then the following cases are considered depending on the root labels l_1 and l_2 of the two subtrees t_1 and t_2 of $t/impO(t)$, resp.:

1. If $l_1 = l_2 = \sigma \in \Delta$, then the σ-decomposition-rule is applied.
2. If, w.l.o.g., $l_1 \in FV$ and $l_2 \in \Delta$, then the occur check for l_1 is applied to the *shell* of t_2, i.e., the set of all occurrences $u \in O(t_2)$ such that there does not exist any occurrence $v \leq u$ which is labelled by a function symbol. If the occur check does not succeed, i.e., l_1 does not occur in the shell of t_2, then the σ-decomposition-rule is applied.
3. If l_1 and l_2 are the same free variable z_i, then $t/impO(t)$ is replaced by z_i.
4. If l_1 and l_2 are different free variables z_i and z_j, resp., then z_i and z_j are bound to a new variable z_k and $t/impO(t)$ is replaced by z_k.
5. If $l_1 \in F$, then one lo-narrowing step is applied to t_1.
6. If $l_1 \in \Delta \cup FV$ and $l_2 \in F$, then one lo-narrowing step is applied to t_2.

In every other case, i.e., l_1 and l_2 are different constructors or the occur check succeeds in Case 2, the narrowing derivation is stopped immediately without yielding an (E_M, Δ)-unifier. Because of the particular structure of the rules' left hand sides, an *lo-narrowing step* applied to a term t has the following form:

If $u \in O(t)$ is the minimal occurrence in t such that $t[u] = f \in F$ and $t[u1] \in \Delta \cup FV$, then we distinguish the following two cases depending on $t[u1]$:

1. If $t[u1]$ is a constructor σ, then the subterm t/u is replaced by the instantiated right hand side of the (f,σ)-rule.
2. If $t[u1]$ is a free variable z, then z is nondeterministically bound to a term $\sigma(z'_1,\ldots,z'_k)$, where $\sigma \in \Delta^{(k)}$ and $z'_1,\ldots,z'_k$ are new variables. After that, Case 1 is applied.

Remark that the situation in Case 2 is the only kind of nondeterminism in $\overset{u}{\leadsto}_{\widehat{M}}$.

We finish this section by recalling the following theorem from [3]. This theorem induces the uu-algorithm which is based on the ulo-narrowing relation.

Theorem 3.2 [3]
Let $t, s \in T\langle F \cup \Delta\rangle(FV)$, and let V be the finite set $\mathcal{V}(t) \cup \mathcal{V}(s)$. Let S be the set of all (FV, Δ)-substitutions φ such that there exists a derivation:

$$(equ(t,s),\varphi_\emptyset) \overset{u}{\leadsto}_{\widehat{M}} (t_1,\varphi_1) \overset{u}{\leadsto}_{\widehat{M}} (t_2,\varphi_2) \overset{u}{\leadsto}_{\widehat{M}} \cdots \overset{u}{\leadsto}_{\widehat{M}} (t_n,\varphi_n),$$

where $t_n \in T\langle\Delta\rangle(FV)$ and $\varphi = \varphi_n|_V$. Then S is a ground complete set of (E_M,Δ)-unifiers of t and s away from V.

4 Implementation of $\overset{u}{\leadsto}_{\widehat{M}}$ on the UU-Machine

In this section we present the implementation of $\overset{u}{\leadsto}_{\widehat{M}}$ on a nondeterministic abstract machine which is called *uu-machine* and which is an extension of the *G-machine* [9]. In the whole section, we commit that an mt $M = (F, \Delta, R)$ and two input terms $t, s \in T\langle F \cup \Delta\rangle(FV)$ are given.

Configurations and Instructions A configuration of the uu-machine consists of the following components:

- The *program store* is a partial function $ps : PA \to Instr$ from the *set of program addresses* $PA = \mathbb{N}$ into the *the set of instructions Instr* which will be defined later. The machine program is only used for the bottom-up creation of the graph representations for t and s and right hand sides of rules in R.
- The *instruction pointer* $ip \in PA \cup \{Gmode\}$ contains the address of the instruction in the program store which has to be executed next, if a graph representation is created. Otherwise, $ip = Gmode$.
- The *graph* is a partial function $G : Adr \to GNodes$ from the *set of graph addresses* $Adr = \mathbb{N}$ into the *set of graph nodes* $GNodes :=$

$$\begin{aligned} &\{\langle FUN, f, reca, a_1,\ldots,a_n\rangle \mid f \in F^{(n+1)}, n \in \mathbb{N}, reca \in Adr, \\ &\qquad\qquad \text{for every } i \in [n] : a_i \in Adr\} \\ \cup\ &\{\langle CON, \sigma, a_1,\ldots,a_k\rangle \mid \sigma \in \Delta^{(k)}, k \in \mathbb{N}, \text{for every } i \in [k] : a_i \in Adr\} \\ \cup\ &\{\langle VAR, a\rangle \mid a \in Adr \cup \{?\}\}\end{aligned}$$

The graph represents the subterms in the first component fc of the current derivation form of $\overset{u}{\leadsto}_{\widehat{M}}$, the father of which are labelled by *equ*. Every free variable is represented by exactly one VAR-node. This is realized by the mechanism of ***sharing*** [9]. If the second component of such a VAR-node contains the symbol "?", then the corresponding free variable is not bound. Otherwise, the second component contains a graph address which points to the corresponding free variable's binding.

- Two ***address stacks*** $rs1, rs2 \in Adr^*$ which refer to the left and right subterms of the occurrences in fc which are labelled by *equ*, respectively.
- The ***number of the active address stack*** $rsn \in [2]$.
- The ***stack of substitutions*** $ss \in (Adr \cup \{?\})^*$, the i-th square of which refers to the graph representation of z_i's binding. Together with the graph, ss represents φ in the second component of the current derivation form of $\overset{u}{\leadsto}_{\widehat{M}}$.
- The ***data stack*** $ds \in Adr^*$ is only used for the creation of the graph.
- The ***output tape*** $ot \in (\Delta \cup Adr)^*$ contains the labels of the occurrences in fc in lexikographical order up to $impO(fc)$. A free variable is represented by the graph address of its binding.

The ***set of configurations*** of the uu-machine, denoted by $Conf$, is the cartesian product over all those components.

The ***set of instructions***, denoted by $Instr$, contains the following instructions:

- $FNODE(f, n+1)$, for every $f \in F^{(n+1)}$ which replaces the $n+1$ topmost graph addresses $a_1, \ldots, a_{n+1}$ by a new address na and which creates a new graph node $\langle FUN, f, a_{n+1}, \ldots, a_1 \rangle$ at na.
- $CNODE(\sigma, k)$, for every $\sigma \in \Delta^{(k)}$ which works very similar to $FNODE$ and which creates a new graph node $\langle CON, \sigma, a_k, \ldots, a_1 \rangle$ at na.
- $PAR\ i$ pushes the i-th address in the function node which is refered by the topmost square of the active address stack, on top of the data stack.
- $RECVAR\ i$ pushes the i-th address in the first argument of the function node which is refered by the topmost square of the active address stack, on top of the data stack.
- $VAR\ i$ pushes the address of z_i's binding on top of the data stack.
- $RESET$ and $INIT$ transfer the address of a generated right hand side or input term from the data stack to the active address stack, respectively.

Initial Configuration, Code Generation, and Graph Creation for Input Terms In the ***initial configuration*** of the uu-machine for M, t, and s, denoted by $input(M, t, s)$, ps contains the machine program, ip contains the address 1, where the code for the creation of the graph for t starts, $rs1$ is active, and ss contains for every free variable in t and s a "?". All the other components are empty. The form of the machine program together with the definition of the translation scheme *trans* is formalized in Figure 2, where the translation of the (f, σ)-rule's right hand side $rhs(f, \sigma)$-rule is labelled by the symbolic address $ca(f, \sigma)$.

machine program:		$trans(t)\ \ INIT;$
		$trans(s)\ \ INIT;$
for every $f \in F$ and $\sigma \in \Delta$	$ca(f,\sigma):$	$trans(rhs(f,\sigma)\text{-rule})\ RESET;$

$$
\begin{aligned}
trans(f_i(t,t_1,\ldots,t_n)) &= trans(t)\ trans(t_1) \cdots trans(t_n)\ FNODE(f_i, n+1);\\
trans(\sigma_j(t_1,\ldots,t_k)) &= trans(t_1) \cdots trans(t_k)\ CNODE(\sigma_j, k);\\
trans(y_i) &= PAR\ i;\\
trans(x_i) &= RECVAR\ i;\\
trans(z_i) &= VAR\ i;
\end{aligned}
$$

Fig. 2. Machine program and translation scheme.

Starting from the initial configuration, the uu-machine creates the graph representation of t. After this, the $INIT$-instruction is executed which transfers the root address from ds to $rs1$. Afterwards, the same is done with s and $rs2$.

Transition Relation The transformation from one configuration into the next one is formalized by the transition relation $\vdash \subseteq Conf \times Conf$.

In the case that ip is a program address, such that ps is defined for ip, $\vdash$ is simply defined by the semantics of the instruction $ps(ip)$.

In the case that ip is *Gmode*, we distinguish the following cases depending on the graph nodes gn_1 and gn_2 which are refered by the topmost squares a and a' of the two address stacks, resp. (We choose the same enumeration as in the explanation of the ulo-narrowing relation in Section 3). Remark that in the case that the considered graph node represents a variable which is bound, i.e., it is a $\langle VAR, a\rangle$-node, we always consider the graph representation of its binding.

1. If gn_1 and gn_2 are σ-nodes, then a and a' are replaced pairwisely on the two address stacks by the addresses of the successors of gn_1 and gn_2, respectively. Furthermore, σ is written to the output tape.
2. If, w.l.o.g., gn_1 is a $\langle VAR, ?\rangle$-node and gn_2 is a σ-node, then the occur check is applied. If it does not succeed, then gn_1 is replaced by a σ-node with $\langle VAR, ?\rangle$-nodes as successors, the addresses of which are stored in ss. After this, the configuration has the form as in Case 1 and the σ-decomposition-rule is applied. Remark that in this case, one derivation step by $\overset{u}{\rightsquigarrow}_{\hat{M}}$ is implemented by two steps of $\vdash$, where $ip = Gmode$.
3. If gn_1 and gn_2 are the same $\langle VAR, ?\rangle$-node, then the address of this node is written to the output tape and a and a' are deleted from the address stacks.
4. If gn_1 and gn_2 are different $\langle VAR, ?\rangle$-nodes, then gn_1 and gn_2 are replaced by a $\langle VAR, nadr\rangle$-node, where the new graph address $nadr$ is stored in ss. Furthermore, a $\langle VAR, ?\rangle$-node is created at $nadr$ and $nadr$ is written to ot.
5. If gn_1 is a function call, then its evaluation is implemented. For that purpose, the graph address ga of the graph representation of the lo-narrowing occurrence is considered. Depending on the root of the recursion argument of the function f at ga, we distinguish the following two cases:

1. If the root is a constructor σ, then the graph representation of the instantiated right hand side of the (f, σ)-rule must be created. For this purpose, the symbolic program address $ca(f, \sigma)$, where the creation starts, is written to ip and $rs1$ becomes active. If the creation of the graph representation gr is finished, then ds contains the address of gr. Next, the instruction $RESET$ has to be executed. Its semantics depends on whether ga is equal to the graph address on the address stack, yes or no. If the addresses are equal, then $RESET$ deletes the topmost address on $rs1$ and transfers the address from the data stack to it. Otherwise, the address of the recursion argument in the f-node is replaced by the address on ds. In both cases, $RESET$ writes *Gmode* to ip.
2. If the root is a $\langle VAR, ?\rangle$-node, then it is replaced nondeterministically by a constructor-node with $\langle VAR, ?\rangle$-node as successors, the addresses of which are stored in ss. After this, the configuration has the same form as in the previous case. Remark that also in this case, one derivation step by $\overset{u}{\rightsquigarrow}_{\widehat{M}}$ is implemented by two steps of $\vdash$, where $ip = Gmode$.

6. If gn_1 is a constructor-node or a $\langle VAR, ?\rangle$-node and gn_2 is a function node, then $rs2$ becomes active and the machine behaves similar to Case 5.

Stop Configuration A computation of the uu-machine is the implementation of a successful derivation by $\overset{u}{\rightsquigarrow}_{\widehat{M}}$ iff it leads to a configuration $conf$ which has no successor with respect to $\vdash$ and which fulfils the following two conditions: (i) ip in $conf$ is *Gmode* and (ii) the two address stacks in $conf$ are empty.

In this case, the result of the computation, denoted by $output(conf)$, is a pair which consists of the term that results from the contents of ot by substituting graph addresses by their bindings, and a substitution that results from the bindings of the graph addresses in ss.

Correctness of the Implementation We present the following theorem which shows the correctness of the implementation and we sketch its proof.

Theorem 4.1 There exists a derivation

$$(equ(t, s), \varphi_\emptyset) \overset{u\ *}{\rightsquigarrow}_{\widehat{M}} (t^*, \varphi) \qquad (1)$$

by the ulo-narrowing relation, such that $t^* \in T\langle\Delta\rangle(FV)$ and $\varphi \in Sub(FV, \Delta)$, iff there exists a computation

$$input(M, t, s) \vdash^* conf \qquad (2)$$

of the uu-machine, such that $(t^*, \varphi) = output(conf)$.

<u>*Proof-Sketch:*</u> We show that every single derivation step by the ulo-narrowing relation can be simulated by a sequence of transition steps of the uu-machine, and vice versa. For this purpose, we use a function $conf\_to\_df$ which transforms configurations of the uu-machine to the corresponding derivation forms of the ulo-narrowing relation. Since configurations, where $ip \in PA$, are only used for intermediate transition steps, $conf\_to\_df$ is only defined on the set $\{conf \in Conf \mid input(M, t, s) \vdash^* conf \text{ and } ip = Gmode\}$ which is denoted by $impConf$.

Direction (1) $\Longrightarrow$ (2) of the theorem is shown by proving that every step in Derivation (1) can simulated by a sequence of transitions. For this purpose, the following claim is used which can be shown by induction over the structure of the first component in the current derivation form:

Claim 1 For every $(t', \varphi'), (\bar{t}, \overline{\varphi}) \in T\langle \widehat{F} \cup \Delta \rangle (FV) \times Sub(FV, \Delta)$: if there exists a derivation $(equ(t,s), \varphi_\emptyset) \overset{u}{\leadsto}^{*}_{\widehat{M}} (t', \varphi') \overset{u}{\leadsto}_{\widehat{M}} (\bar{t}, \overline{\varphi})$, then for every $conf' \in impConf$, where $conf\_to\_df(conf') = (t', \varphi')$, there exists a configuration $\overline{conf} \in impConf$, such that $conf\_to\_df(\overline{conf}) = (\bar{t}, \overline{\varphi})$ and $conf' \vdash^* \overline{conf}$.

Direction (2) $\Longrightarrow$ (1) of the theorem is shown by proving that nearly every sequence of transitions that transforms a configuration in $impConf$ into another configuration in $impConf$ corresponds to one derivation step in Derivation (1). As we have discussed in the explanation of the transition relation $\vdash$, there are two cases, where one derivation step of $\overset{u}{\leadsto}_{\widehat{M}}$ is simulated by a sequence of transition steps with three different configurations in $impConf$. Hence, these situations are considered separately in the following claim which can be shown by induction over the structure of the refered graphs and which is used to show the correctness of the considered direction.

Claim 2 For every $conf_1, conf_2, conf_3 \in impConf$: if there exist transition sequences of one of the following two forms:

1. $input(M,t,s) \vdash^* conf_1 \vdash conf_2$, where the final step is a transition of Case 2) or of the second case in Case 5) or 6) in the explanation of $\vdash$; and a transition sequence $conf_2 \vdash^+ conf_3$, such that there does not exist any configuration $conf' \in impConf$ with $conf_2 \vdash^+ conf' \vdash^+ conf_3$
2. $input(M,t,s) \vdash^* conf_1 \vdash conf_2$, where the final step is a transition of another case in the explanation of $\vdash$; and a transition sequence $conf_2 \vdash^* conf_3$, such that there does not exist any configuration $conf' \in impConf$ with $conf_2 \vdash^+ conf' \vdash^* conf_3$ or $conf_2 \vdash^* conf' \vdash^+ conf_3$

then there exists the derivation step

$$conf\_to\_df(conf_1) \overset{u}{\leadsto}_{\widehat{M}} conf\_to\_df(conf_3).$$

The inductive proofs are omitted because of lack of space. $\oplus$

5 Conclusion and Related Work

In this paper we have described the implementation of the ulo-narrowing relation associated with a modular tree transducer for unrestricted input terms on the uu-machine. Furthermore, we have shown that the implementation is correct.

There exist a lot of implementations of other outermost narrowing relations for functional logic programming languages on abstract machines in the literature (cf., e.g., [6] for an overview). But, since functional logic programming languages have a more complicated structure than modular tree transducers, the implemented narrowing relations are not as efficient as the ulo-narrowing relation. For example, by the lazy narrowing relation [12, 11] also inner occurrences have to be considered and the binding mode and the occur check are not integrated. Hence, lazy narrowing is not a strategy [1] and the considered search space is much bigger. Furthermore, these implementations are not able to take use of the simple structure of modular tree transducers' rewrite rules which leads to problems in deterministic implementations [5] and finally, there does not exist any proof of the correctness of these implementations.

In our current research we extend the presented implementation to an implementation of a deterministic partial uu-algorithm for mt's [4] by adding mechanisms for dealing with backtracking.

References

1. R. Echahed. On completeness of narrowing strategies. In *CAAP'88*, pages 89–101. Springer-Verlag, 1988. LNCS 299.
2. J. Engelfriet and H. Vogler. Modular tree transducers. *Theoretical Computer Science*, 78:267–304, 1991.
3. H. Faßbender and H. Vogler. A universal unification algorithm based on unification-driven leftmost outermost narrowing. *Acta Cybernetica*, 11(3):139–167, 1994.
4. H. Faßbender, H. Vogler, and A. Wedel. Implementation of a deterministic partial E-unification algorithm for macro tree transducers. Technical Report 94-04, University of Ulm, Fakultät für Informatik, D-89069 Ulm, Germany, 1994. accepted for publication in Electronic Journal of Functional and Logic Programming.
5. W. Hans, R. Loogen, and S. Winkler. On the interaction of lazy evaluation and backtracking. In *PLILP'92*, pages 355–369. Springer-Verlag, 1992. LNCS 631.
6. M. Hanus. The integration of functions into logic programming: From theory to practice. *Journal of Logic Programming*, 19,20:583–628, 1994.
7. G. Huet and D.C. Oppen. Equations and rewrite rules: a survey. In R. Book, editor, *Formal Language Theory: Perspectives and Open Problems*. Academic Press, New York, 1980.
8. J.M. Hullot. Canonical forms and unification. In *Proceedings of the 5th conference on automated deduction*, pages 318–334. Springer-Verlag, 1980. LNCS 87.
9. T. Johnsson. Efficient compilation of lazy evaluation. *SIGPLAN Notices*, 6:58–69, 1984.
10. A. Martelli and U. Montanari. An efficient unification algorithm. *ACM Transactions on Programming Languages Systems*, 4:258–282, 1982.
11. J.J. Moreno-Navarro and M. Rodriguez-Artalejo. Logic-programming with functions and predicates: the language BABEL. *Journal of Logic Programming*, 12:191–223, 1992.
12. U.S. Reddy. Narrowing as the operational semantics of functional languages. In *Symposium on Logic Programming*, pages 138–151. IEEE Comp. Soc. Press, 1985.
13. J.H. Siekmann. Unification theory. *Journal of Symbolic Computation*, 7:207–274, 1989.

Dummy Elimination: Making Termination Easier

M. C. F. Ferreira* and H. Zantema
tel: +31-30-532249, fax: +31-30-513791, e-mail: {maria, hansz}@cs.ruu.nl

Utrecht University, Department of Computer Science
P.O. box 80.089, 3508 TB Utrecht, The Netherlands

Abstract. We investigate a technique whose goal is to simplify the task of proving termination of term rewriting systems. The technique consists of a transformation which eliminates function symbols considered "useless" and simplifies the rewrite rules. We show that the transformation is sound, i. e., termination of the original system can be inferred from termination of the transformed one. For proving this result we use a new notion of lifting of orders that is a generalization of the multiset construction.

1 Introduction

Suppose we want to prove termination of the following system

$$f(g(x)) \to f(a(g(g(f(x))), g(g(f(x)))))$$

Intuitively, the function symbol a is created but seems not to have any influence on the reductions. Taking that into account, we can eliminate it and transform the given rule into

$$f(g(x)) \to f(\diamond)$$
$$f(g(x)) \to g(g(f(x)))$$

where $\diamond$ is a fresh constant. Termination of the first system is not easy to prove (since the system is self-embedding orders like *recursive path order (rpo)* cannot be used) while termination of the second system is trivially proven with *rpo* by choosing the precedence $\triangleright$ satisfying $f \triangleright g \triangleright \diamond$. Now if the transformation is sound, i. e., termination of the original system can be inferred from termination of the transformed one, our task is done. In this paper we formally describe this transformation and prove its soundness with respect to termination.

In general, we are interested in simplifying the process of proving termination of term rewriting systems (TRS's). One approach to this goal is to devise sound transformations on TRS's such that the transformed systems are somehow easier to deal with, with respect to termination proofs, than the original ones. As examples of such transformations we have *transformation orderings* [1], *semantic labelling* [10] and *distribution elimination* [11] [2].

* Supported by NWO, the Dutch Organization for Scientific Research, under grant 612-316-041.

[2] For an example of application of some of these techniques, including the one described in this paper, see [12].

The technique we present falls within the same category as distribution elimination, with function symbols occurring only in right-hand-side of rules being eliminated and the rules transformed. As a technical mean to prove our result we make use of trees labelled with terms and of a new construction that lifts an order on a set to an order on trees labelled with labels from that set. This construction is interesting per se and therefore treated separately in section 3.

The rest of the paper is organized as follows. In section 2 we give some basic notions on TRS's and orders. In section 3 we present the tree lifting of an order. It turns out that this lifting is a generalization of the multiset construction, monotone with respect to the order lifted and well-foundedness preserving. This will be the essential tool to be used in the proof of the main result. In section 4 we present the transformation on TRS's and prove its soundeness. The proof is conceptually simple although the technical details may not seem so. Finally in section 5 we make some final remarks.

2 Basic notions

Below we introduce some notation and some basic notions over orders and TRS's. For more information over TRS's the reader is referred to [4].

A *poset* $(S,>)$ is a set S together with a partial order, i. e., an irreflexive and transitive relation $>$. Given a poset $(S,>)$, $M(S)$ denotes the finite multisets over S (see [5]) and $>_{mul}$ denotes the multiset extension of $>$ to $M(S)$, given by $S >_{mul} T \iff T = (S\backslash X) \cup Y$, with $\emptyset \neq X \subseteq S$ and $\forall y \in Y\ \exists x \in X : x > y$. The multiset extension of a partial order is itself a partial order, monotone with respect to the order extended and well-foundedness preserving. We use the parentheses $[\,]$ to denote multisets being $[\,]$ the empty multiset.

Given a non-empty set A, we consider non-empty trees over A, defined by the following data type: $Tr(A) \cong A \times M(Tr(A))$, i. e., if f is the function from sets to sets given by $f(X) = A \times M(X)$, then $Tr(A)$ is the least fixed point of f. Therefore a tree is either a root, represented by $(a,[\,])$, with $a \in A$, or a tree with root $a \in A$ and subtrees $t_1,\ldots,t_n$, represented by $(a,[t_1,\ldots,t_n])$. Since we are not interested in the order of the subtrees, we choose the multiset representation for the subtrees instead of a sequence representation. The *depth* of a tree is given by the function depth and is defined inductively as usual.

Let $\emptyset \neq \mathcal{F}$ be a *signature*, (a set of varyadic function symbols, possibly infinite). Let $\mathcal{X}$ denote a denumerable set of variables with $\mathcal{F} \cap \mathcal{X} = \emptyset$. The function $\mathsf{arity} : \mathcal{F} \cup \mathcal{X} \to \mathcal{P}(\mathbb{N})$, where $\mathcal{P}(\mathbb{N})$ represents the powerset of the natural numbers, gives the possible arities a symbol can have. Constants and variables have arity $\{0\}$. The set of terms over $\mathcal{F}$ and $\mathcal{X}$ is denoted by $\mathcal{T}(\mathcal{F},\mathcal{X})$ and the set of ground terms over $\mathcal{F}$ by $\mathcal{T}(\mathcal{F})$; they are defined as usual. The set $Var(t)$ contains the variables occurring in term t.

A *term rewriting system* (TRS) is a tuple $(\mathcal{F},\mathcal{X},R)$, where R is a subset of $\mathcal{T}(\mathcal{F},\mathcal{X}) \times \mathcal{T}(\mathcal{F},\mathcal{X})$. The elements (l,r) of R are called the rules of the TRS and are usually denoted by $l \to r$. They obey the restriction that the *left-hand-side (lhs)* l must be a non-variable and every variable in the *right-hand-side (rhs)* r

must also occur in l. In the following, unless otherwise specified, we identify the TRS with R, being $\mathcal{F}$ the set of function symbols occurring in R.

A TRS R induces a *rewrite relation* over $\mathcal{T}(\mathcal{F}, \mathcal{X})$, denoted by $\to_R$, as follows: $s \to_R t$ iff $s = C[l\sigma]$ and $t = C[r\sigma]$, for some context C, substitution σ and rule $l \to r \in R$. The transitive closure of $\to_R$ is denoted by $\to_R^+$ and its reflexive-transitive closure by $\to_R^*$. By $\to_R^n$, with $n \in \mathbb{N}$, we denote the composition of $\to_R$ with itself n times (if $n = 0$, $\to_R^n$ is the identity). A TRS is called *terminating* if there exists no infinite sequence of the form $t_0 \to_R t_1 \to_R \ldots$.

3 Ordering trees

We now describe how to lift partial orders on sets to partial orders on trees while preserving well-foundedness. Later we will use this in the context of rewriting.

Definition 1. Let $(A, >)$ be a partially ordered set and consider $Tr(A)$, the finite non-empty trees over A. In $Tr(A)$ we define the following relation $\succ$

$$t = (a, M) \succ (b, M') \iff \begin{cases} a > b \text{ and } \forall u \in M' : (t \succ u) \text{ or } (\exists v \in M : v \succeq u) \\ a = b \text{ and } M \succ_{mul} M' \end{cases}$$

where $\succ_{mul}$ is the multiset extension of $\succ$ and $\succeq = \succ \cup =$. We call the relation $\succ$ the tree lifting of $>$.

The following lemma is an easy induction on the (sum of) depths of the trees.

Lemma 2. *The relation $\succ$ is a partial order on $Tr(A)$.*

Example 1. Let $(A, >)$ be $\mathbb{N}$ with the usual total order. Let

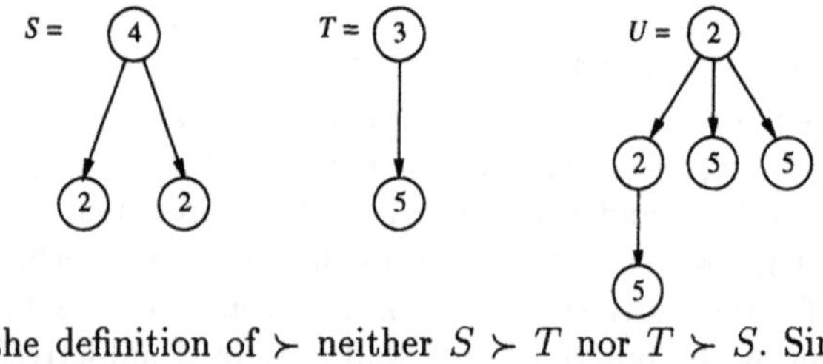

According to the definition of $\succ$ neither $S \succ T$ nor $T \succ S$. Since $S \neq T$, the example shows that the tree lifting does not preserve totality. From definition 1 also follows that $T \succ U$. Note though that the depth of U is greater than the depth of T.

The construction presented in definition 1 has many interesting properties. It is a proper generalization of the multiset construction and it is not difficult to see that it is monotonic with respect to the order lifted. The construction also preserves well-foundedness. Since this property is essential for our purposes, we sketch its proof below.

Theorem 3. *Let $(A, >)$ be a poset. Then $>$ is well-founded on A if and only if $\succ$ is well-founded on $Tr(A)$.*

Proof. (Sketch) For the "if" part, suppose that $>$ is not well-founded in A. Then there is an infinite descending chain $a_0 > a_1 > \cdots$. According to the definition of $\succ$ then $(a_0, [\,]) \succ (a_1, [\,]) \succ \ldots$, is an infinite descending chain in $Tr(A)$, contradicting well-foundedness of $\succ$.

For the "only-if" part we will use *recursive path order*, $>_{rpo}$, on trees, based on $>$ and with multiset status (for a definition of $>_{rpo}$ see for example [2, 3]), given by

$$(a, M) >_{rpo} (b, N) \iff \begin{cases} (a > b) \text{ and } (\forall u \in N : \ (a, M) >_{rpo} u), \text{ or} \\ (a = b) \text{ and } (M >_{rpo,mul} N), \text{ or} \\ \exists u \in M : \ u \geq_{rpo} (b, N), \text{ or} \end{cases}$$

It is well-known that $>_{rpo}$ is well-founded whenever $>$ is well-founded (for a simple proof see [8]), so we only need to check that $\succ \subseteq >_{rpo}$ and this can be achieved by showing that for any trees S, T, $S \succ T \Rightarrow S >_{rpo} T$, using induction on $\mathsf{depth}(S) + \mathsf{depth}(T)$. □

4 Transforming the TRS

In this section we present our transformation and prove its soundness. We establish first some terminology. Let $\mathcal{F}$ be a signature and $\mathcal{X}$ a set of variables with $\mathcal{F} \cap \mathcal{X} = \emptyset$. Let a be a function symbol with non-null arities, i. e., $N > 0$, for all $N \in \mathsf{arity}(a)$, and not occurring in $\mathcal{F}$. Let $\diamond$ be a constant also not occurring in $\mathcal{F}$. We denote by $\mathcal{F}_a$ and $\mathcal{F}_\diamond$ respectively the sets $\mathcal{F} \cup \{a\}$ and $\mathcal{F} \cup \{\diamond\}$. We consider TRS's over $T(\mathcal{F}_a, X)$ such that the function symbol a may only occurs in the rhs of the rules of the TRS. The idea behind the transformation is that the fuction symbol a, not occurring in the lhs of rewrite rules, has no relevant role in the reductions and therefore should not influence the termination behaviour of the TRS. Given a term over $T(\mathcal{F}_a, \mathcal{X})$, we decompose it in its "components" which are the terms over $T(\mathcal{F}_\diamond, \mathcal{X})$ between occurrences of the symbol a, with those occurrences being replaced by the constant $\diamond$. We make this more precise.

Definition 4. Given a term $t \in T(\mathcal{F}_a, \mathcal{X})$, its cap is denoted by $\mathsf{cap}(t)$, where cap: $T(\mathcal{F} \cup \{a, \diamond\}, \mathcal{X}) \to T(\mathcal{F}_\diamond, \mathcal{X})$, is defined inductively as follows:

- $\mathsf{cap}(x) = x$, for any $x \in \mathcal{X}$
- $\mathsf{cap}(f(t_1, \ldots, t_m)) = f(\mathsf{cap}(t_1), \ldots, \mathsf{cap}(t_m))$, if $f \in \mathcal{F}$ and $m \in \mathsf{arity}(f)$
- $\mathsf{cap}(a(t_1, \ldots, t_N)) = \diamond$, with $N \in \mathsf{arity}(a)$

The operation dec collects the caps of the subterms encapsulated between occurrences of the symbol a and adds a symbol $\diamond$ for each occurrence of a encountered except for the topmost.

Definition 5. The decomposition of $t \in T(\mathcal{F}_a, \mathcal{X})$ is denoted by $\mathsf{dec}(t)$, where $\mathsf{dec} : T(\mathcal{F} \cup \{a, \diamond\}, \mathcal{X}) \to M(T(\mathcal{F}_\diamond, \mathcal{X}))$ is defined inductively as follows:

- $\mathsf{dec}(x) = [\,]$

- $\text{dec}(f(t_1,\ldots,t_m)) = \bigcup_{i=1}^{m} \text{dec}(t_i)$, if $f \in \mathcal{F}$ and $m \in \text{arity}(f)$ $(f \neq a)$
- $\text{dec}(a(t_1,\ldots,t_N)) = \bigcup_{i=1}^{N}([\text{cap}(t_i)] \cup \text{dec}(t_i))$, with $N \in \text{arity}(a)$

Example 2. The following term t

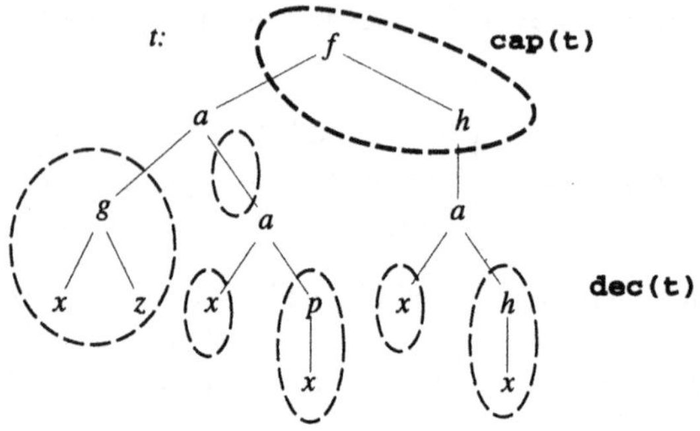

has as decomposition $\text{dec}(t) = [g(x,z), x, \diamond, p(x), x, h(x)]$; its cap is the term $f(\diamond, h(\diamond)))$. The empty ellipse represents the constant $\diamond$.

We define now the transformation on TRS's. As expected we will decompose the rhs of the rules in R and create new rules using this decomposition.

Definition 6. Given a TRS R over $T(\mathcal{F}_a, \mathcal{X})$ where the function symbol a occurs at most in the rhs of the rules in R, $E(R)$ is the TRS over $T(\mathcal{F}_\diamond, \mathcal{X})$ given by

$$E(R) = \{l \to u \mid (l \to r) \in R \text{ and } u = \text{cap}(r) \text{ or } u \in \text{dec}(r)\}$$

Example 3. Let R be given by the rules

$$f(f(x)) \to g(a(f(x), f(x))) \qquad\qquad g(g(x)) \to f(g(x))$$

Then the transformed TRS, $E(R)$ is given by:

$$\begin{array}{ll} f(f(x)) \to g(\diamond) & \qquad g(g(x)) \to f(g(x)) \\ f(f(x)) \to f(x) & \end{array}$$

Note that rules in which a does not occur remain unchanged.

From the definition of E, we see that in general $E(R)$ has more rules but is syntactically simpler than R, so the transformation can be quite useful if we are able to infer termination of R from termination of $E(R)$. Termination however is not preserved Consider the the TRS's R and $E(R)$ given by

$$R : f(x,x) \to f(a(x),x) \qquad\qquad \begin{array}{l} E(R) : f(x,x) \to f(\diamond, x) \\ \qquad\quad\;\; f(x,x) \to x \end{array}$$

The system R is terminating while the system $E(R)$ is not.

The main purpose of this paper is to show that termination of $E(R)$ implies termination of R. Before going into the technical details we give a general idea of the proof. If $E(R)$ is terminating, the relation $\to^{+}_{E(R)}$ is well-founded. If we consider the poset $(Tr(T(\mathcal{F}_\diamond,\mathcal{X})),\succ)$ (where $\succ$ is the tree extension, as defined in 1, of a well-founded extension of $\rightarrow^+_{E(R)}$) then $\succ$ is also well-founded. We now use the trees over $T(\mathcal{F}_\diamond,\mathcal{X})$ to interpret the terms of $T(\mathcal{F}_a,\mathcal{X})$ in such a way that for terms $s,t \in T(\mathcal{F}_a,\mathcal{X})$ if $s \rightarrow_R t$ then $\mathsf{tree}(s) \succ \mathsf{tree}(t)$, where $\mathsf{tree}(u)$ is a tree over $T(\mathcal{F}_\diamond,\mathcal{X})$ associated with the term u. Termination of R follows from well-foundedness of $\succ$.

We introduce some definitions and auxiliary results.

Definition 7. Given a term $t \in T(\mathcal{F}_a,\mathcal{X})$ we associate to it a tree over $T(\mathcal{F}_\diamond,\mathcal{X})$, denoted by $\mathsf{tree}(t)$, where $\mathsf{tree} : T(\mathcal{F}\cup\{a,\diamond\},\mathcal{X}) \rightarrow Tr(T(\mathcal{F}_\diamond,\mathcal{X}))$ is defined as follows:

- $\mathsf{tree}(x) = (x, [])$, for any $x \in \mathcal{X}$
- $\mathsf{tree}(f(s_1,\ldots,s_m)) = (\mathsf{cap}(f(s_1,\ldots,s_m)), \bigcup_{i=1}^m M_i)$, where $\mathsf{tree}(s_i) = (\mathsf{cap}(s_i), M_i)$
- $\mathsf{tree}(a(s_1,\ldots,s_N)) = (\mathsf{cap}(a(s_1,\ldots,s_N)), \bigcup_{i=1}^N [\mathsf{tree}(s_i)])$

Example 4. The following picture shows the same term as in example 2 together with its corresponding tree.

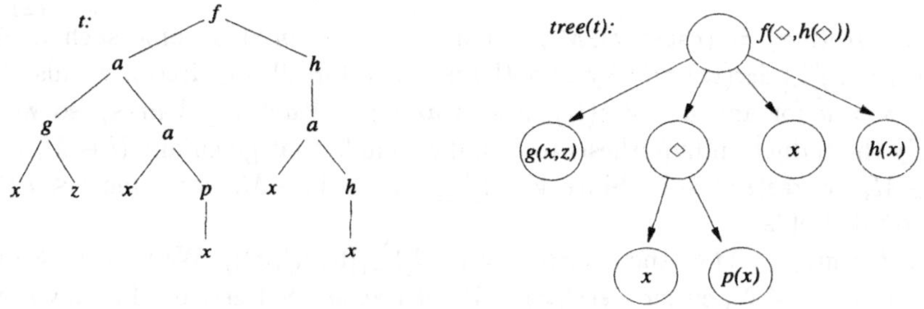

Definition 8. For any arbitrary substitution $\sigma : \mathcal{X} \rightarrow T(\mathcal{F}_a,\mathcal{X})$, the substitution $\mathsf{cap}(\sigma) : \mathcal{X} \rightarrow T(\mathcal{F}_\diamond,\mathcal{X})$ is defined by $\mathsf{cap}(\sigma)(x) = \mathsf{cap}(\sigma(x))$, for all $x \in \mathcal{X}$.

The following lemma is easy to prove using induction on terms.

Lemma 9. *Let $t \in T(\mathcal{F}_a,\mathcal{X})$ and $\sigma : \mathcal{X} \rightarrow T(\mathcal{F}_a,\mathcal{X})$ be an arbitrary substitution. Then $\mathsf{cap}(t\sigma) = \mathsf{cap}(t)\mathsf{cap}(\sigma)$.*

Lemma 10. *Let $t \in T(\mathcal{F},\mathcal{X})$ be non-ground. Let $\sigma : \mathcal{X} \rightarrow T(\mathcal{F}_a,\mathcal{X})$ be any substitution and let $x \in Var(t)$. Let $\mathsf{tree}(t\sigma) = (\mathsf{cap}(t\sigma), M_t)$ and $\mathsf{tree}(\sigma(x)) = (\mathsf{cap}(\sigma(x)), M_x)$. Then $M_x \subseteq M_t$ (being $\subseteq$ multiset inclusion).*

Proof. (Sketch) Since $x \in Var(t)$ we can write t as $C[x]$, for some context $C[\,]$. The lemma is now easily proved by induction on the context. □

Remark. From now on we assume that $E(R)$ is terminating. Consequently the reduction relation $\rightarrow^+_{E(R)}$ is well-founded. If we simply use this reduction relation as base for our tree lifting, we will have problems to establish that $s \rightarrow_R t \Rightarrow \mathsf{tree}(s) \succ \mathsf{tree}(t)$. For this property to hold we need the following extension of $\rightarrow^+_{E(R)}$.

Definition 11. The relation $>$ on $T(\mathcal{F}_\diamond, \mathcal{X})$ is defined as follows: $s > t$ iff $s \neq t$ and $s \rightarrow^*_{E(R)} C[t]$, for some context C.

This relation appeared already in [9]. We note that $C[t] > t$, for any non-trivial context C. Given that $\rightarrow^+_{E(R)}$ is well-founded, the following lemma is also easy to prove.

Lemma 12. *The relation $>$ is a partial well-founded order on $T(\mathcal{F}_\diamond, \mathcal{X})$ extending $\rightarrow^+_{E(R)}$, closed under substitutions.*

We now consider trees over $T(\mathcal{F}_\diamond, \mathcal{X})$ being $\succ$ the tree lifting of $>$.

Lemma 13. *Let $s \in T(\mathcal{F}, \mathcal{X}) \setminus \mathcal{X}$ and $t \in T(\mathcal{F}_a, \mathcal{X})$ such that $Var(t) \subseteq Var(s)$ and $s > v$ for all $v \in \mathsf{dec}(t)$. Let $\sigma : \mathcal{X} \rightarrow T(\mathcal{F}_a, \mathcal{X})$ be any substitution and suppose that $\mathsf{tree}(s\sigma) = (\mathsf{cap}(s\sigma), M_s)$, $\mathsf{tree}(t\sigma) = (\mathsf{cap}(t\sigma), M_t)$. Then for all $U \in M_t$ either $U \in M_s$ or $\mathsf{tree}(s\sigma) \succ U$.*

Proof. By induction on the structure of t. If $t = x \in \mathcal{X}$, the result follows from lemma 10. For $t = f(t_1, \ldots, t_m)$, $\mathsf{tree}(t\sigma) = (\mathsf{cap}(f(t_1\sigma, \ldots, t_m\sigma)), \bigcup_{i=1}^{m} M_i)$, where $\mathsf{tree}(t_i\sigma) = (\mathsf{cap}(t_i\sigma), M_i)$, for all $1 \leq i \leq m$. Fix some such i. Since $\mathsf{dec}(t) = \bigcup_{j=1}^{m} \mathsf{dec}(t_j)$ and by hypothesis $s > v$ for all $v \in \mathsf{dec}(t)$, we also have that $s > u$ for any $u \in \mathsf{dec}(t_i)$. Also $Var(t_i) \subseteq Var(t) \subseteq Var(s)$, so we can apply the induction hypothesis to t_i and conclude that given any $U \in M_i$ either $U \in M_s$ or $\mathsf{tree}(s\sigma) \succ U$. Since $V \in \bigcup_{j=1}^{m} M_j \Rightarrow V \in M_i$, for some $1 \leq i \leq m$, the result holds.

If $t = a(t_1, \ldots, t_N)$ then $\mathsf{tree}(t\sigma) = (\diamond, \bigcup_{i=1}^{N}[\mathsf{tree}(t_i\sigma)])$. We need to see that for any $1 \leq i \leq N$, either $\mathsf{tree}(t_i\sigma) \in M_s$ or $\mathsf{tree}(s\sigma) \succ \mathsf{tree}(t_i\sigma)$. Fix any such i. By lemma 9 we know that $\mathsf{tree}(t_i\sigma) = (\mathsf{cap}(t_i\sigma), M_i) = (\mathsf{cap}(t_i)\mathsf{cap}(\sigma), M_i)$ (see definition 8). Also $\mathsf{tree}(s\sigma) = (\mathsf{cap}(s)\mathsf{cap}(\sigma), M_s) = (s\ \mathsf{cap}(\sigma), M_s)$. By hypothesis $s > u$ for all $u \in \mathsf{dec}(t)$ and since $\mathsf{dec}(t) = \bigcup_{j=1}^{N}([\mathsf{cap}(t_j)] \cup \mathsf{dec}(t_j))$, we can say that $s > u$ for all $u \in \mathsf{dec}(t_i)$. Further $Var(t_i) \subseteq Var(t) \subseteq Var(s)$, so we can apply the induction hypothesis to t_i and conclude that if $U \in M_i$ then either $U \in M_s$ or $\mathsf{tree}(s\sigma) \succ U$. Since by hypothesis $s > \mathsf{cap}(t_i)$, and $>$ is closed under substitutions we conclude that $s\ \mathsf{cap}(\sigma) > \mathsf{cap}(t_i)\mathsf{cap}(\sigma)$ and by definition 1 we have $\mathsf{tree}(s\sigma) \succ \mathsf{tree}(t_i\sigma)$, as we wanted. □

Lemma 14. *Let $s \in T(\mathcal{F}, \mathcal{X}) \setminus \mathcal{X}$ and $t \in T(\mathcal{F}_a, \mathcal{X})$ such that $Var(t) \subseteq Var(s)$ and $s > v$ for all $v \in \mathsf{dec}(t) \cup [\mathsf{cap}(t)]$. Finally let $\sigma : \mathcal{X} \rightarrow T(\mathcal{F}_a, \mathcal{X})$ be any substitution. Then $\mathsf{tree}(s\sigma) \succ \mathsf{tree}(t\sigma)$.*

Proof. By definition 7 and lemma 9, $\mathsf{tree}(s\sigma) = (s\ \mathsf{cap}(\sigma), M_s)$ and $\mathsf{tree}(t\sigma) = (\mathsf{cap}(t)\mathsf{cap}(\sigma), M_t)$. By lemma 13, for any $U \in M_t$ either $U \in M_s$ or $\mathsf{tree}(s\sigma) \succ U$. Since $s > \mathsf{cap}(t)$ and $>$ is closed under substitutions, we have $s\ \mathsf{cap}(\sigma) > \mathsf{cap}(t)\mathsf{cap}(\sigma)$, and by definition 1 we conclude that $\mathsf{tree}(s\sigma) \succ \mathsf{tree}(t\sigma)$. □

Lemma 15. *Let $l \rightarrow r$ be a rule in R and $\sigma : \mathcal{X} \rightarrow T(\mathcal{F}_a, \mathcal{X})$ an arbitrary substitution. Then $\mathsf{tree}(l\sigma) \succ \mathsf{tree}(r\sigma)$.*

Proof. From definition 6, we know that $l \to u$, with $u \in [\mathsf{cap}(r)] \cup \mathsf{dec}(r)$, is a rule in $E(R)$ and so $l > u$ for any $u \in [\mathsf{cap}(r)] \cup \mathsf{dec}(r)$. Also $Var(r) \subseteq Var(l)$ and a does not occur in l, therefore all the hypothesis of lemma 14 are satisfied, so we can apply it to conclude that $\mathsf{tree}(l\sigma) \succ \mathsf{tree}(r\sigma)$. □

The following lemma is not difficult to prove by induction on the definition of reduction. Note that lemma 9 is essential for the base case.

Lemma 16. *Let $s, t \in T(\mathcal{F}_a, \mathcal{X})$. If $s \to_R t$ then $\mathsf{cap}(s) \to^{0,1}_{E(R)} \mathsf{cap}(t)$.*

Lemma 17. *Let $s, t \in T(\mathcal{F}_a, \mathcal{X})$ such that $s \to_R t$ and $\mathsf{tree}(s) \succ \mathsf{tree}(t)$. Then $\mathsf{tree}(C[s]) \succ \mathsf{tree}(C[t])$, for any context C.*

Proof. We proceed by induction on the context. If C is the trivial context, then the result holds by hypothesis. Let then $C = f(s_1, \ldots, \square, \ldots, s_k)$, with $\square$ occurring at fixed position $1 \leq j \leq k$. By definitions 7 and 4 we have

$$\mathsf{tree}(C[s]) = (f(\mathsf{cap}(s_1), \ldots, \mathsf{cap}(s), \ldots, \mathsf{cap}(s_k)), \bigcup_{i=1}^{k} M_i)$$

where $\mathsf{tree}(s_i) = (\mathsf{cap}(s_i), M_i)$ for $1 \leq i \leq k$, $i \neq j$, and $\mathsf{tree}(s) = (\mathsf{cap}(s), M_j)$. Similarly $\mathsf{tree}(C[t]) = (f(\mathsf{cap}(s_1), \ldots, \mathsf{cap}(t), \ldots \mathsf{cap}(s_k)), \bigcup_{i=1}^{k} M_i')$ where $M_i' = M_i$, for $1 \leq i \leq k$, $i \neq j$, and $\mathsf{tree}(t) = (\mathsf{cap}(t), M_j')$.

By hypothesis $\mathsf{tree}(s) \succ \mathsf{tree}(t)$ and therefore either

- $\mathsf{cap}(s) > \mathsf{cap}(t)$ and for all $U \in M_j'$, either $\mathsf{tree}(s) \succ U$ or there is an element $V \in M_j$ such that $V \succeq U$. Since $s \to_R t$ then by lemma 16 we have $\mathsf{cap}(s) \to^{0,1}_{E(R)} \mathsf{cap}(t)$, and due to irreflexivity of $>$ we indeed have $\mathsf{cap}(s) \to_{E(R)} \mathsf{cap}(t)$. Hence $f(\ldots \mathsf{cap}(s) \ldots) \to_{E(R)} f(\ldots \mathsf{cap}(t) \ldots)$ and $f(\ldots \mathsf{cap}(s) \ldots) > f(\ldots \mathsf{cap}(t) \ldots)$. To conclude that $\mathsf{tree}(C[s]) \succ \mathsf{tree}(C[t])$ we only need to see that for any $U \in M_j'$ either $\mathsf{tree}(C[s]) \succ U$ or there is an element $V \in \bigcup_{i=1}^{k} M_i$ such that $V \succeq U$. Take then $U \in M_j'$ and suppose that there is no such element V. Then we must have $\mathsf{tree}(s) \succ U$, since $\mathsf{tree}(s) \succ \mathsf{tree}(t)$ and $\mathsf{cap}(s) > \mathsf{cap}(t)$. But we also have $\mathsf{tree}(C[s]) \succ \mathsf{tree}(s)$, since $M_j \subseteq \bigcup_{i=1}^{k} M_i$ and $f(\mathsf{cap}(s_1), \ldots, \mathsf{cap}(s), \ldots \mathsf{cap}(s_k)) > \mathsf{cap}(s)$. By transitivity of $\succ$ we conclude that $\mathsf{tree}(C[s]) \succ U$.
- $\mathsf{cap}(s) = \mathsf{cap}(t)$; $M_j \succ_{mul} M_j'$. In this case we have $\bigcup_{i=1}^{k} M_i \succ_{mul} \bigcup_{i=1}^{k} M_i'$. Since $f(\mathsf{cap}(s_1), \ldots, \mathsf{cap}(s), \ldots \mathsf{cap}(s_k)) = f(\mathsf{cap}(s_1), \ldots, \mathsf{cap}(t), \ldots \mathsf{cap}(s_k))$, we conclude that $\mathsf{tree}(C[s]) \succ \mathsf{tree}(C[t])$.

Suppose now that $C = a(s_1, \ldots, \square, \ldots, s_N)$, with $\square$ occurring at some fixed position $1 \leq j \leq N$. Then $\mathsf{tree}(C[s]) = (\diamond, \bigcup_{i=1, i\neq j}^{N} [\mathsf{tree}(s_i)] \cup [\mathsf{tree}(s)])$ and $\mathsf{tree}(C[t]) = (\diamond, \bigcup_{i=1, i\neq j}^{N} [\mathsf{tree}(s_i)] \cup [\mathsf{tree}(t)])$.

Since $\mathsf{tree}(s) \succ \mathsf{tree}(t)$ also $\bigcup_{i=1, i\neq j}^{N} [\mathsf{tree}(s_i)] \cup [\mathsf{tree}(s)] \succ_{mul} \bigcup_{i=1, i\neq j}^{N} [\mathsf{tree}(s_i)] \cup [\mathsf{tree}(s)]$ and by definition 1 we conclude that $\mathsf{tree}(C[s]) \succ \mathsf{tree}(C[t])$. □

We have seen that given a TRS R, whenever $E(R)$ terminates, we can lift the well-founded order $\to^+_{E(R)}$ to a well-founded order $\succ$ on $Tr(T(\mathcal{F}_\diamond, \mathcal{X}))$. Furthermore we can associate to each term $t \in T(\mathcal{F}_a, \mathcal{X})$ a tree in $Tr(T(\mathcal{F}_\diamond, \mathcal{X}))$ in such a way that if $s \to_R t$ then $\mathsf{tree}(s) \succ \mathsf{tree}(t)$. Consequently the relation $\to^+_R$ is well-founded. In other words, we have proved our main result.

Theorem 18. *If $E(R)$ terminates then R terminates.*

5 Final remarks

As mentioned before, dummy elimination bears similarities with distribution elimination, and indeed the technique used here to prove the soundness of dummy elimination can also be applied to distribution elimination, namely to solve the conjecture stated in [11]: distribution elimination remains sound when no distributive rules are present and the transformed system is not right-linear.[3] A different proof of our main result can be given by the technique of self-labelling ([6]), which is a particular case of semantic labelling ([10]). Further it has been remarked by Middeldorp and Ohsaki (personal communication) that the whole framework remains valid if dummy symbols are allowed in the lhs. In that case the lhs l in $E(R)$ have to be replaced by $\mathsf{cap}(l)$. In this way still no symbols a occur in $E(R)$.

The two techniques, dummy elimination and distribution elimination, are incomparable. Consider the TRS $R : f(f(x)) \to f(a(f(x)))$, where we want to eliminate the symbol a. Then distribution elimination and dummy elimination result in the systems $E_D(R)$ and $E(R)$, respectively:

$$E_D(R) : f(f(x)) \to f(f(x)) \qquad E(R) : \begin{array}{l} f(f(x)) \to f(\diamond) \\ f(f(x)) \to f(x) \end{array}$$

In this case we have that both R and $E(R)$ terminate but $E_D(R)$ does not, suggesting that the transformation E is stronger than E_D.

Now consider the system $R : f(x,x) \to f(a(0), a(1))$, again with elimination of a. Distribution elimination and dummy elimination result in the systems $E_D(R)$ and $E(R)$, respectively:

$$E_D(R) : f(x,x) \to f(0,1) \qquad E(R) : \begin{array}{l} f(x,x) \to f(\diamond,\diamond) \\ f(x,x) \to 1 \\ f(x,x) \to 0 \end{array}$$

In this case we have that both R and $E_D(R)$ terminate but $E(R)$ does not, suggesting the reverse conclusion. However dummy elimination seems to be more drastic and to produce simpler TRS's. This idea is enforced by the fact that distribution elimination is sound and complete (i. e., termination of R implies termination of $E_D(R)$) with respect to particular kinds of termination like *total*

[3] This same technique can be applied to a whole hierarchy of transformations from which dummy elimination and distribution elimination are particular cases.

termination [7] and (termination proofs using) *recursive path order (rpo)*, while dummy elimination is not. In the first example above, the original system cannot be proven terminating using *rpo*. Furthermore the system is not simply terminating and thus also not totally terminating. The transformed system $E(R)$ is trivially proven terminating by *rpo* taken over a precedence $\triangleright$, satisfying $f \triangleright \diamond$, and therefore is both simply and totally terminating.

Finally instead of using the tree lifting order $\succ$ we could have used *rpo* on trees. However *rpo* has another case in its definition so the inductive proofs would have required the analysis of another case, creating an unnecessary complication.

Dummy elimination could be a useful technique for helping in termination proofs, especially if used in conjunction with automatic tools, since it is very easy to incorporate it as a pre-processing unit to check if the TRS to be proven terminating can be transformed.

References

1. Bellegarde, F., and Lescanne, P. Termination by completion. *Applicable Algebra in Engineering, Communication and Computing 1*, 2 (1990), 79–96.
2. Dershowitz, N. Orderings for term rewriting systems. *Theoretical Computer Science 17*, 3 (1982), 279–301.
3. Dershowitz, N. Termination of rewriting. *Journal of Symbolic Computation 3*, 1 and 2 (1987), 69–116.
4. Dershowitz, N., and Jouannaud, J.-P. Rewrite systems. In *Handbook of Theoretical Computer Science*, J. van Leeuwen, Ed., vol. B. Elsevier, 1990, ch. 6, pp. 243–320.
5. Dershowitz, N., and Manna, Z. Proving termination with multiset orderings. *Communications ACM 22*, 8 (1979), 465–476.
6. Ferreira, M. C. F., Middeldorp, A., Ohsaki, H., and Zantema, H. Transforming termination by self-labelling. in preparation, 1995.
7. Ferreira, M. C. F., and Zantema, H. Total termination of term rewriting. In *Proceedings of the 5th Conference on Rewriting Techniques and Applications* (1993), C. Kirchner, Ed., vol. 690 of *LNCS*, Springer, pp. 213–227. Full version to appear in Applicable Algebra in Engineering, Communication and Computing.
8. Ferreira, M. C. F., and Zantema, H. Well-foundedness of term orderings. To appear at CTRS 94 (Workshop on Conditional and Typed Term Rewriting Systems).
9. Kamin, S., and Lévy, J. J. Two generalizations of the recursive path ordering. University of Illinois, 1980.
10. Zantema, H. Termination of term rewriting by semantic labelling. Tech. Rep. RUU-CS-92-38, Utrecht University, December 1992. Extended and revised version appeared as RUU-CS-93-24, July 1993, accepted for special issue on term rewriting of Fundamenta Informaticae.
11. Zantema, H. Termination of term rewriting: interpretation and type elimination. *Journal of Symbolic Computation 17* (1994), 23–50.
12. Zantema, H., and Geser, A. A complete characterization of termination of $0^p1^q \to 1^r0^s$. In *Proceedings of the 6th Conference on Rewriting Techniques and Applications* (1995), J. Hsiang, Ed., vol. 914 of *LNCS*, Springer, pp. 41–55. Appeared as report UU-CS-1994-44, Utrecht University.

Computing Petri Net Languages by Reductions

Anja Gronewold* and Hans Fleischhack+
*Université de Paris-Sud, LRI, Bât. 490,
F-91405 Orsay Cedex, ag@lri.fr
Fachbereich Informatik, Universität Oldenburg, PF 2503,
D-26111 Oldenburg, fleischhack@informatik.uni-oldenburg.de

Abstract. A method for the computation of (a regular expression for) the language of a safe net is presented. This method is based on net reductions as introduced by Berthelot, and it uses the net's trace language to rebuild the possible interleavings which can get lost during the reduction process. A set of reduction rules preserving the net language is proposed, i. e. in each reduction step, information about the net language is memorized. The result is a regular expression that may be used to decide properties of the net as e. g. mutual exclusion of transitions or possible markings of places.

1 Introduction

One of the benefits of using Petri nets for the specification of concurrent systems is given by the rich variety of analysis methods they offer. Many properties of a system's Petri net model may also be checked by looking at the net language, i. e. the set of all possible transition sequences of the net. The standard algorithm to compute (a regular expression representing) the language of a bounded net involves the construction of the reachability graph which is then interpreted as a finite automaton from which the regular expression is computed. It is well-known that the size of the reachability graph may be exponential in the size of the net - even for safe nets.

We will introduce a new method for the computation of safe net's languages. Our approach is based on Berthelot's work on reductions of Petri nets (cf. [2,3]). His method consists of a set of rules by which a net may be stepwise reduced to a trivial one while preserving the properties of liveness and boundedness. The innovation of our set of reduction rules is the additional preservation of the net language. This is achieved by labelling the transitions of the reduced net with regular expressions over the transitions of the original net. After the reduction, the net language can be computed by stepping backwards through the nets generated by the reduction process, in each step substituting transition names in the regular expressions by the respective labels and applying the independency relation of the net. This method can be seen as an additional approach to solve decision problems for Petri nets without constructing the complete reachability graph. Besides liveness and boundedness, all the properties which are determined by the net language, are preserved and can be checked by looking at the calculated regular expression representing the language (cf. [12]). In the worst case, the analysis can have exponential complexity, but we also have got examples where the reduction is very fast whereas the size of the reachability graph is exponential in the size of the net.

Example 1: We consider a net N for the well-known five dining philosophers problem (cf. Fig. 1(i)). The life of a philosopher process consists of thinking, picking up both forks (n_i), eating, and releasing the two forks (w_i). To determine the language of N, we transform N step by step into a net consisting of one single transition which still has the same language as the original net. Each transition of the reduced net represents a part of the source net. Therefore, it is labelled by a regular expression describing the behaviour of that part. In this introductory example, the transformations are only given informally.

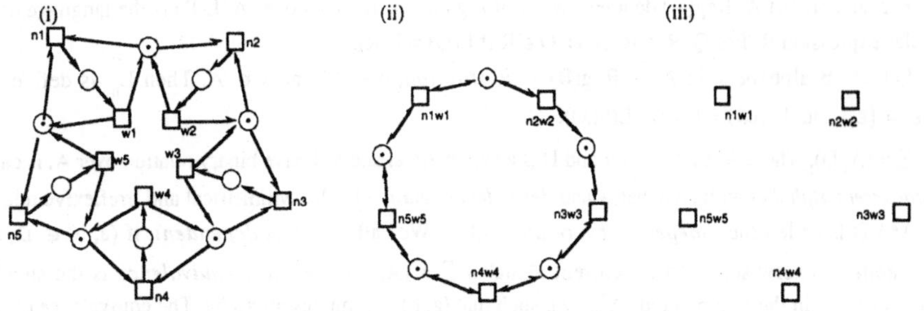

Fig.1. The five dining philosophers

Note that the size of the reachability graph of the philosopher net is exponential in the number of philosophers, but the number of steps needed in the reduction process is linear in that number.

In the net N, each philosopher i, $1 \le i \le 5$, has the sequential behaviour n_iw_i. In a first series of steps, we therefore transform each subnet consisting of n_i and w_i into a single transition labelled with n_iw_i (cf. Fig. 1(ii)). We observe that the (sequential) behaviour of the resulting net is not restricted by its places. So, we apply a transformation which respectively removes one of these places (cf. Fig. 1(iii)). The net now consists of five independent components. Its sequential behaviour may equally be expressed by a single transition which is labelled with the union of the transitions' labels ($n_1w_1 \cup n_2w_2 \cup n_3w_3 \cup n_4w_4 \cup n_5w_5$). Since this last transition is always enabled, it represents the iteration of the label's language. Note that the resulting language L in general is not the net language of N: To compute this, we have to consider the prefix language of the partial commutative closure of L according to N's independency relation.

Example 1 could lead to the assumption that it suffices to take the language of the resulting regular expression and to compute the partial commutative closure according to the independency relation of the start net. But this is not true in the general case which can be seen if an additional run place p is inserted in the net as a side condition with arcs (p, n_3), (n_3, p), (p, n_5) and (n_5, p). The additional place does not restrict the sequential behaviour of the net. So, it may be removed yielding the net of figure 1(i). The reduction process would therefore yield the same result, but the net language of N cannot be generated without taking into account the independency relations of all intermediate nets. Otherwise, we will miss the (still possible) interleavings of n_3 and n_5, because in the beginning, n_3 and n_5 are dependent.

This paper is organized as follows: Section 2 gives some notations and basic definitions. The reduction method is introduced in section 3. We also propose a set of reduction rules and consider correctness and boundedness of the reduction process. In section 4, we are looking at the completeness of the rules and the complexity of the reduction. Finally in section 5, we discuss some open problems and directions for further work. Due to lack of space, we omit all proofs within this paper and refer the reader to [12].

2 Basic notions

In this section, we introduce some notations used throughout the paper. To get an overview of nets, see [4]; the definitions concerning traces are taken from [15,16].

A^* is the *set of words* over an alphabet A. For $L \subseteq A^*$, $\mathrm{Pref}(L) = \{w \in A^* \mid \exists\, u \in A^* \,(wu \in L)\}$ denotes the *prefix closure* of L. Note that Pref(L) is regular, if L is. For $L, L' \subseteq A^*$, we let $L \circ L' = \{uv \mid u \in L, v \in L'\}$.

For an alphabet A, Reg(A) denotes the set of *regular expressions* over A. L(R) is the language of the regular expression R. For $Q, R \in Reg(A)$, $Q \equiv R$ if $L(Q) = L(R)$.

Let A, B alphabets, $h: A \to Reg(B)$ a *substitution*, $u \in A^*$ and $x \in A$. Then L_h is defined by $L_h(\varepsilon) = \{\varepsilon\}$ and $L_h(ux) = L_h(u) \circ L(h(x))$.

$\Sigma = (A, D)$, where A is a finite set and D is a symmetrical and reflexive binary relation over A, is called a *concurrent alphabet* with alphabet A and *dependency relation* D; the symmetrical and irreflexive relation $I = A^2 \setminus D$ is called the *independency relation* of Σ. We call $a, b \in A$ *dependent*, if $(a, b) \in D$, and *independent* otherwise. For a concurrent alphabet $\Sigma = (A, D)$, the *trace equivalence* is the smallest congruence $\equiv_\Sigma$ in the free monoid $(A^*, \circ, \varepsilon)$, such that $(a, b) \in I$ implies $ab \equiv_\Sigma ba$. The equivalence classes $[w]_\Sigma$, $w \in A^*$, of $\equiv_\Sigma$ are called *traces* (the subscript may be dropped, if clear from the context). $[L]_\Sigma = \{[w]_\Sigma \mid w \in L\}$ for every $L \subseteq A^*$. L is called *trace language* over Σ, if $L \subseteq [A^*]_\Sigma$. To give an example, let $\Sigma = (A, D)$ with $A = \{a, b, c\}$ and $D = \{(a, b), (a, c)\}$. Then, $I = \{(b, c)\}$, and the word abcba generates the following equivalence class: [abcba] = {abcba, acbba, abbca}. As a shorthand, we usually drop the symmetrical closure and write e. g. $D = \{(a, a')\}$ instead of $D = \{(a, a'), (a', a)\}$.

For a trace language L over Σ, $String(L) = \bigcup_{[w] \in L} [w] \subseteq A^*$ is called the *string language* of L; for a (string) language $L \subseteq A^*$, $[L]_\Sigma = \{[w]_\Sigma \mid w \in L\}$ is called the trace language of L; $ST_I(L) = String([L]_\Sigma)$ denotes the *partial commutative closure* of the language $L \subseteq A^*$.

$N = (P, T, F, M_0)$ is called a *place/transition system* (P/T-system, Petri net) if P and T are finite, non-empty sets such that $P \cap T = \emptyset$ (the elements of P are called *places* and those of T are called *transitions*), the binary relation $F \subseteq (P \times T) \cup (T \times P)$ is the *flow relation* of N where the elements of F are called *arcs*. and $M_0: P \to \mathbb{N}_0$ is the *initial marking* . For $x \in P \cup T$, ${}^\bullet x = \{y \mid (y, x) \in F\}$ is called the *preset* of x and $x^\bullet = \{y \mid (x, y) \in F\}$ is called the *postset* of x. For $P' \subseteq P$, we define ${}^\bullet P' = \bigcup_{p \in P'} {}^\bullet p$.

A function $M: P \to \mathbb{N}_0$ is called a *marking* of N. A transition t is *enabled* at M iff for all $p \in {}^\bullet t$ holds: $M(p) \geq 1$ (M[t>). A transition t enabled at a marking M can *occur*, yielding a new marking M' defined by $M'(p) = M(p) - 1$ for $p \in {}^\bullet t \setminus t^\bullet$, $M'(p) = M(p) + 1$ for $p \in t^\bullet \setminus {}^\bullet t$, and $M'(p) = M(p)$ for all other places. The changing of the marking M into M''by occurrence of t is denoted by M[t>M'. $w = t_1 \ldots t_r$ ($r \geq 0$, $t_i \in T$) is an *occurrence sequence* leading from M to M' (M[w>M') iff there are markings $M = M_0, M_1, \ldots, M_r = M'$ with $M_{i-1}[t_i>M_i$, $i = 1,\ldots,r$. In this case, w is called *enabled* at M (M[w>).

$[M_0> = \{M' \mid \exists\, w \in T^* \text{ with } M_0[w>M'\}$ is called the *reachability set* of M. The *reachability graph* of N is an edge-labelled graph $G = (V, E)$ with set of vertices $V = [M_0>$ and set of edges $E = \{(M, t, M') \mid M, M' \in V, t \in T, M[t>M'\}$. A marking M is *reachable* from M_0 iff $M \in [M_0>$.

A net is called *live* iff for all $t \in T$ and for all $M \in [M_0>$, there is a marking $M' \in [M>$ with M'[t>. A net is called *n-safe* , $n \in \mathbb{N}$, iff $M(p) \leq n$ for all $p \in P$ and all $M \in [M_0>$, *bounded* iff it is n-safe for some n, and *safe* iff it is 1-safe. Two nodes x, y are called *parallel* in the net N, if ${}^\bullet x = {}^\bullet y$ and $x^\bullet = y^\bullet$.

We let $L_N = \{w \mid w \in T^* \text{ and } M_0[w>\}$ denote the *net language* of the P/T-system N. Note that L_N is prefix closed. For the flow relation F of a net N, $F(x \leftarrow y_1,\ldots,y_n)$ denotes the relation where for all pairs with occurrences of x in F, new pairs are introduced with x replaced by y_i, $i = 1,\ldots,n$. $F \setminus x_1,\ldots,x_n$ (resp. $F \setminus x_i$, $i = 1,\ldots,n$) describes the relation obtained by removing from F all pairs containing one of the x_i.

The *dependency relation* $D_N \subseteq T \times T$ *of a net N* is defined by the reflexive and symmetrical closure of the relation $(t_1, t_2) \in D \iff ({}^\bullet t_1 \cup t_1{}^\bullet) \cap ({}^\bullet t_2 \cup t_2{}^\bullet) \neq \emptyset$. The complementary relation I_N

(sometimes written as I(N)) is called *independency relation of N*; $\Sigma_N = (T, D_N)$ is the *concurrent alphabet of N*. For a safe P/T-system N, $[L_N]_{\Sigma_N}$ denotes the *trace language of N*. It holds (for safe nets) that $ST_{I_N}(L_N) = L_N$.

3 Reduction of safe nets

In this section, we will introduce our method for language preserving reductions of safe nets. This will be done by presenting reduction rules which transform a source net N into a target net N'. N' will consist of less places and/or less transitions than N. During the reduction process, transition sequences of the net may be joined together and represented by one transition whose possible behaviour is then described by a regular expression over transitions. If the reduced net is a single transition, its label can be used directly to calculate the net language. In the other case, the reachability graph of the reduced (and hence smaller) net has to be computed. These considerations lead to the following definition:

Definition 3.1. Let N be a net and $h: T \to Reg(A)$ a mapping of N's transitions into the regular expressions over the alphabet A. We call $\tilde{N} = (N, h)$ a *labelled net*.

A *reduction rule* R is a binary relation on the class S of all safe labelled nets which, for $(\tilde{N}, \tilde{N}') \in R$, satisfies $h': T' \to Reg(T)$. Let $(\tilde{N}, \tilde{N}') \in R$, then $\tilde{N}$ is called *source net* and $\tilde{N}'$ *target net*. For $(\tilde{N}, \tilde{N}') \in R$ we also write $\tilde{N} \to^R \tilde{N}'$ and call $\tilde{N} \to^R \tilde{N}'$ a *reduction step*.

Let $\mathfrak{R}$ be a finite set of reduction rules. We call a sequence $\alpha \equiv (N_0, h_0) \to^{R1} (N_1, h_1) \to^{R2} \ldots \to^{Rr-1} (N_{r-1}, h_{r-1}) \to^{Rr} (N_r, h_r)$ of reduction steps, where $0 \leq r$ and $R_i \in \mathfrak{R}$ for $1 \leq i \leq r$, a *derivation of (N_0, h_0) of length r* in $\mathfrak{R}$. (N_r, h_r) is called *reduced* in $\mathfrak{R}$, if for no $R \in \mathfrak{R}$, there exists a labelled net (N, h) such that $(N_r, h_r) \to^R (N, h)$. ♦

Since parallel processes are sequentialized by the reduction algorithm, we need to consider the independency relation of a net to build the partial commutative closure of the computed regular expressions. (Note that the partial commutative closure is always regular because the language of a safe net is regular.) Unfortunately, it does not suffice to consider the independency relation of the start net, because new independencies can occur during the reduction. So, we get the following definition for the language of a derivation. A derivation of length 0 is a labelled net. In that case, we have to replace each transition name in the net language by the language of the transition label, in order to get the language of the derivation. In the inductive step, we take the language of the tail derivation α' and generate the partial commutative closure according to the independency relation of the start net. In the resulting language, we replace each transition name by the language of the transition's label in the start net and build the prefix closure.

Definition 3.2. Let $\mathfrak{R}$ be a finite set of reduction rules and let α be a derivation in $\mathfrak{R}$. The language L(α) of α is inductively defined by:

(1) $\alpha \equiv (N, h)$: $L(\alpha) = Pref(\bigcup_{w \in L_N} L_h(w))$ and

(2) $\alpha \equiv (N, h) \to^R \alpha'$, where $R \in \mathfrak{R}$ and α' is a derivation in $\mathfrak{R}$: $L(\alpha) = Pref(\bigcup_{w \in ST_{I(N)}(L(\alpha'))} L_h(w))$ ♦

Note that the language of the (unlabelled) net N is given by the language of the derivation (N, id_T). We will call a reduction rule correct, if the language of a derivation step from a source net to a target net corresponds to the language of the source net, i. e., if the rule preserves the language of the source net. So, we get the following definition:

Definition 3.3. A reduction rule R is *correct*, if $(N, h) \rightarrow^R (N', h')$ implies $L_N = \mathrm{Pref}(ST_{I(N)}(L(N', h'))$. A set of reduction rules $\mathfrak{R} = \{R_1,..., R_n\}$ is correct, if each rule R_i, $1 \leq i \leq n$, is correct. ♦

3.4 Theorem. Let $\mathfrak{R} = \{R_1,..., R_n\}$ be correct and α a derivation of (N,h) in $\mathfrak{R}$. Then: $L(\alpha) = L(N,h)$.

3.5 Corollary. Let $\mathfrak{R}$ be correct and α a derivation of (N, id_T) in $\mathfrak{R}$. Then $L(\alpha) = L_N$.

Before defining the reduction rules, let us consider the following example (cf. Fig. 2) to show how the computation of the net language works:

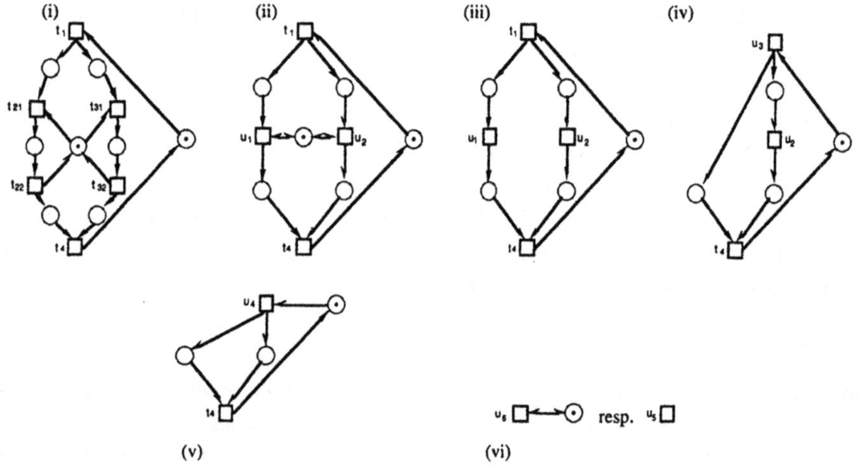

Fig.2.

Example 2: This net satisfies the mutual exclusion of the two critical sections t_{21}, t_{22} and t_{31}, t_{32} where the place in the middle of the net plays the role of a semaphore. The language of the net is $L_{N0} = \mathrm{Pref}\ L((t_1t_{21}t_{22}t_{31}t_{32}t_4 \cup t_1t_{31}t_{32}t_{21}t_{22}t_4)^*)$. We observe that each occurrence of t_{21} (resp.t_{31}) is followed by an occurrence of t_{22} (resp.t_{32}). In a first reduction step, we abstract from those single actions and collapse t_{21}, t_{22} (resp t_{31}, t_{32}) into a single transiton u_1 (resp. u_2) which will be labelled by the respective sequence. The resulting net N^1 is shown in Fig. 2(ii). Here, the semaphore-place does not influence the (string-) language of the net N^1. So, we remove it, thereby getting the net N^2 (cf. Fig. 2(iii)). The following reduction steps consist of combining the transitions t_1 and u_1, resulting in a transition u_3 (cf. Fig. 2(iv)) and combining the transitions u_3 and u_2, resulting in a transition u_4 (cf. Fig. 2(v)). At last, we contract the transitions u_4 and t_4 to a single one named by u_5 and then, we remove the side-condition without affecting the string language of the net (cf. Fig. 2(vi)). For the computation of the net language, we apply definition 3.2 (note that it suffices to apply the prefix closure once at the end of the computation):

$L(N^5, h_5) = \text{Pref}(\bigcup_{w \in L_{N5}} L_{h_5}(w)) = \text{Pref}(\bigcup_{w \in L(u_5^*)} L_{h_5}(w)) = \text{Pref}(\bigcup_{w \in L(u_5^*)} \{h_5(w)\}) = \text{Pref}\, L((u_4t_4)^*)$

$I(N^4) = \varnothing \;\Rightarrow\; ST_{I(N4)}L((u_4t_4)^*) = L(u_4t_4)^*$. Hence

$L((N^4, h_4) \rightarrow (N^5, h_5)) = \text{Pref}(\bigcup_{w \in ST_{I(N4)}(L(N^5,h_5))} L_{h_4}(w)) = \text{Pref}(\bigcup_{w \in ST_{I(N4)}(L(u_4t_4)^*)} L_{h_4}(w))$

$= \text{Pref}(\bigcup_{w \in \text{Pref}L((u_4t_4)^*)} \{h_4(w)\}) = \text{Pref}\, L((u_3u_2t_4)^*)$

$I(N^3) = \varnothing \;\Rightarrow\; ST_{I(N3)}L((u_3u_2t_4)^*) = L(u_3u_2t_4)^*$. Hence

$L((N^3, h_3) \rightarrow^* (N^5, h_5)) = \text{Pref}\, L_{h_3}((u_3u_2t_4)^*) = \text{Pref}\, L((t_1u_1u_2t_4)^*)$

$I(N^2) = \{(u_1,u_2)\} \;\Rightarrow\; ST_{I(N2)}L((t_1u_1u_2t_4)^*) = L(t_1u_1u_2t_4 \cup t_1u_2u_1t_4)^*$. Hence

$L((N^2, h_2) \rightarrow^* (N^5, h_5)) = \text{Pref}\, L_{h_2}((t_1u_1u_2t_4 \cup t_1u_2u_1t_4)^*) = \text{Pref}\, L((t_1u_1u_2t_4 \cup t_1u_2u_1t_4)^*)$

$I(N^1) = \varnothing \;\Rightarrow\; ST_{I(N1)}L((t_1u_1u_2t_4 \cup t_1u_2u_1t_4)^*) = L(t_1u_1u_2t_4 \cup t_1u_2u_1t_4)^*$. Hence

$L((N^1, h_1) \rightarrow^* (N^5, h_5)) = \text{Pref}\, L_{h_1}((t_1u_1u_2t_4 \cup t_1u_2u_1t_4)^*)$

$= \text{Pref}\, L((t_1t_{21}t_{22}t_{31}t_{32}t_4 \cup t_1t_{31}t_{32}t_{21}t_{22}t_4)^*)$

$I(N^0) = \{(t_1, t_{22}), (t_1, t_{32}), (t_{21}, t_4), (t_{31}, t_4)\} \;\Rightarrow\; ST_{I(N0)}L((t_1t_{21}t_{22}t_{31}t_{32}t_4 \cup t_1t_{31}t_{32}t_{21}t_{22}t_4)^*)$

$= L((t_1t_{21}t_{22}t_{31}t_{32}t_4 \cup t_1t_{31}t_{32}t_{21}t_{22}t_4)^*)$. Hence

$L((N^0, h_0) \rightarrow^* (N^5, h_5)) = \text{Pref}\, L_{h_0}((t_1t_{21}t_{22}t_{31}t_{32}t_4 \cup t_1t_{31}t_{32}t_{21}t_{22}t_4)^*)$

$= \text{Pref}\, L((t_1t_{21}t_{22}t_{31}t_{32}t_4 \cup t_1t_{31}t_{32}t_{21}t_{22}t_4)^*) = L_{N0}$

The set *Red* of reduction rules

R1 "Removal of neutral transitions" (Fig. 3)

Let $\tilde{N} = (N, h)$ and $\tilde{N}' = (N', h')$ two safe labelled nets. $(\tilde{N}, \tilde{N}') \in R1 \;\Leftrightarrow$

(i) In N there are a place p and n, $n \geq 1$ (neutral) transitions u_i with $^\bullet u_i = u_i^\bullet = \{p\}$, $i = 1,\ldots,n$ and $M_0(p) = 0$.

(ii) (N', h') is defined by: $T' = T \setminus \{u_1,\ldots,u_n\}$; $P' = P$; $F' = F|u_i$, $i = 1,\ldots,n$; $M_0' = M_0$; $h'(t) = t$ for $t \notin {}^\bullet p$; $h'(t) = t(\bigcup_{i=1}^{n} u_i)^*$ otherwise

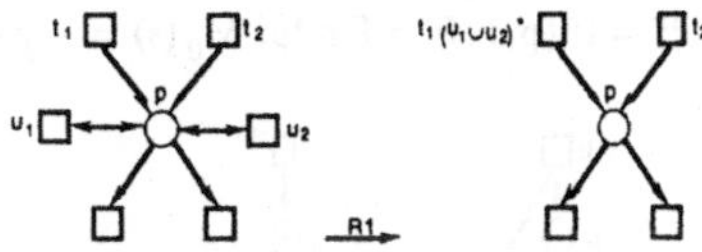

Fig.3. Removal of neutral transitions

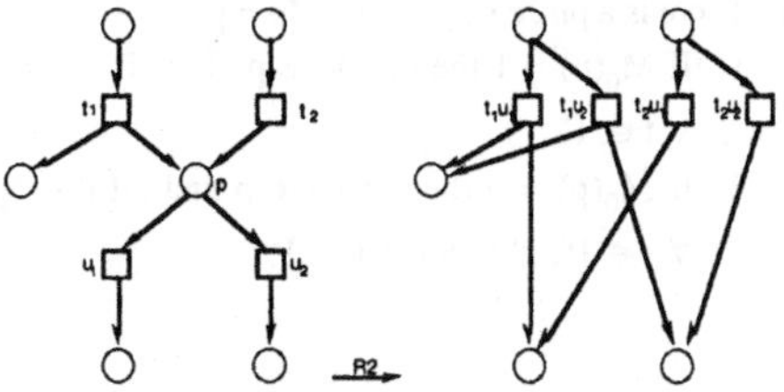

Fig.4. Post-Agglomeration

R2 "Post-Agglomeration" (Fig. 4)

Let $\tilde{N} = (N, h)$ and $\tilde{N}' = (N', h')$ two safe labelled nets. $(\tilde{N}, \tilde{N}') \in R2 \;\Leftrightarrow$

(i) There is a place p whose post-transitions have no other pre-places: $^\bullet(p^\bullet) = \{p\}$. Furthermore, p has the following properties: $^\bullet p = \{t_1,\ldots,t_k\} \neq \varnothing$; $p^\bullet = \{u_1,\ldots,u_n\} \neq \varnothing$; $p^\bullet \cap {}^\bullet p = \varnothing$; $M_0(p) = 0$

(ii) For all i, j with $1 \leq i \leq k$, $1 \leq j \leq n$, a new transition t_{ij} is introduced in the net, connected in the following way: $T' = (T \setminus \{t_1,\ldots,t_k, u_1,\ldots,u_n\}) \cup \{t_{ij} \mid i = 1,\ldots,k,\ j = 1,\ldots,n\}$; $P' = P \setminus \{p\}$; $F' = F(t_i \leftarrow t_{ij},\ u_j \leftarrow t_{ij},\ i = 1,\ldots,k,\ j = 1,\ldots,n) \mid p$; $M_0'(s) = M_0(s)\ \forall s \in P'$; $h'(t) = t$ for $t \neq t_{ij}$, $i = 1,\ldots,k$, $j = 1,\ldots,n$; $h'(t) = t_iu_j$ for $t = t_{ij}$, $i = 1,\ldots,k$, $j = 1,\ldots,n$

R3 "Pre-Agglomeration" (Fig. 5)

Let $\tilde{N} = (N, h)$ and $\tilde{N}' = (N', h')$ two safe labelled nets. $(\tilde{N}, \tilde{N}') \in R3 \Leftrightarrow$

(i) There is a place p whose post-transitions may have several post-places, but whose unique pre-transition t_0 has only unshared pre-places: $({}^\bullet t_0)^\bullet = \{t_0\}$. Furthermore, p has the following properties: ${}^\bullet p = \{t_0\}$; $t_0{}^\bullet = \{p\}$; ${}^\bullet t_0 \neq \emptyset$; $p^\bullet = \{u_1,...,u_n\} \neq \emptyset$; $p^\bullet \cap {}^\bullet p = \emptyset$; $M_0(p) = 0$ and for i = 1,...,n, the following holds: $[\exists\, i : {}^\bullet u_i = \{p\}] \vee [\forall p' \in \bigcup {}^\bullet u_i : (p' \neq p \Rightarrow {}^\bullet t_0 \cap ({}^\bullet p')^\bullet) \neq \emptyset]$

(ii) For all post-transitions u_i with $1 \leq i \leq n$, a new transition is introduced: $T' = (T \backslash \{t_0, u_1,...,u_n\}) \cup \{t_1,...,t_n\}$; $P' = P\backslash\{p\}$; $F' = F(t_0 \leftarrow t_i, u_i \leftarrow t_i, i = 1,...,n)|p$; $M_0'(s) = M_0(s)\ \forall s \in P'$; $h'(t) = t$ for $t \notin \{t_1,..., t_n\}$; $h'(t) = t_0u_i$ for $t = t_i$, i = 1,..., n

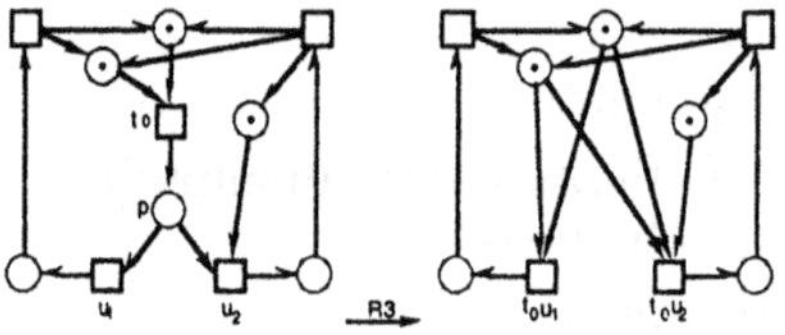

Fig.5. Pre-Agglomeration

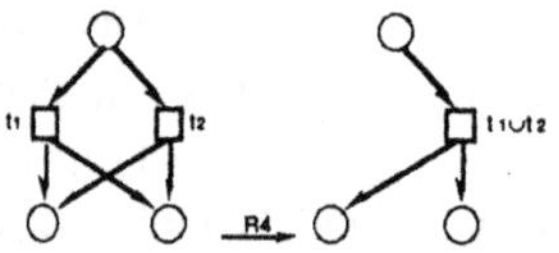

Fig.6. Removal of parallel transitions

R4 "Removal of parallel transitions" (Fig. 6)

Let $\tilde{N} = (N, h)$ and $\tilde{N}' = (N', h')$ two safe labelled nets. $(\tilde{N}, \tilde{N}') \in R4 \Leftrightarrow$

(i) In N, there are two parallel transitions t_1 and t_2.

(ii) Let $t_1, t_2 \in T$ two parallel transitions. Then remove t_1: $T' = T\backslash\{t_1\}$; $P' = P$; $F' = F|t_1$; $M_0'(s) = M_0(s)\ \forall s \in P'$; $h'(t) = t$ for $t \neq t_2$; $h'(t) = t_1 \cup t_2$ for $t = t_2$

R5 "Removal of run-places" (Fig. 7)

Let $\tilde{N} = (N, h)$ and $\tilde{N}' = (N', h')$ two safe labelled nets. $(\tilde{N}, \tilde{N}') \in R5 \Leftrightarrow$

(i) There is a place p such that ${}^\bullet p = p^\bullet$.

(ii) 1. If $M_0(p) = 1$ then remove p: $T' = T$; $P' = P\backslash\{p\}$; $F' = F|p$; $M_0'(s) = M_0(s)\ \forall s \in P'$; $h'(t) = t\ \forall t \in T'$

2. If $M_0(p) = 0$ then remove p and all $t' \in {}^\bullet p$: $T' = T$; $P' = P\backslash\{p\}$; $F' = F|p, {}^\bullet p$; $M_0'(s) = M_0(s)\ \forall s \in P'$; $h'(t) = t\ \forall t \in T'$

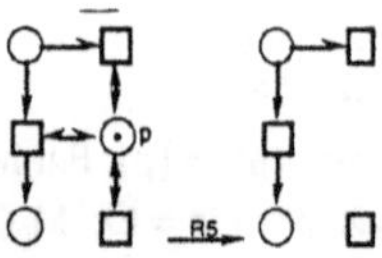

Fig.7. Removal of run-places

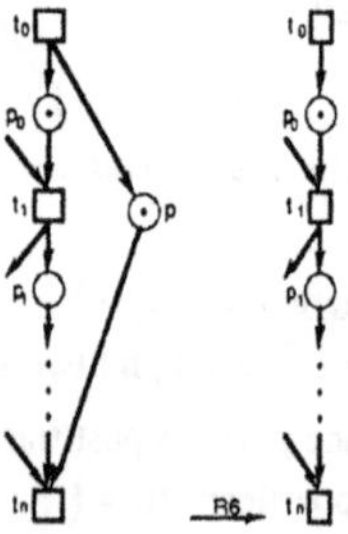

Fig.8. Removal of redundant places

R6 "Removal of redundant places" (Fig. 8)

Let $\tilde{N} = (N, h)$ and $\tilde{N}' = (N', h')$ two safe labelled nets. $(\tilde{N}, \tilde{N}') \in R6 \Leftrightarrow$

(i) There are a place p and two transitions t_0 und t_n in N with the properties: $^\bullet p = \{t_0\}$; $p^\bullet = \{t_n\}$; there is a path $t_0 p_0 t_1 p_1 \ldots p_{n-1} t_n$ from t_0 to t_n with $p_i \neq p$ $(i \in \{0,1,\ldots,n-1\})$, and for all p_i, the following holds: $^\bullet p_i = \{t_i\}$ and [if $M_0(p) = 0$ then $M_0(p_i) = 0$].

(ii) (N', h') is defined by: $T' = T$; $P' = P \backslash \{p\}$; $F' = F|p$; $M_0'(s) = M_0(s)\ \forall s \in P'$; $h'(t) = t\ \forall t \in T'$.

<u>R7 "Lateral fusion"</u> (Fig. 9)

Let $\tilde{N} = (N, h)$ and $\tilde{N}' = (N', h')$ two safe labelled nets. $(\tilde{N}, \tilde{N}') \in R7 \Leftrightarrow$

(i) In N, there are three transitions t_0, t_1, t_2 and two places p_1 and p_2 $(p_1 \neq p_2)$ with the following properties: $^\bullet p_1 = {}^\bullet p_2 = \{t_0\}$; $p_1^\bullet = \{t_1\}$, $p_2^\bullet = \{t_2\}$; $^\bullet t_1 = \{p_1\}$, $^\bullet t_2 = \{p_2\}$; $M_0(p_1) = M_0(p_2) = 0$

(ii) For t_1 and t_2, a new transition t_{12} is introduced: $T' = (T \backslash \{t_1, t_2\}) \cup \{t_{12}\}$; $P' = P$; $F' = F(t_1 \leftarrow t_{12}, t_2 \leftarrow t_{12})$; $M_0'(s) = M_0(s)\ \forall s \in P'$; $h'(t) = t$ for $t \neq t_{12}$; $h'(t) = t_1 t_2 \cup t_2 t_1$ for $t = t_{12}$

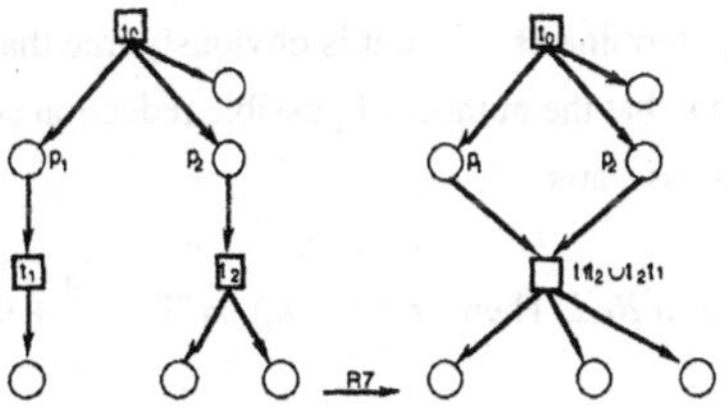

Fig.9. Lateral fusion

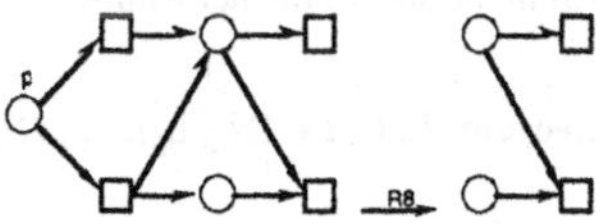

Fig.10. Removal of dead transitions I

<u>R8 "Removal of dead transitions I"</u> (Fig. 10)

Let $\tilde{N} = (N, h)$ and $\tilde{N}' = (N', h')$ two safe labelled nets. $(\tilde{N}, \tilde{N}') \in R8 \Leftrightarrow$

(i) There is $p \in P$ with $^\bullet p = \varnothing$ and $M_0(p) = 0$.

(ii) Remove p and all post-transitions of p: $T' = T \backslash \{t' \mid t' \in p^\bullet\}$; $P' = P \backslash \{p\}$; $F' = F|p,t'$; $M_0'(s) = M_0(s)$ $\forall s \in P'$; $h'(t) = t\ \forall t \in T'$

<u>R9 "Removal of dead transitions II"</u> (Fig. 11)

Let $\tilde{N} = (N, h)$ and $\tilde{N}' = (N', h')$ two safe labelled nets. $(\tilde{N}, \tilde{N}') \in R9 \Leftrightarrow$

(i) There is $p \in P$ with $p^\bullet = \varnothing$ and $M_0(p) = 1$.

(ii) Remove p and all pre-transitions of p: $T' = T \backslash \{t' \mid t' \in {}^\bullet p\}$; $P' = P \backslash \{p\}$; $F' = F \mid p, t'$; $M_0'(s) = M_0(s)\ \forall s \in P'$; $h'(t) = t\ \forall t \in T'$

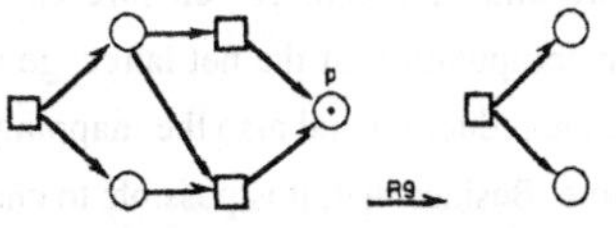

Fig.11. Removal of dead transitions II

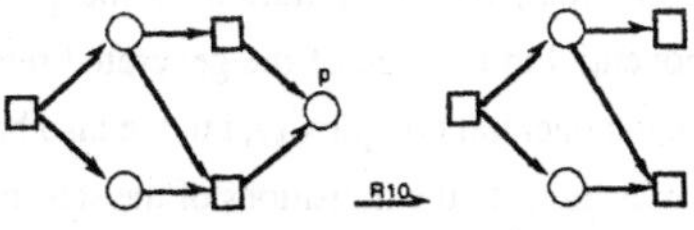

Fig.12. Removal of places with empty postsets

<u>R10 "Removal of places with empty postsets"</u> (Fig. 12)

Let $\tilde{N} = (N, h)$ and $\tilde{N}' = (N', h')$ two safe labelled nets. $(\tilde{N}, \tilde{N}') \in R10 \Leftrightarrow$

(i) There is $p \in P$ with $p^\bullet = \varnothing$ and $M_0(p) = 0$.

(ii) Remove p: $T' = T$; $P' = P \backslash \{p\}$; $F' = F|p$; $M_0'(s) = M_0(s)\ \forall s \in P'$; $h'(t) = t\ \forall t \in T'$

The following theorem states the correctness of the reduction rules, thereby giving a kind of semantical Church-Rosser property of the reduction set *Red* (i. e. the calculated net language is always the same even if the computed regular expressions may differ according to the order of the rule application).

3.6 Theorem. The set of reduction rules *Red* is correct.

The algorithm for the reduction process is applied to a safe net N and returns a regular expression describing the net language. The reduction rules are applied to the net as long as possible. If no further rule is applicable, the net consists either of one single transition, from which the regular expression is taken, or the net consists of several transitions, so that the reachability graph has to be computed for this net, also yielding a regular expression. The net language is computed from the regular expression by applying the partial commutative closure and the substitution h for each reduction step. Finally, the prefix closure is computed for the resulting expression. Due to lack of space, we omit the algorithm here (cf. [12]).

It remains to show that the reduction algorithm always terminates. Since it is obviously true that each application of a reduction rule terminates, it suffices to show that the number of possible reduction steps is bounded in the size of the net with which the reduction process starts.

3.7 Theorem. Let $\alpha \equiv (N_0, h_0), \dots, (N_r, h_r)$ a derivation in *Red*. Then $r \leq G(N_0) := |T_0|^{2^{|S_0|}} + |S_0|$.

4 Completeness and complexity

For our method of language preserving reductions, it is not easy to find a completeness result concerning the reduction rules because - like already for Berthelot - it is very difficult to characterize the class of nets which are completely reducible by the rules (i. e. reducible to a trivial net consisting of one or two vertices). Existing results cannot easily be adopted, because they either include rules which destroy safety of the net (as the rules of Esparza [9] do), or rules that depend on changing the marking during the reduction process (as it is done in the works of Desel [7] and Kovalyov [13]), which changes the net language, too. But nevertheless, it is possible to adapt the reduction method to those rules which change the marking. This is done by slightly changing definition 3.2 for the net language so that possible transition firings during the reduction process are also taken into consideration. This technique will be explained in [11] which yields us the completeness of our method for the nets described in [7] (well-behaved free-choice systems) and those of [13] (subclasses of extended free choice nets).

One problem is the length of the derivation which may grow exponentially in the size of the net. Besides that, the computation of the partial commutative closure and of the prefix closure can be exponential in the size of the generated regular expression. So, the computation of the net language can have exponential complexity, but we also have got examples where the reduction and also the mapping of the transitions to the transitions of the start net is possible in linear time. Besides that, it is possible to check some decision problems without applying the partial commutative closure and the prefix closure ([cf. 12]).

5 Conclusions

We have introduced a new method to compute the language of safe Petri nets. The computation method is based on language preserving net reductions. In many cases, it only needs time polynomial in the size of the net, although the reachability graph is exponential. In this sense, our approach contributes to the fast analysis of Petri nets. We have got the problem that it is difficult to prove the completeness of our method for a nontrivial class of nets or for a nice class of net properties. But we are able to describe a class of reducible nets (cf. [11]) by applying our method to the results of Desel [7] and Kovalyov [13].

Another problem is the complexity of the reduction process: On the one hand, the upper bound given by theorem 3.7 is not satisfactory and might be improved by a more precise inspection. On the other hand, the computation of the partial commutative closure and of the prefix closure can be exponential in the size of the generated regular expression. Both, the length of a reduction and the size of the generated regular expression depend on the order, in which possible reduction rules are applied. So, an important question is that of the existence of optimal strategies for the reduction process. The set of reduction rules *Red* has been implemented in the PEP-Project sponsored by the Deutsche Forschungsgemeinschaft (cf. [5]). We hope that experimental studies based on this implementation will contribute to answer this question.

Language preserving reductions of Petri nets are also semantics preserving reductions. But the net language defines no natural semantics for Petri nets. Reductions preserving the partial order semantics of a net seem to be a promising approach, which hopefully will help to avoid some of the problems mentioned above.

References

[1] IJsbrand Jan Aalbersberg, Grzegorz Rozenberg: Theory of Traces; TCS 60, 1-82, 1988

[2] Gérard Berthelot: Checking Properties of Nets Using Transformations; LNCS 222, 19-40, Springer 1986

[3] Gérard Berthelot: Transformations and Decompositions of Nets; LNCS 254, 359-376, Springer 1987

[4] Eike Best, César Fernández: Notations and Terminology on Petri Net Theory; Arbeitspapiere der GMD 195, Januar 1986

[5] Eike Best, Hans Fleischhack, Hrsg.: Zwischenbericht des Projekts PEP; Hildesheimer Informatikberichte, Institut für Informatik, Universität Hildesheim 1995

[6] Wilfried Brauer: Automatentheorie; Teubner 1984

[7] Jörg Desel: Reduction and Design of Well-behaved Concurrent Systems; CONCUR '90, LNCS 458, 166-181, Springer 1990

[8] Javier Esparza, Manuel Silva: Top-Down Synthesis of Live and Bounded Free Choice Nets; 11th International Conference on Application and Theory of Petri Nets, Paris 1990

[9] Javier Esparza: Reduction and Synthesis of Live and Bounded Free Choice Petri Nets; Hildesheimer Informatikberichte, Institut für Informatik, Universität Hildesheim 1991

[10] Anja Gronewold: Sprachanalyse durch Reduktion von Petri-Netzen; Diploma Thesis, Universität Oldenburg 1993

[11] Anja Gronewold: Language Preserving Reductions of Free-Choice Nets; Paris 1995 (in progress)

[12] Anja Gronewold, Hans Fleischhack: Language Preserving Reductions of Safe Petri-Nets; Berichte aus dem Fachbereich Informatik, Universität Oldenburg 1995

[13] A. V. Kovalyov: On Complete Reducibility of Some Classes of Petri Nets; 11th International Conference on Application and Theory of Petri Nets, Paris 1990

[14] Matthias Jantzen: Language Theory of Petri Nets; LNCS 254, 397-434, Springer 1986

[15] Antoni Mazurkiewicz: Trace Theory; LNCS 255, 279-324, Springer 1987

[16] Antoni Mazurkiewicz: Basic Notions of Trace Theory; Research and Education in Concurrent Systems, REX School / Workshop 1988

Categorial Graphs
(Extended Abstract)

Erik de Haas[1]

ILLC & Department of Mathematics and Computer Science University of Amsterdam, Plantage Muidergracht 24, 1018TV Amsterdam, the Netherlands, email: ehaas@fwi.uva.nl

Abstract. In this paper we present a denotational semantics for a class of database definition languages. We present a language, called categorial graph language, that combines both graphical and textual phrases and is tailored to define databases. The categorial graph language is modeled after a number of practical languages. Its semantics is based on a variant of linear logic, and incorporates directly the notions expressed in the language. We emphasize on the fact that in the semantics presented here, we directly axiomatize the behavior of complex objects, instead of encoding it with help of the traditional mathematical notions. We will argue that this is desirable regarding clearness of semantics and matters of complexity.

1 Introduction

During the last decade a large number of new notions became important in the world of information systems. Many graphical schema techniques entered the syntax definitions of database languages. Some of these schema techniques dealt mainly with pure graph theoretic notions and were introduced with a nice theoretical foundation; examples are IFO of Abiteboul e.a. [AH84] and GOOD of Parendaens e.a. [GPG90]. Some other schema techniques appearing in database languages are borrowed from the field of analysis and design ([dC91]). These last schema techniques involve notions that often do not have a clear theoretical foundation. Moreover they allow wild combinations of graphical and non-graphical constructs. Defining the schema of a database normally involves two tasks: defining the *signature* of the database and defining the *constraints* on this signature. In the field of analysis and design the signatures are often written down graphically, while the constraints are added in a textual form. Moreover it is allowed to cluster complex and constraint signatures and use them for the construction of more complex constraint signatures.

In this paper we examine the semantics of information systems that carry the above notions. For that purpose we designed a language, called '*categorial graph language*', that captures most of the constructions found in the mentioned information system design languages. It contains both graphical and non-graphical phrases and allows the combining of these phrases. We will present a denotational semantics for this language, which is based on a variant of linear logic. The semantics will describe the notions in the language as direct as possible instead of coding them in terms of fundamental mathematical constructions. The main motivation for giving this kind of semantics is that it allows reasoning about the described notions on the same level as in which they are used. This results in a more accessible semantics and, furthermore, it avoids loading the semantics of the described notions with more characteristics then actually used.

2 Related work

In the introduction we described the process of building schemes by constraining types and using the constraint types for building new types. This trade is traditionally treated in the field of type theory, and has proven to be a very useful tool in constructing semantics (e.g. [Mit90]). Most of the type theory done in this context is tailored towards programming semantics and not towards database semantics. We will present calculi that reflect the various notions of type used in database systems. We will also distinguish between signature and constraints.

In a wider perspective this paper presents an attempt to formalize the concepts that became important with the emergence of the paradigm of object orientation, especially of object oriented databases ([ABD89]). There also exist some logics that deal with similar notions, for example F-Logic ([KL89]), O-Logic ([KW93]), LDM ([KV93]) and KR- logic ([Rou91]). Our approach differs from those mentioned, essentially, in the way we deal with the complex structure of the objects. In the logics mentioned only primitive objects have primitive counterparts in the language. Complex objects are constructed. In our approach also the complex objects have a primitive lingual counterpart. The structure of complex objects is given by its axiomatized behavior. Our aim is to have the concepts of interest *intrinsically* present in the logical language instead of coded in more primitive language constructs.

3 Categorial graphs

The basic syntactic structure underlying a categorial graph is an '*arrow graph*'. An arrow graph is a simple generalization of a graph. Consider a collection of *nodes*. In between these nodes we can draw arrows, each having one node as its source and one as its target. Having drawn arrows, we can also draw arrows between the arrows themselves, and between nodes and arrows. If we also imagine the nodes to be arrows, say arrows from one abstract 'empty' arrow, denoted by '**1**', to that same abstract empty arrow **1**, we obtain a structure, which contains only arrows. We will call such a structure an *arrow graph*[1].

Note that in general it is possible that an arrow has itself as its source or has itself as its target, it is even possible that an arrow has itself as the source of its target. We will call these kinds of arrows *unfounded arrows*, and graphs containing such arrows *unfounded arrow graphs.* On the other hand we will say an arrow is *founded* if there is no cyclicity w.r.t its adjacents as with the unfounded Arrows. We call an arrow graph containing, next to the empty arrow, only founded arrows a *founded arrow graph.*

[1] Note that one can describe, or better *encode* the same structure with a normal directed acyclic graph. The notions used will be different though!

We can generalize the concept of arrow graph similarly as done for conventional graphs: Instead of letting an arrow have exactly two adjacents: its source and its target, we can allow an arrow to have an arbitrary number of adjacents. Such an arrow is called a *hyper arrow*. From hyper arrows we can build *hyper arrow graphs*. Hyper arrows come in different flavors; for example *directed hyper arrows*, *undirected* hyper arrows and *set* hyper arrows. For the first flavor, the one of directed arrows, we will consider hyper arrows that have as their adjacency structure an ordered list, so we can talk about the first adjacent, the second adjacent and so forth. Undirected hyper arrows are for which the order of their adjacents is irrelevant. The adjacents of an arrow form a multi-set. In the last case -the set arrow flavor- the adjacents form a set; i.e. we abstract also over the amount of adjacents with identical denotation.

We will not give the formal definitions here due to lack of space. We assume in this paper the existence a simple (syntactic) theory of arrow graphs. For example we can compose two set-hyper arrows by taking the union of their adjacent-sets, and we can compose two undirected hyper arrows by taking the multiset-union of their adjacent-multi-sets. In this paper we will focus on the undirected arrows. Much of the theory will be identical for the other flavors (see for example [Haa94]).

In a *Categorial Graph* an arrow represents a *type* and the structure of the arrow determines the signature of the represented type. Next to the arrows, a categorial graph contains phrases that define *constraints* on the types.

In the framework of databases a categorial graph plays the role of a *database schema*. Instances of such a schema typically consist of objects that obey the structure and the constraints of the categorial graph. Notably, we can write down an instance as an arrow graph in which each arrow represents some object of information, and is assigned a type from the schema. The typing function that assigns these types should be a homomorphism between the instance and its categorial graph. As instances should be considered as databases, we typically require them to have the structure of a founded arrow graph[2]

Example 3.1 Consider figure 3.1. It contains a type arrow graph G. The figure shows that G consists of the simple arrows[3] PERSON and YEAR, one arrow MARRIAGE with adjacents PERSON and PERSON, one arrow CELEBRATION with adjacents MARRIAGE and YEAR and one arrow BIRTH with adjacents PERSON, MARRIAGE and YEAR. For illustration we have written down some plausible constraints at the arrows. It is easy to see that the typed object arrow graph H of figure 3.1 is an instance[4] of G. ♣

In the framework of categorial graphs we make a distinction between the way in which we write down matters of signature, and matters of constraints. The matters of signature are denoted graphically, by arrows and adjacency, while for the constraints we assume the presence of a constraint language in which we write down the restrictions. From a type-theoretic viewpoint this distinction may seem artificial, but

[2] For database objects we require that we can write them down in their full extension, because we want to store them.

[3] We consider an arrow to be simple, if it has only the empty arrow **1** as its adjacent. It is cumbersome to draw it.

[4] In the picture of H we abbreviated the types, e.g. we wrote M instead of MARRIAGE etc.

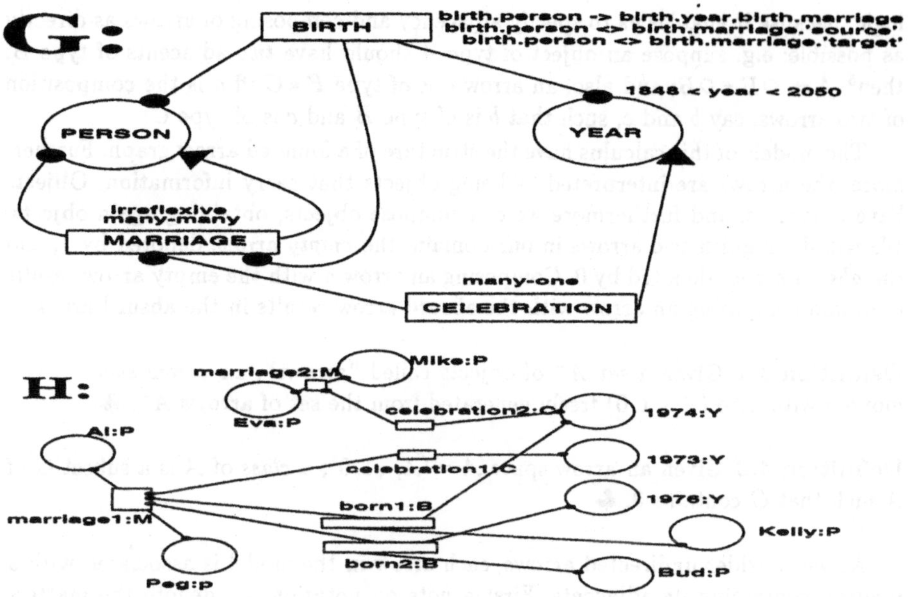

Fig. 1. Example of a type arrow graph: Married with children

in the eyes of a database designer, it is the way it is traditionally done. Our aim is to stay as close as possible to the practitioners intuition as possible. The semantics is constructed as follows: We translate the 'object-language' of categorial graphs, consisting of a graphical *and* a textual part, into a 'meta-language'. In our case this meta-language is some logical language. For the meta-language we have a proper semantics. The minimal requirement for the meta-language is that we have to be able to express in it the structure of the arrow graphs. All the additional features contribute to the expressiveness of the constraint language.

Actually the distinction between signature and constraints becomes a little more vague if we turn to semantics. If we translate the the graphical (signatorial) syntax and the textual (constraint language) syntax to some logical language, we have only *one* logical language. Still we pursue the proposed distinction by designing the logical language in such a way that the graphical ingredients of the object language are *intrinsic* features of the meta-language.

4 Semantics for categorial graphs

In this section we present a variant of linear logic ([Gir87]) tailored for categorial graphs. Recall that we want to be able to conveniently express the structure of an arrow graph in this logic. Furthermore the logic talks about database instances, which can be considered as a collection of resources (objects) with the structure of a founded arrow graph. We chose the multiplicative conjunction as primitive of our logic, because it enables us to specify adjacency and composing of arrows as directly as possible; e.g. suppose an object of type A should have *two* adjacents of type B, then[5] $A \Rightarrow \Diamond B * \Diamond B$; and also, an arrow a is of type $B * C$ iff a is the composition of two arrows, say b and c, such that b is of type B and c is of type C.

The models of this calculus have the structure of a founded arrow graph. Furthermore, the arrows are interpreted as being objects that carry information. Objects have adjacents, and furthermore we can compose objects, obtaining again objects. We will distinguish two arrows in our domain: the *empty arrow*, denoted by 1, and the *absurd arrow*, denoted by 0. Composing an arrow a with the empty arrow results in a, and composing an arrow with the absurd arrow results in the absurd arrow.

Definition 4.1 Given a set A^- of objects called "arrows", an *arrow space* $\mathcal{A}$ is a monoid with zero $(A, \cdot, 1, 0)$ freely generated from the set of arrows A^-. ♣

Definition 4.2 Given an arrow space $\mathcal{A} = (A, \cdot, 1, 0)$; a *class* of $\mathcal{A}$ is a subset C of A such that C contains 0. ♣

As we consider undirected arrows, each arrow in the model is associated with a *multiset* containing its adjacents. First a note on notation: we denote the matters concerning multisets using the usual symbols for sets decorated with a dot; e.g. $\cdot\{a, a, b\}\cdot$ denotes the multiset containing two times a and one time b. Let us associate an arrow with its adjacents via the mapping $f_R : A \rightarrow \mathcal{P}^{\cdot}(A)$ from arrows to multisets of arrows. Observe that this is equivalent to defining a binary multiset-relation R on A.

Definition 4.3 Consider the pair $\mathcal{F} = (\mathcal{A}, f_R)$, where $\mathcal{A} = (A, \cdot, 1, 0)$ is an arrow space and f_R the function of a is a binary (accessibility) multiset relation R[6] on A. We call $\mathcal{F}$ a *pre-undirected-arrow frame* if f_R (accessibility) behaves conjunctive over '$\cdot$' (composition); i.e.

if $f_R(a) = U$ and $f_R(b) = W$ then $f_R(a \cdot b) = U \dot{\cup} W$ ♣

To obtain models for instances of categorial graphs we will have to curtail the class of pre-undirected arrow frames. First of all the the 1 should behave like the empty arrow and the 0 like the absurd arrow. This means that the 0 has all arrows as its adjacent and the only adjacent of 1 is 1. Also the arrows different from 0 or 1 should be founded, which means that if we 'walk down' along the adjacencies of an arrow, we should never encounter the arrow itself, but always encounter the empty arrow at some finite depth. We will call the structure of an adjacent and an adjacent of the adjacent and an adjacent of the adjacent of the adjacent etc. etc. an *adjacency path*. We will require that all arrows have the empty arrow somewhere in their adjacency path. It is sufficient to restrict the accessibility relation R.

[5] Note that we use the notation of Troelstra [Tro92] for the linear logic connectives; i.e. $*$ for multiplicative conjunction, $\sqcap$ for additive conjunction, $\sqcup$ for additive disjunction etc. etc.

[6] R represents the adjacency relation; i.e. $R(b, a)$ if b is an adjacent of a

Definition 4.4 Consider a pre-undirected arrow frame $\mathcal{F} = (\mathcal{A}, R)$ We will define the relation P to be the transitive closure of R. The relation P describes the adjacency paths. We will call $\mathcal{F}$ an *undirected arrow frame* if the following holds:

1. $\forall a \in A[R(a,1) \rightarrow a = 1]$ (i.e. the empty arrow 1 only has itself as its adjacent)
2. $\forall a \in A[R(a,0)]$ (i.e. the absurd arrow has all arrows as adjacent)
3. $\forall a \in A[P(1,a)]$ (all arrows have the empty arrow in their adjacency path)
4. $\forall a \in A[P(a,a) \leftrightarrow a = 1]$ (the adjacency path of a arrow is cyclic iff it is the empty arrow)

We will denote the set of all undirected arrow frames by UA. ♣

Definition 4.5 Let S_{UA} be a similarity type containing two unary modal operators $\Diamond$ and $\wp$, together with the modal constants **1** and **0**. We define the language M_{UA} to be the pair $(S_{\mathsf{UA}}, Q_{\mathsf{UA}})$, where Q_{UA} is a set of propositional variables. The set $\Phi(M_{\mathsf{UA}})$ of formulas in M_{UA} is defined as usual, using next to the modal operators the connectives $\neg\_$, $\_*\_$, $\_\sqcap\_$, $\_\sqcup\_$, ♣

The $\Diamond$ modality will represent the adjacency relation R, the $\wp$ modality represents the path relation P, and the modal constants **0** and **1** respectively the absurd and the empty arrow.

Definition 4.6 Let $\mathcal{F} = (\mathcal{A}, f_R)$ be an undirected arrow frame and let $V := V' \cup \{0\}$, where $V' : Q_{\mathsf{UA}} \mapsto 2^{\mathsf{A}}$, be a mapping from propositions to classes of $\mathcal{A}$. V will be called *valuation* and he pair $(\mathcal{F}, V)$ will be called *undirected arrow model.*

The valuation function V is extended to a function mapping all formulas of M_{UA} to classes by:

$$\begin{aligned}
V(\mathbf{0}) &= \{0\}\\
V(\mathbf{1}) &= \{1\} \cup \{0\}\\
V(\phi \sqcup \psi) &= [V(\phi) \cup V(\psi)]\\
V(\phi \sqcap \psi) &= [V(\phi) \cap V(\psi)]\\
V(\bot) &= V(\mathbf{0})\\
V(\top) &= A\\
V(\neg\phi) &= [A - V(\phi)] \cup \{0\}\\
V(\phi * \psi) &= [V(\phi) \cdot V(\psi)] \cup \{0\}\\
V(\Diamond\phi_1 * \cdots * \Diamond\phi_n) &= \{a | \exists b_1 \in V(\phi), \ldots, \exists b_n \in V(\phi_n)[\{b_1, \ldots, b_n\} \dot{\cdot} \subseteq f_R(a)]\} \cup \{0\}
\end{aligned}$$

Let $\mathcal{M} = (\mathcal{F}, V)$ be a pre-undirected-arrow model, where $\mathcal{F} = (\mathcal{A}, R)$ is a pre-arrow frame, $\mathcal{A} = (A, \cdot, 1, 0)$ is an arrow space and $a \in A$ is an arrow. A formula $\phi \in \Phi(M_{\mathsf{UA}})$ is *true* at arrow a in model $\mathcal{M}$, notation $\mathcal{M}, a \models \phi$ if $a \in V(\phi)$. The formula ϕ is true in $\mathcal{M}$, notation $\mathcal{M} \models \phi$ if $\mathcal{M}, a \models \phi$ for all $a \in A$. The formula ϕ is *valid* in $\mathcal{F}$, notation $\mathcal{F} \models \phi$ if for every valuation V' it holds that $(\mathcal{F}, V') \models \phi$. The formula ϕ is valid for a collection K of frames if $\mathcal{F} \models \phi$ for all $\mathcal{F} \in$ K. A formula ϕ is a *semantic consequence* of a set of formulas Σ *over* a collection of frames K, notation: $\Sigma \models_{\mathsf{K}} \phi$ if for every model M based on a frame in K, and every arrow a in M it holds that $\mathcal{M}, w \models \phi$ if $\mathcal{M}, w \models \sigma$ for all $\sigma \in \Sigma$. ♣

Now we will present a basic calculus for undirected arrows. Axioms and rules will be expressed in Gentzen-style sequent calculus with the restriction that the sequents have, exactly, one formula on the right. For basic connectives $*$, $\sqcup$, $\sqcap$, $\mathbf{0}$, $\mathbf{1}$, $\bot$, and $\top$ we present the usual axioms and rules. We will however incorporated a non-constructive negation. Informally, $\neg A$ means 'not of type A'. We also have modalities in the linear language that are different from the ones that are usually studied (see e.g. [Buc94]). The fundamental difference lies in the fact the the accessibility relations we consider are multiset based. Matters like irreflexivity of the accessibility relations. are thoroughly investigated by the modal logic community (see e.g [Gab81]).

Definition 4.7 *(The sequent calculus Λ_{UA})*

$$(AX) \quad A \Rightarrow A$$

$$(CUT) \quad \frac{\Gamma \Rightarrow A \quad \Gamma', A \Rightarrow B}{\Gamma, \Gamma' \Rightarrow B}$$

$$(EX) \quad \frac{\Gamma, A, B, \Gamma' \Rightarrow C}{\Gamma, B, A, \Gamma' \Rightarrow C}$$

$$(L\sqcap) \quad \frac{\Gamma, A \Rightarrow C}{\Gamma, A \sqcap B \Rightarrow C} \quad \frac{\Gamma, B \Rightarrow C}{\Gamma, A \sqcap B \Rightarrow C} \qquad (R\sqcap) \quad \frac{\Gamma \Rightarrow A \quad \Gamma \Rightarrow B}{\Gamma \Rightarrow A \sqcap B}$$

$$(L*) \quad \frac{\Gamma, A, B \Rightarrow C}{\Gamma, A * B \Rightarrow C} \qquad (R*) \quad \frac{\Gamma \Rightarrow A \quad \Gamma' \Rightarrow B}{\Gamma, \Gamma' \Rightarrow A * B}$$

$$(L\sqcup) \quad \frac{\Gamma, A \Rightarrow C \quad \Gamma, B \Rightarrow C}{\Gamma, A \sqcup B \Rightarrow C} \qquad (R\sqcup) \quad \frac{\Gamma \Rightarrow A}{\Gamma \Rightarrow A \sqcup B} \quad \frac{\Gamma \Rightarrow B}{\Gamma \Rightarrow A \sqcup B}$$

$$(L\mathbf{1}) \quad \frac{\Gamma \Rightarrow A}{\Gamma, \mathbf{1} \Rightarrow A} \qquad (R\mathbf{1}) \quad \Rightarrow \mathbf{1}$$

$$\text{(no } L\top\text{)} \qquad (R\top) \quad \Gamma \Rightarrow \top, \Delta$$

$$(L\mathbf{0}) \quad \mathbf{0} \Rightarrow \qquad \text{(no } R\mathbf{0}\text{)}$$

$$(L\bot) \quad \Gamma, \bot \Rightarrow A \qquad \text{(no } R\bot\text{)}$$

$$(\Diamond I) \quad \frac{A \Rightarrow B}{\Diamond A \Rightarrow \Diamond B}$$

$$(WEAKENING) \quad \Diamond A * B \Rightarrow \Diamond A$$

$$(EMPTY) \quad \mathbf{1} \Rightarrow \Diamond \mathbf{1} \quad \mathbf{1} \Rightarrow \neg \Diamond \neg \mathbf{1}$$

$$(ABSURD) \quad \mathbf{0} \Rightarrow \bot$$

$$(L\neg) \quad \frac{\Gamma \Rightarrow A \sqcup \Delta}{\Gamma \sqcap \neg A \Rightarrow \Delta} \qquad (R\neg) \quad \frac{\Gamma \sqcap A \Rightarrow \Delta}{\Gamma \Rightarrow \neg A \sqcup \Delta}$$

$$(PATH) \quad \Diamond A \sqcup \Diamond \wp A \Leftrightarrow \wp A$$

$$(FOUND) \quad \Gamma \Rightarrow \wp \mathbf{1}$$

$$(IRREFLEXIVE) \quad \frac{\Gamma \Rightarrow \neg(\wp A \rightarrow A) \rightarrow B}{\Gamma \Rightarrow B}$$

♣

Theorem 4.1 *The logic Λ_{UA} is sound and complete with respect to $\models_{UA}$*

proof (sketch): Consider the Lindenbaum algebra L of Λ_{UA}. The algebra L contains sets of equivalent formulas, denoted $(A/\Leftrightarrow)$, especially it contains the set $(\mathbf{0}/\Leftrightarrow)$. Construct an undirected arrow model from L by taking

$$V(A) = (A/\Leftrightarrow) \cup (\mathbf{0}/\Leftrightarrow)$$

It is easily checked that this is a canonical undirected arrow model. Note that the rules and axioms (*FOUND*), (*IRREFLEXIVE*), and (*PATH*) force the accessibility relation R to acyclic and founded. A calculus without these rules and axioms is complete for pre- undirected arrow frames. ♠

Note that the above logic satisfies the our minimal requirement: the ability to express the signature given by the type arrow graphs. Furthermore the logic provides general means to express constraints. We examined several extensions of this logic that increase the expressivity. The most obvious extension concerns adding equality $(=)$ and the inverse adjacency $(\Diamond^{-1})$ ([Rij93]). Also, the strong similarity to linear logic suggest an extension with the 'of course' modality, denoted by '!'. The '!' allows us to construct a type for *sets*. Informally $a \models !A$ means that object a is of type $\mathbf{1}$ or of type A or of type $A * A$ or of type $A * A * A$ etc. etc. . In other words a is composed from an arbitrary number of objects of type A. Another obvious extension we examined is adding (second order) quantification. From type theory we learn that such an extension enables us to express explicit polymorphism, which is a desirable feature for object database definition languages. One has to be careful, though, with adding very general notions to the system, because the complexity of the language will increase enormously. This is especially true considering that, more often than not, we don't need the whole generality of these basic notions to express the notions that are desirable for our application domain.

5 The matter of expressiveness

We will distinguish two kinds of expressiveness: the 'computational expressiveness' and the 'convenience expressiveness'. Computational expressiveness deals with the traditional computation theoretic question of which classes of problems can we express *theoretically* in the language at hand. Convenience expressiveness treats the admittedly vague concept that judges whether some problem can be stated *conveniently* in the considered language. For example theoretically one is able to program a geometrical information system on a Turing machine, but, putting it modestly, it is rather inconvenient for both the programmer to built this system, and also for a user to use it. The main concern in convenience expressiveness is to provide high level notions that enable one to state the problems one considers as direct as possible, involving as little as possible cumbersome encodings.

The main concern of the convenience of expression for the categorial graph formalism was to provide a database model that is at least as flexible as the relational model and which has at least the same richness in type structure as the object oriented databases. Note that the ability to draw pictures for the denoted objects is a purely syntactical matter, but it also adds to convenience. The eye-catcher of the categorial graph approach is the uniform nature of notion of object (arrow). Simple and complex objects are treated in precisely the same manner: as basic entities. We also avoid coding complex objects using structures that are semantically uninteresting (dummies). We will give a few examples of convenience.

Example 5.1 Consider an instance of a categorial graph which contains the types A and B. For a typed relational calculus look at the collection of all objects of some given type as a relation. An object arrow a is an object in the Cartesian product of A and B if $a \models A*B$. In other words the type $A*B$ is the type for the Cartesian product of A and B. Similarly we have union and intersection: respectively $a \models A \sqcup B$ and $a \models A \sqcap B$. Slightly more involved are:

- difference $(A - B)$: $a \models A \sqcap \neg B$;
- projection $(\pi_B(A))$: $a \models \diamond^{-1} A \sqcap B$;
- selection $(\sigma_F(A))$: $a \models A \sqcap F$.

An important type in object oriented systems is the SET-type. An object a represents a set of objects of type A if

$$a \models !A$$

A specialization subtype relation between types is easily formalized as follows:

$$A \leq B \text{ iff } A \Rightarrow B$$

Consider example 3.1. As a last example, one could formulate the following constraints on PERSONs:"With marriage come children", i.e.

$$PERSON \Rightarrow \diamond^{-1}(MARRIAGE \rightarrow \diamond^{-1}(BORN)) \quad \clubsuit$$

Let us now turn the computational expressiveness. There is a natural trade-off between expressiveness and complexity. The more expressive a language is, the more difficult it is to compute with it. Although we like to express much we should strive to keep its complexity as low as possible. For categorial graphs without exponentials (!) we have the following results.

Theorem 5.1 *Given an undirected arrow model with a finite base, an arrow a and a M_{UA} formula ϕ, then deciding whether $M, a \models_{\mathrm{UA}} \phi$ is in* **P**.

Theorem 5.2 *Satisfiability in Λ_{UA} (for non-trivial models) is* **NP**-*hard.*

One easily establishes the above two theorems by observing that the formulas relate directly to resources in the model; deciding whether or not a pre-undirected arrow frame is indeed an undirected arrow frame is in **P**; and furthermore that if a formula is satisfiable, it is so in a poly-size frame.

6 Conclusion and future research

In this paper we presented a mathematical construction called "categorial graph", for the purpose of defining database systems. We presented a theoretically founded language that carries categorial graphs and briefly exemplified its abilities. There is still much work to be done. We think it is useful to look at more sophisticated languages for categorial graphs. In our view the main task is to make the meta-languages as realistic as possible, by which we mean that it has to connect closely to the basic language expressions that are used in practice. This entails that we have to incorporate these basic expressions as intrinsic (high level) features of the meta-language and find a suitable semantic counterpart for them.

One promising aspect of this approach is its impact on the complexity of the language. The emphases on using high level notions instead of encodings of these notions works out favorably on the strive for low complexity. The reason lies in the disqualification of taking unfair advantage of the encodings (see e.g. [AV93]). We hope to gain more results on that aspect in the near future.

References

[ABD89] M. Atkonson, F. Bancilhon, D. DeWitt, K. Dittrich, D. Maier, S. Zdonik, *The Object Oriented Database System Manifesto*; in: *first International Conference on Deductive and Object Oriented Database Systems*, Dec. 1989, pp. 40- 57.

[AH84] Serge Abiteboul, Richard Hull, *IFO: A Formal Semantic Database Model*, In: ACM proc. of Principles of Database Systems (PODS'84), 1984, pp. 119-132.

[AV93] Serge Abiteboul, Victor Vianu, *Computing on Structures*, in: Automata, Languages and Programming (ICALP'93), Springer LNCS 700, 1993, pp.606-620.

[Buc94] Anna Bucalo, *Modalities in Linear Logic Weaker then the Exponential "of Course": Algebraic and Relational Semantics*, in: Journal of Logic, Language and Information 3, pp. 211-232, 1994.

[dC91] D. de Champeaux, *A comparative study of Object Oriented Analysis Methods*, Research Report, HP Laboratories, April 1991.

[Gab81] Dov Gabbay, *An irreflexive lemma with applications to axiomatizations of conditions on linear frames*, In: U. Mönnich (ed.), *Aspects of Philosophical Logic*, Reidel, Dordrect, 1981, pp.67-89.

[Gir87] Jean-Yves Girard, *Linear Logic*, In: Theoretical Computer Science 50 (1987), pp. 10-102.

[GPG90] Marc Gyssens, Jan Parendaens, Dirk van Gucht, *A Graph-Oriented Object Model for Database End-User Interfaces*, in: H. Garciamolina, H.V. Jagadish (eds.), ACM Int. Conf. on Management of Data (SIGMOD), 1990, pp. 24-33.

[Haa94] Erik de Haas, *Categorial graphs: The logical approach*, in: Arthur Nieuwendijk (ed.), ACCOLADE'94 proceedings, Dutch Graduate School in Logic, Department of Philosophy, University of Amsterdam, The Netherlands.

[KL89] M. Kifer, G. Lausen, *F-Logic: A higher-order language for reasoning about objects, inheritance and schema*, in: Proceedings ACM-SIGMOD int. conf. on Management of Data, June 1989, pp. 134-146.

[KV93] Gabriel M. Kuper, Moshe Y. Vardi, *The Logical Data Model*, in: ACM Transactions on Database Systems, Vol. 18, No. 3, September 1993, pp. 379-413.

[KW93] Michael Kifer, James Wu, *A Logic for Programming with Complex Objects*, in: Journal of Computer and System Sciences 47, 1993, pp. 77-120.

[Mit90] John C. Mitchell, *Type systems for programming languages*, in: J. van Leeuwen (ed.), Handbook of Theoretical Computer Science, Elsevier Science Publ., 1990, pp. 365-458.

[Rij93] Maarten de Rijke, *Extending Modal Logic*, PhD Thesis, University of Amsterdam, ILLC Dissertation Series 1993-4.

[Rou91] Bill Rounds, *Situation Theoretic Aspects of Databases*, in: Jon Barwise, Mark Gawron, Gordon Plotkin, Syun Tutiya (eds.), Situation Theory and its Applications, vol. 2, CSLI, 1991, pp. 229-255.

[Tro92] A.S. Troelstra, *Lectures on Linear Logic*, CSLI lecture notes 29, 1992.

Effective Systolic Algorithms for Gossiping in Cycles and Two-Dimensional Grids *

(Extended Abstract)

Juraj Hromkovič [1,**], Ralf Klasing [2], Dana Pardubská [3,***]
Walter Unger [2], Juraj Waczulik [3], Hubert Wagener [2,†]

[1] Institut für Informatik und Praktische Mathematik
Universität zu Kiel, 24098 Kiel, Germany

[2] Department of Mathematics and Computer Science
University of Paderborn, 33095 Paderborn, Germany

[3] Faculty of Mathematics and Physics
Comenius University, 84215 Bratislava, Slovakia

Abstract. The complexity of systolic dissemination of information in one-way (telegraph) and two-way (telephone) communication mode is investigated. The following main results are established:

(i) tight lower and upper bounds on the complexity of one-way systolic gossip in cycles for any length of the systolic period,

(ii) optimal one-way and two-way systolic gossip algorithms in 2-dimensional grids whose complexity (the number of rounds) meets the trivial lower bound (the sum of the sizes of its dimensions).

The second result (ii) shows that we can systolically gossip in grids with the general gossip complexity, i.e., we do not need to pay for the systolization in grids. Opposed to this, the first result (i) shows that the complexity of systolic gossip in cycles is greater than the gossip complexity in cycles for any systolic period length independent of the size of the cycle.

* This work was partially supported by grants Mo 285/9-1 and Me 872/6-1 (Leibniz Award) of the German Research Association (DFG), and by the ESPRIT Basic Research Action No. 7141 (ALCOM II).

** Supported by SAV Grant 2/1138/94 of the Computer Science Institute of the Slovak Academy of Sciences.

*** This author was partially supported by MSV SR and by EC Cooperative Action IC 1000 Algorithms for Future Technologies (ALTEC). Most of the work of this author was done when she was visiting the University of Paderborn.

† This author was supported by the Ministerium für Wissenschaft und Forschung des Landes Nordrhein-Westfalen.

1 Introduction

One of the main quantitative measures characterizing the effectivity (computational power) of interconnection networks is the complexity of the solution of fundamental communication tasks like broadcast and gossip. The complexity of these communication tasks was intensively studied in the last two decades (for a survey see [HHL88, FL94, HKMP95]). This paper continues in the study of the complexity of very regular communication algorithms called "periodic" algorithms in [LR93a, LR93b, LHL94] and "systolic" (or "traffic-light") algorithms in [HKPUW94, KP94]. The main motivation behind this concept corresponds to the idea of Kung [Ku79] who has introduced so-called "systolic computations" as parallel computations with cheap realization due to a very regular, synchronized periodic behaviour of all processors of the interconnection network during the whole execution of the computation (see also some of the first papers about systolic computations in trees and grids [CC84, CGS83, CGS84, CSW84]). Liestman and Richards [LR93a, LR93b] were the first who considered a very regular form of communication algorithms for broadcast and gossip. This form, later called "periodic" communication algorithms [LHL94, LR94], was based on edge coloring of a given interconnection network and on periodic (cyclic) execution of communications via edges with the same colour in one communication step. This periodic gossip based on graph colouring was implicitly used also in [XL92, XL94] in order to solve a completely practical problem – termination detection in synchronous interconnection networks. A little more general concept of regular communication was given by Hromkovič et. al. [HKPUW94] who have introduced k-systolic communication algorithms as a repetition of a given sequence of k communication steps (for some $k \in \mathbb{N}$, k independent of the number of processors of the network). This kind of communication was also considered in [KP94] for the problem of route scheduling under the two-way (telephone) model of communication. Since there is no difference in the complexity of general broadcast algorithms and systolic broadcast algorithms (every broadcast algorithm can be systolized without any increase in the number of communication steps [HKPUW94]), the main research problems formulated in [HKPUW94] are the following two:

1. Which is the k-systolic gossip complexity of fundamental interconnection networks? Which is the most appropriate k (if it exists) for a given network?
2. How much must be paid for the systolization of communication algorithms in concrete interconnection networks (i.e., which is the difference between the complexity of gossip and the complexity of systolic gossip)?

These problems were investigated for one-way (telegraph) and two-way (telephone) communication modes in [HKPUW94], where optimal gossip algorithms for paths and complete d-ary trees are presented. While the optimal two-way gossip algorithm for the paths of odd length is 2-systolic, the complexity of k-systolic one-way gossip in paths is always greater than the complexity of $(k+1)$-systolic one-way gossip in paths for any $k \in \mathbb{N}$. A little surprising result of [HKPUW94] is that we do not need to pay for the systolization in k-ary complete trees. More precisely, for every complete k-ary tree there exists a constant $c_k \in \mathbb{N}$ (depending on k and on the communication mode but independent of the depth of the tree) such that the complexity of c_k-systolic two-way (one-way) gossip is the same as the complexity of two-way (one-way) gossip.

An optimal algorithm for non-systolic gossiping in the two-dimensional grid in the two-way mode of communication was presented in [FP80] matching the diameter lower bound. In [LR93a, KP94], two-way systolic algorithms were described taking only $O(1)$ additional rounds.

Here, we continue in the investigation of the problems 1 and 2 for cycles and for 2-dimensional grids. The organization and the contributions of this paper are as follows. The next section provides the fundamental definitions and notations. The complexity of one-way gossip in the cycle C_n of n nodes is studied in Section 3. It is proved that with growing k, the complexity of k-systolic one-way gossip in C_n decreases and tends to $n/2$. Optimal systolic one-way and two-way gossip algorithms in two-dimensional grids are presented in Section 4. The complexity of these optimal algorithms meet the trivial lower bound given by the diameter (sum of the sizes of the dimensions) i.e., we have a little surprising result (especially for one-way communication mode) that we do not need to pay for the systolization in grids. This improves the result of [LR93a, KP94], where a weaker result with $O(1)$ additional rounds in the more powerful two-way mode is established.

2 Definitions

An (interconnection) network is viewed as a connected undirected graph $G = (V, E)$, where the nodes of V correspond to the processors and the edges correspond to the communication links of the network. An infinite sequence $\{G_i\}_{i=1}^{\infty}$ with $G_i = (V_i, E_i), |V_i| \geq |V_j|$ for $i > j$ is called a class of interconnection networks. Here, we consider the class of cycles $\{C_n\}_{n=1}^{\infty}$ (C_n is the cycle of n nodes) and d-dimensional grids $M(n_1, n_2, \ldots, n_d)$ of the size $n_1 \times n_2 \times \cdots n_d$ for $d, n_1, n_2, \ldots, n_d \in \mathbb{N}$.

The following three fundamental communication problems in networks are defined as follows.

1. Broadcast problem for a network G and a node v of G.
   Let $G = (V, E)$ be a network and let $v \in V$ be a node of G. Let v know a piece of information $I(v)$ which is unknown to all nodes in $V - \{v\}$. The problem is to find a communication strategy such that all nodes in G learn the piece of information $I(v)$.
2. Accumulation problem for a network G and a node v of G.
   Let $G = (V, E)$ be a network, and let $v \in V$ be a node of G. Let each

node $u \in V$ know a piece of information $I(u)$ which is independent of all other pieces of information distributed in other nodes (i.e. $I(u)$ cannot be derived from $\bigcup_{v \in V-\{u\}}\{I(v)\}$). The set $I(G) = \{I(w) | w \in V\}$ is called the cumulative message of G. The problem is to find a communication strategy such that the node v learns the cumulative message of G.

3. Gossip problem for a network G.
   Let $G = (V, E)$ be a network, and let, for all $v \in V$, $I(v)$ be a piece of information residing in v. The problem is to find a communication strategy such that each node from V learns $I(G)$.

Now, it remains to explain what the notion "communication strategy" means. The communication strategy is meant to be a communication algorithm from an allowed set of synchronized communication algorithms. Each communication algorithm is a sequence of simple communication steps called communication rounds (or simply rounds). To specify the set of allowed communication algorithms one defines a so-called communication mode which precisely defines what may happen in one communication step (round). Here, we consider the following two basic communication modes:

a, one-way mode (also called telegraph mode)
   In this mode, in a single round, each node may be active only via one of its adjacent edges either as a sender or as a receiver. This means that if one edge (u, v) is active as a communication link, then the information is flowing only in one direction. Formally, let $G = (V, E)$ be a network, $\vec{E} = \{(v \to u), (u \to v) | (u, v) \in E\}$. A one-way communication algorithm for G is a sequence of rounds $A_1, A_2, \ldots, A_k$, where $A_i \subseteq \vec{E}$ for every $i \in \{1, \ldots, k\}$, and if $(x_1 \to y_1), (x_2 \to y_2) \in A_i$ and $(x_1, y_1) \neq (x_2, y_2)$ for some $i \in \{1, \ldots, k\}$, then $x_1 \neq x_2 \wedge x_1 \neq y_2 \wedge y_1 \neq x_2 \wedge y_1 \neq y_2$ (i.e. each A_i is a matching in the directed graph $(V, \vec{E})$). If $(u \to v) \in A_i$ for some $i \in \{1, \ldots, k\}$, then it is assumed that the whole current knowledge of the node u is known to the node v after the execution of the i-th round A_i.

b, two-way mode (also called telephone mode)
   In two-way mode, in a single round, each node may be active only via one of its adjacent edges and if it is active then it simultaneously sends a message and receives a message through the given, active edge. Formally, let G be a network. A two-way communication algorithm for G is a sequence of rounds $B_1, B_2, \ldots, B_r$, where each round $B_j \subseteq E$, and for each $i \in \{1, \ldots, r\}, \forall (x_1, y_1), (x_2, y_2) \in B_i : (x_1, y_1) \neq (x_2, y_2)$ implies $x_1 \neq x_2 \wedge x_1 \neq y_1 \wedge y_1 \neq y_2 \wedge x_2 \neq y_1$ (i.e. B_i is a matching in G). If $(u, v) \in B_i$ for some i, then it is assumed that the whole current knowledge of u is submitted to v, and the whole current knowledge of v is submitted to u in the i-th round.

Now, we define the systolic communication as introduced in [HKPUW94].

Definition 1. Any one-way (two-way) communication algorithm $A = A_1, A_2, A_3, \ldots, A_m$ for some $m \in \mathbb{N}$ is called k-systolic for some positive integer k, if there exist some $r \in \{1, \ldots, m\}$ and some $j \in \{1, \ldots, k\}$ such that

$$A = (A_1, A_2, \ldots, A_k)^r, A_1, A_2, \ldots, A_j.$$

$P = A_1, \ldots, A_k$ is called the period/cycle of A, k is called the length of P. □

In what follows the complexity of communication algorithms is considered as the number of rounds they consist of.

Definition 2. Let $G = (V, E)$ be a network. Let $A = A_1, A_2, \ldots, A_m$ be a one-way (two-way) communication algorithm on G. The complexity of A is $c(A) = m$ (the number of rounds of A). Let v be a node of V. The one-way complexity of the broadcast problem for G and v is

$$b_v(G) = \min \{c(A) | A \text{ is a one-way communication algorithm solving the broadcast problem for } G \text{ and } v\}.$$

The one-way complexity of the accumulation problem for G and v is

$$a_v(G) = \min \{c(A) | A \text{ is a one-way communication algorithm solving the accumulation problem for } G \text{ and } v\}.$$

We define

$$b(G) = \max \{b_v(G) | v \in V\} \text{ as the broadcast complexity of } G,$$

$$\text{minb}(G) = \min \{b_v(G) | v \in V\} \text{ as the min-broadcast complexity of } G.$$

The one-way gossip complexity of G is

$$r(G) = \min \{c(A) | A \text{ is a one-way communication algorithm solving the gossip problem for } G\},$$

and the two-way gossip complexity of G is

$$r_2(G) = \min \{c(A) | A \text{ is a two-way communication algorithm solving the gossip problem for } G\}.$$ □

Note that we do not define the broadcast (accumulation) complexity for two-way mode because each two-way broadcast (accumulation) algorithm for a network G can be transformed into a one-way broadcast (accumulation) algorithm consisting of the same number of rounds. Observe also that $a_v(G) = b_v(G)$ for any work G and any node v of G (cf. [HKMP95]).

Now, we give the notation for the complexity of systolic gossip. Note that we do not need to define the complexity of k-systolic broadcast and accumulation because it does not differ from the complexity of broadcast and accumulation [HKPUW94].

Definition 3. Let $G = (V, E)$ be a network, and let k be a positive integer. The one-way k-systolic gossip complexity of G is

$$[k]\text{-}sr(G) = \min \{c(A) | A \text{ is a one-way } k\text{-systolic communication algorithm solving the gossip problem for } G\},$$

and the two-way k-systolic gossip complexity of G is

$$[k]\text{-}sr_2(G) = \min \{c(A) | A \text{ is a two-way } k\text{-systolic communication algorithm solving the gossip problem for } G\}.$$ □

3 Systolic Gossip in Cycles

The aim of this section is to investigate the complexity of systolic gossip in cycles in one-way and two-way modes. Similarly as in the non-systolic case [KCV92, HJM94], it is much easier to find optimal systolic two-way gossip algorithms than to get very closed lower and upper bounds on the complexity of one-way systolic gossip in cycles.

We start with the simple two-way case. The next assertion is based on the only fact that the optimal two-way gossip algorithm in cycles can be seen as a systolic one for any even period length.

Theorem 4. *For any even $n \geq 4$, and any even $k \geq 2$:*

$$[k]\text{-}sr_2(C_n) = n/2.$$

If n is odd, the situation is a little more complicated. But also in this case, already period length 4 is enough to get an optimal systolic two-way gossip algorithm for C_n with odd n. We omit the proof of the following assertion in this extended abstract. For the statement of the next theorem, let P_n denote the path of n nodes.

Theorem 5. *For any odd $n \geq 3$:*

$$[2]\text{-}sr_2(C_n) = n = r_2(P_n),$$
$$[3]\text{-}sr_2(C_n) \leq \lceil (2n)/3 \rceil + 2,$$
$$[4]\text{-}sr_2(C_n) = \lceil n/2 \rceil + 1 = r_2(C_n).$$

We call attention to the fact that odd period lengths are not appropriate for systolic gossip in cycles. We can prove for odd k that $[k]$-$sr_2(C_n)$ does not match the trivial lower bound $\lceil n/2 \rceil$ given by the diameter as in the case of even k. But we omit the formulation of these lower bounds in this extended abstract.

Now, we give the results for the one-way case which presents the main contribution of this section. The proof is too technical to be completely presented in this extended abstract.

Theorem 6. *For any $k \geq 6$ and any $n \geq 2$:*

$$\frac{n}{2} \cdot \frac{k+2}{k} - 2 \leq [k]\text{-}sr(C_n) \leq \frac{n}{2} \cdot \frac{k+2}{k} + 2k - 2 \quad \textit{for } k \textit{ even},$$

$$\frac{n}{2} \cdot \frac{k+2}{k} - 2 \leq [k]\text{-}sr(C_n) \leq \frac{n}{2} \cdot \frac{k+1}{k-1} + 2k - 2 \quad \textit{for } k \textit{ odd}.$$

Idea of the Proof. The upper bounds are based on some analysis showing that the best strategy for one-way systolic gossiping in cycles is to totally prefer the information flow in one direction over the flow in the opposite direction. The lower bound is based on a careful analysis of the number of "collisions" between messages flowing in opposite directions in the cycle. □

Finally, we observe that there are some similarities in the results for paths [HKPUW94] and in the results stated above. For both path and cycle we do not need to pay for the systolization in the two-way case, but we have to pay with $const \cdot n$ additional rounds for any period length k in the one-way case.

4 Systolic Gossiping in Two-Dimensional Grids

The main result of this section is an optimal systolic algorithm for the gossip problem in two-dimensional grids. In [KP94], an algorithm was presented for the two–way mode of communication which completes gossiping on an $(m_1 \times m_2)$–grid M (i.e. a grid consisting of $m_1 + 1$ columns and $m_2 + 1$ rows) in $d + O(1)$ rounds, where $d = m_1 + m_2$ is the diameter of the grid. We derive an algorithm for the one–way mode of communication that performs gossiping on M in d rounds, provided m_1 and m_2 are large enough. The period–length of the presented algorithm is 16. Since the diameter is the trivial lower bound on the number of required rounds for any gossip algorithm (even for a non–systolic one in two–way mode of communication), the optimality of our algorithm is obvious. The fact itself that we do not need to pay for the systolization in two–dimensional grids is of special interest, because grids are typical parallel architectures. Note that any solution in the one–way mode of communication is also a solution for the two–way mode. Thus, the possible advantages of using the two–way model solely lie in shorter periods and a weaker restriction on the size of the grids. We have some results of this kind for the two–way mode, but we omit their explicit formulation in this extended abstract.

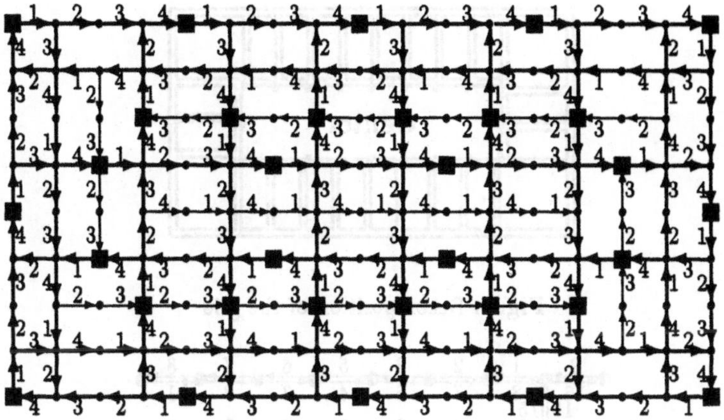

Fig. 1. One-way gossiping in a (16 × 8)-grid (near-optimal scheme)

Theorem 7. *For any $(m_1 \times m_2)$-grid M with m_i even, and $m_i \geq 28$,*

$$[16] - sr(M) \;=\; m_1 + m_2 \,.$$

Idea of the Proof. The proof is realized as follows. First, one designs a near-optimal algorithm S_1 that requires $d + 16$ rounds for gossiping in any 2-dimensional grid. Then, one extends this solution resulting in an optimal algorithm S_2.

To design the near-optimal algorithm we employ the following strategy. We say that a subset K of the grid-nodes forms a *knowledge set* after round t if any information of the system is known to at least one node of K. We specify such a subset K and show that according to our communication scheme K forms a knowledge set after 8 rounds. After additional d rounds any individual node of K knows the cumulative message, i.e. in these d rounds a gossip restricted to K is performed. Thereafter, additional 8 rounds suffice to broadcast the cumulative message to any node of the grid. We can observe that for special values of m_1 and m_2 every fourth node of the border including all corners can be taken into K (see Fig. 1 for a near-optimal gossiping in a (16 × 8)-grid).

To obtain an optimal gossip algorithm we split the grid into a center and a seam as shown in Figure 2.

In the center we use the communication pattern of the near-optimal algorithm. Assume that K_{GC} is the subset of center-nodes among which gossiping in d_{GC} rounds is possible, where d_{GC} denotes the diameter of the center. If we are able to make K_{GC} a knowledge set for the entire grid in $(d - d_{GC})/2$ rounds, and if broadcasting from K_{GC} to the entire grid can be performed within the same number of rounds, we obtain an optimal algorithm using only d rounds of communication. The communication pattern for the seam consequently is

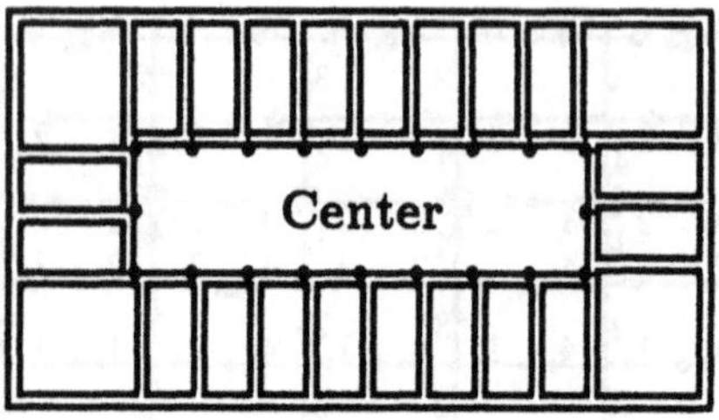

Fig. 2. Decomposition of the grid

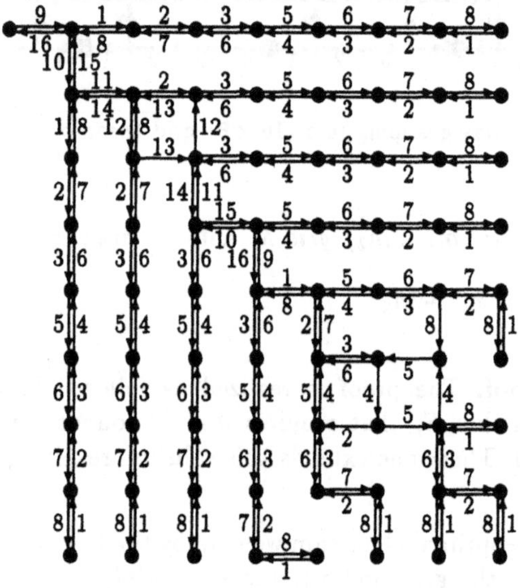

Fig. 3. Accumulation and broadcast in a corner of size 8×9 (optimal scheme)

specialized in the accumulation and broadcast task. Fig. 3 gives an example of an optimal scheme for accumulation and broadcast in a corner of the seam. □

References

[BHMS90] A. Bagchi, S. L. Hakimi, J. Mitchem, E. Schmeichel: *Parallel algorithms for gossiping by mail.* Information Process. Letters 34 (1990), 197-202.

[CC84] C. Choffrut, K. Culik II: *On real-time cellular automata and trellis automata.* Acta Informatica 21 (1984), 393-407.

[CGS83] K. Culik II, J. Gruska, A. Salomaa: *Systolic automata for VLSI on balanced trees.* Acta Informatica 18 (1983), 335-344.

[CGS84] K. Culik II, I. Gruska, A. Salomaa: *Systolic trellis automata. Part I.* Intern. J. Comput. Math. 15 (1984), 195-212.

[CSW84] K. Culik II, A. Salomaa, D. Wood: *Systolic tree acceptors.* R.A.I.R.O. Theoretical Informatics 18 (1984), 53-69.

[FL94] P. Fraigniaud, E. Lazard, *Methods and problems of communication in usual networks.* Discrete Applied Mathematics, special issue on Broadcasting and Gossiping (DAM-BG), Vol. 53 Nos. 1-3, September 1994, pp. 79-134.

[FP80] A.M. Farley, A. Proskurowski, *Gossiping in grid graphs.* J. Combin. Inform. System Sci. 5 (1980), pp. 161-172.

[HJM94] J. Hromkovič, C.-D. Jeschke, B. Monien: *Note on Optimal Gossiping in some Weak-Connected Graphs.* Theoretical Computer Science 127 (1994), No. 2, 395-402.

[HHL88] S. M. Hedetniemi, S. T. Hedetniemi, A. L. Liestmann: *A survey of gossiping and broadcasting in communication networks.* Networks 18 (1988), 319-349.

[HKMP95] J. Hromkovič, R. Klasing, B. Monien, R. Peine: *Dissemination of information in interconnection networks (broadcasting & gossiping).* In: Combinatorial Network Theory (Frank Hsu, Ding-Zhu Du, Eds.), Science Press & AMS, 1994, to appear.

[HKPUW94] J. Hromkovič, R. Klasing, D. Pardubská, W. Unger, H. Wagener: *The complexity of systolic dissemination of information in interconnection networks.* In: Proc. of the 1st Canada-France Conference on Parallel Computing (CFCP '94), Springer LNCS 805, pp. 235-249. R.A.I.R.O. Theoretical Informatics and Applications, Vol. 28 Nos. 3-4, 1994, pp. 303-342.

[KCV92] D.W. Krumme, G. Cybenko, K.N. Venkataraman: *Gossiping in minimal time.* SIAM J. Comput. 21 (1992), pp. 111-139.

[KP94] G. Kortsarz, D. Peleg: *Traffic-light scheduling on the grid.* Discrete Applied Mathematics, special issue on Broadcasting and Gossiping (DAM-BG), Vol. 53 Nos. 1-3, September 1994, pp. 211-234.

[Ku79] H.T. Kung: *Let's design algorithms for VLSI systems.* In: Proc. of the Caltech Conference of VLSI (CL.L. Seifz Ed.), Pasadena, California 1979, pp. 65-90.

[LHL94] R. Labahn, S.T. Hedetniemi, R. Laskar, *Periodic gossiping on trees.* Discrete Applied Mathematics, special issue on Broadcasting and Gossiping (DAM-BG), Vol. 53 Nos. 1-3, September 1994, pp. 235-246.

[LR93a] A.L. Liestman, D. Richards: *Network communication in edge-colored graphs: Gossiping.* IEEE Trans. Par. Distr. Syst. 4 (1993), 438-445.

[LR93b] A.L. Liestman, D. Richards: *Perpetual gossiping.* Parallel Processing Letters 3 (1993), No. 4, 347-355.

[LR94] R. Labahn, A. Raspaud: *Periodic gossiping in back-to-back trees.* Technical Report, Universität Rostock, Germany, 1994.

[XL92] Cheng-Zhong Xu, Francis C.M. Lau: *Distributed termination detection of loosely synchronized computations.* In: Proc. 4th IEEE Symp. on Parallel and Distr. Processing, Texas 1992, pp. 196-203.

[XL94] Cheng-Zhong Xu, Francis C.M. Lau: *Efficient distributed termination detection for synchronous computations in multicomputers.* Technical Report TR-94-04, Department of Computer Science, The University of Hong Kong, March 1994.

Restarting Automata

Petr Jančar*[1], František Mráz[2], Martin Plátek[2], Jörg Vogel[3]

[1] University of Ostrava, Dept. of Computer Science
Dvořákova 7, 701 03 OSTRAVA, Czech Republic
e-mail: jancar@osu.cz

[2] Charles University, Dept. of Computer Science
Malostranské nám. 25, 118 00 PRAHA 1, Czech Republic
e-mail: mraz,platek@kki.mff.cuni.cz

[3] Friedrich Schiller University, Computer Science Institute
07740 JENA, Germany
e-mail: vogel@minet.uni-jena.de

Abstract. Motivated by (natural as well as formal) language analysis, we introduce a special kind of the linear bounded automaton – so called restarting automaton (R-automaton). Its computation proceeds in certain cycles; in each cycle, a (bounded) subsequence of the input word is removed and the computation restarts on the arising shorter word. We consider both nondeterministic and deterministic versions and introduce a natural property of monotonocity. The main result illustrating usefulness of the introduced notions is an elegant characterization of deterministic context-free languages – by deterministic monotonic R-automata. We also show that the monotonocity property is decidable and complete the paper with some related results.

1 Introduction

In [1] we studied forgetting automata, which can be viewed as special linear bounded automata with limited rewriting capability. Here we introduce the notion of restarting automata, which are special forgetting automata with a lucid way of computation.

Our motivation partly comes from the area of natural language analysis and resembles the motivation for Marcus contextual (internal) grammars. These grammars are natural tools for describing the relations of the type "to be a distinguished subsequence (e.g. sentence) of a sentence".

Remark. In [4], [5] such relations are considered as the basic syntactical and semantical criteria for formulation of formal dependency syntax of natural languages. The dependency syntaxes are traditional for free word order natural languages with a considerable flexis, e.g. for Slavonic languages.

Another part of motivation comes from the area of grammar checking of natural as well as programming languages. Having found an error in a sentence,

* The first author was supported by the Grant Agency of the Czech Republic, Grant-No. 201/93/2123

the grammar checker should specify it – often by exhibiting the parts (words) which do not match each other. The method can be based on stepwise leaving out some parts not affecting the (non)correctness of the sentence. E.g. applying it to the sentence

'The little boys I mentioned runs very quickly'

we get successively

'The boys I mentioned runs very quickly'
'boys I mentioned runs very quickly'
'boys runs very quickly'
'boys runs quickly'

coming to the "error core"

'boys runs'.

Being motivated by the mentioned ideas, we define the restarting automaton (R-automaton) which can be roughly described as follows. It has a finite control unit, a head with a lookahead window attached to a tape, and it works in certain cycles. In a cycle, it moves the head from left to right along the word on the tape; according to its instructions, it can at some point "cut off" some symbols out of the lookahead and "restart" – i.e. reset the control unit to the initial state and place the head on the left end of the (shorter) word. The computation halts in an accepting or a rejecting state.

In fact, to make the mentioned "cutting off" more natural, the restarting automata use a (doubly linked) list of items (cells) rather than the usual tape. Here we can mention that (more general) deterministic list automata can serve as a natural basis of "syntax checkers" for programming languages ([7]); the methods used there can also be improved using deterministic restarting automata (as suggested in [8]).

As usual, we define nondeterministic and deterministic versions of the automata. We also consider a natural property of monotonocity (during any computation, "the places of cutting off do not increase their distances from the right end") and show that deterministic monotonic R-automata nicely characterize the class of deterministic context-free languages ($DCFL$). We also show the decidability of the monotonocity property and complete the paper with showing a language outside $DCFL$ which is recognized by both a deterministic (nonmonotonic) R-automaton and a nondeterministic monotonic one.

Section 2 contains definitions, results, and some examples which should illustrate usefulness of the mentioned characterization of $DCFL$. Section 3 contains the main proofs and Section 4 some additional remarks.

2 Definitions and Results

We present the definitions informally; the formal technical details could be added in a standard way of the automata theory.

A *restarting automaton*, or an *R-automaton*, M (with bounded lookahead) is a device with a finite state control unit and one head moving on a finite linear (doubly linked) list of items (cells). The first item always contains a special symbol ¢, the last one another special symbol \$, and each other item contains a symbol from a finite alphabet (not containing ¢, \$). The head has a lookahead "window" of length k (for some $k \geq 0$) – besides the current item, M also scans the next k right neighbour items (or simply the end of the word when the distance to \$ is less than k). In the *initial configuration*, the control unit is in a fixed, initial, state and the head is attached to the item with the left sentinel ¢ (scanning also the first k symbols of the input word).

The *computation* of M is controlled by a finite set of *instructions* of the following two types:

(1) $(q, au) \rightarrow (q', MVR)$
(2) $(q, au) \rightarrow RESTART(v)$

The left-hand side of an instruction determines when it is applicable – q means the current state (of the control unit), a the symbol being scanned by the head, and u means the contents of the lookahead window (u being a string of length k or less if it ends with \$). The right-hand side describes the activity to be performed. In case of (1), M changes the current state to q' and moves the head to the right neighbour item. In case of (2), the activity consists of deleting (removing) some items of the just scanned part of the list (containing au) so that v is left (au is replaced with v where v is a proper subsequence of au) and restarting – i.e. setting the initial state and placing the head on the first item of the list (containing ¢).

We will suppose that the control unit states are divided into two groups – the *nonhalting states* (an instruction is always applicable when the unit is in such a state) and the *halting states* (any computation finishes by entering such a state); the halting states are further divided into the *accepting states* and the *rejecting states*.

In general, an R-automaton can be *nondeterministic*, i.e. there can be two or more instructions with the same left-hand side (q, au). If it is not the case, the automaton is *deterministic*.

An input *word* w *is accepted by* M if there is a computation which starts in the initial configuration with w (bounded by sentinels ¢,\$) on the list and finishes in an *accepting configuration* where the control unit is in one of the accepting states. $L(M)$ denotes the language consisting of all words accepted by M; we say that M *recognizes the language* $L(M)$.

It is natural to divide any computation of an R-automaton into certain phases or *cycles*: in one cycle, the head moves right along the input list (with a bounded lookahead) until a halting state is entered or something in a bounded space is deleted – in that case the computation is resumed in the initial configuration on the new, shorter, word (thus a new cycle starts). It immediately implies that any computation of any R-automaton is finite (finishing in a halting state).

The next two claims (with obvious, and hence omitted, proofs) express certain lucidness of computations of R-automata. The notation $u \longrightarrow_M v$ means that there exists a cycle of M starting in the initial configuration with the word u and finishing in the initial configuration with the word v; the relation $\longrightarrow_M^*$ is the reflexive and transitive closure of $\longrightarrow_M$.

Claim 1 (the error preserving property (for all R-automata)). *Let M be an R-automaton, u, v some words in the alphabet of M. If $u \longrightarrow_M^* v$ and $u \notin L(M)$, then $v \notin L(M)$.*

Claim 2. (the correctness preserving property (for deterministic R-automata)). *Let M be a deterministic R-automaton and $u \longrightarrow_M^* v$ for some words u, v. Then $u \in L(M)$ iff $v \in L(M)$.*

It will turn out useful to consider a natural property of monotonocity. By a *monotonic R-automaton* we mean an R-automaton where the following holds for all computations: all items which appeared in the lookahead window (and were not deleted) during one cycle will appear in the lookahead in the next cycle as well – if it does not finish in a halting state (during any computation, "the places of deleting do not increase their distances from the right endmarker \$").

Considering a deterministic R-automaton M, it is sometimes convenient to suppose it in the *strong cyclic form*; it means that the words of length less than k, k being the length of lookahead, are immediately (hence in the first cycle) accepted or rejected, and that M performs at least two cycles (at least one restarting) for any longer word.
For a nondeterministic R-automaton M, we can suppose the *weak cyclic form* – any word from $L(M)$ longer than k (the length of lookahead) can be only accepted by performing two cycles at least. The cyclic forms are justified by the following claim.

Claim 3. *For any nondeterministic (deterministic) R-automaton M, with lookahead k, there is a nondeterministic (deterministic) R-automaton M', with some lookahead n, $n > k$, such that M' is in the weak (strong) cyclic form and $L(M) = L(M')$. In addition, when M is deterministic and monotonic then M' is deterministic and monotonic as well.*

Proof: First notice that we can easily force M to visit all items of the input list before accepting or rejecting (instead of an "original" accepting (rejecting) state, it would enter a special state which causes moving to the right end and then accepting (rejecting)).

Now suppose that (the modified) M accepts a long word w in the first cycle (without restarting). If w is sufficiently long then it surely can be written $w = v_1auv_2auv_3$ where M enters both occurrences of a in the same state q during the corresponding computation (as above, by a we mean a symbol and by u a string of length k). Then it is clear that the word v_1auv_3 (auv_2 has been deleted) is also accepted. In addition, we can suppose that the length of auv_2auv_3 is less than a fixed (sufficiently large) n.

We sketch a desired M' with lookahead n. Any word w shorter than n is immediately accepted or rejected by M' according to whether $w \in L(M)$ or not.

On a longer w, M' simulates M with the following exception: when $ appears in the lookahead window, M' checks whether M could move to the right end and accept; if so, M' deletes the relevant auv_2 (cf. the above notation) and restarts (recall that n has been chosen so that such auv_2 surely exists). Obviously, $L(M) = L(M')$ holds.

In case M is deterministic, M' can work as above; in addition it can safely delete the relevant auv_2 also when M would reject (due to determinism, the resulting word is also rejected by M).

It should be clear that monotonocity of M implies monotonocity of M' in the deterministic case. □

For brevity, we use the following obvious notation. R denotes the class of all (nondeterministic) restarting automata (with some lookahead). Prefix *det-* denotes the deterministic version, similarly *mon-* the monotonic version. $\mathcal{L}(A)$, where A is some class of automata, denotes the class of languages recognizable by automata from A. E.g. the class of languages recognizable by deterministic monotonic R-automata is denoted by $\mathcal{L}(det\text{-}mon\text{-}R)$.

Now the main theorem of the paper can be expressed as follows.

Theorem 4. $DCFL = \mathcal{L}(det\text{-}mon\text{-}R)$

This is, in fact, the announced characterization of $DCFL$ (the proof is given in Section 3). It can be worth noting that the closure of $DCFL$ under complement is immediately clear when considering R-automata:

Claim 5. *The classes* $\mathcal{L}(det\text{-}mon\text{-}R)$ *and* $\mathcal{L}(det\text{-}R)$ *are closed under complement.*

Proof: Since all computations are finite and in our case also deterministic, it suffices to exchange the accepting and the rejecting states to get a (deterministic) automaton recognizing the complementary language. □

Of course, the last claim only argues for some "lucidness" of deterministic R-automata, it is not a short proof of the closure property of $DCFL$ (the "long" proof is hidden in the proof of Theorem 4). On the other hand, Theorem 4 (or its weaker consequences, together with Claim 3) *is* a means for short proving that some languages are not in $DCFL$. We illustrate it by the next two examples.

Example 1. Consider the language $L_1 = \{ww^R \mid w \in \{a,b\}^*\}$. If it were in $DCFL$, it would be recognized by a deterministic R-automaton M with the length of lookahead k (for some k); M can be supposed in the strong cyclic form. Let us now take a word $a^n b^m b^m a^n$ $(n, m > k)$, on which M performs two cycles at least. In the first cycle, M can only shorten the segment of $b's$. But due to determinism, it would behave in the same way on the word $a^n b^m b^m a^n a^n b^m b^m a^n$, which is a contradiction (to the correctness preserving property).

Example 2. Consider $L_2 = \{a^m b^n c^p \mid m = n \ or \ n = p\}$. If L_2 were in $DCFL$, it would surely be recognized by an R-automaton M with the length of lookahead

k; M can be supposed in the weak cyclic form. Let us take a word $a^r b^r c^{2r}$, $r > k$, which is accepted by performing two cycles at least. In the first cycle, M has to have a possibility to delete m $a's$ and m $b's$ for some $m > 0$. But it could behave in the same way on the word $a^r b^{2r+m} c^{2r}$, which is a contradiction (to the error preserving property).

The natural question of the decidability of monotonocity for a given R-automaton is answered in the affirmative:

Theorem 6. *There is an algorithm which, given an R-automaton M, decides whether M is monotonic or not.*

This theorem holds for all (deterministic and nondeterministic) R-automata. Nevertheless there are other possible versions of monotonocity which can change the situation for nondeterministic automata. E.g. we could only ask all the *accepting* computations to be monotonic (it can be shown that the corresponding classes of languages do not change in that case) or we could ask that any accepted word is (also) accepted by a monotonic computation.

The next theorem shows that the simultaneousness of the conditions of monotonicity and determinism in Theorem 4 cannot be relaxed.

Theorem 7. *There is a language* $L \in \mathcal{L}(det\text{-}R) \cap \mathcal{L}(mon\text{-}R)$ *such that* $L \notin \mathcal{L}(det\text{-}mon\text{-}R)$

3 Proofs

Proof of Theorem 4

The theorem ($\mathcal{L}(det\text{-}mon\text{-}R) = DCFL$) is a consequence of the following two lemmas:

Lemma 8. $\mathcal{L}(det\text{-}mon\text{-}R) \subseteq DCFL$

Proof: Let L be a language recognized by *det-mon-R*-automaton M in strong cyclic form, with lookahead of length k. We show how to construct a deterministic pushdown automaton P which simulates M.

P is able to store in its control unit in a component CSt the current state of M and in a component B a word of length at most $1 + 2k$. P starts by storing the initial state of M in CSt and pushing ¢ – the left endmarker of M – into the first cell of the buffer B and the first k symbols of the input word of M into the next k cells of the buffer B (cells $2, 3, \ldots, k+1$).

During the simulation, the following conditions will hold invariantly:

- CSt contains the current state of M,
- the first cell of B contains the current symbol of M (scanned by the head) and the rest of B contains m right neigbour symbols of the current one (lookahead of length m) where m varies between k and $2k$,
- the pushdown contains the left-hand side (w.r.t. the head) of the list, the

leftmost symbol (¢) being at the bottom. In fact, any pushdown symbol will be composed – it will contain the relevant symbol of the input list and the state of M in which this symbol (this item) was entered last time.

The mentioned invariant will be maintained by the following simulation of instructions of M; the instruction to be simulated (the left-hand side (q, au)) is clear from the information stored in the control unit. The activity to be performed depends on the right-hand side and is dealt with in the next two cases:

(1) (q', MVR) – P puts the contents of the first cell of B and CSt as a composed symbol on the top of the pushdown, stores the new state q' of M in CSt and shifts the contents of B one symbol to the left; if the $(1+k)$-th cell of B is then empty, P reads the next input symbol into it.

(2) $RESTART(v)$ – deleting is simulated in the buffer B (some of the first $k+1$ symbols are deleted and the rest is pushed to the left). Then $k+1$ (composed) symbols are successively taken from the pushdown and the relevant symbols are added from the left to B (shifting the rest to the right). The state parts of k (composed) symbols are forgotten, the state part of the $(k+1)$-th symbol is stored in CSt. Thus not only the $RESTART(v)$-operation is simulated but also the beginning part of the next cycle, the part which surely is the same as in the previous cycle.

It should be clear that due to monotonicity of M the second half of B (cells $k+2, k+3, \ldots, 2k$) is empty at the time of simulating a $RESTART(v)$-operation. Hence the described construction is correct which proves the proposition. □

To show the opposite direction, we use the characterization of deterministic context-free languages by means of $LR(0)$-grammars and $LR(0)$-analyzers; generally $LR(1)$-grammars (lookahead 1) are needed, but it is not necessary when any word is finished by the special sentinel \$.

Lemma 9. *DCFL* $\subseteq \mathcal{L}(det\text{-}mon\text{-}R)$

Proof: Let L' be a deterministic context-free language. Then there is an $LR(0)$-grammar G generating $L = L'\$$ (the concatenation $L' \cdot \{\$\}$ supposing \$ being not in the alphabet of L'), and there is a corresponding $LR(0)$-analyzer P.

For any word $w \in L$ there is only one derivation tree T_w; it corresponds to the analysis of w by the analyzer P. In fact, P simulates constructing T_w in the left-to-right and bottom-up fashion. Due to the standard pumping lemma for context-free languages, there are constants p, q s.t. for any w with length greater than p there are a (complete) subtree T_1 of T_w and a (complete) subtree T_2 of T_1 with the same root labellings; in addition, T_2 has fewer leaves than T_1 and T_1 has q leaves at most. (Cf. Fig. 1; A is a nonterminal of G). Replacing T_1 with T_2, we get the derivation tree for a shorter word w' (w could be written $w = u_1 v_1 u_2 v_2 u_3$ in such a way that $w' = u_1 u_2 u_3$).

Now we outline a *det-mon-R*-automaton M with lookahead of length $k > q$ which recognizes L'.

M stores the contents of the lookahead in a buffer in the control unit. Simulating the $LR(0)$-analyzer P, it constructs (in a bounded space in the control unit) all maximal subtrees of the derivation tree which have all their leaves in

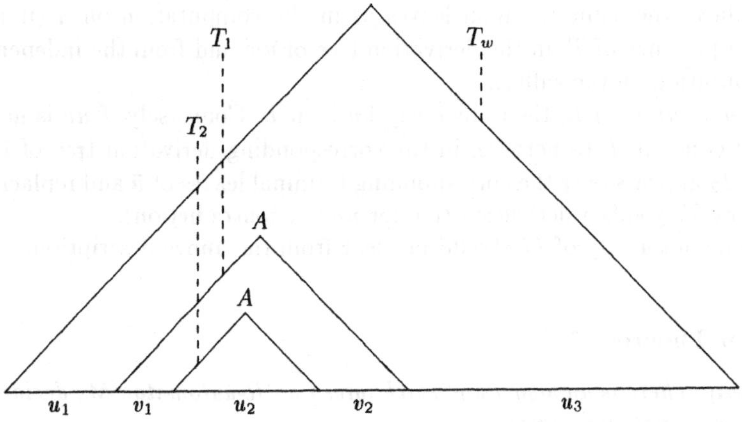

Fig. 1.

the buffer. If one of the subtrees is like the T_1 above, M performs the relevant deleting (of at most two continuous segments) in the input list and restarts. If it is not the case then M forgets the leftmost of these subtrees with all its $n \geq 1$ leaves, and reads n new symbols to the right end of the buffer (shifting the contents left). Then M continues constructing the maximal subtrees with all leaves in the (updated) buffer (simulating P).

In general, the input word w is either shorter than p, such words can be checked using finite memory, or it is longer. If it is longer and belongs to L then M must meet the leftmost above described T_1 (with the subtree T_2); it performs the relevant deleting and restarts on a new, shorter, word. If the (long) input word w does not belong to L, M either meets \$ without restarting and stops in a rejecting state or performs some deleting and restarts. It suffices to show that the resulting shorter word (in both cases) is in L if and only if w is in L.

It can be verified using the following properties of the $LR(0)$–analyzer.

a) For each word $w \in L$ there is exactly one derivation of w in G which corresponds to the analysis of w by P.
b) Let u be the prefix of the input word $w = uv$ which has been already read by the $LR(0)$–analyzer P. If P did not reject the word until now, then there exists a suffix word v' s.t. uv' is in L, and the computation of P on the prefix u is independent w.r.t. the suffix.

Let u be the prefix of the input word, the last symbol of which corresponds to the last symbol in the lookahead window just before M performs a *RESTART*-operation; let $\overline{u}$ be the rest of u after performing the *RESTART*-operation. There exists a suffix v' such that uv' is in L. uv' has a derivation tree in which there is a complete subtree T_1 (with a subtree T_2; as above) corresponding to the place of cutting. Then in the computation of M on $\overline{u}$ the tree T_2 will appear in the buffer above the same terminal leaves as in the computation on u (it follows from the presence of T_2 in the derivation tree of $\overline{u}v'$ and from the independence of computation on the suffix).

Let $w = uv$ is in L, then obviously $\overline{u}v$ is in L. Conversely if uv is not in L then $\overline{u}v$ is not in L (otherwise, in the corresponding derivation tree of $\overline{u}v$, the subtree T_2 appears over the corresponding terminal leaves of $\overline{u}$ and replacing the tree T_2 by T_1 yields a derivation tree for uv – a contradiction).

The monotonicity of M should be clear from the above description. □

Proof of Theorem 6

Theorem. *There is an algorithm which, given an R-automaton M, decides whether M is monotonic or not.*

Proof: We sketch the idea briefly. Consider a given (nondeterministic) R-automaton M; let its lookahead be of length k. Recall that all computations of a monotonic automaton have to be monotonic. The idea is to construct a (nondeterministic) finite automaton which accepts a nonempty language if and only if there is a nonmonotonic computation of M.

Suppose there is a nonmonotonic computation of M. Then there is a word w on which M can perform two cycles (with restarting) where in the second cycle it does not scan all (remaining) items scanned in the first cycle.

Now consider the construction of the mentioned finite automaton A; we can suppose that it has lookahead of length k. A supposes reading the described w. It moves right simulating two consecutive cycles of M simultaneously. At a certain moment, A decides nondeterministically that it has entered the area of the second deleting – it guesses the appropriate contents of the lookahead window which would be encountered in the second cycle. Then it moves right coming to the place of the (guessed) deleting in the first cycle and verifies that the previously guessed lookahead was guessed correctly; if so, A accepts. □

Proof of Theorem 7

Theorem. *There is a language* $L \in \mathcal{L}(det\text{-}R) \cap \mathcal{L}(mon\text{-}R)$ *such that* $L \notin \mathcal{L}(det\text{-}mon\text{-}R)$

Proof: We will show such a language L. The rough idea is as follows. The words of L can be cut in "the middle" by two types of cutting. The information of which type is relevant is placed at the right end of the word and possibly also somewhere between the middle and the right end.

(1) If it is at the middle, the word can be accepted by a deterministic monotonic computation – it keeps cutting at the middle, leaving the information there, and at the end of computation (on a short word) it verifies whether the "middle" information and the "end" information coincide.

(2) If the information is not at the middle, it can be, by another cutting, conveyed stepwise from the right to the middle. This way will be performable for a nonmonotonic *det-R*-automaton but not for any *det-mon-R* one. Nevertheless

a nondeterministic *mon-R*-automaton can guess the information and "write" it at the middle and then proceed deterministically as in (1).

We use two regular languages given by the following regular expressions

$$L_1 = a^*(b_1 + b_1 b_2)^* c_1 \qquad L_2 = a^*(b_2 + b_1 b_2)^* c_2$$

and define

$$L = \{w \mid either\ w \in L_1\ and\ |a|_w = |b_1|_w\ or\ w \in L_2\ and\ 2|a|_w = |b_2|_w\}$$

where $|x|_w$ means the number of occurrences of the symbol x in w. (L is an unambiguous context-free language.)

The details of constructing a (nondeterministic) *mon-R*-automaton M and a *det-R*-automaton M' such that $L(M) = L(M') = L$ as well as the proof that $L \notin \mathcal{L}(det\text{-}mon\text{-}R)$ are omitted. They can be found in the report [3]. □

4 Additional Remarks

Further results on restarting automata are included in [2]. There we also study connections between forgetting (particularly restarting) automata and Marcus grammars (thus following a hint from J.Gruska).

Some techniques used there confirm the similarity between Marcus contextual grammars and restarting automata. E.g. we show there a non-context-free language recognized by a deterministic R-automaton utilizing a method used by Păun for showing a non-context-free language generated by a contextual Marcus grammar with a bounded choice in [6].

References

1. P. Jančar, F. Mráz, M. Plátek: *A taxonomy of forgetting automata;* in Proceedings of MFCS 1993, Gdańsk, Poland; LNCS 711, 527–536, Springer-Verlag, 1993
2. P. Jančar, F. Mráz, M. Plátek, M. Procházka, J. Vogel: *Restarting Automata and Marcus Grammars,* TR Math/95/1 Friedrich Schiller University, Jena, 1995 (accepted for the Second International Conference Developments in Language Theory)
3. P. Jančar, F. Mráz, M. Plátek, J. Vogel: *Restarting Automata,* TR Math/94/8 Friedrich Schiller University, Jena, 1994
4. S. Marcus: *Algebraic Linguistics; Analytical Models,* Academic Press, New York and London, 1967
5. L. Nebeský: *On One Formalization of Sentence Analysis,* Slovo a Slovesnost (*in Czech*), No 2, 104-107, 1962 [Translation to russian: L. Nebeskij: *Ob odnoj formalizacii razbora predlozhenija,* In: Matematitcheskaja lingvistika, Mir 145–149, 1964]
6. G. Păun: *On Some Open Problems about Marcus Contextual Languages,* Intern. J. Computer Math., Vol. 17 9–23, 1985
7. M. Plátek: *Syntactic Error Recovery with Formal Guarantees I.*; TR No. 100, Department of Computer Science, Charles University, Prague, 1992
8. M. Procházka: *Syntax Errors and Their Detection,* Diploma Thesis (in Czech), Charles University, Prague, 1994

Optimal Contiguous Expression DAG Evaluations

Christoph W. Keßler and Thomas Rauber*

Fachbereich 14 Informatik, Universität des Saarlandes, Postfach 151150, D-66041 Saarbrücken, Germany. e-mail: {kessler,rauber}@cs.uni-sb.de

Abstract. Generating evaluations for expression DAGs with a minimal number of registers is NP-complete. We present two algorithms that generate optimal contiguous evaluation for a given DAG. The first is a modification of a complete search algorithm that omits redundant evaluations. The second algorithm generates only the most promising evaluations by splitting the DAG into trees with import and export nodes and evaluating the trees with a modified labeling scheme. Experiments with randomly generated DAGs and large DAGs from real application programs confirm that the new algorithms generate optimal contiguous evaluations quite fast.

1 Introduction

Register allocation is one of the most important problems in compiler optimization. Using fewer registers is essential if the target machine has not enough registers to evaluate an expression without storing intermediate results in the main memory (*spilling*). This is especially important for vector processors that are often used in parallel computers. Vector processors usually have a small number of vector registers (e.g., the CRAY computers have 8 vector register of 64×64 bit) or a register file that can be partitioned into a number of vector registers of a certain length (e.g., the vector acceleration units of the CM5 have register files of length 128×32 bit that can be partitioned into 1, 2, 4 or 8 vector registers, see [5]). A vector operation is evaluated by splitting it into stripes that have the length of the vector registers and computing the stripes one after another. If the register file is partitioned into a small number of vector registers, each of them can hold more elements and the vector operation is split into fewer stripes. This saves initialization costs and results in a faster computation [10].

Scientific programs often contain large basic blocks. Large basic blocks can also result from the application of compiler techniques like loop unrolling [6] and trace scheduling [7]. Therefore, it is important to derive register allocation techniques that cope with large basic blocks [8].

Among the numerous register allocation schemes, register allocation and spilling via graph coloring [3] is generally accepted to yield good results. But register allocation via graph coloring uses a fixed evaluation order within a given basic block B. This is the evaluation order specified in the input program. Often there exists an evaluation order for B that allows to use fewer registers. By using this order, the global register allocation generated via graph coloring could be improved.

The reordering of the operations within a basic block can be arranged by representing the basic block as a set of directed acyclic graphs (DAGs). An

* supported by DFG, SFB 124, TP D4

algorithm to build the DAGs for a given basic block can be found in [2]. A basic block is evaluated by evaluating the corresponding DAGs. For the evaluation of a DAG G the following results are known: (1) If G is a tree, the algorithm of *Sethi* and *Ullman* [13] generates an optimal evaluation in linear time. (In this paper, optimal always means: uses as few registers as possible. Recomputations are not allowed.) (2) The problem of generating an optimal evaluation for G is NP–complete, if G is not restricted [14].

In this paper, we only consider contiguous evaluations. Experiments with randomly generated DAGs and with DAGs that are derived from real programs show that for nearly all DAGs there exists a contiguous evaluation that is optimal. This leads to an algorithm [9] that computes an optimal contiguous evaluation for a given DAG in exponential time. — This paper improves this simple algorithm that performs a rather inefficient complete search by identifying and eliminating redundant evaluations. The second algorithm presented splits the given DAG into a series of trees with import and export nodes and evaluates the trees with a modified labeling scheme. Experiments with DAGs from real applications show that the number of generated evaluations is quite small even for large DAGs. Therefore, the running time of the algorithm remains reasonable.

2 Evaluating DAGs

We assume that we are generating code for a single processor machine with general–purpose registers $\mathcal{R} = \{R_0, R_1, R_2, \ldots\}$ and a countable sequence of memory locations. The arithmetic machine operations are three–address instructions (unary and binary arithmetic operations on registers[2], load and store instructions). Each input program can be partitioned into a number of basic blocks. On the machine instruction level, a *basic block* is a sequence of three–address–instructions that can only be entered via the first statement and only be left via the last statement. The data dependencies in a basic block can be described by a *directed acyclic graph (DAG)*. The leaves of the DAG are the variables and constants occurring as operands in the basic block; the inner nodes represent intermediate results. An example is given in [9].

Definition 1. Let $G = (V, E)$ be a DAG. A subDAG $S = (V', E')$ with $V' \subseteq V$ and $E' \subseteq E \cap (V' \times V')$ is called *complete subDAG* of G with root w, if $V' = \{v \in V : \exists \text{ path in } G \text{ from } v \text{ to } w\}$ and $E' = \{e \in E : e \text{ is an edge on a path in } G \text{ from a node } v \in V' \text{ to } w\}$.

Definition 2. An *evaluation* A of a DAG G is a permutation of the nodes in V such that for all inner nodes $v \in V$, v occurs in A behind all its children.

This implies that the evaluation A is *complete* and contains *no recomputations*, i.e., each node of the DAG appears exactly once in A. Moreover, A is *consistent* because the precedence constraints are preserved. Thus, each topological order of G represents an evaluation, and vice versa.

Definition 3. An evaluation A (represented by the topological order *ord*) of a DAG $G = (V, E)$ is called *contiguous*, if for each node $v \in V$ with children v_1

[2] To simplify the description, we assume w.l.o.g. that the registers participating in an operation are mutually different.

and v_2 the following is true: if w_i is a predecessor of v_i, $i = 1, 2$, and $ord(v_1) < ord(v_2)$, then $ord(w_1) < ord(w_2)$.

A contiguous evaluation of a node v first evaluates the complete subDAG with one of the children of v as root before evaluating any part of the remaining subDAG with root v. — While general evaluations can be generated by variants of topological-sort[3], contiguous evaluations are generated by variants of *depth-first search* (*dfs*). From now on, we restrict our attention to contiguous evaluations to reduce the number of generated evaluations. By doing so, we may not always get the evaluation with the least register need. There are some DAGs for which a general evaluation exists that uses fewer registers than every contiguous evaluation. However, these DAGs are usually quite large and do very rarely occur in real programs. The smallest DAG of this kind that we could construct so far has 14 nodes and is given in [9]. Note that for larger DAGs, it is quite difficult to decide whether there exists a general evaluation that uses fewer registers than every contiguous evaluation. This is because of the enormous running time of the algorithms that generate general evaluations.

Definition 4. [14] Let $num: \mathcal{R} \rightarrow \{0, 1, 2 \ldots\}$, $num(R_i) = i$ denote the numbering of registers. A mapping $reg: V \rightarrow \mathcal{R}$ is called a (consistent) *register allocation* for A, if for all nodes $u, v, w \in V$ the following holds: If u is a child of w, and v appears in A between u and w, then $reg(u) \neq reg(v)$.

$$m(A) = \min_{reg \text{ is reg. alloc. for } A} \{ \max_{v \text{ appears in } A} \{num(reg(v)) + 1\}\}$$

is called the *register need* of the evaluation A. An evaluation A for a DAG G is called *optimal* if for all evaluations A' of G holds $m(A') \geq m(A)$.

Sethi proved in [14] that the problem of computing an optimal evaluation for a given DAG is NP-complete. Assuming **P** $\neq$ **NP**, we expect an optimal algorithm to require nonpolynomial time.

3 Counting Evaluations

In [9], we give the following definitions and prove the following lemmata:

Definition 5. Each leaf is a *tree node*. An inner node is a *tree node* iff all its children are tree nodes and none of them has outdegree > 1.

Definition 6. For a leaf v, $label(v) = 1$. For a unary node v, $label(v) = \max\{label(child(v)), 2\}$. For a binary node v, $label(v) = \max\{3, \max\{label(lchild(v)), label(rchild(v))\} + q\}$ where $q{=}1$ if $label(lchild(v)) = label(rchild(v))$ and $q{=}0$ otherwise.

Let *new_reg*() be a function that returns an available register and marks it 'busy'. Let *regfree*(*reg*) be a function that marks register *reg* 'free' again. The Labeling-algorithm *labelfs* of *Sethi* and *Ullman* [13] generates optimal evaluations for a tree with labels by first evaluating the child with the greater label value for each binary node.

Definition 7. A *decision node* is a binary node which is not a tree node.

[3] See [11] for a summary on topological sorting.

Thus, all binary nodes that have at least one predecessor with more than one parent are decision nodes. In a tree, there are no decision nodes.

Lemma 8. *For a tree T with one root and b binary nodes, there exist exactly 2^b different contiguous evaluations. For a DAG with one root and k binary nodes, there exist at most 2^k different contiguous evaluations.*

Lemma 9. *Let G be a DAG with d decision nodes and b binary tree nodes which form t (disjoint) subtrees $T_1, \ldots, T_t$. Let b_i be the number of binary tree nodes in T_i, $i = 1 \ldots t$, with $\sum_{i=1}^{t} b_i = b$. If we fix an evaluation A_i for T_i, then there remain at most 2^d different contiguous evaluations for G.*

The following simple algorithm performs a complete search to create all 2^d contiguous evaluations for G, provided that a fixed contiguous evaluation (supplied by *labelfs*) for the tree nodes of G is used:

```
algorithm simple_complete_search
Let v_1, ..., v_d be the decision nodes of a DAG G.
forall 2^d different β ∈ {0,1}^d do
      start dfs(root) with each β, such that for 1 ≤ i ≤ d
      if β_i = 0 in the call dfs(v_i),
      then the left child of v_i is evaluated first
      else the right child of v_i is evaluated first fi od
end simple_complete_search;
```

This algorithm has exponential running time, since a DAG with n nodes can have up to $d = n - 2$ decision nodes [9]. The running time of the algorithm can be reduced by exploiting the following observation (consider the example DAG in Fig. 1): Assume that the algorithm to generate a contiguous evaluation decides to evaluate the left child f of the root h first (i.e., the decision bit of h is set to zero). Then node e appears in the evaluation before g, since e is in the subDAG of f, but g is not. Therefore, there is no real decision necessary when node g is evaluated, because the child e of g is already evaluated. But because g is a decision node, the algorithm generates bitvectors containing 0s and 1s for the decision bit of g, although bitvectors that only differ in the decision bit for g describe the same evaluation.

We say that g is *excluded* from the decision by setting the decision bit of h to 0, because the child e (and c) are already evaluated when the evaluation of g starts. We call the decision bit of g *redundant* and mark it by an asterisk (∗).

The following algorithm computes only those bitvectors that yield different evaluations. We suppose again that tree nodes are evaluated by *labelfs*. Fig. 1 shows the application of *descend* to an example DAG.

```
Let v_1, ..., v_d be the decision nodes in reverse topological order (the root comes first)
We call descend(Θ, 1) where Θ is a bitvector that contains d 0's.
function descend ( bitvector β, int pos )
while β_pos = * and pos < d do pos ← pos + 1 od
if pos ≥ d
then if β_pos = * then print β   // new evaluation found //
      else β_d = 0; print β;  β_d = 1; print β;   // 2 new evaluations found //   fi
else β_pos = 0;
      mark exclusions of nodes v_j, j ∈ {pos + 1, ..., d} through lchild(v_pos) by β_j ← *;
      descend( β, pos + 1);
      β_pos = 1;
      mark exclusions of nodes v_j, j ∈ {pos + 1, ..., d} through rchild(v_pos) by β_j ← *;
      descend( β, pos + 1);
fi end descend;
```

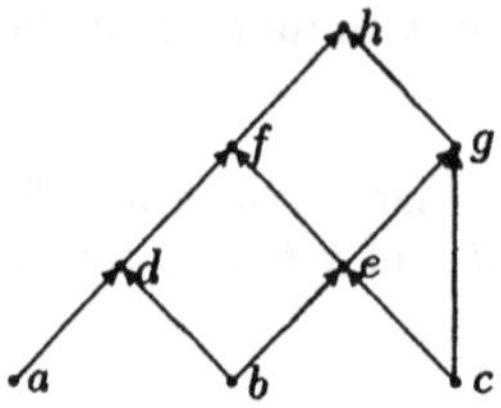

Fig. 1. Example DAG. For this DAG, algorithm *descend* executes the above evaluation steps. Only 7 instead of $2^5 = 32$ contiguous evaluations are generated.

decision nodes $v_1, v_2, \ldots, v_5$:	h	f	g	d	e	
start at the root: preset first bit	**0**		*			
propagate bits and asterisks to next stage:	0	**0**	*		*	
all bits set: first evaluation found	0	0	*	**0**	*	A_1
	0	0	*	**1**	*	A_2
'backtrack':	0	**1**	*	*		
	0	1	*	*	**0**	A_3
	0	1	*	*	**1**	A_4
'backtrack':	**1**	*		*		
	1	*	**0**	*		
	1	*	0	*	**0**	A_5
	1	*	0	*	**1**	A_6
'backtrack':	1	*	**1**	*	*	A_7

4 Reducing the Number of Evaluations

We now construct an algorithm that reduces the number of generated evaluations further. The reduction is based on the following observation: Let v be a decision node with two children v_1 and v_2. Let $G(v) = (V(v), E(v))$ be a DAG with root v, $G(v_i)$ the complete subDAG with root v_i, $i = 1, 2$. By deciding to evaluate v_1 before v_2, we decide to evaluate all nodes of $G(v_1)$ before the nodes in $G_{rest} = (V_{rest}, E_{rest})$ with $V_{rest} = V(v) - V(v_1), E_{rest} = E(v) \cap (V_{rest} \times V_{rest})$. Let $e = (u, w) \in E(v)$ be an edge with $u \in V(v_1), w \in V_{rest}$. The function *descend* marks w with a $*$. This can be considered as eliminating e: at decision node w, we do not have the choice to evaluate the child u first, because u has already been evaluated and will be held in a register until w is evaluated. Therefore, *descend* can be considered as splitting the DAG G into smaller subDAGs. We will see later that these subDAGs are trees after the splitting has been completed. The root of each of these trees is a decision node.The trees are evaluated in reverse of the order in which they are generated. For the example DAG of Fig. 1, there are 7 possible ways of carrying out the splitting. The splitting steps that correspond to evaluation A_1 from Fig. 1 are shown in Fig. 2.

If we look at the subDAGs that are generated during the splitting operation, we observe that even some of the intermediate subDAGs are trees which could be evaluated without a further splitting. E.g., after the second splitting step ($\beta_2 = 0$) in Fig. 2, there is a subtree with nodes a, b, d which does not need to be split further, because an optimal contiguous evaluation for the subtree can be found by a variant of *labelfs*.. By stopping the splitting operations in these cases, the number of generated evaluations can be reduced from 7 to 3 for the example DAG.

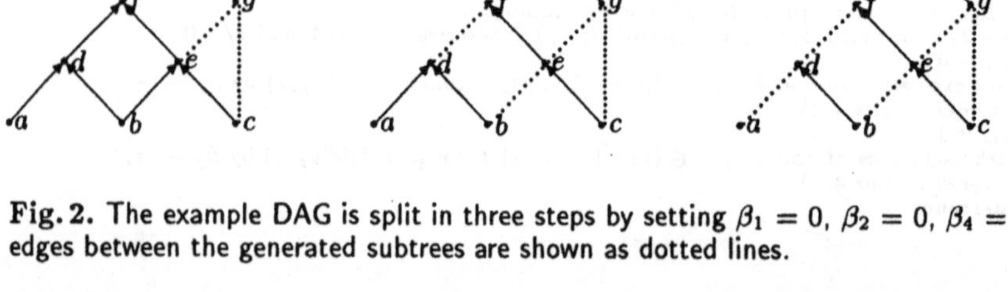

Fig. 2. The example DAG is split in three steps by setting $\beta_1 = 0$, $\beta_2 = 0$, $\beta_4 = 0$. The edges between the generated subtrees are shown as dotted lines.

Depending on the structure of the DAG, the number of generated evaluations may be reduced dramatically when splitting the DAG into trees. To evaluate the generated trees we need a modified labeling algorithm *labelfs2* that is able to cope with the fact that some nodes of the trees must be held in a register until the last reference from any other tree is resolved. Before applying the new labeling algorithm, we explicitly split the DAG in subtrees $T_1 = (V_1, E_1), \ldots, T_k = (V_k, E_k)$. We suppose that these subtrees must be evaluated in this order. The splitting procedure is described in detail in the next section. After the splitting, we introduce additional import nodes which establish the communication between the trees. The resulting trees to the second DAG in Fig. 2 are given in Fig. 3.

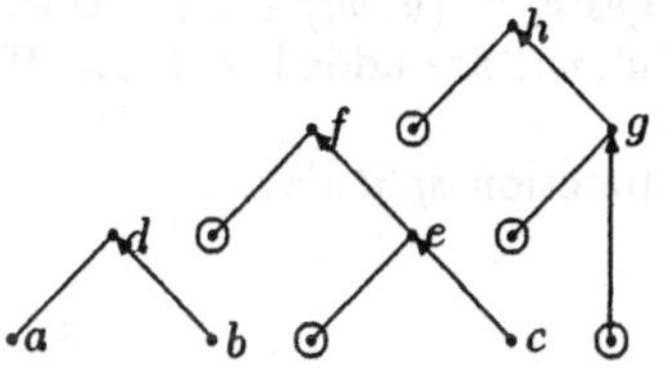

Fig. 3. The example DAG is split into 3 subtrees by setting $\beta_1 = 0, \beta_2 = 0, \beta_4 = 0$. The newly introduced import nodes are marked with a circle. They are all non-permanent.

We present *labelfs2* in Section 5 with the notion of import and export nodes: An *export node* of a tree T_i is a node which has to be left in a register because another tree $T_j (j > i)$ has a reference to v, i.e., T_j has an import node which corresponds to v. An *import node* of T_i is a leaf which is already in a register R because another tree $T_j (j < i)$ that has been evaluated earlier has left the corresponding export node in R. Therefore, an import node need not to be loaded in a register and does not appear again in the evaluation. For each import node, there exists a corresponding export node. Two import nodes $v_1 \neq v_2$ may have the same corresponding export node.

We distinguish between permanent and non-permanent import nodes: A *permanent* import node v can be evaluated without being loaded in a register. v can*not* be removed from the register after the parent of v is evaluated, because there is another import node of T_i or of another tree T_j that has the same corresponding export node as v and that has not been evaluated yet. A *non-permanent* import node v can also be evaluated without being loaded into a register. But the register that contains v can be *freed* after the parent of v has been evaluated, because all other import nodes that have the same corresponding export node as v are already evaluated.[4]

Let the DAG nodes be $V = V_1 \cup \ldots \cup V_k$. We describe the import and export nodes by the following characteristic functions:
$exp : V \rightarrow \{0, 1\}$, $exp(v) = 1$ if v is an export node, and 0 otherwise;
$imp_p : V \rightarrow \{0, 1\}$, $imp_p(v) = 1$ if v is a permanent import node, and 0 otherwise;
$imp_{np} : V \rightarrow \{0, 1\}$, $imp_{np}(v) = 1$ if v is a non-permanent import node, and 0 otherwise;
$corr : V \rightarrow V$, $corr(v) = u$ if u is the corresponding export node to v.
These definitions imply $exp(v) + imp_p(v) + imp_{np}(v) \leq 1$ for each $v \in V_i$.

The recursive procedure *descend2*, a modification of *descend*, generates a set of evaluations for a given DAG G by splitting G into subtrees and evaluating the subtrees with *labelfs2*. Among the generated evaluations are all optimal ones.

[4] This partitioning of the import nodes is well-defined as the order of the T_i is fixed.

Let d be the number of decision nodes. The given DAG is split into at most d subtrees to generate an evaluation. After each split operation, export nodes are determined and corresponding import nodes are introduced as follows: Let $v = v_{pos}$ be a decision node with children v_1 and v_2 and let $G(v), G(v_1)$ and G_{rest} be defined as in the previous section. We consider the case that v_1 is evaluated before v_2 ($\beta_{pos} = 0$). Let $u \in V(v_1)$ be a node for which an edge $(u, w) \in E(v)$ with $w \in V_{rest}$ exists. Then u is an export node in $G(v_1)$. A new import node u' is added to G_{rest} by setting $V_{rest} = V_{rest} \cup \{u'\}$ and $E_{rest} = E_{rest} \cup \{(u', w)\}$. u' is the corresponding import node to u. If u has already been marked in $G(v_1)$ as export node, then u' is a permanent import node, because there is another reference to u (from another tree) that is evaluated later. Otherwise, u' is a non-permanent import node. If there are other edges $e_i = (u, w_i) \in E(v)$ with $i = 1, \ldots, k$ and $w_i \in V_{rest}$, then new edges $e'_i = (u', w_i)$ are added to E_{rest}. If $k \geq 1$, G_{rest} is not a tree and will be split later on.

One splitting step is executed by the following function *split_dag*:

```
function split_dag(node v, v1, v2, dag G = (V, E)) : dag;
// v is a decision node with children v1 and v2 //
u1 = new_node(); V = V ∪ {u1}; E = E ∪ {(u1, v)};
if exp(v1) == 0 then imp_np(u1) = 1 else imp_p(u1) = 1; fi;
exp(v1) = 1; corr(u1) = v1; delete (v1, v) from E;
for each edge e = (v1, w) ∈ E do
    u1 = new_node(); V = V ∪ {u1}; E = E ∪ {(u1, w)};
    imp_p(u1) = 1; corr(u1) = v1; delete e from E; od;
Let G(v) = (V(v), E(v)) be the subDAG of G with root v, and
let G(v1) = (V(v1), E(v1)) be the subDAG of G with root v1
build G_rest = (V_rest, E_rest) with V_rest = V(v) - V(v1), E_rest = E(v) ∩ (V_rest × V_rest);
for each u ∈ V(v1) do
    if ∃w1, ..., wn ∈ V_rest with (u, wi) ∈ E(v)
    then u1 = new_node(); V = V ∪ {u1}; E = E ∪ {(u1, wi), 1 ≤ i ≤ n};
         if exp(u) == 0 then imp_np(u1) = 1 else imp_p(u1) = 1; fi
         exp(u) = 1; corr(u1) = u; delete (u, wi) from E, 1 ≤ i ≤ n; fi; od;
return the subDAG of G with root v; end split_dag;
```

new_node returns a new node x and sets $exp(x), imp_p(x)$ and $imp_{np}(x)$ to 0. *descend2* visits the decision nodes in reverse topological order (in the same way as *descend*). For each decision node v with children v_1 and v_2, *descend2* executes two possible split operations by using the complete subDAGs with roots v_1 and v_2. For each split operation, two subDAGs G_{left} and G_{right} are built. If one of these is a tree, all decision nodes in the tree are marked with a $*$ so that no further split is executed for these decision nodes. The root of the tree is stored in *roots*, a set of nodes that is empty at the beginning. If all decision nodes are computed, the trees rooted in *roots* are evaluated according to *ord* with the modified labeling scheme *labelfs2*. — To evaluate a DAG G, we start *descend2*$(\Theta, 1, G)$ where Θ is a bitvector with 0's at all positions. The decision nodes $v_1, \ldots, v_d$ are supposed to be sorted in reversed topological order (the root first).

```
function descend2 ( bitvector β, int pos, dag G )
while β_pos = * and pos ≤ d do pos = pos + 1 od;
if pos == d + 1
then ord = top_sort(roots); for i = 1 to d do labelfs2(ord(i)) od;
else β_pos = 0; G1 = copy(G);
     mark exclusions of nodes v_j, j ∈ {pos + 1, ..., d} through lchild(v_pos) with β_j = *;
     G_left = complete subDAG of G1 with root lchild(v_pos)
     if is_tree(G_left)
     then mark all decision nodes in G_left with a *; roots = roots ∪ {lchild(v_pos)} fi;
     G_right = split_dag(v_pos, lchild(v_pos), rchild(v_pos), G1);
     if is_tree(G_right)
     then mark all decision nodes in G_right with a *; roots = roots ∪ {v_pos} fi;
```

```
        descend2( β, pos + 1, G_1);
        β_pos = 1; G_2 = copy(G);
        mark exclusions of nodes v_j, j ∈ {pos + 1, ..., d} through rchild(v_pos) with β_j = *;
        G_right = complete subDAG of G_2 with root rchild(v_pos)
        if is_tree(G_right)
        then mark all decision nodes in G_right with a *; roots = roots ∪ {rchild(v_pos)} fi;
        G_left = split_dag(v_pos, rchild(v_pos), lchild(v_pos), G_2);
        if is_tree(G_left)
        then mark all decision nodes in G_left with a *; roots = roots ∪ {v_pos} fi;
        descend2( β, pos + 1, G_2);
fi end descend2;
```

By fixing the evaluation order of the trees, we also determine the type of the import nodes and thus which import nodes return a free register after their evaluation. An import node is non-permanent if it is the last reference to the corresponding export node. Otherwise it is permanent: The register cannot be freed until the last referencing import node is computed.

5 Evaluating trees with import and export nodes

The split operation executed by *descend2* yields a series of trees $T_1 = (V_1, E_1)$, ..., $T_k = (V_k, E_k)$ with import and export nodes. Now we describe how an optimal evaluation is generated for these trees. We define two functions:

$$occ : V \rightarrow \{0,1\} \qquad with \qquad occ(v) = \sum_{w \text{ is a proper predecessor of } v} exp(w)$$

counts the number of export nodes in the subtree $T(v)$ with root v (excluding v), i.e. the number of registers that remain occupied after $T(v)$ has been evaluated.

$$freed : V \rightarrow \{0,1\} \qquad with \qquad freed(v) = \sum_{w \text{ is a proper predecessor of } v} imp_{np}(w)$$

counts the number of import nodes of the second type in $T(v)$, i.e. the number of registers that are freed after $T(v)$ has been evaluated.

We now define for each node v of a tree $T_i (1 \leq i \leq k)$ a label $label(v)$ which specifies the number of registers required to evaluate v as follows: If v is a leaf, then $label(v) = 2 - 2 \cdot (imp_p(v) + imp_{np}(v))$. Otherwise, let v be an inner node with two children v_1 and v_2. Let S_i be the subtree with root $v_i, i = 1, 2$. We have two possibilities to evaluate v, when we use contiguous evaluations: If we evaluate S_1 before S_2, we use $m_1 = \max(label(v_1), label(v_2) + occ(v_1) + 1 - freed(v_1))$ registers, provided that v_1 (v_2) can be evaluated with $label(v_1)$ ($label(v_2)$) registers. After S_1 is evaluated, we need $occ(v_1)$ registers to hold the export nodes of S_1 and one register to hold v_1. On the other hand, we free $freed(v_1)$ registers, when evaluating S_1. If we evaluate S_2 before S_1, we use $m_2 = \max(label(v_2), label(v_1) + occ(v_2) + 1 - freed(v_2))$ registers. We suppose that the best evaluation order is chosen by *labelfs2* and set $label(v) = \min(m_1, m_2)$.[5]

Theorem 10. *labelfs2 generates a contiguous evaluation that uses no more registers than any other contiguous evaluation.*

---

[5] Until now, we have assumed that two different import nodes of a tree T_i have different corresponding export nodes. We now explain what has to be done if this is not true. Let $A = \{w_1, \ldots, w_n\} \subseteq V_i$ be a set of import nodes of T_i with the same corresponding export node that is stored in a register r. As described above we have set $imp_p(w_1) = \ldots = imp_p(w_n) = 1$ and $imp_{np}(w_1) = \ldots = imp_{np}(w_n) = 0$. But r can be freed, after the last node of A is evaluated. By choosing an appropriate node $w \in A$ to be evaluated last, T_i eventually can be evaluated with one register less than the label of the root specifies. We determine w by a top–down traversal of T_i.

n	d	N_{simple}	$N_{descend}$	$N_{descend2}$
24	12	4096	146	5
25	14	16384	1248	3
27	17	131072	744	15
28	19	524288	630	32
33	21	2097152	1148	98
36	24	16777216	2677	312
39	27	134217728	1280	358
42	29	536870912	6072	64
42	31	2^{31}	2454	152
46	34	2^{34}	4902	707
54	39	2^{39}	30456	592
56	43	2^{43}	21048	4421

n	d	N_{simple}	$N_{descend}$	$N_{descend2}$
20	14	16384	160	10
28	16	65536	784	8
30	21	2097152	1040	64
37	23	8388608	13072	24
38	24	16777216	11924	56
45	27	134217728	100800	18
41	29	536870912	74016	364
41	31	2^{31}	3032	142
44	33	2^{33}	40288	435
46	34	2^{34}	40244	1008
48	37	2^{37}	21488	1508
53	42	2^{42}	79872	3576

Table 1. Some examples from a test series for large random DAGs. The number of contiguous evaluations generated by the algorithms *simple*, *descend* and *descend2* are given for typical examples. The tests confirm the large improvements of *descend* and *descend2*.

6 Experimental Results

We have implemented *descend* and *descend2* and have applied them to a great variety of randomly generated test DAGs with up to 150 nodes and to large DAGs taken from real application programs, see Tables 1 and 2. We observed:

- *descend* reduces the number of different contiguous evaluations considerably.
- *descend2* often leads to a large additional improvement over *descend*, especially for DAGs where *descend* is not so successful in reducing the number of different contiguous evaluations. *descend2* works even better for DAGs from real application programs than for random DAGs.
- Only one of the considered DAGs with $n \leq 25$ nodes has a non-contiguous evaluation that uses fewer registers than the computed contiguous one.[6]
- In almost all cases, the computational effort of *descend2* seems to be justified. This means that, *in practice*, an *optimal* contiguous evaluation (and thus, contiguous register allocation) can be computed in acceptable time even for large DAGs.

7 Conclusions

We have presented two variants of the simple algorithm that evaluates only the tree nodes by a labeling algorithm and generates 2^d contiguous evaluations

Let v be an inner node of T_i with children v_1 and v_2. Let S_j be the subtree with root v_j, $j = 1, 2$. If only one of S_1 and S_2 contains nodes of A, we descend to the root of this tree. If both S_1 and S_2 contain nodes of A, we examine, whether we can decrease the label value of v by choosing S_1 or S_2. Let be $a = label(v_1) + occ(v_2) - freed(v_2)$ and $b = label(v_2) + occ(v_1) - freed(v_1)$ If $a > b$, this can only be achieved by searching w in S_1. If $a < b$, this can only be achieved by searching w in S_2. If $a = b$, we cannot decrease the register need and can search in S_1 or S_2. — We repeat this process until we reach a leaf $w \in A$. We set $imp_p(w) = 0$, $imp_{np}(w) = 1$.

6 For a subDAG of MDG with $n = 24$ nodes, there is a non-contiguous that uses 6 registers. The computed contiguous evaluation takes 7 registers. The program to compute the non-contiguous evaluation has run for about 7 days, the corresponding program for the contiguous evaluation took less than 0.1 seconds. For DAGs with $n > 25$ nodes it is not possible to compute the best non-contiguous evaluation because of the runtime of the program that computes them is growing too fast.

where d is the number of decision nodes of the DAG. The first variant excludes redundant decision nodes. The second variant splits the DAG into subtrees and evaluates these by a modified labeling algorithm. The experimental results confirm that this variant generates only a small number of contiguous evaluations; thus, this method finds the optimal contiguous evaluation in a reasonable time even for large DAGs. making it suitable for usage in optimizing compilers, especially for time–critical regions of the source program.

References

1. Aho, A.V., Johnson, S.C.: *Optimal Code Generation for Expression Trees*, J. ACM **23:3**, 1976, 488–501
2. Aho, A.V., Sethi, R., Ullman, J.D.: *Compilers: Principles, Techniques, and Tools.* Addison–Wesley, 1986
3. Chaitin, G.J., Auslander M.A., Chandra A.K., Cocke J., Hopkins M.E., Markstein P.W.: *Register allocation via coloring.* Computer Languages Vol. **6**, 47–57, 1981
4. Chaitin, G.J.: *Register allocation & spilling via graph coloring.* ACM SIGPLAN Notices **17:6**, 201–7, 1982
5. *The Connection Machine CM–5 Technical Summary*, Thinking Machines Corporation, Cambridge, MA, 1991
6. Dongarra, J.J., Jinds, A.R.: *Unrolling Loops in Fortran*, Software Practice and Experience, **9**:3, 219–26, 1979
7. Fisher, J,: *Trace Scheduling: A Technique for Global Microcode Compaction*, IEEE Transactions on Computers, **C–30**:7, 1981
8. Goodman J.R., Hsu Wei–Chung: *Code Scheduling and Register Allocation in Large Basic Blocks*, ACM Int. Conf. on Supercomputing, 1988, 442–52
9. Keßler, C.W., Paul, W.J., Rauber, T.: *A Randomized Heuristic Approach to Register Allocation.* Proc. of PLILP'91 3rd Int. Symp. on Programming Language Implem. and Logic Programming, Aug. 26–28, 1991. Springer LNCS 528, 195–206
10. Keßler, C.W., Paul, W.J., Rauber, T.: *Scheduling Vector Straight Line Code on Vector Processors.* in: R. Giegerich, S.L. Graham (Ed.): Code Generation — Concepts, Tools, Techniques. Springer Workshops in Computing Series (WICS), 1992
11. Mehlhorn, K.: *Data Structures and Algorithms 2.* Springer, 1984
12. Rauber, T.: *An Optimizing Compiler for Vector Processors.* Proc. ISMM Int. Conf. on Par. and Distr. Computing and Systems, New York 1990, Acta press, 97–103
13. Sethi, R., Ullman, J.D.: *The generation of optimal code for arithmetic expressions.* J. ACM, Vol. **17**, 715–28, 1970
14. Sethi, R.: *Complete register allocation problems.* SIAM J. Comput. **4**, 226–48, 1975

Source	DAG	n	d	N_{simple}	$N_{descend}$	$N_{descend2}$	$T_{descend}$	$T_{descend2}$
LL 14	second loop	19	10	1024	432	18	0.1 sec.	< 0.1 sec.
LL 20	inner loop	23	14	16384	992	6	0.2 sec.	< 0.1 sec.
MDG	calc. $\cos(\theta), \sin(\theta), \ldots$	26	15	32768	192	96	< 0.1 sec.	< 0.1 sec.
MDG	calc. forces, first part	81	59	2^{59}	—	7168	—	13.6 sec.
	subDAG of this	65	45	2^{45}	—	532	—	0.9 sec.
	subDAG of this	52	35	2^{35}	284672	272	70.2 sec.	0.8 sec.
	subDAG of this	44	30	2^{30}	172032	72	42.9 sec.	0.3 sec.
SPEC77	mult. FFT analysis	49	30	2^{30}	131072	32768	20.05 sec.	21.1 sec.

Table 2. Some measurements for DAGs taken from real programs (LL = Livermore Loop Kernels; MDG = Molecular Dynamics, and SPEC77 = atmospheric flow simulation, both from the Perfect Club Benchmark Suite). The run times of the algorithms *descend* and *descend2*, implemented on a SUN SPARC station SLC, show that for large DAGs *descend* is too slow, but the run times required by *descend2* remain really acceptable.

Communication as Unification in the Petri Box Calculus

Hanna Klaudel[1] and Elisabeth Pelz[1,2]

[1] Université Paris-Sud, L.R.I. bât. 490, F-91405 Orsay
[2] Université Paris-Val-de-Marne, Av. du G^l. de Gaulle, F-94010 Créteil

Abstract. A way of handling abstract data types in the Petri Box Calculus [BDH92] is proposed. Semantic objects are A-nets, a kind of algebraic nets with Boxe like interfaces. The synchronization operation on A-nets is discussed and some basic (non-trivial) algebraic properties of the synchronization with syntactic unification are shown.

1 Introduction

The Petri Box Calculus (PBC) is a formalism developed in [BDH92, BDE93] in order to apply the Petri net theory to the verification of concurrent algorithms and also to address the compositional semantics of languages needed to express them. The PBC syntax yields *box expressions*; their compositional semantics is given by classes of labeled place/transition nets, called *Boxes*. Boxes are nets with two kinds of interface: an entry/exit interface (places) and a communication interface (transitions). Boxes can be composed with each other across these interfaces. Furthermore, the PBC specific action structure allows partial synchronizations between multisets of actions.

PBC was applied to the semantics of parallel languages, and in particular of $B(PN)^2$ (Basic Petri Net Programming Notation) introduced in [BH93] for the specification of concurrent algorithms. A formal semantics in terms of Boxes has been proposed in [BH93] by associating a box expression to every meaningful sub-construct of a concurrent $B(PN)^2$ program, and in turn, providing a compositional Petri net semantics [BDH92]. Anyway, the size of the nets obtained was a problem, which has been solved in [BF$^+$95a] by proposing the algebra of M-nets in order to provide a compositional high level semantics for $B(PN)^2$ programs [BF$^+$95b]. M-nets are a fairly powerful Petri net model; however the lack of handling complex data structures is a serious drawback if one wants to use them as a semantic domain for a real programming language.

Our aim is to generalize M-nets by giving them a full abstract data type orientation. In fact, concurrent systems often use various, sometimes complex data structures, whose abstract formal specification is most appropriately expressed by an algebraic specification [AR91]. This approach provides a formal way of describing properties of the values and operations, independently of any particular implementation [GTW78]. It also allows to reason about programs using such data types and to prove implementations.

We propose a model, called A-nets, based on algebraic nets as proposed by Vautherin in [Vau87]. Algebraic nets are a version of colored nets where the tokens of different color are represented by elements of different sorts in the initial model of the specification of an abstract data type, and the structure of the net is given as usual. We define an algebra of A-nets, with operators similar to those of PBC and M-nets. Here we stress on synchronization, a crucial operation, which is based on a mechanism of syntactic unification. Basic, non-trivial, algebraic properties of this synchronization are shown. Future work will include the application of the A-net algebra (extended by refinement and recursion, see [DK95]) as semantic counterpart of parallel languages, and in particular for some extensions of $B(PN)^2$ including complex data structures.

2 Abstract data types, terms and substitutions

An *abstract data type*, shortly adt, can be considered as a many-sorted algebraic theory which is given by a *presentation* (S, F, E, X), including: a set of *sorts* (type names) S; a *signature* F, i.e. a set of operation (function) names with their corresponding typed arities $ar : F \rightarrow S^+$; and a set of *axioms* E, i.e. equations between terms composed with operations and free variables from a set of typed variables X.

The sets $\mathbf{T} = \bigcup_{s \in S} \mathbf{T}_s$ of terms with variables, and $\mathbf{G}$ of *ground* terms are defined inductively as usual.

A *model* of the adt is a many-sorted algebra D which satisfies each axiom in E. In particular, $\mathbf{G}/_E$, the quotient algebra of $\mathbf{G}$ modulo the smallest congruence generated by the set of axioms E, is a model of the adt. Moreover, it is initial in the class of all its models.

We consider finite multisets m over a set A as functions $m : A \rightarrow N$, such that the set $\{a \in A \mid m(a) \neq 0\}$ is finite. We denote by $\mathcal{M}_f(A)$ the set of all finite multisets over A, and by $\subseteq_{\mathcal{M}}$ the multiset inclusion[3]. Let $m : A \rightarrow N$ be a finite multiset and let $\sim$ be an equivalence relation on A. We define a finite multiset $m/_\sim : A/_\sim \rightarrow N$, by setting $m/_\sim([a]) = \sum_{a' \in [a]} m(a')$. Notice that $m/_\sim$ is indeed a finite multiset. Let us call $\sim_E$ the smallest congruence relation on $\mathbf{T}$ generated by the set of axioms E. Then, two finite multisets m and n over $\mathbf{T}$ are said to be *E-equal*, denoted $m =_E n$, if $m/_{\sim_E} = n/_{\sim_E}$. Sometimes, we use for multisets an extended set notation (e.g. $\{a, a, b, c, c, c\} = m$ for $m(a) = 2$, $m(b) = 1$, $m(c) = 3$, and $m(x) = 0$ for each $x \in A \setminus \{a, b, c\}$); also when no confusion is possible, we will represent singleton multisets by its unique element (allowing to write a instead of $\{a\}$). The symbols $\oplus$ and $\ominus$ are used then for multiset sum and difference, respectively.

We will use standard definitions for renamings and substitutions. A *substitution* is an expression $[x_1/t_1, \ldots, x_n/t_n]$ where all x_i are distinct variables, all t_i are terms of corresponding sort, and for no i $t_i = x_i$. Formally, substitutions are mappings (almost everywhere identity) from the set of variables X to $\mathbf{T}$,

[3] $\forall m, n \in \mathcal{M}_f(A) \;\; [m \subseteq_{\mathcal{M}} n \Leftrightarrow \forall a \in A \;\; m(a) \leq n(a)]$.

extended in a canonical way to terms, vectors, sets and multisets of terms. A substitution is called *ground* if its co-domain is **G**. The composition of substitutions is denoted by juxtaposition and for any object[4] B and any substitutions η and θ : $B(\eta\theta) = (B\eta)\theta$. A particular case of substitution is a *renaming* $[x_1/t_1, \ldots, x_n/t_n]$ where all t_i are distinct variables. A substitution θ is called a *unifier* of two objects B_1 and B_2 if $B_1\theta = B_2\theta$. A *most general unifier* (mgu) for B_1 and B_2 is a unifier σ such that every unifier of B_1 and B_2 is of the form $\sigma\theta'$ for some substitution θ'. Also, if B_1 and B_2 are unifiable then an mgu of B_1 and B_2 exists and is unique up to renaming.

We call an *equation* any pair (u, v) of terms of the same sort. A *system of equations* is a finite set $\{(u_k, v_k) \mid 1 \leq k \leq n\}$ of equations (for instance, the one given by the set of axioms of the used adt). If $\vec{u} = u_1, \ldots, u_n$ and $\vec{v} = v_1, \ldots, v_n$ are two vectors of terms of the same length, then we denote by $E(\vec{u}, \vec{v})$ the equation system $\{(u_k, v_k) \mid 1 \leq k \leq n\}$. A solution of a system of equations E is a substitution (unifier) σ such that $u_k\sigma$ and $v_k\sigma$ for $1 \leq k \leq n$ are equal up to renaming. If σ is a solution of E and it is the most general, we write $\sigma = mgu(E)$. Obviously, if E has a solution, then the mgu of E exists.

3 Nets: syntax, semantics and equivalences

Let (S, F, E, X) be a presentation of an adt. Suppose that *bool* $\in S$ and a constant *true* $\in F_{bool}$. We consider a set of action symbols (port names) A with a typed arity function, which will allow to handle parameterized actions, $ar: A \rightarrow S^*$, and a *conjugation* bijection $^-: A \rightarrow A$, such that for all $a \in A$ we have $\bar{a} \neq a$, $\bar{\bar{a}} = a$ and $ar(a) = ar(\bar{a})$. We also define $A_w = ar^{-1}(w)$ with $w \in S^*$; a particular case is the set $A_\emptyset$ of action symbols whose typed arities are the empty word: $A_\emptyset$ coincides with the set of (low-level) PBC action names.

The action symbols whose arities are non zero can take parameters which are terms of appropriate sort from the many sorted term algebra **T**. The set

$$A_{HL} = A_\emptyset \cup \bigcup_{w \in S^+} \{(a; \vec{u}) \mid a \in A_w, w = s_1 \cdots s_n, \vec{u} \in \mathbf{T}_{s_1} \times \ldots \times \mathbf{T}_{s_n}\}$$

is the set of *parameterized actions*.

We consider a kind of algebraic Petri nets, called A-nets. Their main difference with respect to the algebraic nets defined in [Vau87, Rei91] is a particular labeling on places and transitions due to the desired correspondence with the PBC model.

Definition 3.1 An A-net N is a triple $(P, T; \lambda)$ where P is a set of places, T is a set of transitions, $(P \times T) \cup (T \times P)$ is the set of arcs, and λ is a labeling function on places, transitions and arcs.

[4] An *object* is any variable, term, vector, set or multiset of terms which may be substituted.

1. $\forall p \in P\ \lambda(p) = (\alpha_p, \beta_p) = (\mathbf{status}, \mathbf{sort})$ where $\alpha_p \in \{e, \emptyset, x\}$ and $\beta_p \in S$; as in the PBC, the places can have one of the three allowed statuses: *entry* - label e, *internal* - label $\emptyset$, or *exit* - label x;
2. $\forall t \in T\ \lambda(t) = (\alpha_t, \beta_t) = (\mathbf{label}, \mathbf{condition})$ where α_t is a finite multiset of parameterized actions, i.e. $\alpha_t \in \mathcal{M}_f(A_{HL})$, and β_t is a finite set of boolean terms (predicates), i.e. $\beta_t \subseteq \mathbf{T}_{bool}$;
3. $\forall t \in T, \forall p \in P$ the arc inscriptions are finite multisets of terms with variables of adequate sort, i.e. $\lambda((t,p)), \lambda((p,t)) \in \mathcal{M}_f(\mathbf{T}_{\beta_p})$. ◇

A concrete semantics of A-nets depends on a particular model D of the used abstract data type. For instance, let us take $D = G/_E$. A marking of an A-net $N = (P, T, \lambda)$ associates to each place $p \in P$ a multiset of ground terms of its sort β_p: $m(p) \in \mathcal{M}_f((\mathbf{G})_{\beta_p})$. Its behavior is defined by giving the usual transition rule for algebraic nets [Vau87] or [Rei91], as follows:

A transition $t \in T$ is enabled at marking m if there exists a (sort preserving) ground substitution[5] $\sigma : X \to \mathbf{G}$ such that for each $b \in \beta_t\sigma\ \ b =_E true$, and for each $p \in P$ there exists $U_p \subseteq_{\mathcal{M}} m(p)$ such that $U_p =_E (\lambda(p,t))\sigma$. If t is enabled at m, each such U_p and σ leads to a marking m' resulting from the occurrence of t at m as follows:

$$m'(p) = m(p) \ominus U_p \oplus (\lambda(t,p))\sigma.$$

For our purpose, we restrict the class of A-nets to those which satisfy usual (for PBC) constraints concerning their structure and labeling of entry/exit interfaces, see [BDH92]. In particular, it is required that there is at least one entry and at least one exit place, and no in-coming arcs to entry places nor out-going arcs from exit places. The transitions are required to have at least one input place and at least one output place. The *sort* of entry and exit places must be of a particular kind, i.e. of sort "token", as in the case of M-nets [BF+95a]. The motivation of the above is the coherence with PBC requirements.

We may consider some equivalences identifying in a structural way A-nets with the same behavior. We may take, for instance, net *isomorphism* allowing to change the *name* (or *identity*) of nodes. Another possibility is *renaming equivalence* allowing to consistently change the names of the term variables occurring in the inscriptions of each transition and in the arcs going in and out of it, since these variables only have a local meaning. But also *duplication/renaming equivalence* may be considered (denoted $\equiv_T$) allowing to add/remove duplicate transitions, i.e., with the same, up to renaming, inscriptions and connectivity.

We will denote by $var(t)$ the set of variables occurring in the inscriptions of a transition t and in the arcs going in and out of it, i.e., in the whole *area* of t (following the terminology of [KP95]). The symbol $\approx$ will be used for the *renaming equivalence* on transitions.

[5] σ extends to $\mathbf{T} \to \mathbf{G}$

4 Algebra of A-nets

We can distinguish two kinds of operators: those, like sequence, parallel composition, choice, iteration, which deal with the entry/exit interface - their definitions are analogous to those in the M-net model, see [BF+95a, KP95] and also [DK95]; and those, like synchronization and restriction, which deal with the communication interface - their definitions should support parameterized actions which are more general w.r.t. those of the M-net model.

Thus, the restriction operation being particularly easy to generalize, we will principally stress on the crucial definition of synchronization.

4.1 Synchronization

The intuitive idea behind the synchronization operator is to group multisets of transitions and to connect them through communication links. The transitions then yield new ones, which can still take part to other synchronizations, even during the same synchronization operation.

The definition of synchronization is given in two steps. First, we introduce a general *schema* of synchronization on the net level. More precisely, we propose two versions: a symmetrical one (Def. 4.1), and an asymmetrical one (Def. 4.2). They have to be completed by a *basic synchronization*, determining the condition when two high level transitions can give rise to a synchronization link (Def. 4.3).

We will see later that the two synchronization schemas (4.1 and 4.2) can be shown equivalent, cf. Theorem 1. Each of them has its own advantage for the proofs we need to establish (not only for A-nets, but also for M-nets): The symmetrical definition will be essential for the proof of commutativity of synchronization (Theorem 2) while the asymmetrical one is more amenable for induction arguments, as used in the inductive proof of coherence of unfolding for M-nets [BF+95a].

Definition 4.1 (Symmetrical synchronization)
Let $N = (P, T, \lambda)$ be an A-net and a an action symbol. The net $N' = N\mathbf{sy}\ a = (P, T', \lambda')$ is the smallest[6] A-net having the same set of places as N, with the same inscriptions, and satisfying:

1. Every transition of N is also in N', with the same surrounding arcs.
2. If t_1 and t_2 are transitions of N', then any transition t (and its surrounding arcs) arising through a *basic synchronization* out of t_1 and t_2 over a is also in N'. ◇

Definition 4.2 (Asymmetrical Synchronization)
Let $N = (P, T, \lambda)$ be an A-net and a an action symbol. The net $N' = N\mathbf{sy}'\ a = (P, T', \lambda')$ is the smallest A-net having the same set of places as N, with the same inscriptions, and satisfying:

1. Every transition of N is also in N', with the same surrounding arcs.

[6] In the sense of inclusion of the transition sets of $\equiv_T$-equivalent A-nets.

2. If t_1 is a transition of N and t_2 a transition of N', then any transition t (and its surrounding arcs) arising through a *basic synchronization* out of t_1 and t_2 over a is also in N'. ◇

In the low level model, where a value domain and an interpretation are fixed, the condition to create a synchronization link between two transitions t_1 and t_2 is clear: it is only required that the labels of t_1 and t_2 contain a conjugate pair of actions. Consider, for instance, in the low level model, the action a_1 whose meaning could be "the value of a is set to 1". Let t_1 and t_2 be two low-level transitions. If $(\lambda(t_1))(a_1) \geq 1$ and $(\lambda(t_2))(\bar{a}_1) \geq 1$, then t_1 and t_2 can synchronize w.r.t. the action a_1.

In the high level approach, because of parameters, it is more difficult to determine when two transitions can synchronize. An analogous example in the high level model is that of two transitions t_1 and t_2 with parameterized actions $(a; u_1)$ in the label of t_1, and $(\bar{a}; u_2)$ in the label of t_2 where u_1, u_2 are terms with variables (assuming that $ar(a) = 1$). The problem now consists in determining the conditions under which a communication link can be created (in other words, when a *basic synchronization* of t_1 and t_2 is allowed). One could desire to allow an arbitrary interpretation, or on the contrary, argue only for a fixed one, or favor a syntactical view. As a consequence, at least four different view points on basic synchronization (BS) can be considered:

• **Semantic BS** consists in taking a model of the adt, and synchronizing if the evaluation of u_1 and u_2 may give the same element in the model.

As consequence, for a given model, only expected transitions are created. But in general, we could obtain by synchronization different net "skeletons" for different models. Thus, the main drawback of this approach is the lack of abstraction.

• **Greedy BS** consists in only requiring corresponding action symbols to be conjugate, but nothing about terms. Formally, this can be expressed by adding a new condition (like, for instance, $u_1 = u_2$) which is satisfied if u_1 and u_2 evaluate to the same element in the model.

This approach satisfies the abstraction feature; and indeed, it is worthwhile to consider that an A-net has a meaning without a particular choice of interpretation and value domain. Otherwise, we would fall back on a semantical synchronization.

On the other hand, it is clear that all transitions for all models are anticipated, but unfortunately most of them (often infinitely many) will be "dead" (the condition being always false) for almost all models. In particular, if (at least) two conjugate parameterized actions occur in the label of some transition, this synchronization would provide an infinite (w.r.t. the number of transitions) net. From a theoretical point of view (because of this explosion phenomenon) we should avoid this alternative. But, in practice (at least in the case of the $B(PN)^2$ semantics) this approach could be satisfactory.

• **BS with semantic unification** consists in requiring u_1 and u_2 to unify modulo the set of axioms E of the adt.

This approach is particularly interesting when the intended model is the quotient algebra $\mathbf{G}/_E$ (or any isomorphic one). In that case, exactly the expected transitions are created. For instance, for the usual arithmetic axioms

$$E = \{\ldots, *(succ(x), y) = +(y, *(x, y)),\ *(x, succ(0)) = x, \ldots\}$$

the terms $u_1 = +(succ(0), succ(0))$ and $u_2 = *(succ(succ(0)), succ(0))$ unify, i.e., $u_1 =_E u_2$, and give rise to a communication link. But very often such a unification is non effective or even undecidable. This is a serious drawback because it means that this can only be used for very particular algebraic theories

• **BS with syntactic unification** consists in requiring u_1 and u_2 to unify syntactically. This approach coincides with the unification modulo the empty set of axioms.

In general, it presents some drawbacks, namely some terms which cannot syntactically unify may however evaluate to the same value, and as consequence, some expected (for a model) transitions cannot be created. For instance, u_1 and u_2 from the previous example do not unify syntactically, but evaluate to the same value for the standard interpretation of $+$ and $*$ in the naturals.

However, this approach has the advantage to be decidable and could be particularly satisfactory in at least two interesting cases:

- if the intended model is (isomorphic to) $\mathbf{G}$, or
- if the parameterized actions only involve terms being either variables or constants; notice that is the case for M-nets. Indeed, in that case the semantic unification coincides with the syntactic one. Moreover, unlike the previous approach, we are not bound to a specific model.

Thus, according to the above arguments we shall define the synchronization *syntactically*, which means with syntactic unification.

Definition 4.3 (Basic synchronization with syntactic unification)
Let t_1 and t_2 be transitions of N' as above. A transition t (and its surrounding arcs) *arises out of* t_1 *and* t_2 *through a basic synchronization over* a, if the label of t_1 contains[7] a parameterized action $(a; \vec{u_1})$ for some $\vec{u_1}$, the label of t_2 contains a parameterized action $(\bar{a}, \vec{u_2})$ for some $\vec{u_2}$, and there exist a substitution θ and two renamings ρ_1, ρ_2, such that the five following conditions hold[8]:

1. $var(t_1)\rho_1 \cap var(t_2)\rho_2 = \emptyset$
2. $\theta = mgu(E(\vec{u_1}\rho_1, \vec{u_2}\rho_2))$
3. $\forall p \in P \quad \lambda'((p,t)) = \lambda'((p,t_1))\rho_1\theta \oplus \lambda'((p,t_2))\rho_2\theta$ (sim. for $\lambda'((t,p))$)
4. $\alpha_t = (\alpha_{t_1}\rho_1\theta \oplus \alpha_{t_2}\rho_2\theta) \ominus \{(a; \vec{u_1}\rho_1\theta), (\bar{a}; \vec{u_2}\rho_2\theta)\}$
5. $\beta_t = \beta_{t_1}\rho_1\theta \cup \beta_{t_2}\rho_2\theta$ ◇

Notice that the mgu, and the pair of substitutions $\rho_1\theta, \rho_2\theta$, are unique (up to renaming) making the number of created transitions as small as possible.

[7] The label of t_1 *contains* a parameterized action $(a; \vec{u_1})$ if $\alpha_{t_1}((a; \vec{u_1})) > 0$.
[8] Note that this definition is symmetrical w.r.t. the choice of t_1 and t_2.

Notice also that Def. 4.3 is symmetrical w.r.t. the conjugation of actions. As a direct corollary we have that for every A-net N and every action name $a \in A$: $N\mathbf{sy}\ a = N\mathbf{sy}\ \bar{a}$ and $N\mathbf{sy}'\ a = N\mathbf{sy}'\ \bar{a}$.

4.2 Restriction

The restriction operation $N\mathbf{rs}\ a$ removes from an A-net N all transitions whose labels contain parameterized actions $(a; \vec{u})$ or $(\bar{a}; \vec{v})$, for any $\vec{u}$ and $\vec{v}$. One easily gets that this operation is commutative and idempotent [KP95].

5 Properties

The results mentioned in this section are basic algebraic properties of the syntactic synchronization operator on A-nets. Detailed proofs are given in [KP95], they use a common technique based on BS-trees[9], i.e., labeled binary complete trees whose nodes are labeled by transitions, edges are labeled by substitutions, and all internal nodes are typed by action symbols from A, as shown for instance in Figure 1. We will say that a BS-tree τ is *associated* to a transition t if t labels the root of τ. In that case, all internal nodes of τ correspond to basic synchronizations leading to t.

Lemma 5.1 *(local commutativity) Let $a, b \in A$, $N = (P, T; \lambda)$ be an A-net resulting from an arbitrary number of synchronizations over a and b in any order, and $t, t_1, t_2, t_3 \in T$. If τ shown in Figure 1(a) is a BS-tree associated to t, then at least one of the two following facts holds:*

1. *$\exists t' \in T$ such that the labeled tree τ' shown in Figure 1(b) is a BS-tree associated to t', and $t' \approx t$, or*
2. *$\exists t'' \in T$ such that the labeled tree τ'' shown in Figure 1(c) is a BS-tree associated to t'', and $t'' \approx t$.*

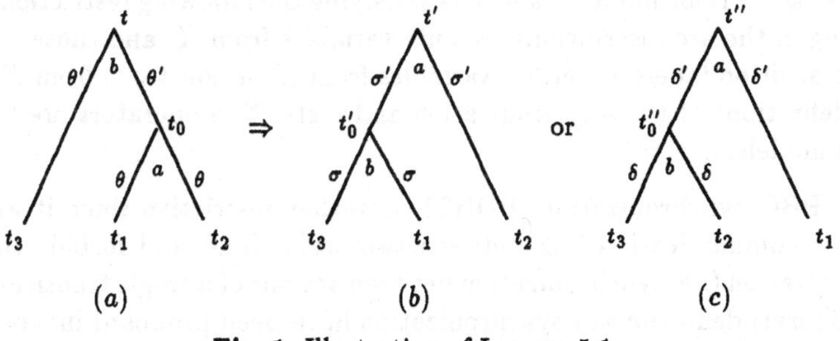

Fig. 1. Illustration of Lemma 5.1

Lemma 5.1 is crucial for the results given below. It says that if t is a transition of an A-net N coming from two successive synchronizations (first over a and second over b) of transitions t_1, t_2 and t_3 (as shown in Figure 1(a)), then there always exists a (renaming equivalent) transition t' (or resp. t'') coming from

[9] *Basic Synchronization* trees.

synchronizations done first over b and then over a. The non-trivial part of the proof consists in showing that t and t' (or resp. t'') are renaming equivalent. This follows essentially from the fact that variables in areas of transitions have only a local meaning, and that concerned (composed) substitutions (mgu's) can be shown equivalent.

Theorem 1.
The symmetrical and asymmetrical synchronization schemes are equivalent.

Proof. Let N be an A-net, and $a \in A$. The proof of Theorem 1 reduces to show that for each transition t of $N\mathbf{sy}\ a$ there exists a (renaming equivalent) transition t' in $N\mathbf{sy}'\ a$. Let τ be a BS-tree associated to t. By induction on the number of basic synchronizations, and by successively applying Lemma 5.1 (with $a = b$) we show that t and t' are renaming equivalent.

Theorem 2.
Synchronization with syntactic unification is commutative and idempotent.

Proof. Let N be an A-net, and $a, b \in A$. *Idempotence* is straightforward.
For *commutativity*, we have to show that $(N\mathbf{sy}\ a)\mathbf{sy}\ b \equiv_T (N\mathbf{sy}\ b)\mathbf{sy}\ a$. Since $(N\mathbf{sy}\ a)\mathbf{sy}\ b$ and $(N\mathbf{sy}\ b)\mathbf{sy}\ a$ have the same sets of places and inscriptions on places, proving commutativity reduces to show some property of transitions (and their areas). For each (non-trivial) BS-tree associated to a transition t of the first A-net, there exists a cut such that all nodes of the cut and above it are typed by b's and all nodes below it are typed by a's. By iterated application of Lemma 5.1 on the "frontier" between a's and b's we get the existence of a transition t' (renaming equivalent to t) whose associated BS-tree has a cut such that all nodes of the cut and above it are typed by a's and all nodes below it are typed by b's. Thus, t' is a transition of $(N\mathbf{sy}\ b)\mathbf{sy}\ a$.

Corollary 5.1 An important consequence of Theorem 2 is the commutativity and idempotence of the synchronization operator for M-nets [BF+95a], since:
- M-nets can be reformulated as A-nets satisfying the following restrictions: terms occurring in the arc inscriptions are only variables from X, and those occurring in the transition labels are either variables from X or constants from F.
- The definitions of the synchronization and restriction operators are the same in both models.

The PBC synchronization [BDH92] is rather restrictive since it allows at most one communication link between two transitions, and forbids the auto-synchronization (i.e. synchronization between actions of a single transition). Two (more liberal) definitions of synchronization have been proposed independently in [Dev93, FK94] where an arbitrary number of communication links or auto-synchronization links have been considered and the problem of finiteness has been discussed. The full paper [KP95] introduces an operator allowing auto-synchronization in the context of A-nets and it is shown that it has the expected properties of commutativity and idempotence, too.

Some other features of the A-net algebra (refinement and recursion) were developed in [DK95].

6 Concluding remarks

We proposed a way of handling abstract data types in the Petri Box Calculus, by introducing algebraic Box-like nets, called A-nets. Between various ways of defining synchronization operators on A-nets, we took an approach based on the syntactic unification of terms: it has the advantage to coincide with M-net synchronization, which has been proved coherent with the low level Boxes [BF+95a]. Also, we show that it can be defined as well symmetrically, as asymmetrically, what offers real facilities for proofs. The results of this paper are mandatory if we want to use A-nets in a compositional context.

Acknowledgment: This work is partly supported by ESPRIT WG CALIBAN, no 6067. Our thanks go to Raymond Devillers for helpful comments on this work.

References

[AR91] E. Astesiano and G. Reggio. Algebraic specification of concurrency. In *Recent Trends in Data Type specification*, LNCS 655, 1993.

[BDE93] E. Best, R. Devillers, and J. Esparza. General refinement and recursion operators for the Petri Box Calculus. In *STACS'93*, LNCS 665, 1993.

[BDH92] E. Best, R. Devillers, and J. Hall. The box calculus: a new causal algebra with multi-label communication. In *Advances in PN-92*, LNCS 609, 1992.

[BF+95a] E. Best, H. Fleischhack, W. Fraczak, R.P. Hopkins, H. Klaudel, and E. Pelz. A Class of Composable High Level Petri Nets. To appear in *ICPN-95*, Torino, LNCS, June 1995.

[BF+95b] E. Best, H. Fleischhack, W. Fraczak, R.P. Hopkins, H. Klaudel, and E. Pelz. A High Level Petri Net Semantics of $B(PN)^2$. To appear in WiC Springer, *STRICT-95*, Berlin, May 1995.

[BH93] E. Best and R.P. Hopkins. $B(PN)^2$ - A Basic Petri Net Programming Notation. In *PARLE-93*, LNCS 694, 1993.

[Dev93] R. Devillers. On a more liberal synchronization operator for the Petri Box Calculus. LIT TR 281, Université Libre de Bruxelles, May 1993.

[DK95] R. Devillers and H. Klaudel. Refinement and Recursion in a High Level Petri Box Calculus. To appear in WiC Springer, *STRICT-95*, Berlin, May 1995.

[FK94] W. Fraczak and H. Klaudel. A multi-action synchronization schema and its application to the PBC. In *ESDA-94*, ASME, July 1994.

[GTW78] J.A. Goguen, J.W. Thatcher, and E.G. Wagner. An initial algebra approach to the specification, correctness and implementation of abstract data types. In *Current Trends in Programming Methodology*, vol. 4, Prentice-Hall, 1978.

[KP95] H. Klaudel and E. Pelz. Communication as Unification in the Petri Box Calculus. LRI TR 967, Université Paris-Sud, Orsay, April 1995.

[Rei91] W. Reisig. Petri nets and algebraic specifications. *TCS*, vol. 80, 1991.

[Vau87] J. Vautherin. Parallel systems specification with colored Petri nets and algebraic specification. In *Advances in PN 1987*, LNCS 266, 1987.

Distributed Catenation and Chomsky Hierarchy

Manfred Kudlek[1] and Alexandru Mateescu[2,3] *

[1] Fachbereich Informatik, Universität Hamburg, Germany
email : kudlek@rz.informatik.uni-hamburg.de
[2] Faculty of Mathematics, University of Bucharest, Romania
[3] Academy of Finland and Department of Mathematics, University of Turku, Finland
email : mateescu@sara.utu.fumail.fi

Abstract. We introduce a new operation between words and languages, called *distributed catenation.* The distributed catenation is a natural extension of the well known catenation operation from the theory of formal languages. As for partial shuffle operation the introduction of this operation is strongly motivated by the theory of concurrency. At the same time the distributed catenation is a powerful operation. For instance, any Turing machine can be simulated by a pushdown automaton that uses distributed catenation for the pushdown memory.

1 Introduction

In [4] *partial shuffle* has been introduced as a new operation on sets of words, based on the normal shuffle operation. Here we introduce another new operation on words, *(left) distributed catenation*, similar to partial shuffle, but based on normal catenation. This operation between words, extended to sets of words, is situated between normal catenation and partial shuffle. It also defines a special parallel composition of concurrent processes.

Assume the following situation :

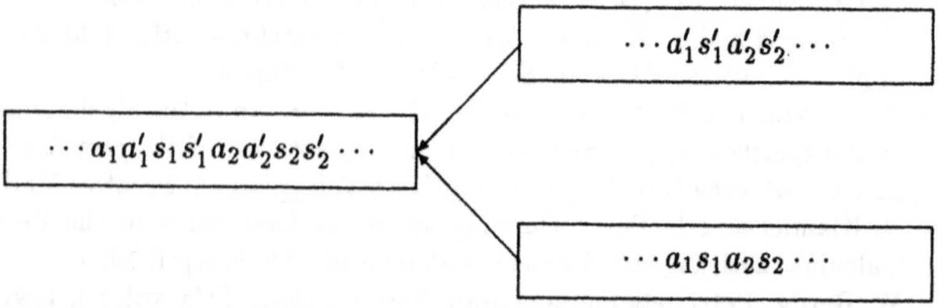

¿From two sources letters or messages with destination either Australia or South America arrive at some node where they are sent to the next node.

* Research supported by the Academy of Finland, Project 11281.

Note that at the node messages are collected by destination (with order of the sources) until messages with another destination arrive.

Throughout this paper we will omit *left* and use only ***distributed catenation***. For all unexplained notions of formal languages we refer to monograph [7].

2 Basic Properties of Distributed Catenation

We start by introducing the basic definitions of distributed catenation and considering the basic properties of that operation.

Let Σ be a finite and nonempty set, an alphabet. The empty word is denoted by λ, the set of all nonempty words over Σ by Σ^+, and the set of all words over Σ by $\Sigma^* = \Sigma^+ \cup \{\lambda\}$.

For general results on the theory of formal languages the reader may consult the monograph [7].

Let Σ be an alphabet, $\Delta \subseteq \Sigma$ and $\Gamma \subseteq \Sigma$ such that $\Sigma = \Gamma \cup \Delta$ and $\Gamma \cap \Delta = \emptyset$. Note that Γ or Δ can be empty.

Notations.
For $\Delta \subseteq \Sigma$ let

$$M_0 = \Gamma^* \quad , \quad M_{k+1} = \Gamma^*(\Delta^+\Gamma^+)^k\Delta^+\Gamma^* \cup \{\lambda\}, k \geq 0.$$

Remark 1. Note that :

(*i*) $\bigcup_{k\geq 0} M_k = \Sigma^*$.
(*ii*) if $i \neq j$, then $M_i \cap M_j = \{\lambda\}$.
(*iii*) for any $w \in \Sigma^+$, there exists a unique $k, k \geq 0$, such that $w \in M_k$.

□

Definition 2. Let w be in Σ^+. The *Δ-degree* of w is :

$$deg_\Delta(w) = k, \text{ where } w \in M_k.$$

By definition, $deg_\Delta(\lambda) = 0$. □

Comment. Note that for any word $w \in \Sigma^*$, $deg_\Delta(w)$ has a unique value (see Remark 1).

Any $w \in \Sigma^*$ can be represented in the following *canonical decomposition with respect to Δ*, or shortly, *canonical decomposition*, if Δ is clear from the context :
$w = u_0v_1u_1 \cdots v_ku_k$ with
$u_i \in \Gamma^*, i = 0, \cdots, k$ and $v_j \in \Delta^+, j = 1, \cdots, k$.

Definition 3. The length of the first Δ-block v_1 in the canonical decomposition of a word $w = u_0v_1u_1 \cdots v_ku_k \in \Sigma^*$ will be called the *Δ-length* of w, denoted by $||w||_\Delta = |v_1|$. □

Comment. Note that $||w||_\Delta = 0$ for all $w \in \Gamma^*$.

For arbitrary $k \geq 0$ a binary operation on M_k is defined by

Definition 4. If $x \in M_k$, $y \in M_k$, $x = u_0v_1u_1 \dots v_ku_k, y = u'_0v'_1u'_1 \dots v'_ku'_k$, with $u_0, u'_0, u_k, u'_k \in \Gamma^*, u_i, u'_i \in \Gamma^+, i = 1, \cdots, k-1, v_i, v'_i \in \Delta^+, i = 1, \dots, k$, then :

$$x \circ_{\Delta,k} y = (u_0u'_0)(v_1v'_1)(u_1u'_1)\cdots(v_kv'_k)(u_ku'_k).$$

By definition,

$$x \circ_{\Delta,k} \lambda = \lambda \circ_{\Delta,k} x = x.$$

$\circ_{\Delta,k}$ is called the k-Δ-distributed catenation of x with y. □

Remark 5. Note that for the special cases $\Delta = \emptyset$ and $\Delta = \Sigma$ we get the normal catenation $x \circ_{\Delta,k} y = xy$.

Trivially,

Lemma 6. For any Σ, $\Delta \subseteq \Sigma$ and $k \geq 0$, the operation $\circ_{\Delta,k}$ is associative, and the triple $\mathcal{M}_{\Delta,k} = (M_k, \circ_{\Delta,k}, \lambda)$ is a monoid. □

Proof. Assume $x, y, z \in M_k$ with canonical decompositions

$$x = u_0v_1u_1\cdots v_ku_k\ ,\ y = u'_0v'_1u'_1\cdots v'_ku'_k\ ,\ x = u''_0v''_1u''_1\cdots v''_ku''_k.$$

Then

$$(x \circ_{\Delta,k} y) \circ_{\Delta,k} z = (u_0u'_0u''_0)(v_1v'_1v''_1)(u_1u'_1u''_1)\cdots(v_kv'_kv''_k)(u_ku'_ku''_k)$$
$$= x \circ_{\Delta,k} (y \circ_{\Delta,k} z)$$

and λ is the unit element. □

The binary operation $\circ_{\Delta,k}$ can be extended to subsets of M_k in a natural way.

Let $\mathcal{P}(M_k)$ be the set of all subsets of M_k. Then a binary operation on $\mathcal{P}(M_k)$ is defined as follows :

Definition 7. Let A, B be subsets of M_k. The *distributed $k - \Delta$-catenation* or shortly, *$k - \Delta$-catenation* of A with B, is by definition :

$$A \circ_{\Delta,k} B = \bigcup_{x \in A, y \in B} x \circ_{\Delta,k} y,$$

□

Then, again :

Lemma 8. For any Σ, $\Delta \subseteq \Sigma$, and $k \geq 0$ the operation $\circ_{\Delta,k}$ is associative and the triple $\mathcal{T}_{\Delta,k} = (\mathcal{P}(M_k), \circ_{\Delta,k}, \{\lambda\})$ is a monoid. □

Moreover,

Lemma 9. $\circ_{\Delta,k}$ is distributive with $\cup$ and $\mathcal{S}_{\Delta,k} = (\mathcal{P}(M_k), \cup, \circ_{\Delta,k}, \emptyset, \{\lambda\})$ is a (noncommutative) ω-complete semiring. □

For the theory of semirings one may consult the monographs [1] and [3].

Definition 10. If $A \subseteq M_k$, $k \geq 0$, then the *Δ-catenation closure* of A is :

$$A^{\circ_{\Delta,k}} = \bigcup_{i \geq 0} A^{(i)_{\Delta,k}}, \text{ where}$$

$$A^{(0)_{\Delta,k}} = \{\lambda\} \text{ and } A^{(i+1)_{\Delta,k}} = A \circ_{\Delta,k} A^{(i)_{\Delta,k}}.$$

□

Comment. Note that $A^{\circ_{\Delta,k}}$ is the submonoid generated by A, with respect to $\circ_{\Delta,k}$, in the monoid $\mathcal{T}_{\Delta,k}$ and in the semiring $\mathcal{S}_{\Delta,k}$.

The Δ-catenation operation will now be extended to arbitrary words from Σ^*. The new operation will be denoted by $\circ_\Delta$.

Definition 11. Let x, y be in Σ^+ such that $deg_\Delta(x) = n$ and $deg_\Delta(y) = m$. Assume that

$$x = u_0 v_1 u_1 \cdots v_n u_n \ , \ y = u'_0 v'_1 u'_1 \cdots v'_m u'_m,$$

with $u_0, u'_0, u_n, u'_m \in \Gamma^*, u_i \in \Gamma^+, i = 1, \cdots, n-1, v_i \in \Delta^+, i = 1, \cdots, n,$ $u'_i \in \Gamma^+, i = 1, \cdots, m-1, v'_i \in \Delta^+, i = 1, \cdots, m.$

Then the *distributed Δ catenation* of x with y is :

$$x \circ_\Delta y = \begin{cases} (u_0 u'_0)(v_1 v'_1)(u_1 u'_1) \cdots (v_n v'_n)(u_n u'_n) v'_{n+1} u'_{n+1} \cdots v'_m u'_m, & if\ n \leq m, \\ (u_0 u'_0)(v_1 v'_1)(u_1 u'_1) \cdots (v_m v'_m)(u_m u'_m) v_{m+1} u_{m+1} \cdots v_n u_n, & otherwise. \end{cases}$$

By definition : $x \circ_\Delta \lambda = \lambda \circ_\Delta x = x$. □

Remark 12. Note that actually we have defined a *left* distributed catenation. A *right* one can be defined similarly.

For the special cases $\Delta = \emptyset$ and $\Delta = \Sigma$ we get the normal catenation $x \circ_\Delta y = xy$.

Example 13. Assume that $\Sigma = \{a, b, c, d\}$ and consider the following words x, y over Σ : $x = acdbccac$ and $y = badad$.

(i) Consider that $\Delta = \{c, d\}$ and note that $deg_\Delta(x) = 3$, $deg_\Delta(y) = 2$, $||x||_\Delta = 2$, and $||y||_\Delta = 1$. By Definition 11 we get $x \circ_\Delta y = abacddbaccdac$.

(ii) Assume now that x, y are the same, but $\Delta = \{b, c, d\}$. It follows that $deg_\Delta(x) = 2$, $deg_\Delta(y) = 3$, $||x||_\Delta = 5$, $||y||_\Delta = 1$, and $x \circ_\Delta y = acdbccbaacdad$.

(iii) As above, except that $\Delta = \{d\}$. Note that $deg_\Delta(x) = 1$, $deg_\Delta(y) = 2$, and $||x||_\Delta = ||y||_\Delta = 1$. In this case $x \circ_\Delta y = acbaddbccacad$. □

Definition 14. A word $w \in \Sigma^*$ is called *Δ-prime* if

$$w = w_1 \circ_\Delta w_2 \Rightarrow (w_1 = \lambda \vee w_2 = \lambda).$$

Let $Pr_\Delta \subseteq \Sigma^*$ denote the set of all Δ-prime words. □

Lemma 15. Let $\Delta \subseteq \Sigma$. Then the set of all Δ-prime words is given by :

$$Pr_\Delta = \{\lambda\} \cup \Sigma \cup \Delta\Gamma\Sigma^*.$$

Proof. $w = \lambda$ and $w = x \in \Sigma$ are trivial cases. Therefore assume $|w| > 1$.

(i) If $w \notin \Delta\Gamma\Sigma^*$ then either $w = xyw'$ with $x, y \in \Delta$ or $w = xw'$ with $x \in \Gamma$. In the first case $w = x \circ_\Delta yw'$, and in the second $w = x \circ_\Delta w'$, both not giving Δ-prime words. Therefore $Pr_\Delta \subseteq \Delta\Gamma\Sigma^*$.

(ii) If $w = w_1 \circ_\Delta w_2$ then there are the following possibilities for $w_1 = u_1 v_1$ and $w_2 = u_2 v_2$:

$u_1 \in \Delta^+, v_1 \notin \Delta\Sigma^*$ $v_2 \in \Delta^+, v_2 \notin \Delta\Sigma^*$ $w = u_1 u_2 w'$
$u_1 \in \Delta^+, v_1 \notin \Delta\Sigma^*$ $v_2 \in \Gamma^+, v_2 \notin \Gamma\Sigma^*$ $w = u_2 u_1 w'$
$u_1 \in \Gamma^+, v_1 \notin \Gamma\Sigma^*$ $v_2 \in \Delta^+, v_2 \notin \Delta\Sigma^*$ $w = u_1 w'$
$u_1 \in \Gamma^+, v_1 \notin \Gamma\Sigma^*$ $v_2 \in \Gamma^+, v_2 \notin \Gamma\Sigma^*$ $w = u_1 u_2 w'$

In all cases $w = w_1 \circ_\Delta w_2 \notin \Delta\Gamma\Sigma^*$.
Therefore $\Delta\Gamma\Sigma^* \subseteq Pr_\Delta$. □

Remark 16. Note that a word $w \in \Sigma^*$ usually can not be uniquely factorized into Δ-prime words with respect to $\circ_\Delta$. This can be seen from the following example :

Let $\Sigma = \{a, b, c, d\}$ and $\Delta = \{a, b\}$. Then $deg_\Delta(abcd) = 1$, $||abcd||_\Delta = 2$, and

$$abcd = a \circ_\Delta bcd = ac \circ_\Delta bd = acd \circ_\Delta b.$$

□

Lemma 17. For any Σ and $\Delta \subseteq \Sigma$ the operation $\circ_\Delta$ is associative, and the triple $\mathcal{M}_\Delta = (\Sigma^*, \circ_\Delta, \lambda)$ is a (noncommutative) monoid.

Proof. First, observe that λ is the unit element. It remains to show that $\circ_\Delta$ is an associative operation on Σ^*. Let $x, y, z \in \Sigma^+$ with $deg_\Delta(x) = i, deg_\Delta(y) = j$, and $deg_\Delta(z) = k$. Assume that x, y, z have canonical decompositions:

$$x = u_0 v_1 u_1 \cdots v_i u_i \ , \ y = u'_0 v'_1 u'_1 \cdots v'_j u'_j \ , \ z = u''_0 v''_1 u''_1 \cdots v''_k u''_k.$$

Observe that :

$$(x \circ_\Delta y) \circ_\Delta z = (u_0 u'_0 u''_0)(v_1 v'_1 v''_1)(u_1 u'_1 u''_1) \cdots (v_s v'_s v''_s)(u_s u'_s u''_s)\cdot$$

$$\cdot(v^{(1)}_{s+1} v^{(2)}_{s+1})(u^{(1)}_{s+1} u^{(2)}_{s+1}) \cdots (v^{(1)}_r v^{(2)}_r)(u^{(1)}_r u^{(2)}_r) v^{(2)}_{r+1} u^{(2)}_{r+1} \cdots v^{(2)}_t u^{(2)}_t = x \circ_\Delta (y \circ_\Delta z),$$

where $s = min(i, j, k), r = min(\{i, j, k\} - \{s\}), t = max(i, j, k)$

and, moreover :

if $i \leq j \leq k$, then $v^{(1)}_p = v'_p, u^{(1)}_p = u'_p, v^{(2)}_q = v''_q, u^{(2)}_q = u''_q$, else
if $i \leq k \leq j$, then $v^{(1)}_p = v''_p, u^{(1)}_p = u''_p, v^{(2)}_q = v'_q, u^{(2)}_q = u'_q$, else
if $j \leq i \leq k$, then $v^{(1)}_p = v_p, u^{(1)}_p = u_p, v^{(2)}_q = v''_q, u^{(2)}_q = u''_q$, else

if $j \leq k \leq i$, then $v_p^{(1)} = v_p'', u_p^{(1)} = u_p'', v_q^{(2)} = v_q, u_q^{(2)} = u_q$, else
if $k \leq i \leq j$, then $v_p^{(1)} = v_p, u_p^{(1)} = u_p, v_q^{(2)} = v_q', u_q^{(2)} = u_q'$, else
if $k \leq j \leq i$, then $v_p^{(1)} = v_p', u_p^{(1)} = u_p', v_q^{(2)} = v_q, u_q^{(2)} = u_q$,
where $p = s+1, \ldots, r$ and $q = s+1, \ldots, t$.
It follows that $\circ_\Delta$ is an associative operation. □

Remark 18. Note that for any $x, y \in \Sigma^*$:
$deg_\Delta(x \circ_\Delta y) = max(\{deg_\Delta(x), deg_\Delta(y)\})$ and $||x \circ_\Delta y||_\Delta = ||x||_\Delta + ||y||_\Delta$. □

Again, $\circ_\Delta$ can be extended to $\mathcal{P}(\Sigma^*)$ by

Definition 19. Let A, B be subsets of Σ^*. The *distributed Δ-catenation* or shortly, *Δ-catenation* of A with B, is by definition :

$$A \circ_\Delta B = \bigcup_{x \in A, y \in B} x \circ_\Delta y,$$

□

Then, trivially

Lemma 20. For any Σ and $\Delta \subseteq \Sigma$, the triple $\mathcal{T}_\Delta = (\mathcal{P}(\Sigma^*), \circ_\Delta, \{\lambda\})$ is a (noncommutative) monoid, and the quintuple $\mathcal{S}_\Delta = (\mathcal{P}(\Sigma^*), \cup, \emptyset, \circ_\Delta, \{\lambda\})$ is an ω-complete (noncommutative) semiring. □

Definition 21. If $A \subseteq \Sigma^*$ then the *Δ-catenation closure* of A is :

$$A^{\circ_\Delta} = \bigcup_{i \geq 0} A^{(i)_\Delta}, \text{ where } A^{(0)_\Delta} = \{\lambda\} \text{ and } A^{(i+1)_\Delta} = A \circ_\Delta A^{(i)_\Delta}.$$

□

Comment. Note that A°_Δ} is the submonoid generated by A, with respect to $\circ_\Delta$, in the monoid $\mathcal{T}_\Delta$ and in the semiring $\mathcal{S}_\Delta$.

Notation. For any $k \geq 0$ let

$$H_k = \bigcup_{i \leq k} M_i.$$

Lemma 22. For any alphabet Σ, for any subset $\Delta \subseteq \Sigma$ and for any $k \geq 0$,

$$\mathcal{H}_{\Delta,k} = (\mathcal{P}(H_k), \cup, \emptyset, \circ_{\Delta,k}, \{\lambda\})$$

is also a (noncommutative) semiring.

Proof. This fact follows from Lemma 20. □

3 Closure under Distributed Catenation

Let REG, CF, CS, RE denote the classes of regular, context-free, context-sensitive and recursively enumerable languages, respectively.

Theorem 23. REG and CF are not closed under distributed catenation closure.

Proof. Let $\Sigma = \{a, b, c\}$ and $\Delta = \{b\}$. Then $L = \{abc\} \in \underline{REG}$, but

$$L^{\circ_\Delta} = \{a^n b^n c^n \mid n \geq 0\} \notin \underline{CF}.$$

□

Theorem 24. CF is not closed under distributed catenation.

Proof. Let $\Sigma = \{a, b, c, d\}$ and $\Delta = \{b, d\}$. Then $L_1 = \{a^m b^m \mid m \geq 0\} \in \underline{CF}$, $L_2 = \{c^n d^n \mid n \geq 0\} \in \underline{CF}$, $L_1 = \{a, b\}^{\circ_\Delta}$, $L_2 = \{c, d\}^{\circ_\Delta}$, but

$$L_1 \circ_\Delta L_2 = \{a^m c^n b^m d^n \mid m, n \geq 1\} \notin \underline{CF}.$$

□

Theorem 25. REG is closed under distributed catenation.

Proof. Assume that $L_i \in \underline{REG}$ are accepted by some deterministic finite automata $A_i = (Q_i, \Sigma, \delta_i, q_0^i, F_i)$, $i = 1, 2$. We will define a nondeterministic finite automaton $A = (Q, \Sigma, \delta, Q_0, F)$ such that $L(A) = L_1 \circ_\Delta L_2$. Define $Q = Q_1 \times Q_2 \times \{1, 2, 3, 5\} \times \{1, 2\}$, $Q_0 = \{(q_0^1, q_0^2, i, j) \mid 1 \leq i \leq 5, j = 1, 2\}$, $F = F_1 \times F_2 \times \{1, 2, 3, 4, 5\} \times \{1, 2\}$. An element of Q will be denoted by $[p, q, i, j]$ instead of (p, q, i, j). The transition function δ is defined as follows :

$$\delta([p, q, 1, 1], a) = \{[\delta_1(p, a), q, 1, 1], [p, \delta_2(q, a), 1, 2]\}\ , \ \forall a \in \Gamma$$
$$\delta([p, q, 1, 2], a) = \{[p, \delta_2(q, a), 1, 2]\}\ , \ \forall a \in \Gamma$$
$$\delta([p, q, 1, 1], b) = \{[\delta_1(p, b), q, 2, 1]\}\ , \ \forall b \in \Delta$$
$$\delta([p, q, 1, 2], b) = \{[p, \delta_2(q, b), 2, 1]\}\ , \ \forall b \in \Delta$$
$$\delta([p, q, 2, 1], b) = \{[\delta_1(p, b), q, 2, 1], [p, \delta_2(q, b), 2, 2]\}\ , \ \forall b \in \Delta$$
$$\delta([p, q, 2, 2], b) = \{[p, \delta_2(q, b), 2, 2]\}\ , \ \forall b \in \Delta$$
$$\delta([p, q, 2, 2], a) = \{[\delta_1(p, a), q, 3, 1]\}\ , \ \forall a \in \Gamma$$
$$\delta([p, q, 3, 1], a) = \{[\delta_1(p, a), q, 3, 1], [p, \delta_2(q, a), 3, 1]\}\ , \ \forall a \in \Gamma$$
$$\delta([p, q, 3, 2], a) = \{[p, \delta_2(q, a), q, 3, 2]\}\ , \ \forall a \in \Gamma$$
$$\delta([p, q, 3, 2], b) = \{[p, \delta_2(q, b), 2, 1], [\delta_1(p, b)q, , 4, 1], [p, \delta_2(q, b); 4, 2]\}\ , \ \forall b \in \Delta$$
$$\delta([p, q, 4, 1], b) = \{[\delta_1(p, b), q, 4, 1]\}\ , \ \forall b \in \Delta$$
$$\delta([p, q, 4, 1], a) = \{[\delta_1(p, a), q, 5, 1]\}\ , \ \forall a \in \Gamma$$
$$\delta([p, q, 5, 1], a) = \{[\delta_1(p, a), q, 5, 1]\}\ , \ \forall a \in \Gamma$$
$$\delta([p, q, 5, 1], b) = \{[\delta_1(p, b), q, 4, 1]\}\ , \ \forall b \in \Delta$$
$$\delta([p, q, 4, 2], b) = \{[p, \delta_2(q, b), 4, 2]\}\ , \ \forall b \in \Delta$$
$$\delta([p, q, 4, 2], a) = \{[p, \delta_2(q, b), 5, 2]\}\ , \ \forall a \in \Gamma$$

$$\delta([p,q,5,2],a) = \{[p,\delta_2(q,b),5,2]\}\ ,\ \forall a \in \Gamma$$
$$\delta([p,q,5,2],b) = \{[p,\delta_2(q,b),4,2]\}\ ,\ \forall b \in \Delta$$

Then it is easy to observe that $L(A) = L_1 \circ_\Delta L_2$. □

Theorem 26. If $R \in \underline{REG}$ and $L \in \underline{CF}$, then $R \circ_\Delta L \in \underline{CF}$ and $L \circ_\Delta R \in \underline{CF}$.

Proof. The proof is similar to the proof of Theorem 25, except that one automaton is a pushdown automaton. Hence, one component must simulate the behaviour of a pushdown automaton. Thus, the resulting pushdown automaton uses its finite control to simulate the finite deterministic automaton that accepts the language R. □

Theorem 27. $\underline{CS}$ is closed under distributed catenation.

Proof. To prove $L_1, L_2 \in \underline{CS} \Rightarrow L_1 \circ_\Delta L_2 \in \underline{CS}$, note that L_1, L_2 are accepted by LBA's M_1, M_2. Define a LBA M guessing for an input w if there are scattered subwords u, v of w with $w = u \circ_\Delta v$, and then checking by M_1 and M_2 if $u \in L_1, v \in L_2$. □

Theorem 28. $\underline{RE}$ is closed under distributed catenation.

Proof. Similar to the proof of Theorem 27. □

4 A Characterization of Recursively Enumerable Languages Using Distributed Catenation

The main result of this section is that a language L is recursively enumerable if and only if it is accepted by a pushdown automaton which uses distributed catenation on its pushdown memory.

Definition 29. A *Δ-distributed (nondeterministic) pushdown automaton* (*Δ-PDA*) is an ordered system $A = (Q, \Sigma, \Theta, \Delta, \delta, q_0, B, F)$ such that the system $A' = (Q, \Sigma, \Theta, \delta, q_0, B, F)$ is a pushdown automaton and $\Delta \subseteq \Theta$.

The *transition function* of A is defined as follows :

$(p, aw, z\gamma) \underset{A}{\rightarrow} (q, w, \alpha \circ_\Delta \gamma)$ iff $(q, \alpha) \in \delta(p, a, z)$

where $p, q \in Q$, $a \in \Sigma \cup \{\lambda\}$, $w \in \Sigma^*$, $z \in \Theta$, and $\alpha, \gamma \in \Theta^*$.

The language accepted by A is :

$L(A) = \{w \in \Sigma^* \mid \exists q_f \in F \exists \gamma \in \Gamma^* : (q_0, w, B) \overset{*}{\underset{A}{\rightarrow}} (q_f, \lambda, \gamma)\}$.

Note that a Δ-PDA A behaves like a classical pushdown automaton, except of using Δ-distributed catenation for storing on the pushdown memory instead of the classical catenation. This increases the power of a pushdown automaton as can be seen in the next theorem.

Theorem 30. For any language L holds : $L \in \underline{RE}$ if and only if L is accepted by some Δ-PDA.

Proof.

a) If L is accepted by some Δ-PDA A then it is obvious that there exists some (nondeterministic) one tape Turing machine M accepting L. M just simulates on its tape the behaviour of A according to the operation $\circ_\Delta$.

b) Let the deterministic one tape Turing machine $M = (Q, \Sigma, q_0, q_f, \delta)$ with blank $B \in \Sigma$ accept L.

Define a Δ-PDA $\quad A = (Q', \Sigma \cup \{\#\}, \Theta, \Delta, \delta', q_0', \overline{B}, F)$ with
$Q' = \{q_0', q_0'', q_{01}, q_{02}\}$
$\cup \{q'', q_{11}, q_{12}, q_{13} \mid q \in Q\} \cup \{[qxR], [qxL], [qR], [qL] \mid q \in Q, x \in \Sigma\}$
$F = \{q_f''\}$, $\Theta = \Sigma \cup \{\#\} \cup \overline{\Sigma} \cup \{\overline{\#}\}$, $\Delta = \overline{\Sigma} \cup \{\overline{\#}\}$

Any configuration $By_m \cdots y_1 q x_1 \cdots x_k$ of M will be represented as a configuration $(q'', \lambda, Bx_m \cdots x_2[x_1 q]\overline{y_1} \cdots \overline{y_m}\overline{B})$ of A.

For the initial configuration $Bx_1 \cdots x_k B$ of M this is achieved from the initial configuration $(q_0', x_1 \cdots x_k, \overline{B})$ of A by the transitions

$$\delta(q_0', \lambda, \overline{B}) = \{(q_0', \overline{B}\overline{B}), (q_{02}, [Bq_0]\overline{B})\} \ , \ \delta(q_0', x, \overline{B}) = \{(q_{01}, [xq_0])\}$$
$$\delta(q_{01}, y, [xq_0]) = \{(q_{01}, y[xq_0])\} \ , \ \delta(q_{01}, y, x) = \{(q_{01}, yx)\}$$
$$\delta(q_{01}, \lambda, [xq_0]) = \{(q_{02}, B[xq_0])\} \ , \ \delta(q_{01}, \lambda, x) = \{(q_{02}, Bx)\}$$
$$\delta(q_{02}, \lambda, [Bq_0]) = \{(q_{02}, B[Bq_0])\} \ , \ \delta(q_{02}, \lambda, B) = \{(q_{02}, BB), (q_0'', BB)\}$$

This causes

$$(q_0', x_1 \cdots x_k, \overline{B}) \xrightarrow[A]{*} (q_0'', \lambda, B^s x_k \cdots x_2[x_1 q_0]\overline{B}^r)$$
$$(q_0', x_1, \overline{B}) \xrightarrow[A]{*} (q_0'', \lambda, B^s[x_1 q_0]\overline{B}^r)$$
$$(q_0', \lambda, \overline{B}) \xrightarrow[A]{*} (q_0'', \lambda, B^s[Bq_0]\overline{B}^r)$$

for the initial configurations $Bq_0x_1 \cdots x_k B$, Bq_0x_1B, and Bq_0B of M, respectively, and with $r, s \geq 1$ just representing enough working space for M.

Note that A is nondeterministic in this initial part.

A transition of M with right move

$$By_m \cdots y_1 p a x_1 \cdots x_k B \xrightarrow[M]{} By_m \cdots y_1 b q x_1 \cdots x_k B$$

where some of the x_i, y_j at the ends may be equal to B, is simulated by A in the following way :

$$(p'', \lambda, Bx_k \cdots x_1[ap]\overline{y_1} \cdots \overline{y_m}\overline{B}) \xrightarrow[A]{*} (q'', \lambda, Bx_k \cdots [x_1 q]\overline{b}\overline{y_1} \cdots \overline{y_m}\overline{B}).$$

This is achieved by the transitions

$$\delta(p'', \lambda, x) = \{(q_{11}, \overline{\#}x)\}$$
$$\delta(q_{11}, \lambda, x) = \{(q_{11}, \overline{\#}x)\} \ , \ \delta(q_{11}, \lambda, [xp]) = \{([qbR], \lambda)\}$$
$$\delta([qbR], \lambda, \overline{\#}) = \{([qbR], \lambda)\} \ , \ \delta([qbR], \lambda, \overline{y}) = \{([qR], \overline{b}\overline{y})\}$$
$$\delta([qR], \lambda, \overline{y}) = \{([qR], \overline{\#}\#\overline{y})\} \ , \ \delta([qR], \lambda, \overline{\#}) = \{([qR], \lambda)\}$$
$$\delta([qR], \lambda, \#) = \{([qR], \lambda)\} \ , \ \delta([qR], \lambda, x) = \{(q_{12}, \overline{\#}[xq])\}$$
$$\delta(q_{12}, \lambda, x) = \{(q_{12}, \overline{\#}x)\} \ , \ \delta(q_{12}, \lambda, \overline{\#}) = \{(q_{12}, \lambda)\} \ , \ \delta(q_{12}, \lambda, \overline{y}) = \{(q_{13}, \overline{\#}\#\overline{y})\}$$
$$\delta(q_{13}, \lambda, \overline{\#}) = \{(q_{13}, \lambda)\} \ , \ \delta(q_{13}, \lambda, \overline{y}) = \{(q_{13}, \overline{\#}\#\overline{y})\}$$

$$\delta(q_{13},\lambda,\#) = \{(q_{13},\lambda)\}\ ,\ \delta(q_{13},\lambda,x) = \{(q'',x)\}$$

A transition of M with left move

$$By_m\cdots y_1pax_1\cdots x_kB \underset{M}{\rightarrow} By_m\cdots y_2qy_1bx_1\cdots x_kB$$

is simulated by A in the following way :

$$(p'',\lambda,Bx_k\cdots x_1[ap]\overline{y_1}\cdots\overline{y_m}\overline{B}) \overset{*}{\underset{A}{\rightarrow}} (q'',\lambda,Bx_k\cdots x_1b[y_1]\overline{y_2}\cdots\overline{y_m}\overline{B}).$$

In this case the transitions of A are the same as for the right move, except that $[qbR],[qR]$ are replaced by $[qbL],[qL]$ and the transitions from $[qbR]$ to $[qR]$ and from $[qR]$ to q_{12} are changed into

$$\delta([qbL],\lambda,\overline{y}) = \{([qL],\overline{\#}[y_1q]b)\}\ ,\ \delta([qL],\lambda,[xq]) = \{(q_{12},\overline{\#}[xq])\}$$

Note that A is deterministic in this part.

Now it is obvious that A arrives at q''_f if and only if M arrives at q_f. □

5 References

1. J.S. Golan : The Theory of Semirings with Application in Mathematics and Theoretical Computer Science, Longman Scientific and Technical, 1992.
2. H. Jürgensen : Syntactic Monoids of Codes, Report 327, Dept.Comp.Sci., The University of Western Ontario, 1992.
3. W. Kuich, A. Salomaa : Semirings, Automata, Languages, EATCS Monographs on Theoretical Computer Science, Springer-Verlag, Berlin, 1986.
4. A. Mateescu : On (Left) Partial Shuffle, Proceedings of 'Results and Trends in Theoretical Computer Science', LNCS 812, Springer-Verlag, (1994) 264-278.
5. C. Reutenauer : Free Lie Algebras, Clarendon Press, Oxford, 1993.
6. J. Sakarovitch, I. Simon : Subwords, in Combinatorics on Words, M. Lothaire (ed.), Addison-Wesley, Read. Mass., 1983.
7. A. Salomaa : Formal Languages, Academic Press, New York, 1973.

The Power of Frequency Computation
(Extended Abstract)

Martin Kummer and Frank Stephan *

Universität Karlsruhe, Institut für Logik, Komplexität und Deduktionssysteme,
D-76128 Karlsruhe, Germany. {kummer; fstephan}@ira.uka.de

Abstract. The notion of *frequency computation* concerns approximative computations of n distinct parallel queries to a set A. A is called (m,n)-recursive if there is an algorithm which answers any n distinct parallel queries to A such that at least m answers are correct. This paper gives natural combinatorial characterizations of the fundamental inclusion problem, namely the question for which choices of the parameters m, n, m', n', every (m,n)-recursive set is (m',n')-recursive. We also characterize the inclusion problem restricted to recursively enumerable sets and the inclusion problem for the polynomial-time bounded version of frequency computation. Furthermore, using these characterizations we obtain many explicit inclusions and noninclusions.

1 Introduction

Frequency computation is a classical approximation notion introduced by Rose [19] in the early sixties and subsequently developed (in historical order) by Trakhtenbrot [20], Kinber [11], and Dëgtev [8] until the early eighties. Recently an upsurge of interest occurred in connection with the investigation of bounded query computations in complexity theory [1, 2, 3, 6] and computability theory [4, 5, 9]. Very recently frequency computations were also studied in inductive inference [7, 13, 17]. In the present paper we characterize the fundamental inclusion problem of frequency computation by a combinatorial property. This yields a short proof of McNicholl's new result [18] that the inclusion problem is decidable. Similar characterizations are provided for the restriction of the inclusion problem to r.e. sets and for a polynomial time version of frequency computation.

We now discuss the definitions and results in more detail. For natural numbers m, n $(1 \leq m \leq n)$ a set A is called (m,n)-recursive (in short $A \in \Omega(m,n)$) if any n distinct parallel queries to A can be effectively answered such that at least m of the answers are correct (i.e., at most $(n-m)$ errors are made). More formally, $A \in \Omega(m,n)$ if there is a recursive function $f : \omega^n \to \{0,1\}^n$, mapping n-tuples of numbers to n-tuples of bits, such that for all n pairwise distinct numbers $x_1, \ldots, x_n$:

$$f(x_1, \ldots, x_n) = (b_1, \ldots, b_n) \Rightarrow |\{i : A(x_i) = b_i\}| \geq m.$$

* Supported by the Deutsche Forschungsgemeinschaft (DFG) grant Me 672/4-2.

Trivially, every recursive set is (m,n)-recursive for all m,n, and every (n,n)-recursive set is recursive. Trahktenbrot [20] proved that there exist nonrecursive $(1,2)$-recursive sets, and, answering a question of Myhill, that if $\frac{m}{n} > \frac{1}{2}$ then every (m,n)-recursive set is already recursive – see [10, 14] for a recent survey and related results. Intuitively, no nonrecursive set can be approximated with frequency greater than $\frac{1}{2}$, and this bound is tight. This result is nontrivial as Kinber [11] showed that for $n > 1$ there can be no uniform method to extract a decision procedure from an $(n-1,n)$-algorithm. In particular, Trahktenbrot's inclusions $\Omega(m,n) \subseteq \Omega(1,1)$ for $\frac{1}{2} < \frac{m}{n} < 1$ cannot be shown by local modifications of the (m,n)-algorithm.

In general, the inclusion problem concerns the power of (m,n)-computations with different parameters m,n, namely the question whether, for any given m,n,h,k, every (m,n)-recursive set is (h,k)-recursive. As we have seen above, Trakhtenbrot's result answers this for $\frac{h}{k} > \frac{1}{2}$. For frequency less than or equal to $\frac{1}{2}$, Dëgtev [8] characterized the special case $\{(1,n,h,k) : \Omega(1,n) \subseteq \Omega(h,k)\}$ and stated first results on explicit inclusions and noninclusions. Additional results were obtained in [15], but the question whether the full inclusion problem $\{(m,n,h,k) : \Omega(m,n) \subseteq \Omega(h,k)\}$ is decidable was left open. Very recently, McNicholl [18] answered it positively. In his proof he also uses a finite combinatorial property, but this is rather implicit and the proof is very lenghty (however, due to its greater generality, it could also be useful for other related types of inclusion problems which appear in the literature).

In this paper, the inclusion problem is characterized by natural combinatorial conditions. This yields a new and much shorter proof of McNicholl's result. We first sketch Dëgtev's approach of local (m,n)-computations and apply it to obtain a characterization of the polynomial-time version of frequency computation. Then we weaken Dëgtev's condition to obtain our characterization of the inclusion problem. A new proof, which has no counterpart in Dëgtev's approach is required to show that our condition is sufficient. In section 4 it is shown that the recursively enumerable (r.e.) case "When is every *r.e.* (m,n)-recursive set (h,k)-recursive?" is also decidable. Finally, using our characterizations, we obtain several explicit inclusions and noninclusions.

Definitions and Notation. We are using standard notation. $\omega = \{0,1,2,\ldots\}$. We identify a set A with its characteristic function; so we write $A(x) = 1$ for $x \in A$ and $A(x) = 0$ for $x \notin A$.

The concatenation of strings v,w is denoted by $v \cdot w$ or simply by vw; $V_1 \cdot V_2 = \{u \cdot v : u \in V_1, v \in V_2\}$. For $v = (b_1,\ldots,b_n) \in \{0,1\}^n$ and $i, i_1, \ldots, i_j$ such that $1 \leq i \leq n$ and $1 \leq i_1 \leq \ldots \leq i_j \leq n$, $v[i]$ denotes the i-th component b_i, and $v[i_1,\ldots,i_j]$ denotes the projection $(b_{i_1},\ldots,b_{i_j})$. $V[i_1,\ldots,i_j] = \{v[i_1,\ldots,i_j] : v \in V\}$, $\#v = |\{i : v[i] = 1\}|$. If $X = \{i_1,\ldots,i_j\}$ then $V[X]$ abbreviates $V[i_1,\ldots,i_j]$. We write $\sigma \preceq \tau$ if σ is an initial segment of τ. A *tree* is a set of strings closed under initial segments.

φ_i is the i-th function in a standard enumeration of all partial recursive functions of one argument. $\varphi_{i,s}(x)$ denotes the result, if any, of performing s steps in the computation of $\varphi_i(x)$. A set is called k-r.e. for $k = 1,2,\ldots$ if there is a recursive function $g(x, s)$ such that $A(x) = \lim_{s\to\infty} g(x, s)$, $|\{s : g(x, s) \neq g(x, s+1)\}| \leq k$ and $g(x, 0) = 0$ for all x. Intuitively, the characteristic function of a k-r.e. set can be computed with at most k mindchanges. Clearly a set is r.e. iff it is 1-r.e.

2 A Local Version of (m, n)-Computation

Dëgtev [8] (and implicitly already Kinber [11]) introduced the following local version of (m, n)-computability in order to study the combinatorial aspect of frequency computation.

Definition 1. [8] Let $s \geq n$. A set $V \subseteq \{0, 1\}^s$ of binary vectors is (m, n)-admissible if for all $x_1 < \ldots < x_n \in \{1, \ldots, s\}$ there is $(b_1, \ldots, b_n) \in \{0, 1\}^n$ such that

$$|\{j : b_j = v[x_j]\}| \geq m \text{ for all } v \in V.$$

We extend the definition to the case $s < n$ by appending $(n - s)$ zeros to each of the vectors. (In other words, if $s < n$, i.e., $t := n - s > 0$, then V is (m, n)-admissible iff either $t \geq m$ or V is $(m-t, n-t)$-admissible in the original sense.)

Example: (1) $\{0, 1\}^s$ is (m, n)-admissible iff $n - m \geq s$.
(2) Let $2m \leq n$. $\{0, 1\}^{n-2m} \times \{0^{2m}, 1^{2m}\}$ is (m, n)-admissible and not $(m+1, n)$-admissible.

This notion leads to a combinatorial condition that is sufficient for inclusions.

Fact 1. [8] *If every (m, n)-admissible set $V \subseteq \{0, 1\}^k$ is (h, k)-admissible, then $\Omega(m, n) \subseteq \Omega(h, k)$.*

Trakhtenbrot's Theorem implies that the condition is not necessary when $m > \frac{n}{2}$. It is open whether it is necessary when $m \leq \frac{n}{2}$. However, Dëgtev settled the case $m = 1$.

Theorem 2. [8] (1) *If there is a $(1, n)$-admissible set $V \subseteq \{0, 1\}^k$ which is not (h, k)-admissible, then there exists a 2-r.e. set $A \in \Omega(1, n) - \Omega(h, k)$.*
(2) *If in addition $\{0^k, 1^k\} \subseteq V$ then A can be chosen to be an r.e. set.*

For the polynomial-time version of frequency computation, the admissibility condition is also a necessary condition for inclusions, as we now show. Let $\Omega_p(m, n)$ denote the class of all recursive sets which are (m, n)-recursive via an algorithm which runs in polynomial time w.r.t. to the length of the input (i.e., the sum of the lengths of the binary representations of the x_i). Kinber [11] (see also Amir, Gasarch [1]) proved that there are arbitrarily complex sets in $\Omega(n-1, n)$. Thus, the analog of Trahktenbrot's Theorem (which would be that every set in $\Omega(m, n)$ with $\frac{m}{n} > \frac{1}{2}$ is in **P**) fails badly. On the other hand, the lack of nonuniform inclusions enables the following characterization

Theorem 3. *Let $1 \leq m \leq n, 1 \leq h \leq k$. $\Omega_p(m,n) \subseteq \Omega_p(h,k)$ iff every (m,n)-admissible set $V \subseteq \{0,1\}^k$ is (h,k)-admissible.*

From results in [17, Proposition 3.3] and Theorem 3, it follows that for $\frac{m}{n} < \frac{2}{3}$ and any h, k, we have $\Omega_p(m,n) = \Omega_p(h,k)$ iff $(m,n) = (h,k)$. Thus, in this case there are only trivial equalities. From a result of Kinber [11] (see [17, Fact 3.12]) it follows that $\Omega_p(n,n+1) = \Omega(2,3)$ for all $n \geq 2$. An explicit characterization of all equalities for $\frac{m}{n}, \frac{h}{k} \geq \frac{2}{3}$ is not known, but Kinber conjectures that $\Omega_p(m,n) = \Omega_p(h,k)$ iff $n - m = k - h$.

3 A Characterization of the Inclusion Problem

The following definition slightly strengthens Definition 1 and will lead to a combinatorial characterization of the inclusion problem for $\Omega(m,n)$ classes. The condition was found by looking for a property such that Dëgtev's construction in the proof of Theorem 2 can be carried out for all $m \geq 1$.

Definition 4. A set $V \subseteq \{0,1\}^s$ of vectors is strongly (m,n)-admissible if for every suitable t, every projection of V on $n-t$ distinct components is $(m-t, n-t)$-admissible via a vector *with at least m zeros*. Formally, for every t with $0 \leq t < m$ and $n-t \leq s$, and all $x_1 < \ldots < x_{n-t} \in \{1,\ldots,s\}$ there is $(b_1,\ldots,b_{n-t}) \in \{0,1\}^{n-t}$ such that

- $|\{j : b_j = 0\}| \geq m$ and
- $|\{j : b_j = v[x_j]\}| \geq m-t$ for all $v \in V$.

Note that every $(1,n)$-admissible set $V \subseteq \{0,1\}^k$ with $0^k \in V$ is also strongly $(1,n)$-admissible. Thus, every $(1,n)$-admissible set can be transformed into a strongly $(1,n)$-admissible set by interchanging 0 and 1 in some columns. But this does not hold in general: Kinber [11] found a $(3,5)$-admissible set which is not $(4,6)$-admissible. On the other hand, every strongly $(3,5)$-admissible set V is also $(4,6)$-admissible: every projection $V[X]$ onto three coordinates is $(1,3)$-admissible via 0^3, thus no $v \in V$ contains more than two 1s and V is $(4,6)$-admissible via 0^6.

The following theorem establishes a necessary condition for inclusions. Note that the case $m = 1$ already follows from Dëgtev's Theorem 2, (1)

Theorem 5. *If $\frac{m}{n} \leq \frac{1}{2}$ and if there is a strongly (m,n)-admissible set $V \subseteq \{0,1\}^k$ which is not (h,k)-admissible then $\Omega(m,n) \not\subseteq \Omega(h,k)$.*

Proof. Let m, n, h, k, V be as in the hypothesis of the theorem. The proof is based on Dëgtev's interval construction [8]. $\{I_j\}_{j\in\omega}$ denotes a partition of ω into the intervals $I_j = \{jk+1, jk+2, \ldots, jk+k\}$ of length k and $A(x,s)$ is a 2-r.e. approximation of the value $A(x)$ where A is the witness for $\Omega(m,n) \not\subseteq \Omega(h,k)$. We satisfy the following requirements P_i for all $i \in \omega$ to ensure that $A \notin \Omega(h,k)$.

$$(P_i) \quad A \text{ is not } (h,k)\text{-recursive via } \varphi_i.$$

Construction
Stage 0: Let $A(x,0) = 0$, for all x. Let $p(i,0) = i$, for all i.
Stage $s{+}1$: If there is an i such that

- $p(i,s) < s$,
- $\varphi_{i,s}(p(i,s)k+1, p(i,s)k+2, \ldots, p(i,s)k+k) = (b_1, b_2, \ldots, b_k)$ and
- $|\{l : A(p(i,s)k+l) = b_l\}| \geq h$,

then we let the interval $I_{p(i,s)}$ for the least such i *act*, i.e., we define

- $A(p(i,s)k+l, s{+}1) = v[l]$ for some $v \in V$ with $|\{l : v[l] = b_l\}| < h$; note that v exists because V is not (h,k)-admissible.
- $A(x, s{+}1) = 0$ for all $x > p(i,s)k+k$. ("Reset action")
- $p(j, s{+}1) = s+j$ for all $j > i$; note that the intervals $I_{p(j,s+1)}$ for $j > i$ are by now unused.

Else do nothing.
End of construction

Every interval acts at most once, in this case all intervals with larger index are reset to zero. The intervals with smaller index are not changed. If $A(x, s+1)$ is reset from 1 to 0 then x is also thrown out of any diagonalization interval (by the way in which $p(j, s+1)$ is updated), thus $A(x, s') = 0$ for all $s' > s$. This proves that A is a 2-r.e. set. By induction on i it is easy to show that each i is chosen at only finitely many stages and consequently (P_i) is satisfied.

It remains to construct the (m,n)-algorithm for A. For given $x_1 < \ldots < x_n$ let $s = x_n$. Determine j such that $x_m \in I_j$ and choose numbers $a \geq 0$, $b \geq 1$ such that $I_j \cap \{x_1, \ldots, x_n\} = \{x_{a+1}, \ldots, x_{a+b}\}$; let $c = n - a - b$. In this way we partition the input vector into three blocks: The initial segment of a components, the middle segment of b components which all belong to the interval which contains x_m, and the end segment of c components. Note that if A changes on one block in some stage $s' > s$ then all blocks to the right of it are definitely set to zero. The (m,n)-algorithm $f(x_1, \ldots, x_n)$ is defined according to one of the following three cases:

1.1 $b \geq n - m$ and $A(x_l, s) = 0$ for all $x_l \in I_j$.
As V is strongly (m,n)-admissible there is a vector $v \in \{0,1\}^b$ with at least m zeros such that $V[x_{a+1}-jk, x_{a+2}-jk, \ldots, x_{a+b}-jk]$ is $(m-a-c, b)$-admissible via v.
Let $f(x_1, \ldots, x_n) = (A(x_1, s), \ldots, A(x_a, s), v[1], \ldots, v[b], 0, \ldots, 0)$.
1.2 $b \geq n - m$ and $A(x_l, s) \neq 0$ for some $x_l \in I_j$.
Let $f(x_1, \ldots, x_n) = (A(x_1, s), \ldots, A(x_m, s), 0, \ldots, 0)$.
2 $b < n - m$, i.e., $a + c > m$.
Let $f(x_1, \ldots, x_n) = (A(x_1, s), \ldots, A(x_m, s), 0, \ldots, 0)$.

Let p be the least index such that I_p acts after stage s. Now it follows by case distinction that A is (m,n)-recursive via f.

For the converse we need the following combinatorial lemma which shows that if a certain extension of a set V – by appending a block of ones and including the all zero vector – is still (m,n)-admissible, then V is even strongly (m,n)-admissible.

Lemma 6. *Let $V \subseteq \{0,1\}^s$. If $W = \{1^n v : v \in V\} \cup \{0^{n+s}\}$ is (m,n)-admissible then V is strongly (m,n)-admissible.*

Proof. Consider any $t < m$ such that $n-t \le s$ and any distinct $x_1,\ldots,x_{n-t} \in \{1,2,\ldots,s\}$. The projection $W[1,2,\ldots,t,n+x_1,n+x_2,\ldots,n+x_{n-t}] = \{0^n\} \cup 1^t V[x_1,\ldots,x_{n-t}]$ is (m,n)-admissible via a vector u. W.l.o.g. the x_i are ordered such that $u = 0^b 1^{t-b} 0^a 1^{n-t-a}$. If $a \ge m$ let $w = 0^a 1^{n-t-a}$; otherwise let $w = 0^m 1^{n-t-m}$. w has at least m zero components.

By hypothesis, for each $v \in V[x_1,\ldots,x_{n-t}]$, at least m components of $1^{m-t}v$ and u agree. Thus at least $m-(t-b)$ components of v and $0^a 1^{n-t-a}$ agree. If $a \ge m$, then $w = 0^a 1^{n-t-a}$ has at least $m-t$ correct components. If $a < m$ then w differs from $0^a 1^{n-t-a}$ in $m-a$ components and at least $m-(t-b)-(m-a) = a+b-t$ are correct. Since u agrees with 0^n in at least m components, $a+b \ge m$, i.e., $a+b-t \ge m-t$.

Theorem 7. *If every strongly (m,n)-admissible $V \subseteq \{0,1\}^k$ is (h,k)-admissible then $\Omega(m,n) \subseteq \Omega(h,k)$.*

Proof. Assume that every strongly (m,n)-admissible set $V \subseteq \{0,1\}^k$ is (h,k)-admissible. Let A be (m,n)-recursive via an (m,n)-algorithm f. We define the tree T of characteristic functions that are consistent with f, and then we construct a recursive (h,k)-algorithm from T in a nonuniform way. This idea goes back to Trakhtenbrot [20], but McNicholl [18] realized that it can also be fruitful for frequencies below $\frac{1}{2}$. It is not known whether the nonuniformity can be eliminated in this case.

$$\begin{aligned} T &= \{\sigma \in \{0,1\}^* : \sigma \text{ is consistent with } f\} \\ &= \{\sigma \in \{0,1\}^* : (\forall x_1 < x_2 < \ldots < x_n < |\sigma|) \\ &\qquad [f(x_1,x_2,\ldots,x_n) = (b_1,b_2,\ldots,b_n) \Rightarrow |\{i : \sigma[x_i] = b_i\}| \ge m]\}. \end{aligned}$$

T is a recursive tree and A is an infinite branch of T. Trakhtenbrot [20] observed that if all infinite branches are a finite variant of A, then all infinite branches are recursive, in particular A is (h,k)-recursive.

Now assume that there is an infinite branch B which differs from A in at least n places, say w.l.o.g. $1^n \preceq A$ and $0^n \preceq B$. In particular, for all $x > n$ there is $\tau \in T$ with $0^n \preceq \tau$ and $|\tau| = x$. We define an (h,k)-algorithm g for A as follows:

Given $x_1,\ldots,x_k$ such that $n \le x_1 < x_2 < \ldots < x_k$, let $x = x_k + 1$. Take any string $\tau \in T$ with $0^n \preceq \tau$ and $|\tau| = x$, and let

$$V = \{\sigma[x_1,x_2,\ldots,x_k] : \sigma \in T \wedge 1^n \preceq \sigma \wedge |\sigma| = x\}.$$

By definition of T, the set $1^n V \cup \{0^n \tau[x_1,x_2,\ldots,x_k]\}$ is (m,n)-admissible and contains $A[x_1,x_2,\ldots,x_k]$. Let U be obtained from V by interchanging 0 and 1 in all columns i with $\tau(x_i) = 1$. Then $W = 1^n U \cup \{0^{n+k}\}$ is (m,n)-admissible and hence, by Lemma 6, U is strongly (m,n)-admissible. Thus, by hypothesis, U is (h,k)-admissible. Then also V is (h,k)-admissible, say via w. Let $g(x_1,\ldots,x_k) = w$.

If $x_1 < x_2 < \ldots < x_l < n \leq x_{l+1}$, then let $w = g(x_{l+1},\ldots,x_k,x_k+1,\ldots,x_k+l)$ and define $g(x_1,\ldots,x_k) = 1^l w[1,\ldots,k-l]$.

These results can be summarized as follows:

Theorem 8. *$\Omega(m,n) \subseteq \Omega(h,k)$ iff $\frac{m}{n} > \frac{1}{2}$ or every strongly (m,n)-admissible set $V \subseteq \{0,1\}^k$ is (h,k)-admissible.*

Corollary 9. [McNicholl, 1994] *The inclusion problem for frequency classes is decidable.*

4 The Inclusion Problem for R.E. Sets

We now present a combinatorial characterization of the inclusion problem restricted to r.e. sets, i.e., of $\{(m,n,h,k) : \Omega(m,n) \cap \mathcal{E} \subseteq \Omega(h,k)\}$ where $\mathcal{E}$ is the class of all r.e. sets. This is based on a further strengthening of the notion strongly admissible which we call *special (m,n)-admissible*. Intuitively, a set V was called *strongly admissible* if it is admissible via vectors which contain a certain amount of zeros. For *special admissible* it is required that they contain in addition a certain amount of ones. One is led to this condition by looking for a property which allows the set A in the construction of Theorem 5 to be made r.e. (instead of merely 2-r.e.).

Definition 10. A set $V \subseteq \{0,1\}^k$ is called special (m,n)-admissible iff for every t with $0 \leq t < m$ and $n-t \leq k$, for all $a,c \geq 0$ with $a+c = t$ and for all $x_1 < \ldots < x_{n-t} \in \{1,\ldots,k\}$ there is $(b_1,\ldots,b_{n-t}) \in \{0,1\}^{n-t}$ such that

- $|\{j : b_j = 0\}| \geq m-a$,
- $|\{j : b_j = 1\}| \geq m-c$ and
- $|\{j : b_j = v[x_j]\}| \geq m-t$ for all $v \in V$.

Note that every special (m,n)-admissible set is also strongly (m,n)-admissible. Furthermore, if V is special (m,n)-admissible so is $V \cup \{0^k, 1^k\}$.

Theorem 11. *Let $\frac{m}{n} \leq \frac{1}{2}$. $\Omega(m,n) \cap \mathcal{E} \subseteq \Omega(h,k)$ iff every special (m,n)-admissible set $V \subseteq \{0,1\}^k$ is (h,k)-admissible.*

Sketch. We restrict ourselves to the direction ($\Rightarrow$). We show the contrapositive. Let m,n,h,k,V be as in the hypothesis of the theorem, i.e., $m \leq \frac{n}{2}$ and V is special (m,n)-admissible but not (h,k)-admissible. The proof is again based on Dëgtev's interval construction [8]. In fact, the case $m = 1$ already follows from Theorem 2, (2). We use almost the same construction as in Theorem 5. Only the "Reset action" is modified: this time we reset to ones (instead of zeros) so that

$A(x, s)$ never changes from one to zero, i.e., A is r.e. Formally, with the notation from the proof of Theorem 5, the "Reset action" is done as follows:

$$A(x, s+1) = 1 \text{ for } x = p(i, s)k + k + 1, p(i, s)k + k + 2, \ldots, sk.$$

As before it follows by induction on i that each requirement (P_i) is satisfied. Thus A is an r.e. set not in $\Omega(h, k)$. Similariy as before it can be shown that $A \in \Omega(m, n)$.

5 Explicit Inclusions and Noninclusions

In this section we present some explicit results on inclusions and noninclusions for frequencies below $\frac{1}{2}$. These are obtained by applying the combinatorial characterizations of the previous sections.
Dëgtev [8] noted the following three easy inclusions: $\Omega(m, n) \subseteq \Omega(m, n+1)$; $\Omega(m+1, n+1) \subseteq \Omega(m, n)$; $\Omega(m, n) \cap \Omega(h, k) \subseteq \Omega(m+h, n+k)$. The next result shows that there are also inclusions which are nontrivial:

Theorem 12. $\Omega(2,4) \subsetneq \Omega(3,6)$, $\Omega(2,5) \subsetneq \Omega(3,8)$ *and* $\Omega(3,6) \subsetneq \Omega(4,9)$.

We have obtained these inclusions by showing that every $(2,4)$-admissible set is $(3,6)$-admissible and so on. Further it is possible to show that every class $\Omega(m, 2m)$ is contained in almost all classes $\Omega(h, 2h)$; therefore we conjecture that $\Omega(m, 2m) \subseteq \Omega(h, 2h)$ for all $h \geq m$.

Theorem 13. $(\forall m)(\forall^\infty h)[\Omega(m, 2m) \subseteq \Omega(h, 2h)]$.

The next fact shows that every $(1, k)$-recursive set is "almost" $(h, 2h)$-recursive for large h.

Fact 2. [16, Fact 2.7(b)] *The quotient* $q_k(n) = max\{\frac{m}{n} : \Omega(1, k) \subseteq \Omega(m, n)\}$ *converges to* $\frac{1}{2}$ *for* $n \to \infty$ *and fixed* k.

Dëgtev showed that in general it is impossible to reduce the number of errors made by an (m, n)-algorithm:

Theorem 14. [8] *If* $n - m > k - h$ *and* $\frac{m}{n} \leq \frac{1}{2}$ *then* $\Omega(m, n) \not\subseteq \Omega(h, k)$.

This noninclusion is witnessed by the set $\{0, 1\}^{n-m}$ which is (h, k)-admissible iff $k - h \leq n - m$. Further noninclusions are given by the following theorem:

Theorem 15. *The set* $V = \{0^n, 1^n\} \cup \{0^i 1 0^j : i+j = n-1\}$ *is (strongly)* (m, n)*-admissible iff* $\frac{m}{n} < \frac{1}{2}$.

The next theorem generalizes Dëgtev's example of a $(1,3)$-admissible set which is not $(2,5)$-admissible [8].

Theorem 16. *The set* $V = \{0^{2h+1}, 1^{2h+1}, 10^{2h-1}1\} \cup \{0^i10^j : i+j = 2h\} \cup \{0^i110^j : i+j = 2h-1\}$ *is strongly* $(m, 2m+1)$*-admissible for* $m < h$ *but not* $(h, 2h+1)$*-admissible.*

Corollary 17. $\Omega(m,n) \not\subseteq \Omega(h,k)$ *if* $\frac{m}{n} \leq \frac{1}{2}$ *and one of the following conditions holds:* (a) $n - m > k - h$; (b) $n - 2m > k - 2h$; (c) $n - 2m = k - 2h > 0$ *and* $m \neq h$.

Trakhtenbrot's Theorem and Corollary 17, (a), (b) yield an explicit solution of the equality problem for frequency classes:

Theorem 18. $\Omega(m,n) = \Omega(h,k)$ *iff* $(m,n) = (h,k)$ *or* $\frac{m}{n} > \frac{1}{2} \wedge \frac{h}{k} > \frac{1}{2}$.

Trivially, $\Omega(1,n) \subseteq \Omega(2,2n)$. Is this optimal, i.e., does $\Omega(1,n) \not\subseteq \Omega(2,2n-1)$ hold? The case $n = 2$ follows from Trakhtenbrot's Theorem. Dëgtev proved it for $n = 3$. The case $n = 4$ is witnessed by the closure of {0000000, 0000011, 0000111, 0001101, 0011101, 1111111} under rotational shifts.

All noninclusions $\Omega(m,n) \not\subseteq \Omega(h,k)$ in this section are witnessed by special (m,n)-admissible sets which are not (h,k)-admissible, thus in each case there are r.e. sets $A \in \Omega(m,n) - \Omega(h,k)$. In particular the equality problem restricted to r.e. sets has the same solution as the general equality problem.

6 Conclusion

We have provided combinatorial characterizations for the inclusion problem of frequency computation. These conditions are derived from a local version of (m,n)-computation and look natural. In particular, it follows that the various inclusion problems are decidable and all computability issues are now clearly separated from the combinatorics. While from the point of view of computability theory the problem is completely solved, it would be desirable to obtain a better understanding of the underlying combinatorics which would lead to further explicit results. (An explicit solution for the full inclusion problem, say by a simple formula, is not to be expected, for similar reasons as in [5].) For instance, there is a connection with the "covering problem" from coding theory, cf. [15], but this seems to be rather superficial. Another obvious question is, whether the three criteria which we presented are really different for frequencies less than or equal to $\frac{1}{2}$. We conjecture that they are mutually distinct, but they agree on all values which we were able to test by now.

Acknowledgments: We would like to thank Valentina Harizanov, Susanne Kaufmann, and Tim McNicholl for comments.

References

1. A. Amir, W. I. Gasarch. Polynomial terse sets. *Information and Computation*, 77:37–56, 1988.

2. R. Beigel. Bi-immunity results for cheatable sets. *Theoretical Computer Science*, 73:249–263, 1990.
3. R. Beigel. Bounded queries to SAT and the boolean hierarchy. *Theoretical Computer Science*, 83:199–223, 1991.
4. R. Beigel, W. I. Gasarch, J. Gill, J. C. Owings, Jr. Terse, superterse, and verbose sets. *Information and Computation*, 103:68–85, 1993.
5. R. Beigel, M. Kummer, F. Stephan. Quantifying the amount of verboseness. To appear in: Information and Computation. (A preliminary version appeared in *Logical Foundations of Computer Science – Tver'92*, Lecture Notes in Computer Science 620, pp. 21–32, 1992).
6. R. Beigel, M. Kummer, F. Stephan. Approximable sets. To appear in: Information and Computation. (A preliminary version appeared in *Proceedings Structure in Complexity Theory, Ninth Annual Conference*, pp. 12–23, IEEE Press, 1994.)
7. J. Case, S. Kaufmann, E. Kinber, M. Kummer. Learning recursive functions from approximations. In *Proceedings EuroCOLT'95*, Lecture Notes in Computer Science 904, pp. 140–153, 1995. Springer-Verlag
8. A. N. Dēgtev. On (m,n)-computable sets. In *Algebraic Systems (Edited by D.I. Moldavanskij)*. Ivanova Gos. Univ. 88–99, 1981. (Russian) (MR 86b:03049)
9. W. I. Gasarch. Bounded queries in recursion theory: a survey. In *Proceedings of the Sixth Annual Structure in Complexity Theory Conference*, pp. 62–78, IEEE Press, 1991.
10. V. Harizanov, M. Kummer, J. C. Owings, Jr. Frequency computation and the cardinality theorem. *J. Symb. Log.*, 57:677–681, 1992.
11. E. B. Kinber. Frequency-computable functions and frequency-enumerable sets. Candidate Dissertation, Riga, 1975. (Russian)
12. E. B. Kinber. On frequency real-time computations. In *Teoriya Algorithmov i Programm*, Vol. 2 (Edited by Ya. M. Barzdin). Latv. Valst. Gos. Univ. pp. 174–182, 1975. (Russian) (MR 58:3624, Zbl 335:02023)
13. E. Kinber, C. Smith, M. Velauthapillai, R. Wiehagen. On learning multiple concepts in parallel. In *Proceedings COLT'93*, pp. 175–181, ACM Press, 1993.
14. M. Kummer. A proof of Beigel's cardinality conjecture. *J. Symb. Log.*, 57:682–687, 1992.
15. M. Kummer, F. Stephan. Some aspects of frequency computation. Technical Report No. 21/91, Fakultät für Informatik, Universität Karlsruhe, 1991.
16. M. Kummer, F. Stephan. Recursion theoretic properties of frequency computation and bounded queries. To appear in: Information and Computation (A preliminary version appeared in *Third Kurt Gödel Colloquium*, Lecture Notes in Computer Science 713, pp. 243–254,1993).
17. M. Kummer, F. Stephan. Inclusion problems in parallel learning and games. In *Proceedings COLT'94*, pp. 287–298, ACM Press, 1994.
18. T. McNicholl. A solution to the inclusion problem for frequency classes. Manuscript, 66 pp., Dept. of Mathematics, George Washington University, Washington DC, June 1994.
19. G. F. Rose. An extended notion of computability. In *Abstr. Intern. Congr. for Logic, Meth., and Phil. of Science*, Stanford, California, 1960.
20. B. A. Trakhtenbrot. On frequency computation of functions. *Algebra i Logika*, 2:25–32, 1963. (Russian)

Randomized Incremental Construction of Simple Abstract Voronoi Diagrams in 3-Space

(Extended Abstract)

Ngọc-Minh Lê *

Praktische Informatik VI
FernUniversität Hagen
D-58084 Hagen, Germany

ngoc-minh.le@fernuni-hagen.de

Abstract

The simple abstract Voronoi diagram in 3-space is introduced as an abstraction of the usual Euclidean Voronoi diagram. We show that the three-dimensional simple abstract Voronoi diagram of n sites can be computed in $O(n^2)$ expected time using $O(n^2)$ expected space by a randomized algorithm.

1 Preliminaries

The Voronoi diagram is an important data structure in computational geometry. Given n sites in $\mathbb{R}^D$, the Voronoi diagram partitions D-space into n regions, one for each site. The region of a site p consists of all points in D-space that lie closer to p than to any of the other sites. For a survey on Voronoi diagrams and their applications we refer to Aurenhammer [2].

Klein [6] recognized that many types of planar Voronoi diagrams can be regarded as specializations of only one type of diagram, which he calls *abstract Voronoi diagrams.* He proposed an axiomatic approach that is based on the concept of bisecting curves rather than the concept of proximity derived from a concrete distance measure. For each pair of sites p and q, he assumes the existence of a bi-infinite bisecting curve that divides the plane into a p-region and a q-region. The Voronoi region of a site p is defined to be the intersection of all p-regions for different q's. Voronoi regions are assumed to be path-connected and to partition the plane. Klein develops a divide-and-conquer algorithm to compute the planar abstract Voronoi diagrams in $O(n \log n)$ time.

Mehlhorn, Meiser, Ó'Dúnlaing, and Klein [11, 7] show how to construct the planar abstract Voronoi diagram in $O(n \log n)$ expected time incrementally by

*This work was partly supported by Deutsche Forschungsgemeinschaft grant Kl 655/2-2.

adding the sites one by one in random order and maintaining the diagram constructed so far. Their algorithm is based on the randomized incremental construction technique of Clarkson and Shor [4] and its improvement [3].

In this paper we present a "low-technology" generalization of the concept of abstract Voronoi diagram to 3-space, "low-technology" in the sense that we don't try to keep the set of axioms to be minimal and we allow them to contain redundancy; moreover we assume, by analogy with the case of the Euclidean metric, that the intersection of bisectors satisfy certain specific conditions. We develop an algorithm for computing such diagrams.

Similar to the axiomatic approach of Klein [6] we postulate, for each pair of sites p and q, the existence of a bisecting surface $B(p,q)$ that divides 3-space into a p-region and a q-region. We define the Voronoi region of a site p to be the intersection of all p-regions for different q's. Further, we assume that Voronoi regions are either empty or homeomorphic to a 3-ball and partition 3-space. Since the non-degenerate Voronoi diagram in the Euclidean metric should serve as a model for our axioms, we assume that, given 5 sites p,q,r,s,t, the set $B(p,q)\cap B(q,r)$ is a bi-infinite curve, the set $B(p,q)\cap B(q,r)\cap B(r,s)$ is a point, and the set $B(p,q)\cap B(q,r)\cap B(r,s)\cap B(s,t)$ is empty; in addition, we require that bisecting surfaces should intersect transversely.

The diagram arising this way is called the *(3-dimensional) simple abstract Voronoi diagram*. We present an algorithm that constructs the simple abstract Voronoi diagram in $O(n^2)$ expected time using $O(n^2)$ expected space. The algorithm generalizes the approach for the planar case [11, 7] and is an instance of the randomized incremental construction of Clarkson and Shor, but the history graph [3] is used instead of the original conflict graph.

The proofs of the results in this paper can be found in [10]. We use *bd* A, *int* A, and $\overline{A}$ to denote the boundary, interior, and closure of a subset $A\subset \mathbf{R}^3$.

2 Simple abstract Voronoi diagrams in 3-space

We extend the notion of planar abstract Voronoi diagrams [6] to 3-space as follows. Let $n\in\mathbf{N}$, and let $S=\{1,\dots,n-1\}$. We define a *dominance system over* S to be a family $\mathcal{D}=\{D(p,q)\mid p,q\in S \text{ and } p\neq q\}$ of subsets of 3-space satisfying

1. $D(p,q)$ is a non-empty open subset of 3-space.
2. $D(p,q)\cap D(q,p)=\emptyset$ and $bd\ D(p,q)=bd\ D(q,p)$.
3. $B(p,q):=bd\ D(p,q)$ is homeomorphic to a plane in 3-space.

The elements of S are *sites*, the surface $B(p,q)$ is the *bisector of* p *and* q, and $D(p,q)$ is the *region of dominance of* p *over* q.

Definition 1 We set

$$R(p,q) = \begin{cases} D(p,q)\cup B(p,q) & \text{if } p<q \\ D(p,q) & \text{if } p>q \end{cases} \qquad (1)$$

and define the *extended Voronoi region of p with respect to S*, denoted $EVR(p,S)$, by

$$EVR(p,S) \quad = \quad \bigcap_{\substack{q \in S \\ q \neq p}} R(p,q). \tag{2}$$

The *Voronoi region of p with respect to S* is defined to be

$$VR(p,S) \quad = \quad int\, EVR(p,S). \tag{3}$$

The *abstract Voronoi diagram of S*, denoted $V(S)$, is defined by

$$V(S) \quad = \quad \bigcup_{p \in S} bd\; VR(p,S). \tag{4}$$

A dominance system is *admissible* if

1. The intersection of any set of bisectors can be described as the union of a finite number of faces, curves and points.

2. For all non-empty subsets S' of S

   (a) $VR(p,S')$ is either empty or homeomorphic to an open 3-ball.

   (b) $\mathbf{R}^3 = \bigcup_{p \in S'} EVR(p,S')$.

Definition 2 A *face* f of $V(S)$ is a maximal connected subset of $V(S)$ so that every point $x \in f$ lies on the boundary of exactly two Voronoi regions. An *edge* e (resp. *vertex* v) of $V(S)$ is a 1-dimensional (resp. 0-dimensional) maximal connected subset of $V(S)$ so that every point $x \in e$ (resp. v) lies on the boundary of exactly k Voronoi regions with some $k \geq 3$.

It is convenient to assume that each Voronoi region $VR(p,S)$, with $p \in S$, is bounded. To ensure this, we use a "bounding sphere trick". The trick consists in introducing a symbolic huge bounding sphere that encloses all finite features of the system of bisectors in an appropriate way:

Definition 3 A *symbolic bounding sphere*, denoted Γ, is a surface homeomorphic to the 2-sphere so that for all distinct $p,q \in S$

1. Γ and $B(p,q)$ intersect transversely in a simple closed curve.

2. Let $\mathcal{A}(S)$ denote the cell complex induced by the set of all bisectors $B(p,q)$ for all distinct $p,q \in S$. Then the part of the restriction of the cell complex $\mathcal{A}(S)$ onto $B(p,q)$ that lies outside Γ consists only of 1-cells and 2-cells that are homeomorphic to a halfline or a halfspace, respectively, see Figure 1(a).

We add a site ∞ to S and define $B(p,\infty) = B(\infty,p) = \Gamma$, for any $p \in S$. Let $D(p,\infty)$ and $D(\infty,p)$ denote the inner and outer domain bounded by Γ, respectively. We add $D(p,\infty)$ and $D(\infty,p)$ to $\mathcal{D}$, for all $p \in S \setminus \{\infty\}$.

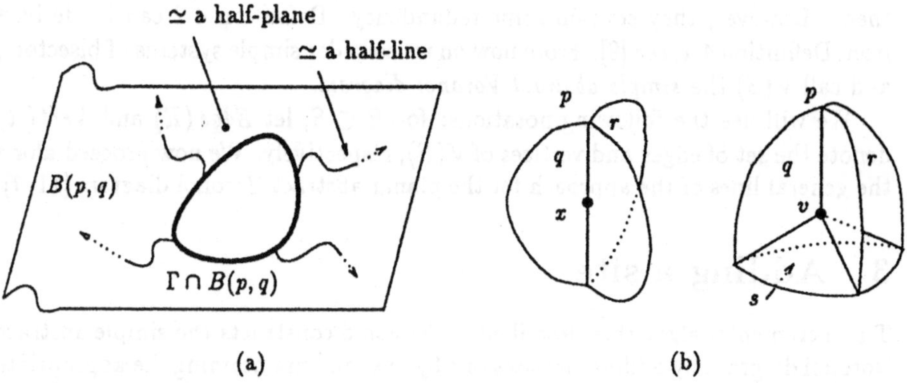

Figure 1: (a) Restriction of the part of $\mathcal{A}(S)$ that lies outside Γ onto a bisector. (b) Local shapes of the simple abstract Voronoi diagram.

Definition 4 The system of the bisectors derived from $\mathcal{D}$ is called *simple* if

1. For all $p, q, r, s, t \in S \setminus \{\infty\}$,

   (a) $B(p,q)$ and $B(p,r)$ intersect transversely. Let $B(p,q,r) = B(p,q) \cap B(p,r)$, we require that $B(p,q,r)$ does not depend on the order of p, q, r in the triple.

   (b) $B(p,q,r)$ and $B(p,s)$ intersect transversely. We set $B(p,q,r,s) = B(p,q,r) \cap B(p,s)$ and require that $B(p,q,r,s)$ does not depend on the order of p,q,r,s in the quadruple.

   (c) $B(p,q,r)$ is homeomorphic to a line, $B(p,q,r,s)$ consists of exactly one point, and $B(p,q,r,s)$ and $B(p,t)$ have empty intersection.

2. With respect to any non-empty subsets S' of S hold:

   (a) Each bounded Voronoi region is either empty or homeomorphic to an open 3-ball. The closure of any non-empty bounded Voronoi region is homeomorphic to a closed 3-ball. The boundary of any non-empty bounded Voronoi region is homeomorphic to a 2-sphere.

   (b) All edges and vertices of $V(S')$ are locally (topologically) equivalent to that of the non-degenerate Euclidean Voronoi diagram; i.e., for any point x of an edge (resp. for any vertex v) of $V(S')$, there is a neighborhood U of x (resp. v) so that $U \cap V(S')$ looks like as illustrated in Figure 1(b).

Definitions 4.1 and 4.2 are consistent; indeed, it can be easily seen that non-degenerate systems of bisectors under the Euclidean distance is a model for them. However, they contain some redundancy: Definition 4.2 can be derived from Definition 4.1, see [9]. From now on we consider simple systems of bisectors, and call $V(S)$ the *simple abstract Voronoi diagram.*

We will use the following notations: for $R \subset S$, let $Edge(R)$ and $Vert(R)$ denote the set of edges and vertices of $V(R)$, respectively. We now proceed along the general lines of the approach for the planar abstract Voronoi diagram [11, 7].

3 Adding a site

The incremental algorithm described in Section 5 constructs the simple abstract Voronoi diagram by adding the sites one by one, and maintaining the appropriate data structures. In this section, we investigate the changes of the current simple abtract Voronoi diagram when a new site is inserted. Throughout this section let $R \subset S$ with $\infty \in R, |R| \geq 4$, and $u \in S \setminus R$.

Lemma 5 *Let* $\mathcal{U} := VR(u, R \cup \{u\})$. *If* $\mathcal{U} \neq \emptyset$ *then* $V(R) \cap \overline{\mathcal{U}}$ *is a non-empty connected set that intersects* $bd\,\mathcal{U}$.

Next we investigate the topology of the intersection of the new region with a face of $V(R)$.

Lemma 6 *Let* $\mathcal{U} := VR(u, R \cup \{u\})$, *and let* f *be a face of* $V(R)$. *Assume that* f *is simply-connected. If* $f \cap \overline{\mathcal{U}} \neq \emptyset$ *then*

1. *The intersection of* $\overline{f}$ *and* $\overline{\mathcal{U}}$ *is homeomorphic to a closed disk.*
2. *The intersection of* $bd\,f$ *and* $\overline{\mathcal{U}}$ *is homeomorphic to a closed interval or a simple closed curve.*

By a simple induction argument, using Lemma 6, we see that all faces of $V(R)$ are simply-connected. We call the *skeleton* of the Voronoi diagram of R the set

$$Skel(R) = \Big(\bigcup_{e \in Edge(R)} e\Big) \bigcup \Big(\bigcup_{x \in Vert(R)} x\Big).$$

Lemma 7 *If* $\mathcal{U} \neq \emptyset$ *then*

1. *The intersection of* $Skel(R)$ *and* $\overline{\mathcal{U}}$ *is a non-empty connected set,*
2. *The intersection of* $Skel(R)$ *and* $bd\,\mathcal{U}$ *is non-empty and is not just a single point.*

The following lemma describes the possible forms of the intersection of any edge e of $V(R)$ with $\overline{\mathcal{U}}$.

Lemma 8 *Let* $e \in Edge(R)$. *If* $e \cap \overline{\mathcal{U}} \neq \emptyset$ *then* $e \cap \overline{\mathcal{U}}$ *is a single component, and* $e \setminus \overline{\mathcal{U}}$ *is also a single component (possibly empty).*

With the results above, we are now in a position to estimate the complexity of the simple abstract Voronoi diagram in 3-space.

Lemma 9 *The complexity of the 3-dimensional simple abstract Voronoi diagram of* n *sites is* $O(n^2)$.

4 The intersection of a new region with an edge

As in the previous section, let $R \subset S$ with $\infty \in R$, $|R| \geq 4$, and $u \in S \setminus R$.

Definition 10 Let e be an edge of $V(R)$.

1. We say that u *intersects e with respect to R* if $e \cap \overline{VR(u, R \cup \{u\})} \neq \emptyset$.
2. Let v be an endpoint of e. We say that u *clips e at v with respect to R* if $e \cap \overline{VR(u, R \cup \{u\})}$ contains a connected component incident to v.

4.1 Characterizing the edges

Definition 11 Let p, q, r, and s be four distinct sites in R. We call a vertex x of $V(R)$ a *pqrs-vertex* if x is incident to the p-, q-, r-, and s-region, and x is the startpoint of the oriented edge defined by p, q, r where the orientation is by the mathematical "opening bottle" rule.

Lemma 12 *Let p, q, r, and s be four distinct sites in R. Then $V(R)$ contains at most one pqrs-vertex and at most one edge incident to the p-, q-, r-region and to that vertex.*

We now adapt the basic result [7, Lemma 4] to our 3-space case.

Lemma 13 *Let e be a pqrst-edge of $V(R)$. Then, for all $R' \subset R$ containing the sites p, q, r, s, t, the point set e is also a pqrst-edge of $V(R')$. Furthermore, for any $u \in S \setminus R$, holds*

$$e \cap \overline{VR(u, R \cup \{u\})} = e \cap \overline{VR(u, R' \cup \{u\})}.$$

Lemmas 12 and 13 state that a *pqrst*-edge of $V(R)$ is also the unique *pqrst*-edge of $V(\{p, q, r, s, t\})$, and that the intersection of the *pqrst*-edge and the u-region with respect either to $R \cup \{u\}$ or to $\{p, q, r, s, t, u\}$ are identical. This observation is fundamental in defining the basic operation for the incremental algorithm in the next section.

4.2 The basic operation

Input: A 6-tuple (p, q, r, s, t, u) so that $V(\{p, q, r, s, t, u\})$ contains a *pqrst*-edge e and $u \notin \{p, q, r, s, t\}$.
Output: The structure of $e \cap \overline{VR(u, \{p, q, r, s, t, u\})}$, which is one of the following:

1. *EMPTY*: intersection is empty
2. intersection is non-empty and consists of a single component:
   (a) *ENTIRE_EDGE*: e itself
   (b) *SEGMENT_1*: a segment of e incident to the *pqrs*-endpoint
   (c) *SEGMENT_2*: a segment of e incident to the *prqt*-endpoint.

Note that, by Lemma 8, the output of the basic operation covers all the cases that may occur.

5 The Algorithm

To describe the algorithm we need some more notations. If e is a $pqrst$-edge then we call the tuple $pqrst$ the *description* of e, denoted $D(e)$; let $u \in S \setminus \{p, q, r, s, t\}$, then we say that site u *intersects* $D(e)$ if $e \cap \overline{VR(u, \{p, q, r, s, t, u\})} \neq \emptyset$. The algorithm constructs the simple abstract Voronoi diagram incrementally. It starts with four sites ∞, p, q and r, where p, q, r are chosen from $S \setminus \{\infty\}$ randomly. The algorithm then adds the remaining sites in random order. We maintain the following data structures.

1. The Voronoi diagram $V(R)$. To store $V(R)$, we use the incidence graph.
2. The history graph $\mathcal{H}(R)$. The history graph is a rooted directed acyclic graph. Its vertex set consists of a particular vertex, called *source*, and the descriptions of all the Voronoi edges that have appeared at some stage of the algorithm.

We maintain the following history-graph invariants.

1. Each vertex of $\mathcal{H}(R)$ has outdegree of at most 4. The vertices that correspond to the edges of $V(R)$ have outdegree 0, i.e., they are the leaves of $\mathcal{H}(R)$.
2. Each edge e of $V(R)$ is linked to its description $D(e)$.
3. For each site $u \in S \setminus R$ and each leaf D of $\mathcal{H}(R)$ that is intersected by u, there is a path from the source of $\mathcal{H}(R)$ to D so that the path visits only vertices that are intersected by u.

5.1 Determining the intersected edges

Let E_u be the set of edges of $V(R)$ that are intersected by u. Let A denotes the set of vertices of $\mathcal{H}(R)$ that are intersected by u. The following lemma shows that, using the history graph, E_u can be found quickly.

Lemma 14 *The set E_u can be found in time $O(|A|)$.*

Proof. Starting from the source of $\mathcal{H}(R)$, we search all vertices of $\mathcal{H}(R)$ that are intersected by u. Observe that, by the history-graph invariant 1, the outdegree of $\mathcal{H}(R)$ is bounded by 4, and that, using the basic operation, we can test in constant time whether a vertex of $\mathcal{H}(R)$ is intersected by u. Hence the search needs $O(|A|)$ time. Moreover, the history-graph invariant 3 ensures that the set of all leaves of $\mathcal{H}(R)$ that are intersected by u are found. By the history-graph invariant 2 and Lemma 13, this set is the set E_u. □

5.2 Updating the simple abstract Voronoi diagram

Define
$$V_{del} = \{x \in Vert(R) \mid \text{all edges incident to } v \text{ are clipped at } v \text{ by } u\},$$
$$V_{unch} = \{x \in Vert(R) \mid \text{no edge incident to } x \text{ is clipped at } x \text{ by } u\},$$
$$V_{new} = \{x \mid x \notin Vert(R) \text{ and } x \text{ is endpoint of } e \setminus \overline{U} \text{ for some } e \in E_u\}.$$

Lemma 15 *$Vert(R \cup \{u\}) = V_{unch} \cup V_{new}$.*

Lemma 16 *Let v be a pqrst-vertex of $V(R)$ so that, after adding site u, holds $v \in V_{unch}$. Then, for all sufficiently small neighborhoods of vertex v, we have $U \cap V(R) = U \cap V(R \cup \{u\})$. In particular, v is a pqrst-vertex of $V(R \cup \{u\})$.*

Lemma 17 *Given E_u, then $V(R \cup \{u\})$ can be constructed from $V(R)$ in time $O(|E_u|)$.*

Proof. Clearly the sets V_{del} and V_{new} can be determined from E_u in time $O(|E_u|)$. Let v be a new vertex that lies in an edge e intersected by u. Suppose e is defined by sites p, q, and r. We first show how to construct the face separating p- and u-region. Suppose e borders face f separating p- and q-region. Starting with e, we trace the sequence of edges of f that are intersected by u until we reach another new vertex v' that lies in an edge e' of f; Lemma 6 ensures that we find v'. Knowing v and v', we can construct the new edge e^* joining v and v' (if e^* was not already created by this procedure performed at some stage before), and consequently the new face f', which is $f \setminus \overline{U}$. We repeat the procedure above for the next face separating p-region and some s-region and adjacent to the face previously considered until the vertex v is reached, which is ensured because the face separating p and u is homeomorphic to a disk. In this way, we obtain the boundary, and thus, the new face separating p- and u-region.

Hence, we can build $V(R \cup \{u\})$ visiting each edge intersected by u at most three times (see also Lemma 7). □

5.3 Updating the history graph

To characterize the new vertices of $\mathcal{H}(R \cup \{u\})$, i.e., vertices that are not present in $\mathcal{H}(R)$, we need some definitions. An edge e of $V(R \cup \{u\})$ is said to be *new* if it is not a subset of any edge of $V(R)$, *shortened* if it is a proper subset of some edge of $V(R)$.

The next two lemmas are needed for proving the correctness of the procedure for updating the history graph (Lemma 20).

Lemma 18 *Let e be a shortened edge of $V(R \cup \{u\})$, let e' be the edge of $V(R)$ with $e \subset e'$, and let $w \in S \setminus (R \cup \{u\})$ intersect e with respect to $R \cup \{u\}$. Then w intersects e' with respect to R.*

Let e be a new edge of $V(R \cup \{u\})$ with endpoints x_1 and x_2. Then x_1 and x_2 are in V_{new}. Let $p, q \in R$ such that e is incident to p-, q- and u-region in $V(R \cup \{u\})$. Let f be the face of $V(R)$ separating p- and q-region. Call e_1 and e_2 the edges of $V(R)$ that border f and contain x_1 and x_2, resp. Let $\mathcal{P}$ be the path in $bd\, f \cap \overline{U}$ connecting x_1 and x_2.

Lemma 19 *Let $w \in S \setminus (R \cup \{u\})$, and let w intersect e with respect to $R \cup \{u\}$. Then w intersects either e_1 or e_2 or one of the edges of $\mathcal{P}$ with respect to R.*

Lemma 20 *Given E_u, then $\mathcal{H}(R \cup \{u\})$ can be constructed from $V(R)$ and $\mathcal{H}(R)$ in time $O(|E_u|)$.*

Proof. Suppose E_u is given, we construct $\mathcal{H}(R \cup \{u\})$ from $V(R)$ and $\mathcal{H}(R)$ as follows. Case 1, for each shortened edge e of $V(R \cup \{u\})$, let e' be the edge of $V(R)$ with $e \subset e'$, we add to the history graph the edge $(D(e'), D(e))$. Case 2, for each new edge e of $V(R \cup \{u\})$, let the set $\{e_1, e_2\} \cup \mathcal{P}$ be defined as for Lemma 19, we add to the history graph the edges $(D(e'), D(e))$ for all e' in $\{e_1, e_2\} \cup \mathcal{P}$.

We show that the history graph invariants hold for $\mathcal{H}(R \cup \{u\})$. Updating the history graph as above, only leaves of $\mathcal{H}(R)$ can get edges to new children. We show that each leaf of $\mathcal{H}(R)$ can get at most four edges to new children. Suppose e' is an edge of $V(R)$. Case 1, if e' remains an edge of $V(R \cup \{u\})$, then the construction above did not add to the history graph any edge out of $D(e')$. Case 2, there is a shortened edge e of $V(R \cup \{u\})$ so that $e \subset e'$. Since e' is contained in sets of the form $\{e_1, e_2\} \cup \mathcal{P}$, as e_1 or e_2, three times, and $D(e')$ is parent of $D(e)$, the outdegree of $D(e')$ is four. Case 3, $e' \subset \overline{\mathcal{U}}$. Since e' is contained in sets of the form $\mathcal{P}$ three times, the outdegree of $D(e')$ is three. □

5.4 Expected running time and space

Lemmas 9, 14, 17, and 20 enable us to apply the performance analysis of randomized algorithms in [3, 4]. Hence, we get the main result of this paper.

Theorem 21 *The simple abstract Voronoi diagram $V(S)$ of n sites in 3-space can be computed by a randomized algorithm in $O(n^2)$ expected time and $O(n^2)$ expected space. Furthermore, the r-th site can be inserted in $O(r)$ expected time.*

6 Applications

1. *Power diagrams.* Power diagrams in higher dimensions have been thoroughly investigated by Aurenhammer [1]. In 3-space, our algorithm is just another one to compute power diagrams within $O(n^2)$ resource bounds.
2. *Ellipsoid convex distances.* Previously, to compute diagrams under ellipsoid distances one first transforms the sites to new ones using a certain affine mapping, computes the Euclidean diagram of the new sites, and then inversely transforms the computed diagram to one under the corresponding symmetric ellipsoid distance, from which the final diagram can be obtained. In contrast to this method, our algorithm constructs diagrams under ellipsoid convex distances "directly".

The following result strongly indicates that exactly ellipsoid convex distances give rise, *in general*, to simple systems of bisectors.

Lemma 22 *Let d be a convex distance in D-space whose unit-sphere is not an ellipsoid, with $D \geq 3$. Then there are $D+1$ affinely independent points $p_1, \ldots, p_{D+1}$ so that the cardinality of $B(p_1, \ldots, p_{D+1})$ is ≥ 2.*

3. *Hausdorff distance.* The *Hausdorff distance* from a point $x \in \mathbf{R}^3$ to a segment s is defined to be $d_H(x, s) = \max\{|x - y| \; ; \; y \in s\}$; see [2]. Let S be a set of parallel segments all having the same length, and let S be non-degenerate under d_H. Then it can be shown that the arising system of bisectors under d_H is simple. Thus, our algorithm computes the Voronoi diagram of S under the Hausdorff distance in expected time $O(n^2)$ using $O(n^2)$ expected space.

7 Conclusion

Is it possible to extend our approach to diagrams in which $B(p, q, r)$ consists of an arbitrary number of bi-infinite curves, and $B(p, q, r, s)$ consists of an arbitrary number of points? Such diagrams include Voronoi diagrams of points in general position under convex distance functions that are different from the ellipsoid ones.

Acknowledgements. I thank Rolf Klein, Kurt Mehlhorn and Stephan Meiser for having directly or indirectly contributed some initial ideas. Thanks also go to Marisa Mazon Calpena and Francisco Santos for some hints concerning ellipsoid convex distances. I wish to thank the anonymous referees for suggestions that helped to improve the presentation of this abstract.

References

[1] F. Aurenhammer: *Power diagrams: properties, algorithms and applications.* SIAM Journal of Computing 16 (1987), pp. 78–96.

[2] F. Aurenhammer: *Voronoi Diagrams — A Survey of a Fundamental Geometric Data Structure.* ACM Computer Surveys 23(3), 1991.

[3] J. D. Boissonnat, O. Devillers, R. Schott, M. Teillaud, and M. Yvinec: *Applications of random sampling to on-line algorithms in computational geometry.* Discrete & Comput. Geom. 8, pp. 51–71, 1992.

[4] K. L. Clarkson and P. W. Shor: *Applications of Random Sampling in Computational Geometry, II.* Discrete & Comput. Geom. 4, pp. 387–421, 1989.

[5] C. Icking, R. Klein, N.-M. Lê, L. Ma: *Convex Distance Functions in 3-Space are Different.* Proceedings 9th ACM Symposium on Computational Geometry, 1993.

[6] R. Klein: *Concrete and Abstract Voronoi Diagrams.* LNCS 400, Springer, 1989.

[7] R. Klein, K. Mehlhorn, and S. Meiser: *Randomized Incremental Construction of Abstract Voronoi Diagrams.* Comput. Geometry: Theory and Applications 3 (1993), pp. 157–184.

[8] N.-M. Lê: *On Voronoi diagrams in the L_p-metric in higher dimensions.* Proc. 11th STACS, P. Enjalbert *et al.* (Eds.), LNCS 775, Springer, pp. 711–722, 1994.

[9] N.-M. Lê: *An axiomatic approach to Voronoi diagrams in 3-space.* Manuscript, 1994.

[10] N.-M. Lê: *Randomized incremental construction of simple abstract Voronoi diagrams in 3-space.* TR 174, Dep. of Comp. Science, FernUniv. Hagen, Germany.

[11] K. Mehlhorn, S. Meiser, and C. Ó'Dúnlaing: *On the Construction of Abstract Voronoi Diagrams.* Discrete & Comput. Geom. 6, pp. 211–224, 1991.

Properties of Probabilistic Pushdown Automata

(Extended abstract)

Ioan I. Macarie and Mitsunori Ogihara

Department of Computer Science
University of Rochester
Rochester, NY 14627

Abstract. Properties of probabilistic as well as "probabilistic plus nondeterministic" pushdown automata and auxiliary pushdown automata are studied. These models are analogous to their counterparts with nondeterministic and alternating states. Complete characterizations in terms of well-known complexity classes are given for the classes of languages recognized by polynomial time-bounded, logarithmic space-bounded auxiliary pushdown automata with probabilistic states and with "probabilistic plus nondeterministic" states. Also, complexity lower bounds are given for the classes of languages recognized by these automata with unlimited running time. It follows that, by fixing an appropriate mode of computation, the difference between classes of languages such as P and PSPACE, NL and SAC^1, PL and $Diff_>(\#SAC^1)$ is characterized as the difference between the number of stack symbols; that is, whether the stack alphabet contains one versus two distinct symbols.

1 Introduction

The notion of deterministic as well as nondeterministic auxiliary pushdown automata[1] was introduced by Cook [Coo71]. Cook proved that the class of languages recognized by these automata with work space bound $S(n)$ is exactly the class of those languages recognized by $2^{S(n)}$ time-bounded deterministic Turing machines. Properties of auxiliary pushdown automata have been studied since then. Especially, as logarithmic space-bound yields a characterization of P, we could hope to deepen our knowledge about the internal structure of P by investigating the computational power of time-bounded, logarithmic space-bounded auxiliary pushdown automata. Several tight connections between these automata and boolean circuits as well as arithmetic circuits have been obtained [Ruz80, Ven91,Vin91,AJ93].

As a natural analog to alternating Turing machines introduced by Chandra, Kozen, and Stockmeyer [CKS81], one could consider the notion of alternating auxiliary pushdown automata, which has been introduced and studied in the

[1] An auxiliary pushdown automaton is a pushdown automaton that additionally has a worktape (usually logarithmically space-bounded in the input) and two-way access to the input tape.

seminal paper by Ladner, Lipton, and Stockmeyer [LSL84]. Interestingly, with the aid from alternation, auxiliary pushdown automata gain immense computational power. Ladner, Lipton, and Stockmeyer showed that the languages recognized by $S(n)$ space-bounded alternating auxiliary pushdown automata are exactly those languages recognized by $2^{2^{S(n)}}$ time-bounded deterministic Turing machines. Thus, while only single exponentiation is achieved by nondeterminism, double exponentiation is achieved by alternation. One should observe here that, by employing logarithmic space-bound, nondeterministic and alternating auxiliary pushdown automata respectively characterize P (i.e., polynomial-time languages) and EXP (i.e. exponential-time languages).

This observation gives rise to question of how much computational power the space-bounded auxiliary pushdown automata will gain if different computation modes, such as probabilistic computation and "probabilistic plus nondeterministic" computation, are employed.

The purpose of the present paper is to study the computational power of auxiliary pushdown automata with probabilistic computation as well as "probabilistic plus nondeterministic" computation, where by "probabilistic plus nondeterministic" we mean that the automata have both probabilistic states and nondeterministic states and nondeterministic choices are made to maximize accepting probability. The concept of "probabilistic plus nondeterministic" computation is an interpretation of Papadimitriou's "games against nature" [Pap83] and closely related to the notion of Arthur-Merlin games (see for example, [GS89,Con89].) We are particularly interested in the power of those automata with logarithmic space-bound. We study the computational power of these models in the following three settings: (i) with no work tapes, (ii) with logarithmic space-bound plus polynomial time-bound, and (iii) with logarithmic-space but no time-bound.

Firstly, regarding case (i), we show that nondeterministic pushdown automata are strictly less powerful than probabilistic pushdown automata and that probabilistic pushdown automata are strictly less powerful than "probabilistic plus nondeterministic" pushdown automata.

Secondly, as to case (ii), we obtain that probabilistic auxiliary pushdown automata are less powerful than both alternating and "probabilistic plus nondeterministic" auxiliary pushdown automata. We show that logspace polynomial-time auxiliary pushdown automata recognize precisely PSPACE (i.e., polynomial-space) languages with either alternating or "probabilistic plus nondeterministic" states. In contrast, the languages recognized by probabilistic auxiliary pushdown automata are characterized as the difference between two $\#SAC^1$ functions being positive, and consequently, they are contained in TC^1.

Finally, regarding case (iii), we show that "probabilistic plus nondeterministic" pushdown automata are at least as powerful as alternating ones. In fact, we show that all of EXP are recognized by "probabilistic plus nondeterministic" pushdown automata with logarithmic space-bound.

2 Definitions and notations

We will use a short-hand 'pda' to denote pushdown automata and add to word 'pda' prefixes d-, n-, p-, a-, and (p+n)- to denote deterministic, nondeterministic, probabilistic, alternating, and "probabilistic plus nondeterministic" pushdown automata, respectively. As to d-pda's and n-pda's we use the standard definitions (see for example [HU79]), which employ acceptance with final states (stack need not be empty when halting). A configuration of a pda is a triple consisting of its state, stack content, and input head position. It is worth noticing that the input head is one-way. Without loss of generality, we assume for an n-pda, there exist exactly two reachable configurations from a non-halting configuration. Thus, the computation of an n-pda on an input can be viewed as a binary tree with the initial configuration as the root and the halting configuration as the leaves. We say that an n-pda accepts an input if there is a leaf corresponding to an accepting configuration in the computation tree associated with the computation of the n-pda.

Definitions for a-pda's are from [CKS81]. The computation of an a-pda can be viewed as a binary tree, where non-leaf nodes are either universal nodes or existential nodes. Acceptance of an a-pda is recursively defined as follows. (I) A leaf is accepting if it corresponds to an accepting configuration. (II) A universal node is accepting if its children are all accepting. (III) An existential node is accepting if some of its children are accepting. (IV) The computation tree is accepting if its root is accepting. We say that the a-pda accepts an input if the root of the associated computation tree is accepting.

A probabilistic pushdown automaton (p-pda) is a pda with probabilistic states. There are two configurations reachable from a non-halting configuration, and one of the two reachable configuration is chosen with probability a half by flipping a fair coin. Again, the computation of a p-pda can be viewed as a binary tree, where each path from the root to a leaf, which represents a particular computation path, is associated with the probability that the corresponding computation path is chosen. The probability is equal to $1/2^k$, where k is the number of coin-tosses occuring along that computation. As we can assume that there are exactly two reachable configuration from a non-halting configuration, the number k equals the length of the computation path. The acceptance probability of a p-pda on an input is the sum of the probability associated with accepting computation paths. We say that a p-pda accepts an input if the acceptance probability is greater than a half.

A "probabilistic plus nondeterministic" pushdown automaton ((p+n)-pda) is a pda with nondeterministic and probabilistic states. A strategy of a (p+n)-pda on an input string is a function that maps each nondeterministic configuration to one of its successors. We assume that a (p+n)-pda chooses one of its strategies. Therefore, the computation of a (p+n)-pda can be viewed as a collection of binary trees, each representing the computation as a p-pda for a fixed strategy. We say that a (p+n)-pda accepts an input if there exists a binary tree whose acceptance probability (as a computation tree for a p-pda) is greater than a half. The concept of (p+n)-pda's corresponds to "games against nature" (defined in [Pap83]) with probabilistic pushdown automata as verifiers. A (p+n)-pda is said to have bounded-error property if there is a fixed constant $\epsilon > 0$ such that for any nondeterministic strategy, the acceptance probability does not fall between $1/2$ and $1/2 + \epsilon$. Bounded-error (p+n)-pda's correspond to Arthur-Merlin games with probabilistic pushdown automata as verifiers.

For each of the computation modes above, the notion of auxiliary pushdown automata (auxpda's) is naturally defined by allowing each pda to have access to single worktape and by allowing two-way access on the input tape. A configuration of an auxpda is a quintuple consisting of its state, worktape content, stack content, input head position, worktape head position. In the same way as we did for pda's we define acceptance criteria for auxpda's. (Formal definitions for deterministic, nondeterministic, and alternating auxiliary pushdown automata can be found in [Coo71,LSL84].) For a function $S(n)$, we say that an auxpda is $S(n)$ space-bounded if for any input x and for any computation path, the auxpda uses at most $O(S(n))$ worktape cells. Although many of our results in this paper hold for a general space-bounding function, we will focus on logarithmic space-bounded auxpda's.

In what follows we list the notations that are used throughout this paper.

- x-pda, x $\in$ { a, d, n, p, (p+n)}, denote the sets of alternating, deterministic, nondeterministic, probabilistic, and "probabilistic plus nondeterministic" pushdown automata, respectively;
- (X)CFL, X $\in$ { A, N, P, P+N}, denote the classes of languages recognized by alternating, nondeterministic, probabilistic, and "probabilistic plus nondeterministic" pushdown automata, respectively;
- x-auxpda, x $\in$ { a, d, n, p, p+n}, denote the sets of alternating, deterministic, nondeterministic, probabilistic, and "probabilistic plus nondeterministic" auxiliary pushdown automata, respectively;
- (X)AuxPDA(logspace), X $\in$ { A, N, P, P+N}, denote the classes of languages recognized by alternating, nondeterministic, probabilistic, and "probabilistic plus nondeterministic" logspace auxiliary pushdown automata, respectively;
- (X)AuxPDA(logspace, T-time), X $\in$ { A, N, P, P+N}, denote the classes of languages recognized by alternating, nondeterministic, probabilistic, and "probabilistic plus nondeterministic" T-time logspace auxiliary pushdown automata, respectively, where T-time will be either poly-time (polynomial-time) or exp-time (exponential-time);
- Bounded(P+N)AuxPDA(logspace, T-time) denotes the class of languages recognized by bounded-error "probabilistic plus nondeterministic" T time-bounded logspace auxiliary pushdown automata;
- P (EXP) is the class of languages recognized by deterministic Turing machines in polynomial (exponential) time;
- NL (RPL, PL) is the class of languages recognized by logspace nondeterministic Turing machines (logspace one-sided bounded-error probabilistic Turing machines, logspace unbounded-error probabilistic Turing machines);
- PSPACE is the class of languages recognized by polynomial space-bounded deterministic Turing machines;

- #L is the class of functions that count the number of accepting computation paths of logspace nondeterministic Turing machines;
- $\#SAC^1$ is the class of functions that count the number of accepting computations of a logspace-uniform family of semi-unbounded fan-in circuits;
- $\text{Diff}_>(\#X)$, $X \in \{L, SAC^1\}$, are the classes that consist of the languages L for which there exist $f, g \in \#X$ such that for every x, it holds that

$$x \in L \Leftrightarrow f(a) > g(a);$$

- ALT(poly-time), ALT(poly-space), ALT(logspace) are respectively the classes of languages recognized by polynomial time-bounded, polynomial space-bounded, and logarithmic space-bounded alternating Turing machines;
- AM(logspace) is the class of languages recognized by Arthur-Merlin games with probabilistic logarithmic space-bounded verifiers.

3 Results

It is well-known that the d-pda's are less powerful than the n-pda's. We supplement this result by showing that the n-pda's are less powerful than the p-pda's, and the p-pda's are less powerful than the (p+n)-pda's.

Theorem 1 $\text{(N)CFL} \subsetneqq \text{(P)CFL} \subsetneqq \text{(P + N)CFL}$.

Next we study relations among the closures under logspace reductions of these classes. Regarding nondeterministic computation, Sudborough [Sud78] showed that (N)AuxPDA(logspace, poly-time) = LOG(N)CFL and Venkateswaran [Ven91] showed that $\text{LOG(N)CFL} = SAC^1$. For other computation modes, probabilistic and "probabilistic plus nondeterministic," and alternating computation, we obtain the following theorem.

Theorem 2

$$\begin{aligned}
\text{LOG(P)CFL} &= \text{(P)AuxPDA(logspace, poly-time)} &&= \text{Diff}_>(\#SAC^1),\\
\text{LOG(P + N)CFL} &= \text{(P + N)AuxPDA(logspace, poly-time)} &&= \text{PSPACE, and}\\
\text{LOG(A)CFL} &= \text{(A)AuxPDA(logspace, poly-time)} &&= \text{PSPACE}.
\end{aligned}$$

Remark 1 The second equality of the first relation contrasts with the relation: PL= $\text{Diff}_>(\#L)$ [Vin91]. Also note that, from $\text{Diff}_>(\#SAC^1) \subseteq TC^1$ [AJ93], it follows that $\text{LOG(P)CFL} \subseteq TC^1$. From the last two relations, we obtain that PSPACE can be defined as the class of languages that are logspace reducible to languages recognized by (p+n)-pda's (or by a-pda's), i.e., by one-head one-way one-stack automata with "probabilistic plus nondeterministic" states (or alternating states). In [Mac95], it is shown that P can be defined as the class of languages that are logspace reducible to languages recognized by one-head one-way one-counter automata with "probabilistic plus nondeterministic" states. A similar result can be obtained for alternating computation by adapting a technique from [Kin88]. As a counter can be regarded as a stack with one stack symbol, we can say that, in terms of one-head one-way one-stack automata with either alternating states or "probabilistic plus nondeterministic" states, the difference between P and PSPACE is characterized as the difference between the number of stack symbols: whether one or two symbols are allowed on the stack.

Galil and Sudborough [Gal74,Sud78] proved that a language L is recognized by a poly-time logspace nondeterministic auxpda if and only if L is logspace many-one reducible to a language recognized by a nondeterministic pda. The statement holds for other modes of computation.

MetaTheorem 1 *Let μ denote a computation mode chosen from nondeterministic, probabilistic, alternating, "probabilistic plus nondeterministic". Then a language L is recognized by a polynomial-time logspace μ-auxpda if and only if L is logspace many-one reducible to a language recognized by a μ-pda; i.e., by a one-head, one-way, one-stack, finite-state μ-automaton.*

Sudborough [Sud75] showed that the class NL coincides with the class of languages logspace reducible to languages recognized by a one-head, one-way, one-counter, finite-state nondeterministic-automaton. Similar results have been proven for probabilistic as well as "probabilistic plus nondeterministic" computation modes [Mac95]. We summarize all these in the next Metatheorem.

MetaTheorem 2 *Let μ denote a computation mode chosen from nondeterministic, probabilistic, alternating, and "probabilistic plus nondeterministic." Then a language L is recognized by a polynomial time-bounded μ-type Turing machine if and only if L is logspace many-one reducible to a language recognized by a one-head, one-way, one-counter, finite-state μ-automaton.*

From Theorem 2 and Metatheorem 2 it follows that the difference between NL and SAC^1 = LOG(N)CFL, and between PL = $\mathrm{Diff}_>(\#\mathrm{L})$ and LOG(P)CFL = $\mathrm{Diff}_>(\#\mathrm{SAC}^1)$, can be regarded as the difference between using one symbol versus using two symbols in a stack, in appropriate models of computation. This observation parallel the observation from Remark 1 concerning the difference between P and PSPACE.

We obtain the following theorem for the unbounded time computation.

MetaTheorem 3 *Let μ denote a computation mode chosen from nondeterministic, probabilistic, alternating, "probabilistic plus nondeterministic", bounded-error and one-sided-error probabilistic. Then a language L is recognized by a logspace μ-auxpda if and only if L is logspace many-one reducible to a language recognized by a one-sweeping-head, one-stack, finite-state μ-automaton.*

For the nondeterministic mode, this was proven in [Sud78], and it can be easily extended to other cases.

We attempt to characterize the classes of languages recognized by logspace auxiliary pushdown automata. As to the deterministic and nondeterministic aux-pda's, Cook [Coo71] showed that they are equal to P; that is,

$$\text{(D)AuxPDA(logspace)} = \text{(N)AuxPDA(logspace)} = \text{P}.$$

As to the alternating auxpda's, Ladner, Lipton, and Stockmeyer [LSL84] showed that the class equals exponential-time; that is,

$$\text{(A)AuxPDA(logspace)} = \text{ALT(poly-space)} = \text{EXP}.$$

Since nondeterministic acceptance criteria can be modified to probabilistic criteria with a simple modification of the computation, we have $\text{P} \subseteqq$ (P)AuxPDA(logspace). So far, this is the largest lower bound we have obtained for logspace bounded probabilistic auxpda's.

In order to obtain lower bounds for (P + N)AuxPDA(logspace), the following lemma, which is proven by some counting technique, is useful.

Lemma 1. [Ang80] *Deterministic logspace auxiliary pushdown automata can simulate deterministic exponential counters.*

This lemma provides an interesting example in which deterministic logspace auxiliary pushdown automata are better than polynomial-time deterministic Turing machine, in spite of the fact that both types of devices recognize the same languages.

Now based on the above simulation lemma, we obtain the following theorem.

Theorem 3 $\text{EXP} \subseteqq$ (P + N)AuxPDA(logspace, exp-time) *and* $\text{EXP} \subseteqq$ Bounded(P + N)AuxPDA(logspace, double-exp-time).

The proof of the second inclusion in the above can be regarded as an extension of the proof for $\text{P} = \text{ALT(logspace)} \subseteqq \text{AM(logspace)}$ by Condon [Con89], which is an extension of the proof for $\text{NL} \subseteqq \text{RPL}$ by Gill [Gil77].

4 Proofs

Proof of Theorem 1 First we show that $\text{(N)CFL} \subsetneqq \text{(P)CFL}$. In order to simulate an n-pda, our p-pda chooses one of the two possible successors by flipping a fair coin, accepts if the n-pda reaches an accepting configuration, and accepts or rejects with probability 1/2 if it reaches a rejecting configuration. This gives $\text{(N)CFL} \subseteqq \text{(P)CFL}$. In order to prove the properness, recall that $L = \{a^n b^n c^n \mid n \in \mathbf{N}\} \notin \text{(N)CFL}$, that $L' = \{a^n b^n \mid n \in \mathbf{N}\}$ is recognized by a one-head, two-way, bounded-error probabilistic finite-state automaton [Fre81], and that one-head, two-way, probabilistic finite-state automata can be simulated by one-head, one-way, probabilistic finite-state automata [Kan89]. Thus, L' is in (P)CFL. Consider a probabilistic pda that, on input $u = a^k b^l c^m$, checks probabilistically whether $a^k b^l$ is in L' using $O(1)$ working space, and pushes b^l on the stack; if the p-pda finds that $a^k b^l$ is not L' then it rejects, else it checks deterministically if $l = m$ and accepts or rejects depending on whether this equality is true or false. It follows that L is in (P)CFL.

Next we show that $\text{(P)CFL} \subsetneqq \text{(P + N)CFL}$, where inclusion holds trivially. The properness follows from Theorem 2 by observing that

$$\text{Diff}_{>}(\#\text{SAC}^1) \subseteq \text{Dspace}(\log^2 n) \subsetneqq \text{PSPACE}. \blacksquare$$

Proof of MetaTheorem 1 and MetaTheorem 3 Omitted. ∎

Proof of Theorem 2 The first equalities in each relation follows from MetaTheorem 1. We will prove the second equalities.
[(P)AuxPDA(logspace, poly-time) = Diff$_>$(#SAC^1):] ($\subseteq$) For each polynomial-time logspace p-auxpda we build two polynomial-time logspace n-auxpda's so that the acceptance probability of the p-pda is $> 1/2$ if and only if the difference between the numbers of accepting paths of the two n-auxpda's is > 0. Note that each polynomial-time logspace p-auxpda P can be modified (by adding some extra states and some extra space) into a p-auxpda P' that makes probabilistically branches at each step and all its paths have the same length. So, we have only to define M_1 (M_2) to be an n-auxpda that simulates probabilistic choices of P' and accepts if and only if P' accepts (rejects).

($\supseteq$) For any two n-auxpda's, we build complete (and of equal length) computation trees so that the number of accepting computations is preserved and each rejecting configuration in the initial n-auxpda's branches into two complete subtrees of equal size (one accepting and the other rejecting on any path). Additionally, the second n-auxpda has its final configurations complemented. By replacing the nondeterministic states by probabilistic states we obtain two p-auxpda's. The p-auxpda we are looking for, probabilistically chooses to simulate one of these two p-auxpda's.

[PSPACE $\subseteq$ (A)AuxPDA(logspace, poly-time):] Recall that PSPACE = ALT(poly-time) [CKS81]. In order to prove that ALT(poly-time) $\subseteq$ (A)AuxPDA(logspace, poly-time), we adapt a technique from [LSL84], which was used to show that (A)AuxPDA(logspace) = ALT(poly-space). The details are omitted.

[(A)AuxPDA(logspace, poly-time) $\subseteq$ (P + N)AuxPDA(logspace, poly-time):] Adapt a standard technique used to prove NP $\subseteq$ PP. We omit the details. ∎

Proof of Theorem 3 Ladner, Lipton, and Stockmeyer [LSL84] showed that EXP = (A)AuxPDA(logspace, exp-time). Consequently, it suffices to prove that (A)AuxPDA(logspace, exp-time) $\subseteq$ (P + N)AuxPDA(logspace, exp-time) and that (A)AuxPDA(logspace, exp-time) $\subseteq$ Bounded(P + N)AuxPDA(logspace, double-exp-time). But the proof for (A)AuxPDA(logspace, poly-time) $\subseteq$ (P + N)AuxPDA(logspace, poly-time) does not seem to be applicable for this case, for, the running time of the simulating device is exponential here, and apparently we do not seem to have enough space to count coin tosses and to generate low probability events. This difficulty is solved by using the simulation from Lemma 1.

To prove the first relation, we use the method from the proof of Theorem 2 combined with Lemma 1. For each a-auxpda M, processing a long enough length-n input string x in time less than 2^{n^k}, we build an equivalent (p+n)-auxpda N that has "similar" states and simulate "closely" the computation of M on x, as follows: at the beginning of the simulation, N tosses 2^{n^p} fair coins using the technique from Lemma 1 (where $p > k$), and accepts x if the outcome is "all heads", an event that occurs with probability $1/2^{2^{n^p}}$. If this event does not occur, then N simulates the computation of M on x, as in the proof of Theorem 2.

In order to establish the second relation, we combine the counting technique from Lemma 1 with the standard technique to prove that NL $\subseteq$ RPL [Gil77] and that Alt(logspace) $\subseteq$ AM$(\log n)$ [Con89]. The analysis of the algorithm below can be done as it is done in [Gil77]. So, we omit it. We prove that for each automaton M, processing a long enough length-n input string x in time less than 2^{n^k}, we build the following equivalent bounded-error (p+n)-auxpda B that operates in double-exponential time:

- Associated with the nondeterministic states of M, B has nondeterministic states (so existential branches in M are replaced by existential branches in B).
- Associated with the universal states of M, B has probabilistic states (so universal branches in M are replaced by probabilistic branches in B).
- When M reaches an accepting state, then B tosses a fair coin 2^{n^p} times using the counting technique from Lemma 1 (where $p > k$), and accepts if tosses are all heads. If not, B restarts the simulation of M from the very beginning.
- When M enters a rejecting state, then B rejects.

The crucial effect of B's computation is B drastically decreases the weights of M's accepting paths, leaving the weights of the rejecting paths unmodified. Suppose that M accepts x. Then there exists a strategy in which all paths are accepting, so B will accept x. Suppose that M rejects x. As the computation tree of M has height less than 2^{n^k} and the number of paths is less than $2^{2^{n^k}}$ we note that, between two consecutive "simulation rounds", the weight of a rejecting path of B is greater than $1/2^{2^{n^k}}$, which is greater than $2^{2^{n^k}}/2^{2^{n^p}}$, which is grater than the weights of all the accepting paths. By the definition of B, B halts with probability 1, so B rejects x. It follows that M and B are equivalent and the computation time of B is double-exponential. ∎

5 Acknowledgments

The first author is grateful to Eric Allender for enjoyable discussions on this topic.

References

[AJ93] E. Allender and J. Jiao. Depth reduction for noncommutative arithmetic circuit. In *Proceedings of the 25th Symposium on Theory of Computing*, pages 515–522. ACM Press, 1993.

[Ang80] D. Angluin. On Relativising Auxiliary Pushdown Machines. *Mathematical Systems Theory*, 13(4):283–299, 1980.

[CKS81] A. Chandra, D. Kozen, and L. Stockmeyer. Alternation. *Journal of the Association for Computing Machinery*, 28(1):114–133, January 1981.

[Con89] A. Condon. *Computational models of games.* M.I.T. Press, 1989.

[Coo71] S. Cook. Characterizations of pushdown machines in terms of time-bounded computers. *Journal of the Association for Computing Machinery*, 18(1):4–18, January 1971.

[Fre81] R. Freivalds. Probabilistic two-way machines. In *Proceedings of the 6th Symposium on Mathematical Foundations of Computer Science*, pages 33–45. Springer-Verlag *Lecture Notes in Computer Science #118*, 1981.

[Gal74] Z. Galil. Two-way deterministic pushdown automata and some open problems in the theory of computation. In *Proceedings of the 15th Symposium on Switching and Automata Theory*, pages 170–177, 1974.

[Gil77] J. Gill. Computational complexity of probabilistic Turing machines. *SIAM Journal on Computing*, 6(4):675–695, 1977.

[GS89] S. Goldwasser and M. Sipser. Private coins versus public coins in interactive proof systems. In S. Micali, editor, *Randomness and Computation*, pages 73–90. Advances in Computing Research # 5, JAI Press Inc., Greenwich, CT, 1989.

[Har72] J. Hartmanis. On non-determinancy in simple computing devices. *Acta Informatica*, 1:336–344, 1972.

[HU79] J. Hopcroft and J. Ullman. *Introduction to Automata Theory, Languages, and Computation.* Addison-Wesley, Reading, MA, 1979.

[Kan89] J. Kaneps. Stochasticity of the languages recognized by 2-way finite probabilistic automata. *Diskretnaya Mtematika*, 1(4):63–77, 1989. (Russian).

[Kin88] K. King. Alternating multihead finite automata. *Theoretical Computer Science*, 61:149–174, 1988.

[LSL84] R. Ladner, L. Stockmeyer, and R. Lipton. Alternating pushdown and stack automata. *SIAM Journal on Computing*, 13(1):135–155, February 1984.

[Mac95] I.I. Macarie. On the structure of log-space probabilistic complexity classes. In *Proceedings of the 12th Symposium on Theoretical Aspects of Computer Science*, pages 583-596. Springer-Verlag *Lecture Notes in Computer Science #900*, 1995.

[Mon76] B. Monien. Transformational methods and their application to complexity problems. *Acta Informatica*, 6:95–108, 1976. Corrigendum, 8:383–384, 1977.

[Pap83] C. Papadimitriou. Games against nature. In *Proceedings of the 24th Symposium on Foundations of Computer Science*, pages 446–450. IEEE Computer Society Press, 1983.

[Ruz80] W. Ruzzo. Tree-size bounded alternation. *Journal of Computer and System Science*, 21:218–235, 1980.

[Sud75] I. Sudborough. On tape-bounded complexity classes and multihead finite automata. *Journal of Computer and System Science*, 10:62–76, 1975.

[Sud78] I. Sudborough. On the tape complexity of deterministic context-free languages. *Journal of the Association for Computing Machinery*, 25(3):405–414, July 1978.

[Ven91] H. Venkateswaran. Properties that characterize LOGCFL. *Journal of Computer and System Science*, 43:380–404, 1991.

[Vin91] V. Vinay. Counting auxiliary pushdown automata and semi-unbounded arithmetic circuits. In *Proceedings of the 6th Conference on Structure in Complexity Theory*, pages 270–284. IEEE Computer Society Press, 1991.

Formal Parametric Equations

G.S. Makanin, Steklov Mathematical Institut
H. Abdulrab, LIR, INSA de Rouen
M.N. Maksimenko, LIR, INSA de Rouen

Abstract

The goal of this article is to present formal parametric equations, used as a new tool for the computation of the general solution *of word equation. The computation of the general solution is reduced here to the computation of a finite graph whose nodes are families of formal parametric equations, and whose arcs are labelled by some powerful transformations that can replace arbitrary long sequences of elementary transformations.*

The main concepts (tables of polarisation, formal parametric transformations) and ideas of our approach are presented and demonstrated for the computation of a finite graph describing the general solution of word equations with three variables.

1 Introduction, definitions, and notations

Let Π be a free monoid with a countable alphabet of generators

$$a_1, a_2, \ldots, a_k, \ldots \tag{1}$$

A *coefficientless* equation in Π is given by an alphabet of word variables

$$x_1, x_2, \ldots, x_n \tag{2}$$

and a left noncancellable equality

$$\varphi(x_1, x_2, \ldots, x_n) = \psi(x_1, x_2, \ldots, x_n) \tag{3}$$

(the two words $\varphi(x_1, \ldots, x_n)$ and $\psi(x_1, \ldots, x_n)$ do not start by the same letter).

A list $X_1, X_2, \ldots, X_n$ of words on the alphabet (1) is called a *solution* of the equation (2), (3) if the words $\varphi(X_1, X_2, \ldots, X_n)$ and $\psi(X_1, \ldots, X_n)$ coincide.

By the *general solution* of a word equation we mean a description of the set of all the solutions of the equation by a finite set of formulas depending on the word variables and on some specially organised collections of natural parameters.

If $\varphi(x_1, \ldots, x_n)$ and $\psi(x_1, \ldots, x_n)$ are empty words, the equation (2), (3) is called *trivial*.

Two types of *elementary transformations* of the equation (2), (3), called *nondegenerated* and *degenerated* elementary transformations, are defined:

- The nondegenerated transformation $x_p \to x_q x_p$, where $p \neq q$, can be applied to the equation (2), $x_p\varphi_1(x_1, x_2, \ldots, x_n) = x_q\psi_1(x_1, x_2, \ldots, x_n)$. The result of the application of this nondegenerated transformation is given by the alphabet (2) and the equality $x_p(\varphi_1)^{x_p \to x_q x_p} = (\psi_1)^{x_p \to x_q x_p}$.

- The degenerated transformation $x_p \to 1$, can be applied to the equation (2), $x_p\varphi_1(x_1, x_2, \ldots, x_n) = \psi(x_1, x_2, \ldots, x_n)$. The result of the application of this degenerated transformation is the equation given by the alphabet $x_1, \ldots, x_{p-1}$, $x_{p+1}, \ldots, x_n$ and the equality $(\varphi_1)^{x_p \to 1} = (\psi)^{x_p \to 1}$, after all the possible left cancellations.

A sequence of equations $E \to E_1 \to, \ldots, \to E_r$ connected by the elementary transformations $a_1, \ldots, a_r$ is called a *finished sequence* associated with E, if E_r is trivial.

The transformations

$$\begin{cases} x_1 \to W_1(x_1, x_2, \ldots, x_n) \\ \ldots \\ x_n \to W_n(x_1, x_2, \ldots, x_n) \end{cases} \tag{4}$$

resulting from all the successive applications of the elementary transformations in any finished sequence of the equation (2), (3) is called a *principal solution* of the equation (2), (3).

It is well-known [Len 72, Lot 83] that:

Theorem

For each solution $X_1, \ldots, X_n$ of the equation (2), (3) there exists a principal solution (4) and a list $L_1, \ldots, L_n$ of words on the alphabet (2) such that $X_i = W_i(L_1, \ldots, L_n)$, for all $i = 1, \ldots, n$. On the other hand, for each principal solution (4) the words $\varphi(W_1(x_1, \ldots, x_n), \ldots, W_n(x_1, \ldots, x_n))$ and $\psi(W_1(x_1, \ldots, x_n), \ldots, W_n(x_1, \ldots, x_n))$ coincide. □

Thus, the set of all the solutions of any equation in Π can be described by the set of all the finished sequences of elementary transformations of this equation.

An *elementary* equation is defined by the alphabet (2) and the equality

$$x_1 x_2 \ldots x_n = x_{i_1} x_{i_2} \ldots x_{i_n}$$

where $\{i_1, \ldots, i_n\}$ is a permutation of $\{1, \ldots, n\}$, and $i_1 \neq 1$, $i_n \neq n$.

The process [Len 72; Plo 72] generating a set of all the principal solutions (i.e. a *complete* set of *minimal unifiers* [Sie 89]) by means of elementary transformations does not generally terminate. It gives rise, in general, to an infinite graph. A finite graph (called *PigPid* graph by Lentin) is ensured only in the case where each variable of the equation has at most two occurrences. This is for example the case of elementary equations.

We denote the length of the word X_i by $\partial(X_i)$.

Coincidence of two words P and Q will be denoted by $P == Q$.

We denote by (x), where x is a word, the set $\{1, x\}$.

By a *directed equation* in Π, we mean an equation of the form $x_t\varphi(x_1, \ldots, x_n) = x_s\psi(x_1, \ldots, x_n)$ with the additional *condition* $\partial(x_t) > \partial(x_s)$, (i.e. the *relation* $\partial(X_t) > \partial(X_s)$ is satisfied by each solution of the directed equation). This directed equation will be written as follows:

$$x_t\varphi(x_1,\ldots,x_n) \rightarrow x_s\psi(x_1,\ldots,x_n). \quad (5)$$

An expression of the form $\alpha(x_1,\ldots,x_n)\ldots = \beta(x_1,\ldots,x_n)\ldots$ will be called the *prefix-equation* of the equation $\alpha(x_1,\ldots,x_n)\varphi(x_1,\ldots,x_n) = \beta(x_1,\ldots,x_n)\psi(x_1,\ldots,x_n)$, if $\alpha(x_1,\ldots,x_n)$ and $\beta(x_1,\ldots,x_n)$ are minimal nonempty words, such that every variable x_i occurs in $\alpha(x_1,\ldots,x_n)$ iff it occurs in $\beta(x_1,\ldots,x_n)$. If each variable of (2) occurs in the prefix-equation, it is called *nondegenerated*. Otherwise, it is called *degenerated*. For example, the equation $x_1x_2^2x_3^3x_2 = x_3x_1^2x_2^2x_1$ has the nondegenerated prefix-equation $x_1x_2^2x_3\ldots = x_3x_1^2x_2\ldots$. And the equation $x_1x_2^2x_3^3 = x_2x_1^2x_2^2x_3$ has the degenerated prefix-equation $x_1x_2\ldots = x_2x_1\ldots$.

We suppose that each variable of (2) occurs in each hand-member of the equation (otherwise, it is sufficient to concatenate the word $x_1x_2\ldots x_n$ at the right of each hand-member, and to use the new equation obtained after the concatenation). Consequently, the equation (2), (3) can be written (after a possible renaming of variables) in the form:

$$x_1\varphi(x_1,\ldots,x_n) = \alpha(x_2,\ldots,x_n)x_1\psi(x_1,\ldots,x_n), \quad (6)$$

where the first occurrence of x_1 in the right hand-member is extracted.

2 Main Goal

We are interested in finding some precise and explicit functions that allow to describe the general solution of word equations. These functions depend on the word variables and on some specially organised collections of natural parameters. For examples:

1) The general solution of the equation $x_1x_2 = x_2x_1$ in (1) is given by: the free word variable u, and all the values of the natural parameters α, β in: $X_1 = u^\alpha, X_2 = u^\beta$.

2) The general solution of *mirror equations* with n variables is given in [Mak 91], [Abd-Maks 94] using words *with vectors of parameters*.

We suppose in our approach that the general solution can be computed by induction on the number of variables. Equations with one variable are isomorphic to systems of linear diophantine equations. The general solution of such a system can be effectively computed [Mak 78], [Abd-Maks 94].

Our main goal in the project of the computation of the general solution of word equations [Mak-Abd 94] is to prove that the use of some new nondegenerated transformations (parametric, Nielsen's, Rouen's transformations,...) that both replace some arbitrary long sequences of elementary transformations and simplify the structure of transformed equations, allows to find a finite process that reduces the computation of the general solution to that of elementary equations (for which elementary transformations are sufficient, as mentioned above). This process can be represented by a finite graph whose nodes are sets of equations, and whose arcs are labelled by the transformations, from which the general solution can be deduced.

Then, for each node of such a graph, one can apply degenerated transformations. The application of such transformations produces equations with less than n variables. In which case the computation of the general solution is supposed known by induction.

The obtained graph should then be used to find a new transformation allowing to replace the sequences of elementary transformations where n variables can be transformed unboundedly, in order to be used in the graph of equations with $n+1$ variables.

In this framework, the general solution of equations with coefficients (or with constants) $\varphi(x_1,\ldots,x_n,a_1,a_2,\ldots,a_m) = \psi(x_1,\ldots,x_n,a_1,a_2,\ldots,a_m)$ is reduced to the case of coefficientless equations. In fact, it is sufficient to replace each coefficient a_i by a new variable y_i, and to respect the following three rules:

1) An equation of the form $y_i\varphi(x_1,\ldots,x_n,y_1,y_2,\ldots,y_m) = y_j\psi(x_1,\ldots,x_n, y_1,y_2,\ldots,y_m)$, with $i \neq j$, has no solution.

2) No degenerated transformation of the form $y_i \rightarrow 1$ is allowed.

3) Nondegenerated transformations to each equation of the form $x_i\varphi(x_1,\ldots,x_n,y_1,y_2,\ldots,y_m) = y_j\psi(x_1,\ldots,x_n,y_1,y_2,\ldots,y_m)$, are only applied to the directed equation $x_i\varphi(x_1,\ldots,x_n,y_1,y_2,\ldots,y_m) \rightarrow y_j\psi(x_1,\ldots,x_n,y_1,y_2,\ldots,y_m)$.

3 Main Methods and results

We propose two methods to characterise the nodes of our desired finite graph.

3.1 Prefix-equations method

In this method, each node will be characterised by its prefix-equation. The prefix-equation p associated with an equation e is the minimal part of e that defines the transformations. The other part of the equation only receives the application of the transformation. Remark that if p is an elementary equation, then the set of the solutions of e is a subset of the set of the solutions of p.

Before showing a complete example using this method for equations with two variables, we first introduce parametric transformations.

It is easy to observe that in each solution of (6), we have $X_1 = (\alpha(X_2,\ldots, X_n))^p\alpha'(X_2,\ldots,X_n)$, where $\alpha'(X_2,\ldots,X_n)$ is a prefix of $\alpha(X_2,\ldots,X_n)$, $p \in N$. And in *PigPid* graph, all the sequences of transformations of the form

$$x_1 \rightarrow (\alpha(x_2,\ldots,x_n))^q\alpha_1(x_2,\ldots,x_n)x_1 \qquad q = 0,1,2,\ldots \tag{7}$$

followed by $x_i \rightarrow x_1x_i$, where

$$\alpha(x_2,\ldots,x_n) == \alpha_1(x_2,\ldots,x_n)x_i\alpha_2(x_2,\ldots,x_n), \tag{8}$$

are used. The sequence (7) is replaced by the unique parametric transformation:

$$x_1 \rightarrow (\alpha(x_2,\ldots,x_n))^\lambda\alpha_1(x_2,\ldots,x_n)x_1, \tag{9}$$

followed by the elementary transformation

$$x_i \rightarrow x_1 x_i, \tag{10}$$

where λ is a natural parameter that can take any natural value (natural parameters will be denoted by Greek letters).

The graph of prefix-equations of equations with two variables is shown here.

Example 1

$$\varphi(x_1, x_2) \quad = \quad \psi(x_1, x_2)$$

$$\downarrow$$

$$x_1^{p+1} x_2 \ldots \quad = \quad x_2^{q+1} x_1 \ldots$$

$$\swarrow x_1 \rightarrow x_2 x_1$$

$x_1 x_2 \ldots = x_2^{q+1} x_1 \ldots$ Similar to left

$\swarrow x_1 \rightarrow (x_2)^\lambda x_1$, followed by $x_2 \rightarrow x_1 x_2$

$x_1 x_2 \ldots = x_2 x_1 \ldots \longleftarrow$ *Elementary prefix-equation.*

□

For equations with three variables, we prove in [Mak-Abd 93] the following:

Theorem 3.1 *Each prefix-equations with three variables can be reduced via a finite sequence of parametric and Nielsen's transformation into elementary prefix-equations with two and three variables.* □

A finite graph using this method is given in [Mak-Abd 93], where a new transformation, called Rouen's transformation, allowing to replace any sequence where three variables can be transformed unboundedly, is deduced from this graph. It is to be used in the construction of the graph describing the general solution of word equations with four variables.

3.2 Formal Parametric Equations method

In this method each node of the desired finite graph is captured by some precedence relations among the transformed variables, called a *table of polarisation*; and some words called *atoms*. Let us now introduce these two concepts.

Parametric transformations (9): $x_1 \rightarrow (\alpha(x_2, \ldots, x_n))^\lambda \alpha_1(x_2, \ldots, x_n) x_1$, and (10): $x_i \rightarrow x_1 x_i$, give new constraints on the positions of the variables. After the application of (9), each occurrence of x_1 (except, of course, the first one) is always preceded by the last letter x_j of $\alpha_1(x_2, \ldots, x_n)$. This constraint will be denoted by $x_j \leftarrow x_1^+$, where x_1^+ is x_1, but the label " + " points out that it has here an occurrence as a successor.

After the application of (10), each occurrence of x_i (which will be denoted by x_i^+ for the same reason as before) is always preceded by an occurrence (which will be denoted by x_1^*) of the previously transformed variable x_1. This new constraint is denoted by $x_1^* \leftrightarrow x_i^+$.

The transformation: $x_1 \rightarrow (\alpha(x_2,\ldots,x_n))^\lambda \alpha_1(x_2,\ldots,x_n)x_1^+$ followed by $x_i \rightarrow x_1^* x_i^+$ will be called a *formal parametric transformation*. The result of the application of this formal parametric transformation will be called a *formal parametric equation*.

Define formally now, by a joint induction, formal parametric equations and formal parametric transformations :

- Equation (2), (3) is a formal parametric equation.

- If a formal parametric equation has the form (6): $x_1\varphi(x_1,\ldots,x_n) = \alpha(x_2, \ldots,x_n)x_1\psi(x_1,\ldots,x_n)$, then the formal parametric transformation is given by the two successive transformations

$$x_1 \rightarrow (\alpha(x_2,\ldots,x_n))^\lambda \alpha_1(x_2,\ldots,x_n)x_1^+, \tag{11}$$

$$x_i \rightarrow x_1^* x_i^+, \tag{12}$$

where $\alpha(x_2,\ldots,x_n) == \alpha_1(x_2,\ldots,x_n)x_i\alpha_2(x_2,\ldots,x_n)$ for some $\alpha_2(x_2,\ldots,x_n)$, and λ is a natural parameter which does not occur in (6).

The result of the application of the transformation (11), (12) to the equation (6) is the formal parametric equation:

$\varphi((\alpha(x_1^*x_2^+,\ldots,x_n))^\lambda \alpha_1(x_1^*x_2^+,\ldots,x_n)x_1^+, x_1^*x_2^+,\ldots,x_n) = x_2^+\alpha_2(x_1^*x_2^+,\ldots, x_n)\alpha_1(x_1^*x_2^+,\ldots,x_n)x_1^+\psi((\alpha(x_2,\ldots,x_n))^\lambda\alpha_1(x_1^*x_2^+,\ldots,x_n)x_1^+, x_1^*x_2^+,\ldots,x_n)$

The variable x_i having everywhere in the equation the labels " + " or " * " will be called *polarised*.

Precedence relations among the variables of a formal parametric equation can be done by the notion of the *table of polarisation of variables*, as follows:

$x_i^* \rightarrow x_j$ denotes that there exists an occurrence of x_i^* directly before x_j,

$x_i \leftarrow x_j^+$ denotes that there exists an occurrence of x_j^+ directly after x_i,

$x_i^* \leftrightarrow x_j^+$ denotes that we have both $x_i^* \rightarrow x_j^+$ and $x_i^* \leftarrow x_j^+$.

Example 2

The table of polarisation of variables of the word $(x_1^*x_2^+)^{p+1}x_1^+$ is

$x_1^* \leftrightarrow x_2^+ \leftarrow x_1^+$. □

Formal parametric equations will be written on a base of words called *atoms*. An atom is essentially a word in which no variable labelled by " + " precedes a variable labelled by " * ". In the tables of polarisation, the atoms can be seen as the sequences of letters $x_s^* \ldots x_t^+$ in the paths of the form: $x_s^* \leftarrow \ldots \leftarrow x_t^+$.

Formally, a sequence of variables of the form $y_1y_2\ldots y_k$ (y_i can be polarised or not) will be called an atom, if it satisfies the following condition:

if y_1 is not polarised, then $k = 1$;

otherwise, $y_1 \neq y_1^+$ (i.e. y_1 is not a latter labelled by " + "), $y_k = y_k^+$ (i.e. y_k is a latter labelled by " + "), and no y_j^+ precedes y_i^*.

Example 3

- x_1, $x_1x_2^+$ are atoms, and $x_2^+x_1^+$, $x_2^+x_1^*$, are not.

- if $p = 0$, the word $(x_1^*x_2^+)^{p+1}x_1^+$ seen in the last example is the atom $x_1^*x_2^+x_1^+$, otherwise it can be written on the set of two atoms $\{x_1^*x_2^+, x_1^*x_2^+x_1^+\}$. □

Note that, in the last case of this example, the fact that the atom $x_1^*x_2^+x_1^+$ succeeds the atom $x_1^*x_2^+$ is not reflected in the table of polarisation of variables.

That is why we will introduce the notion of the *table of polarisation of atoms*, in which we numerate all the atoms and precise their precedence relations. Thus, if $x_1^* x_2^+ x_1^+$ is the first atom and $x_1^* x_2^+$ is the second atom, then we have the following table of polarisation of atoms: $2 \leftarrow 1$, if $p > 0$ (there is no table of polarisation, if $p = 0$).

Now, we show the graph of formal parametric equations with two variables $\varphi(x_1, x_2) = \psi(x_1, x_2)$.

Example 4

Consider (without loss of generality) the directed equation: $x_1 \varphi(x_1, x_2) \rightarrow x_2 \psi(x_1, x_2)$. After extracting the first occurrence of x_1, we obtain: $x_1 \varphi(x_1, x_2) \rightarrow x_2^{p+1} x_1 \psi(x_1, x_2)$.

We can apply the transformation $x_1 \rightarrow x_2^{\lambda+1} x_1^+$. Thus, the transformed equation becomes: $x_1^+ \varphi(x_2^{\lambda+1} x_1^+, x_2) \leftarrow x_2^{p+1} x_1^+ \psi(x_2^{\lambda+1} x_1^+, x_2)$. The letter that precedes x_1^+ in this formal parametric equation is always x_2. This relation gives us the table of polarisation: $x_2 \leftarrow x_1^+$.

Reduce now our directed equation by the transformation $x_2 \rightarrow x_1^* x_2^+$. This transformation leads to $\varphi((x_1^* x_2^+)^{\lambda+1} x_1^+, x_1^* x_2^+) = x_2^+ (x_1^* x_2^+)^p x_1^+ \psi((x_1^* x_2^+)^{\lambda+1} x_1^+, x_1^* x_2^+))$, and to the following table of polarisation: $x_1^* \leftrightarrow x_2^+ \leftarrow x_1^+$.

Describing these formal parametric equations by their atoms gives:

$$\varphi(x_1^* x_2^+ x_1^+, x_1^* x_2^+) = x_2^+ (x_1^+) \psi(x_1^* x_2^+ x_1^+, x_1^* x_2^+), \tag{13}$$

with the same table of polarisation. This is a closed set of formal parametric equations with two polarised variables: the application of formal parametric transformations to (13) gives rise to formal parametric equations of (13).

After such a closed set is obtained, we interest in transforming its subsets that have different prefix-equations. Here, each equation of (13) has the elementary prefix-equation $x_1^* x_2^+ = x_2^+ x_1^+$, or $x_1^* x_2^+ = x_2^+ x_1^*$. That is $x_1 x_2 = x_2 x_1$.

Thus, we obtain the following graph:

$$\begin{array}{ccc} x_1 \varphi(x_1, x_2) & \rightarrow & x_2 \psi(x_1, x_2) \\ & \swarrow & x_1 \rightarrow x_2^{\lambda+1} x_1^+ \textit{ followed by } x_2 \rightarrow x_1^* x_2^+. \end{array}$$

$\varphi(x_1^* x_2^+ x_1^+, x_1^* x_2^+) = x_2^+ (x_1^+) \psi(x_1^* x_2^+ x_1^+, x_1^* x_2^+) \longleftarrow$ *Elementary prefix-equation.*

□

We can observe in this example that there is only one table of polarisation for two variables.

We applied this method in the case of word equations with three variables. We started from $\varphi(x_1, x_2, x_3) = \psi(x_1, x_2, x_3)$. We obtained a finite and closed set (D) of formal parametric equations in which two variables are polarised, in a similar way as (13) of example 4. We transformed this new set by polarising the third variable. We obtained a finite and closed set of all the formal parametric equations with three polarised variables. It is denoted by (T).

There exists only the following five tables of polarisation for formal parametric equations with three variables, denoted by (P_1), (P_2), (P_3), (P_4), (P_5):

$$x_2^* \leftrightarrow x_3^+ \leftarrow x_2^+$$
$$\downarrow \qquad (P_1)$$
$$x_3^* \leftrightarrow x_1^+$$

$$x_2^* \leftrightarrow x_3^+$$
$$\downarrow \qquad (P_2)$$
$$x_3^* \leftrightarrow x_1^+ \leftarrow x_2^+$$

$$x_2^* \rightarrow x_3^* \leftrightarrow x_1^+ \leftarrow x_2^+ \leftarrow x_3^+ \qquad (P_3)$$

$$x_2^* \rightarrow x_3^* \leftrightarrow x_1^+ \leftarrow x_3^+ \leftarrow x_2^+ \qquad (P_4)$$

$$x_3^+$$
$$\swarrow$$
$$x_2^* \rightarrow x_3^* \leftrightarrow x_1^+ \qquad (P_5)$$
$$\nwarrow$$
$$x_2^+$$

The set (T) is divided into some subsets, called ***bunches*** $T_{i,j}$, where i is the index of the table P_i and j is the index of the j-th bunch, associate with P_i, as follows:

$$x_3^+\varphi(x_1^*x_2^+x_1^+, x_1^*x_2^+, x_2^*x_3^+, (x_1^*x_2^+x_3^+)) =$$
$$(x_1^+)\psi(x_1^*x_2^+x_1^+, x_1^*x_2^+, x_2^*x_3^+, (x_1^*x_2^+x_3^+)) \qquad (T_{1,1})$$
$$x_2^+(x_1^+)\varphi(x_3^*x_2^+x_1^+, x_3^*x_2^+, x_1^*x_3^+, (x_1^*x_3^*x_2^+)) =$$
$$\psi(x_3^*x_2^+x_1^+, x_3^*x_2^+, x_1^*x_3^+, (x_1^*x_3^*x_2^+)) \qquad (T_{2,1})$$
$$x_2^+(x_1^+)\varphi(x_3^*x_2^+x_1^+, x_3^*x_2^+, x_1^*x_3^+, (x_1^*x_3^*x_2^+x_1^+)) =$$
$$\psi(x_3^*x_2^+x_1^+, x_3^*x_2^+, x_1^*x_3^+, (x_1^*x_3^*x_2^+x_1^+)) \qquad (T_{2,2})$$
$$x_1^+(x_2^+)\varphi(x_3^*x_1^+, x_2^*x_3^+, x_2^*x_3^*x_1^+, x_2^*x_3^*x_1^+x_2^+) =$$
$$\psi(x_3^*x_1^+, x_2^*x_3^+, x_2^*x_3^*x_1^+, x_2^*x_3^*x_1^+x_2^+) \qquad (T_{2,3})$$
$$x_1^+(x_2^+)\varphi(x_3^*x_1^+x_2^+, x_2^*x_3^+, x_2^*x_3^*x_1^+, x_2^*x_3^*x_1^+x_2^+) =$$
$$\psi(x_3^*x_1^+x_2^+, x_2^*x_3^+, x_2^*x_3^*x_1^+, x_2^*x_3^*x_1^+x_2^+) \qquad (T_{2,4})$$
$$x_2^+(x_1^+)\varphi(x_3^*x_2^+x_1^+, x_3^*x_2^+, x_3^*x_2^+x_1^+x_3^+, x_1^*x_3^*x_2^+x_1^+) =$$
$$\psi(x_3^*x_2^+x_1^+, x_3^*x_2^+, x_3^*x_2^+x_1^+x_3^+, x_1^*x_3^*x_2^+x_1^+) \qquad (T_{3,1})$$
$$x_2^+(x_1^+)\varphi(x_3^*x_2^+x_1^+, x_3^*x_2^+, x_1^*x_3^*x_2^+x_1^+x_3^+, (x_1^*x_3^*x_2^+x_1^+)) =$$
$$\psi(x_3^*x_2^+x_1^+, x_3^*x_2^+, x_1^*x_3^*x_2^+x_1^+x_3^+, (x_1^*x_3^*x_2^+x_1^+)) \qquad (T_{3,2})$$
$$x_2^+(x_1^+)\varphi(x_3^*x_2^+, x_3^*x_2^+x_1^+, x_3^*x_2^+x_1^+x_3^+, x_1^*x_3^*x_2^+) =$$
$$\psi(x_3^*x_2^+, x_3^*x_2^+x_1^+, x_3^*x_2^+x_1^+x_3^+, x_1^*x_3^*x_2^+) \qquad (T_{3,3})$$
$$x_1^+\varphi(x_3^*x_1^+, x_2^*x_3^*x_1^+, x_2^*x_3^*x_1^+x_2^+, x_2^*x_3^*x_1^+x_2^+x_3^+) =$$
$$\psi(x_3^*x_1^+, x_2^*x_3^*x_1^+, x_2^*x_3^*x_1^+x_2^+, x_2^*x_3^*x_1^+x_2^+x_3^+) \qquad (T_{3,4})$$
$$x_1^+x_2^+x_3^+\varphi(x_3^*x_1^+, x_3^*x_1^+x_2^+, x_3^*x_1^+x_2^+x_3^+, x_2^*x_3^*x_1^+x_2^+x_3^+) =$$
$$\psi(x_3^*x_1^+, x_3^*x_1^+x_2^+, x_3^*x_1^+x_2^+x_3^+, x_2^*x_3^*x_1^+x_2^+x_3^+) \qquad (T_{3,5})$$
$$x_1^+x_2^+\varphi(x_3^*x_1^+x_2^+, x_2^*x_3^*x_1^+, x_2^*x_3^*x_1^+x_2^+, x_2^*x_3^*x_1^+x_2^+x_3^+) =$$
$$\psi(x_3^*x_1^+x_2^+, x_2^*x_3^*x_1^+, x_2^*x_3^*x_1^+x_2^+, x_2^*x_3^*x_1^+x_2^+x_3^+) \qquad (T_{3,6})$$
$$x_1^+x_2^+x_3^+\varphi(x_3^*x_1^+x_2^+x_3^+, x_2^*x_3^*x_1^+, x_2^*x_3^*x_1^+x_2^+, x_2^*x_3^*x_1^+x_2^+x_3^+) =$$
$$\psi(x_3^*x_1^+x_2^+x_3^+, x_2^*x_3^*x_1^+, x_2^*x_3^*x_1^+x_2^+, x_2^*x_3^*x_1^+x_2^+x_3^+) \qquad (T_{3,7})$$
$$x_1^+x_2^+x_3^+\varphi(x_3^*x_1^+x_2^+, x_3^*x_1^+x_2^+x_3^+, x_2^*x_3^*x_1^+, x_2^*x_3^*x_1^+x_2^+) =$$
$$\psi(x_3^*x_1^+x_2^+, x_3^*x_1^+x_2^+x_3^+, x_2^*x_3^*x_1^+, x_2^*x_3^*x_1^+x_2^+) \qquad (T_{3,8})$$
$$x_3^+\varphi(x_1^*x_2^*x_3^+x_2^+x_1^+, x_1^*x_2^*x_3^+x_2^+, x_2^*x_3^+, (x_1^*x_2^*x_3^+)) =$$
$$(x_1^+)\psi(x_1^*x_2^*x_3^+x_2^+x_1^+, x_1^*x_2^*x_3^+x_2^+, x_2^*x_3^+, (x_1^*x_2^*x_3^+)) \qquad (T_{4,1})$$

$$x_3^+(x_1^+)\varphi(x_2^*x_3^+x_2^+x_1^+, x_2^*x_3^+x_2^+, x_2^*x_3^+, x_1^*x_2^*x_3^+) =$$
$$(x_1^+)\psi(x_2^*x_3^+x_2^+x_1^+, x_2^*x_3^+x_2^+, x_2^*x_3^+, x_1^*x_2^*x_3^+) \qquad (T_{4,2})$$
$$x_1^+x_3^+(x_2^+)\varphi(x_3^*x_1^+, x_3^*x_1^+x_3^+, x_3^*x_1^+x_3^+x_2^+, x_2^*x_3^*x_1^+x_3^+x_2^+) =$$
$$(x_2^+)\psi(x_3^*x_1^+, x_3^*x_1^+x_3^+, x_3^*x_1^+x_3^+x_2^+, x_2^*x_3^*x_1^+x_3^+x_2^+) \qquad (T_{4,3})$$
$$x_1^+x_3^+(x_2^+)\varphi(x_3^*x_1^+, x_3^*x_1^+x_3^+, x_3^*x_1^+x_3^+x_2^+, x_2^*x_3^*x_1^+x_3^+) =$$
$$(x_2^+)\psi(x_3^*x_1^+, x_3^*x_1^+x_3^+, x_3^*x_1^+x_3^+x_2^+, x_2^*x_3^*x_1^+x_3^+) \qquad (T_{4,4})$$
$$x_1^+x_3^+\varphi(x_3^*x_1^+, x_3^*x_1^+x_3^+, x_2^*x_3^*x_1^+x_3^+, x_2^*x_3^*x_1^+x_3^+x_2^+) =$$
$$(x_2^+)\psi(x_3^*x_1^+, x_3^*x_1^+x_3^+, x_2^*x_3^*x_1^+x_3^+, x_2^*x_3^*x_1^+x_3^+x_2^+) \qquad (T_{4,5})$$
$$x_1^+x_3^+(x_2^+)\varphi(x_3^*x_1^+x_3^+, x_3^*x_1^+x_3^+x_2^+, x_2^*x_3^*x_1^+, x_2^*x_3^*x_1^+x_3^+) =$$
$$(x_2^+)\psi(x_3^*x_1^+x_3^+, x_3^*x_1^+x_3^+x_2^+, x_2^*x_3^*x_1^+, x_2^*x_3^*x_1^+x_3^+) \qquad (T_{4,6})$$
$$x_1^+x_3^+(x_2^+)\varphi(x_3^*x_1^+x_3^+, x_3^*x_1^+x_3^+x_2^+, x_2^*x_3^*x_1^+, x_2^*x_3^*x_1^+x_3^+x_2^+) =$$
$$(x_2^+)\psi(x_3^*x_1^+x_3^+, x_3^*x_1^+x_3^+x_2^+, x_2^*x_3^*x_1^+, x_2^*x_3^*x_1^+x_3^+x_2^+) \qquad (T_{4,7})$$
$$x_1^+x_3^+\varphi(x_3^*x_1^+x_3^+, x_2^*x_3^*x_1^+, x_2^*x_3^*x_1^+x_3^+, x_2^*x_3^*x_1^+x_3^+x_2^+) =$$
$$(x_2^+)\psi(x_3^*x_1^+x_3^+, x_2^*x_3^*x_1^+, x_2^*x_3^*x_1^+x_3^+, x_2^*x_3^*x_1^+x_3^+x_2^+) \qquad (T_{4,8})$$
$$x_1^+x_3^+x_2^+\varphi(x_3^*x_1^+x_3^+x_2^+, x_2^*x_3^*x_1^+, x_2^*x_3^*x_1^+x_3^+, x_2^*x_3^*x_1^+x_3^+x_2^+) =$$
$$\psi(x_3^*x_1^+x_3^+x_2^+, x_2^*x_3^*x_1^+, x_2^*x_3^*x_1^+x_3^+, x_2^*x_3^*x_1^+x_3^+x_2^+) \qquad (T_{4,9})$$
$$x_1^+x_3^+x_2^+\varphi(x_3^*x_1^+, x_3^*x_1^+x_3^+x_2^+, x_2^*x_3^*x_1^+x_3^+, x_2^*x_3^*x_1^+x_3^+x_2^+) =$$
$$(x_2^+)\psi(x_3^*x_1^+, x_3^*x_1^+x_3^+x_2^+, x_2^*x_3^*x_1^+x_3^+, x_2^*x_3^*x_1^+x_3^+x_2^+) \qquad (T_{4,10})$$
$$x_2^+(x_1^+)\varphi(x_3^*x_2^+x_1^+, x_3^*x_2^+, x_3^*x_2^+x_3^+, x_1^*x_3^*x_2^+x_1^+) =$$
$$\psi(x_3^*x_2^+x_1^+, x_3^*x_2^+, x_3^*x_2^+x_3^+, x_1^*x_3^*x_2^+x_1^+) \qquad (T_{5,1})$$
$$x_2^+(x_1^+)\varphi(x_1^*x_3^*x_2^+, x_3^*x_2^+x_1^+, x_3^*x_2^+, x_3^*x_2^+x_3^+) =$$
$$\psi(x_1^*x_3^*x_2^+, x_3^*x_2^+x_1^+, x_3^*x_2^+, x_3^*x_2^+x_3^+) \qquad (T_{5,2})$$
$$x_2^+(x_1^+)\varphi(x_3^*x_2^+x_1^+, x_3^*x_2^+, x_1^*x_3^*x_2^+x_3^+, (x_1^*x_3^*x_2^+)) =$$
$$\psi(x_3^*x_2^+x_1^+, x_3^*x_2^+, x_1^*x_3^*x_2^+x_3^+, (x_1^*x_3^*x_2^+)) \qquad (T_{5,3})$$
$$x_1^+x_3^+\varphi(x_3^*x_1^+, x_3^*x_1^+x_2^+, x_3^*x_1^+x_3^+, x_2^*x_3^*x_1^+x_3^+) =$$
$$\psi(x_3^*x_1^+, x_3^*x_1^+x_2^+, x_3^*x_1^+x_3^+, x_2^*x_3^*x_1^+x_3^+) \qquad (T_{5,4})$$
$$x_1^+\varphi(x_3^*x_1^+, x_2^*x_3^*x_1^+, x_2^*x_3^*x_1^+x_2^+, x_2^*x_3^*x_1^+x_3^+) =$$
$$\psi(x_3^*x_1^+, x_2^*x_3^*x_1^+, x_2^*x_3^*x_1^+x_2^+, x_2^*x_3^*x_1^+x_3^+) \qquad (T_{5,5})$$
$$x_1^+x_3^+\varphi(x_3^*x_1^+, x_3^*x_1^+x_3^+, x_2^*x_3^*x_1^+, x_2^*x_3^*x_1^+x_2^+) =$$
$$\psi(x_3^*x_1^+, x_3^*x_1^+x_3^+, x_2^*x_3^*x_1^+, x_2^*x_3^*x_1^+x_2^+) \qquad (T_{5,6})$$
$$x_1^+x_3^+\varphi(x_3^*x_1^+x_3^+, x_2^*x_3^*x_1^+, x_2^*x_3^*x_1^+x_2^+, x_2^*x_3^*x_1^+x_3^+) =$$
$$\psi(x_3^*x_1^+x_3^+, x_2^*x_3^*x_1^+, x_2^*x_3^*x_1^+x_2^+, x_2^*x_3^*x_1^+x_3^+) \qquad (T_{5,7})$$
$$x_1^+x_2^+\varphi(x_3^*x_1^+x_2^+, x_2^*x_3^*x_1^+, x_2^*x_3^*x_1^+x_2^+, x_2^*x_3^*x_1^+x_3^+) =$$
$$\psi(x_3^*x_1^+x_2^+, x_2^*x_3^*x_1^+, x_2^*x_3^*x_1^+x_2^+, x_2^*x_3^*x_1^+x_3^+) \qquad (T_{5,8})$$

Then, we transformed the subset of (T) divided according to their different prefix-equations, in the same manner as example 4. This leads [Mak-Abd-Maks, 94] to new formal parametric equations whose prefix-equations coincide with those of the low levels of the graph of word equations with three variables mentioned in the theorem 3.1. That leads to:

Theorem 3.2. Each prefix-equations of (T) can be reduced via a finite sequence of parametric and Nielsen's transformation into elementary equations with two and three variables. □

Conclusion and perspectives

We showed in this article a new method (using formal parametric equations) used to compute a finite graph describing the general solution of word equation. We demonstrated this method in the case of equations with three variables, and introduce the idea of its generalisation: the set (D) of formal parametric equations with *deux* polarised variables allowed to produce the closed set (T) of formal parametric equations with *trois* polarised variables. We are using now (T), in the same way, as a base for the creation of the set (Q) of formal parametric equations with *quatre* polarised variables, in order to obtain a finite graph describing the general solution of word equations with four variables.

References

[Abd-Maks 93] *General Solution of Mirror Equations.* H. Abdulrab et M. Maksimenko. Proceedings of FCT'93 (Fundamentals of Computation Theory). Springer Verlag, LNCS. n: 710, Edited by Prof. Esik, p.p. 133-142.

[Abd-Maks 94] *General Solution of Systems of Linear Diophantine Equations and Inequations.* H. Abdulrab and M. Maksimenko, Rewriting Techniques and Applications. Springer Verlag, LNCS. n: 914, Edited by Jieh Hsiang, p.p. 339-351.

[Len 72] *Equations dans le Monode Libre.* A. Lentin, Gauthier-Villars, Paris 1972.

[Lot 83]*Combinatorics on Word.* M. Lothaire, chapter 9, by C. Choffrut, Addisson-Wesley, 1983.

[Mak 78] *The Systems of Standard Word Equations in the n -layer Alphabet of Variables.* (in Russian) G.S. Makanin, Sibirski matematictheski journal, AN SSSR, Sibirskoje otdelenije, vol. XIX, N 3, 1978.

[Mak 91] *On general solution of equations in a free semigroup.* G.S. Makanin, Proceedings of IWWERT'91, Rouen, 667, LNCS. Edited by H. Abdulrab and J.P. Pecuchet, 1991.

[Mak-Abd 93] *Transformations of Word Equations with Three Variables: Rouen's Function.* G.S. Makanin, and H. Abdulrab, Rapport LITP 93.44, July 93.

[Mak-Abd 94] *On General Solution of Word Equations.* G.S. Makanin, and H. Abdulrab, Proceedings of "Important Results and Trends in Theoretical Computer Science", LNCS, N. 812. P.P. 251-263.

[Mak-Abd-Maks 94] *Bunches of Formal Parametric Equations with Three Variables.* G.S. Makanin, H. Abdulrab and M.N. Maksimenko, LIR, Research rapport, 1994.

[Plo 72] *Building-in Equational Theories.*, G. Plotkin, Machine Intelligence 7, P. 73-90, 1972.

[Sie 89] *Unification Theory: a survey.* J. Siekmann, in C. Kirchner (Ed.), Special Issue on Unification, JSC 7, 1989 7, P. 73-90, 1972.

PRAM's Towards Realistic Parallelism: BRAM's

Rolf Niedermeier[1] and Peter Rossmanith[2]

[1] Wilhelm-Schickard-Institut für Informatik, Universität Tübingen, Sand 13, D-72076 Tübingen, Fed. Rep. of Germany, niedermr@informatik.uni-tuebingen.de
[2] Fakultät für Informatik, Technische Universität München, Arcisstr. 21, D-80290 München, Fed. Rep. of Germany, rossmani@informatik.tu-muenchen.de

Abstract. Due to its many idealizing assumptions, the well-known parallel random access machine (PRAM) is not a very practical model of parallel computation.

As a more realistic model we suggest the BRAM. Here each of the p processors gets a piece of length n of the input, which thus has size pn in total. Access to global memory has to be data-independent, block-wise, and has to obey the owner restriction. Assuming different global memory sizes, BRAM's are suitable for modeling various parallel computers ranging from bounded degree networks to completely connected parallel machines, while abstracting from architectural details.

We present optimal BRAM algorithms requiring different global memory sizes and different numbers of block communications for the longest common subsequence and the sorting problem.

1 Introduction

It is a "well-known pragmatic rule that any parallel machine can be used efficiently, provided it is used to solve large enough problem sizes" [12]. The purpose of this paper is to specify the above claim more precisely and to put it into a formal framework drawn from parallel complexity theory. Herein, we make use of the most favorite model in parallel algorithm and complexity theory, the parallel random access machine (PRAM). The two features that decide the PRAM's conceptual simplicity (and, thus, its popularity) are its synchronous mode of operation and its unit time access to global shared memory.

The obvious dilemma between theory (PRAM's) and practice (asynchronous, distributed memory machines) is the seemingly large gap between ideal model and real machine. There roughly are two main ways to bridge the gap. The first is to try to preserve the full PRAM on the level of algorithm design and to show up ways how to implement PRAM's on bounded degree networks [16, 18]. The second one is to abandon the intact world of PRAM's and to use models of computation closer to really existing parallel machines [6]. Whereas the first approach still suffers at least from large constant factors for PRAM simulations on existing machines, the second approach destroys the conceptual simplicity of

This research was partially supported by the Deutsche Forschungsgemeinschaft, Sonderforschungsbereich 0342, TP A4 "KLARA."

the underlying model and makes it harder or even impossible to create a real complexity theory upon it. So we subsequently try to find a middle-way between both the extremes.

A typical, existing and foreseeable parallel computer is characterized by a relatively small number of processors compared to a relatively large number of data-items per processor [6]. In our eyes, the deficiency of newly presented models is that they either lack *conceptual simplicity* or they stick to close to the classical PRAM model and do not take into consideration an important side condition like small number of processors paired with large number of data-items. So our main goal is to present a new model without the above deficiencies. In this sense, the contribution of this paper is first of all to methodology in parallel computation and only second to concrete algorithm design.

To introduce our new model of a parallel machine, called *BRAM*[1] (Block-RAM), our general plan of attack is to use known concepts from the PRAM world, to combine them in a new way, and to add a further demand concerning the input/output convention. More precisely, the BRAM model, which is formally introduced in Section 3, evolves informally speaking by putting together the concept of *data-independent* (or oblivious) computations [15] with the *owner read, owner write* restriction for PRAM's [17]. In addition, we assume that *each* of the p processors of a BRAM holds n data-items of an input of total size pn in its *local memory* and that the number of global memory cells may range from p to p^2. Note that, for example, BRAM's with global memory size $O(p)$ in essence model bounded degree networks of processors and that BRAM's with global memory size p^2 in fact model completely connected parallel machines. Thus, in a sense, BRAM's model parallel computers ranging from bounded degree to completely connected networks from a more abstract point of view, ignoring architectural details.

One prospect of the BRAM is that it allows the use of sequential algorithms in the design of parallel ones in a systematic way. Often we are able to *prove* the *work-optimality* of our algorithms. In addition, algorithms designed for the BRAM, due to their static communication pattern known at compile time, due to their owner read, owner write communication protocol inter alia avoiding contention effects, and due to their small number of blocked communications enabling the bypassing of latency effects and requiring only simple synchronization mechanisms and a strong use of *locality*, should allow an easy and efficient mapping to existing distributed memory machines. Nevertheless the BRAM still provides enough conceptual simplicity necessary to build a structural theory of BRAM algorithms.

As to the algorithmic problems studied in this work, let us only mention that the sorting problem can be solved work-optimally with a speedup factor between $p/2$ and $p/3$ compared to the best sequential algorithm.

Due to the lack of space several details had to be omitted. They will be in the full paper.

---

[1] Van Emde Boas [19] uses the terminus "BRAM" in order to refer to a kind of Boolean RAM, which should not lead to any confusion with our model.

2 Preliminaries

A *PRAM* (Parallel Random Access Machine) is a set of Random Access Machines, called processors, that work *synchronously* and communicate via a *global, shared memory.* Each computation step takes *unit time* regardless whether it performs a local or a global (i.e., remote) operation. Each processor has a local memory. The input to a PRAM computation is initially given in global memory. A PRAM processor is identified by its processor number. We call processors *neighbors* if their processor numbers differ by one.

The most restricted PRAM variant with respect to access possibilities to shared memory is the *owner mechanism* introduced by Dymond and Ruzzo [7]. Later on, OROW-PRAM's (Owner read, Owner write) were introduced as a still more restricted variation of EREW-PRAM's [17]. Here, a uniquely determined read-owner, resp. write-owner, processor is assigned to each global memory cell. The read owner is the only processor with read and the write owner is the only processor with write permission for this cell. More formally, there are functions called *write-owner*(i, n) resp. *read-owner*(i, n) that map global memory cells into the set of processors and if $j =$ *write-owner*(i, n), then j is the only processor allowed to write into cell i. Both functions have to be easily computable, that is, for example, by a logarithmically space bounded Turing machine. In this way, global memory is deteriorated to a set of directed channels between pairs of processors and, in general, we may view OROW-PRAM's just as completely connected, synchronous processor networks.

Recently, the concept of data-independence for PRAM's was studied from a structural complexity theoretic point of view [15]. The central idea behind *data-independence*, also called *obliviousness* by other authors, is the demand for control and communication structures that are independent of the input. Thus, e.g., the global memory access pattern, that is, the addresses used, may only depend on the *size* of the input to the PRAM computation, but not on the concrete input word. The communication structure is static along these lines. The demand for data-independent communication appears to be an important prerequisite for the efficient implementation of PRAM algorithms on distributed memory machines [9]. Note that data-independence allows to optimize parallel algorithms at compile time.

The most important resource bounds for PRAM's are the number of processors and the running time. We call a parallel algorithm *work-optimal* if it performs up to a constant factor the same amount of *work* (that is, the number of processors multiplied with the running time) as the best sequential algorithm.

In the rest of this paper for the ease of presentation we always assume that any occurring fractions or roots of integers shall yield integer values.

3 The Model: BRAM's

The two main points of criticism against the PRAM model concern its synchronous mode of operation and its unit time access to global shared memory.

Further on, it is often criticized that PRAM's do not impose any structure on the communication pattern, that most PRAM algorithms are excessively fine-grained, that PRAM's do not allow to model locality in computations, that even for EREW-PRAM's contention may occur during memory access to the same memory bank, and so on [4, 6, 11]. On the other hand, exactly its *conceptual simplicity* is the reason why the PRAM is so popular.

The idea behind the introduction of the BRAM model as an adherent of the PRAM is that we still want to preserve the two most important points of the PRAM's simplicity (synchronism and shared memory), but put forward some natural requirements for communication structure and protocol (data-independent, OROW), for input size (*each* processor gets locally n input items) and the global memory size (at most quadratic in the number of processors). Altogether, we may define the BRAM model using known concepts from the world of PRAM's.

Definition 1. A p-processor-*BRAM (Block-RAM)* is an OROW-PRAM with p processors obeying the following restrictions:

1. The size of global shared memory lies between p and p^2.
2. The input of total size pn is equally distributed among the local memory banks of the p processors, that is, *each* processor has local input of size n.
3. The communication structure is data-independent, it only depends on p and n.

The term "block" refers to two important characteristics of BRAM's — the partitioning of the input into p pieces each of size n and the use of blocking techniques for communication. So blocking of communications means that data items are only exchanged in large portions between processors and generally *not* item-wise.

We chose the owner read, owner write (OROW) mechanism with a number of memory cells lying between p and p^2, because this exactly allows to model a broad range of parallel computers, going from bounded degree to completely connected networks. Eventually, the OROW restriction compared to other conflict resolution rules like EREW, in particular prevents contention conflicts due to the "channel character" of its communication mechanism.

The requirement for a local input piece of size n has many faces. First, it enables a good modeling of locality and of the fact that in most practical situations the size of the input exceeds the number of processors by large. It also models the observation that one advantage of using parallel machines is that they increase the total main memory size at hand. Our input size demands also lead in a natural way to the use of relatively few communications with relatively large data blocks to be transferred. So latency problems might be done away with. Finally, it also makes a direct comparison with sequential algorithms possible. We can use the best sequential algorithms during local computations. Since we may use sequential algorithms in some sense as parameters, we can often say that our parallel algorithms are *provably* work-optimal up to a certain constant factor of small size (e.g., in the case of sorting, see Section 6).

At last, the request for data-independent communications, as pointed out in the previous section, puts the missing structure on PRAM communications and admits optimizations at compile time [9, 15].

Typical BRAM algorithms are clearly separated in global communication and local computation phases. The blocking of communications and the in general small number of communication phases facilitates the implementation of BRAM algorithms on existing asynchronous distributed memory machines. In particular, synchronization costs remain quite modest. A useful, additionally restricting demand for BRAM algorithms also might be that the local computation time has to be significantly greater than the communication time (measured in the size of the data to be transferred). Several of our BRAM algorithms possess this property, as the subsequent sections show.

There is good reason to assume that the BRAM model may still be used to build a structural theory in analogy to the one for PRAM's. This is due to the definition of BRAM's in the line of known PRAM concepts and the use of *only* two parameters, number of processors p and local input size n.

4 Related Work

There is a lot of literature dealing with more practical models of parallel computation. See the papers of Chin [4] and Heywood and Leopold [11] for recent surveys. We only mention some papers with close relations to our work.

Perhaps the closest relationship is with Gottlieb and Kruskal [10]. They also study the phenomenon of a more efficient use of parallel machines through enlargement of problem sizes. For example, they introduce the so-called supersaturation limit, which, informally speaking, asks how the relation of input size to number of processors has to be in order to get an asymptotically optimal speedup of $\Theta(p)$. If there exists a supersaturation limit, they call a problem *completely parallelizable.* In three tables they present lower and upper bounds for several problems. Gottlieb and Kruskal seemingly do not use parallelization by using sequential algorithms.

Kruskal, Rudolph, and Snir [12] started to build a complexity theory of efficient parallel algorithms. They emphasize speedup *and* work-optimality as two crucial aspects of parallel algorithm design. Using the conventional PRAM model, they compare the performance of a parallel algorithm with the best sequential algorithm. As there is not always a provably best sequential algorithm known, it is not possible to formally define complexity classes with notions like reducibility and completeness. Moreover, they do not work under the proviso that the input size exceeds the number of processors by large.

Aggarwal, Chandra, and Snir introduced Block-PRAM's (BPRAM's, not to be confused with out BRAM's) [1] and Local-memory PRAM's (LPRAM's) [2] to model latency restrictions and locality aspects. Their basis models are "conventional" EREW- resp. CREW-PRAM's. In their framework, the input is still given in global memory and they do not include the use of sequential algorithms.

Culler *et al.* [6] presented the LogP model as a more realistic model of parallel computation. They say that "parallel algorithms will need to be developed under the assumption of a large number of data elements per processor." The LogP model, however, is quite far away from the PRAM model and has as many as four main parameters: number of processors and memory modules, per-processor communication bandwidth, communication delay or latency, and per-processor communication overhead. The large number of parameters makes algorithm design and analysis rather difficult. Culler *et al.* also emphasize that current PRAM simulation techniques on existing parallel machines suffer from large constant factors, context switching costs, and hardware support requests for synchronization.

Heywood and Leopold [11] underline the importance of input and output, which is often neglected in algorithm design. Further on, they criticize that the PRAM imposes no structure on communication and that it assumes a "worst case" scenario for locality. Heywood and Leopold also stress that the LogP model only is designed for short term because it has not enough conceptual simplicity. According to them a computational model needs to offer the opportunity to use locality, but also has to provide synchronous communication and modularity.

Vitányi [22] analyzes and discusses in detail the interplay between locality of computation, communication, and physical realization of multicomputers. In particular, he provides lower bounds on the average interconnect length of physical embeddings of various parallel architectures. As a consequence, he asks for "*realistic* formal models of nonsequential computation." So he concludes that "mesh-connected architectures may be the ultimate solution for interconnecting the extremely large (in numbers) computer complexes of the future."

In an announcement for a workshop on suggesting computer science agendas for high-performance computing [21], several requirements are enumerated. Here, questions and problems like latency, synchronization, and portability are of major importance. One of the focal questions there is how the transition from serial to parallel computation may occur.

5 Longest Common Subsequence

The longest common subsequence problem is, given two sequences of same length, to find a maximum length common subsequence of both sequences. The longest common subsequence problem possesses a well-known, quadratic time dynamic programming solution. To the authors' knowledge only for some special cases more efficient sequential solutions are known. The decisive fact we will also make use of here is that to compute the longest common subsequence, one basically has to compute the entries of a table $c[0..n, 0..n]$. The only thing of interest for the time being is that table entries $c[0, i]$ and $c[i, 0]$, $i \geq 0$, are initialized with zero and that for $i, j > 0$, entry $c[i, j]$ only depends on the entries $c[i-1, j]$, $c[i, j-1]$, and $c[i-1, j-1]$ and whether the ith element of the first sequence is the same as the jth element of the second sequence.

We now briefly discuss a BRAM parallelization of the dynamic programming approach that exhibits the fundamental importance of how to partition the algorithm's input, output and intermediate data structures in order to enable blocked communications and bounded degree communication, i.e., communication only with neighboring processors.

Our modus operandi is to exhaust table c by using right angled triangles of increasing size, having two equal sides each time. The pairs of triangles represent the areas of local computations. In general, each processor is assigned to a pair of triangles each time. The right-hand picture shows a partitioning of table c for four processors. The advantage of this approach is that it allows for local computations and thus also for the blocking of communications. Due to the lack of space, we only state the result.

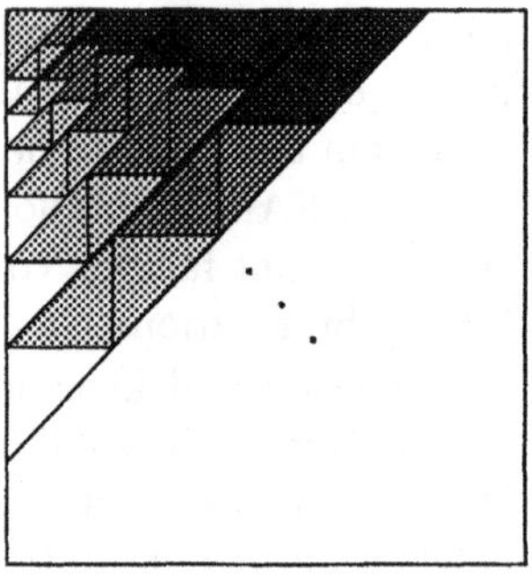

Theorem 2. *The longest common subsequence of two strings of length pn each can be computed by a p-processor BRAM in time $O(pn^2)$ using $2p$ global memory cells and performing $O(p \log p)$ blocked communications per processor.*

6 Sorting

Cole [5] gave a work-optimal merge-sort algorithm that sorts in $O(\log n)$ time using n processors of an EREW-PRAM. From a purely theoretical point of view, this result is the best possible. From a more practical point of view, large constants and the model of computation ("full" EREW-PRAM) are still disturbing factors. We present an efficient, work-optimal sorting algorithm for BRAM's. For this purpose, we make use of Kunde's technique of sorting with all-to-all mappings on grids of processors [13]. To combine a certain number of grid processors into sub-grids called blocks and to sort "locally" within blocks is a basic idea of Kunde. In our case a single processor and its local memory bank play the rôle of a block.

Algorithm 1 (Sorting)

1. Locally sort each processor's n items.
2. Distribute the sorted data items among all processors as "equally" as possible: Assume that the items of the ith processor ($1 \leq i \leq p$) are called $x_1^i, \ldots, x_n^i$. Then x_j^i goes to the kth processor, where $k = 1 + (i + j) \bmod p$.
3. Locally sort each processor's n items.
4. Send all items stored in a processor's local memory from cell $((i-1)n/p) + 1$ to cell $in/p + p$ to the ith processor.
5. Locally sort each processor's $n + p^2$ items.
6. Finally throw out overlaps of sorted arrays of neighboring processors.

In Algorithm 1 local sorting predominates the overall complexity. So we may sort pn data-items in parallel in nearly the same (up to a constant factor between 2 and 3) time as we need to sort n data-items sequentially.

Theorem 3. *Sorting pn items can be done by a p-processor-BRAM in time $2t(n) + O((n + p^2) \log p)$, using p^2 global memory cells and performing $2p + 2$ blocked communications per processor, where $t(n)$ denotes the time needed to sort n elements sequentially.*

Proof. (Sketch) Since we mainly transferred Kunde's sorting methodology for grids [13] to the BRAM model, the correctness of Algorithm 1 follows similar to there. With respect to the underlying BRAM model, note that in Algorithm 1 communication partners are statically determined, they only depend on the number of processors p. Even the size of the data blocks to be transferred only depends on n and p. Clearly, one communication channel between each pair of processors is sufficient.

Let us shortly analyze the time and communication complexity of Algorithm 1. The first and third step simply can be done using some sequential sorting algorithm of time complexity $t(n)$. The second and fourth step consist of all-to-all inter-processor data-exchange using blocked communications in a static (data-independent) fashion. In the second step we transmit pn data-items and in the fourth step we transmit $pn + p^2$ data-items. So we transmit in these two steps $2pn + p^2$ items. The sixth step simply needs the exchange of constant size messages between neighboring processors.

The fifth step could also be done using a local sorting as in the first and third step. But it is also possible to make use of the fact that the given data is already "pre-sorted." Remember that each processor gets a sorted sub-array of size $n/p + p$ from each other processor. The idea is to keep a sorted "minimum-list" of size p, where in the beginning we put the first element of each of the above sub-arrays. To get all data-items sorted, we do the following: Take away the first element of the minimum-list and subsequently insert into the minimum-list the currently first element from the sub-array where the taken element came from. Doing this, use insertion sort. Then again take away the new first element of the minimum-array and so on. Altogether, this takes time $O((n + p^2) \log p)$ with a small constant, which, for $p = o(\sqrt{n})$, beats the application of general local sorting in step 5.

In total, we get the following bounds for Algorithm 1: The per-processor computation costs are $2t(n) + O((n + p^2) \log p))$ and the per-processor communication costs are $2n + p^2 + O(1)$. Altogether that yields time complexity $2t(n) + O((n + p^2) \log p)$. □

In Algorithm 1 the computation costs are predominant compared to the communication costs, because sequential sorting needs time $\Theta(n \log n)$. It is easy to see that the memory space needed on principle is bounded by $p(n + p^2)$ provided that we use a sequential algorithm that sorts in place. Under the common assumption of a very small number of processors compared to the number of data-items that means that Algorithm 1 "nearly sorts in place."

Gottlieb and Kruskal [10, Table III] state that sorting is completely parallelizable (cf. Section 4) on bounded degree networks if for the relation between number n of per-processor local data-items and number p of processors it holds $n = p^{\Omega(\log p)}$. In their terminology, Theorem 3 says that in our framework sorting is completely parallelizable for $n = p^2$. This still is improved in the subsequent Theorem 6. Kruskal, Rudolph, and Snir [12, pages 106–107] point out that based on a bitonic sorting network one can get an efficient sorting algorithm that is optimal provided that $\log^2 p = O(\log n)$ in their setting. It is thus in their class *ANC* (almost efficient *NC* fast). Translated into our setting that means that they get a completely parallelizable sorting algorithm under the same conditions as already Gottlieb and Kruskal [10] do.

Note that although our BRAM used has unbounded degree, that is, global memory size p^2, due to the small number of large, blocked communications these data exchanges can also be performed on bounded degree networks without time loss. In essence, this already follows from Kunde's realization of all-to-all mappings on grids of processors [13]. On the other hand, observe that in a sense Ajtai, Komlós, and Szemerédi's NC^1 sorting network [3] on principle shows that sorting is completely parallelizable on bounded degree networks even for $n = 1$. This result, however, is of purely theoretical interest since it at least suffers from huge constant factors (also cf. [12]).

Corollary 4. *For $p = O(\sqrt{n})$ sorting pn items can be done by a p-processor-BRAM in time $O(n \log n)$, which is work-optimal up to a constant factor close to 3.*

For integer sorting (that is, the values of the items are drawn from a polynomial range), we still get work-optimality for $p = O(\sqrt{n})$.

Corollary 5. *If the items are drawn from a polynomial range, then sorting pn items can be done by a p-processor-BRAM in time $O(n + p^2)$.*

The following result owing to its larger constants has a somewhat less practical flavor. It shows that we can even increase the number of processors up to an arbitrary polynomial in n and still get an asymptotically work-optimal parallel sorting algorithm for BRAM's. In Gottlieb's and Kruskal's terms, it says that sorting is completely parallelizable for $n = p^\epsilon$ for any $\epsilon > 0$.

Theorem 6. *For $p = n^{O(1)}$ sorting pn items can be done by a p-processor-BRAM in time $O(n \log n)$, which is work-optimal.*

Proof. (Sketch) In order to increase the number of processors in comparison to Corollary 4, the basic idea is to make use of an "hierarchy of localities." For the time being, assume that we want to improve Corollary 4 from $p = O(\sqrt{n})$ to $p = O(n)$. Then partition the p processors (and, thus, the input) into $\sqrt{p}$ groups of $\sqrt{p}$ processors each time. Each group of processors shall play the rôle of a single processor in Algorithm 1. To sort within such a group, we again make use of Algorithm 1, but now really with individual processors. We now need 9 applications of a sequential sorting algorithm.

The step from $p = O(n)$ to $p = n^{O(1)}$ is an easy generalization of the above method. The number of applications of sequential sorting algorithms and communication phases grows exponentially with the degree of the polynomial. □

Recently, it has been shown that *average case* sorting on grids for uniformly distributed inputs can be done with only one instead of two all-to-all mappings [14]. This implies that for average case sorting on BRAM's we can save one local sorting and one global communication phase (i.e., steps 1 and 2 in Algorithm 1) and thus Theorem 3 improves to time $t(n) + O((n + p^2)\log p)$.

7 Conclusion and Open Questions

The purpose of this paper was to introduce the BRAM model as a natural adherent of the PRAM. The two main points of interest in the model formation were *conceptual simplicity* on the one hand and *realistic modeling* of most aspects of existing parallel machines on the other hand. Some of the advantages of the BRAM model we see are as follows. *Locality* of computation is a natural part of BRAM computations, leading to the possible (re)use of sequential algorithms, sometimes enabling results of provable work-optimality. The clear separation into local computation and into in general few phases of blocked, data-independent communications simplifies the transfer to asynchronous distributed memory machines with communication delay. BRAM's take into account that one aspect of parallel computers is the increase of main memory size and thus larger problems may be solved exclusively within the main memories of a parallel machine, whereas a sequential machine possibly had to make use of secondary storage (think of thrashing effects!).

We also studied BRAM algorithms for the matrix multiplication, the selection, and the closest pair problem. Here, we also get BRAM algorithms optimal up to small constant factors. The complexity of the list ranking problem remains open. Also work-optimal BRAM algorithms for several graph problems, e.g., depth-first tree of an undirected graph, (strongly) connected components, transitive closure etc. are of interest.

References

1. A. Aggarwal, A. K. Chandra, and M. Snir. On communication latency in PRAM computations. In *Proc. of 1st SPAA*, pages 11–21, 1989.
2. A. Aggarwal, A. K. Chandra, and M. Snir. Communication Complexity of PRAMs. *Theoretical Comput. Sci.*, 71:3–28, 1990.
3. M. Ajtai, J. Komlós, and E. Szemerédi. Sorting in $c \log n$ parallel steps. *Combinatorica*, 3:1–19, 1983.
4. A. Chin. Complexity models for all-purpose parallel computation. In Gibbons and Spirakis [8], chapter 14, pages 393–404.

5. R. Cole. Parallel merge sort. *SIAM J. Comput.*, 17(4):770–785, Aug. 1988.
6. D. Culler et al. LogP: Towards a realistic model of parallel computation. In *4th ACM SIGPLAN Symposium on Principles and Practice of Parallel Programming*, pages 1–12, May 1993.
7. P. W. Dymond and W. L. Ruzzo. Parallel RAMs with owned global memory and deterministic language recognition. In *Proc. of 13th ICALP*, number 226 in LNCS, pages 95–104. Springer-Verlag, 1986.
8. A. Gibbons and P. Spirakis, editors. *Lectures on parallel computation*. Cambridge International Series on Parallel Computation. Cambridge University Press, 1993.
9. D. Gomm, M. Heckner, K.-J. Lange, and G. Riedle. On the design of parallel programs for machines with distributed memory. In A. Bode, editor, *Proc. of 2d EDMCC*, number 487 in LNCS, pages 381–391, Munich, Federal Republic of Germany, Apr. 1991. Springer-Verlag.
10. A. Gottlieb and C. P. Kruskal. Complexity results for permuting data and other computations on parallel processors. *J. ACM*, 31(2):193–209, April 1984.
11. T. Heywood and C. Leopold. Models of parallelism. Technical Report CSR-28-93, The University of Edinburgh, Department of Computer Science, July 1993.
12. C. P. Kruskal, L. Rudolph, and M. Snir. A complexity theory of efficient parallel algorithms. *Theoretical Comput. Sci.*, 71:95–132, 1990.
13. M. Kunde. Block gossiping on grids and tori: Sorting and routing match the bisection bound deterministically. In T. Lengauer, editor, *Proc. of 1st ESA*, number 726 in LNCS, pages 272–283, Bad Honnef, Federal Republic of Germany, Sept. 1993. Springer-Verlag.
14. M. Kunde, R. Niedermeier, K. Reinhardt, and P. Rossmanith. Optimal Average Case Sorting on Arrays. In E. W. Mayr and C. Puech, editors, *Proc. of 12th STACS*, number 900 in LNCS, pages 503–514. Springer-Verlag, 1995.
15. K.-J. Lange and R. Niedermeier. Data-independences of parallel random access machines. In R. K. Shyamasundar, editor, *Proc. of 13th FST&TCS*, number 761 in LNCS, pages 104–113, Bombay, India, Dec. 1993. Springer-Verlag.
16. W. F. McColl. General purpose parallel computing. In Gibbons and Spirakis [8], chapter 13, pages 337–391.
17. P. Rossmanith. The Owner Concept for PRAMs. In C. Choffrut and M. Jantzen, editors, *Proc. of 8th STACS*, number 480 in LNCS, pages 172–183, Hamburg, Federal Republic of Germany, Feb. 1991. Springer-Verlag.
18. L. G. Valiant. General purpose parallel architectures. In van Leeuwen [20], chapter 18, pages 943–971.
19. P. van Emde Boas. Machine models and simulations. In van Leeuwen [20], chapter 1, pages 1–66.
20. J. van Leeuwen, editor. *Algorithms and Complexity*, volume A of *Handbook of Theoretical Computer Science*. Elsevier, 1990.
21. U. Vishkin. Workshop on "Suggesting computer science agenda(s) for high-performance computing" (Preliminary announcement). Announced via electronic mail on "TheoryNet", January 1994.
22. P. M. B. Vitányi. Locality, Communication, and Interconnect Length in Multicomputers. *SIAM J. Comput.*, 17(4):659–672, August 1988.

Some Results Concerning Two-Dimensional Turing Machines and Finite Automata *

H.Petersen †

Institut für Informatik der Universität Stuttgart
Breitwiesenstraße 20–22, D-70565 Stuttgart
e-mail: petersen@informatik.uni-stuttgart.de

Abstract

We show that emptiness is decidable for three-way two-dimensional nondeterministic finite automata as well as the universe problem for the corresponding class of deterministic automata. Emptiness is undecidable for three-way (and even two-way) two-dimensional alternating finite automata over a single-letter alphabet. Consequently inclusion, equivalence, and disjointness for these automata are undecidable properties.

We establish a hierarchy result for space bounded two-dimensional alternating Turing machines above logarithm where the languages witnessing the hierarchy are over single-letter alphabets. Below logarithm we prove that an infinite hierarchy of languages over larger alphabets exists.

The results rely mainly on a translational technique from one to two dimensions. Using this technique we can also show some connections between open problems of two-dimensional automata theory and one-dimensional complexity theory.

1 Introduction and Notation

We study problems arising in the theory of space bounded two-dimensional Turing machines and two-dimensional finite automata. We solve several of the open problems posed in the survey [4, 5] and link other problems to questions about one-dimensional automata.

For a very extensive bibliography as well as for the exact definitions of the models investigated we refer to [4, 5]. Note that the Turing machines are equipped with a two-dimensional rectangular read-only *input* tape and a one-dimensional read-write work-tape. The input is bordered by special endmarkers.

We will denote the classes of one-dimensional languages accepted by $s(n)$ space bounded deterministic, nondeterministic, and alternating Turing machines

*The main part of this work was done at the University of Hamburg

†Supported in part by ESPRIT Basic Research Action WG 6317: Algebraic and Syntactic Methods in Computer Science (ASMICS 2)

by DSPACE($s(n)$), NSPACE($s(n)$), and ASPACE($s(n)$). For classes of two-dimensional automata and languages we adopt the notation from [4, 5]. The classes of two-dimensional finite deterministic, nondeterministic, and alternating automata are denoted by DFA, NFA, and AFA. If the automata cannot move their head up they are called three-way which is denoted by a prefix 'T' (TDFA, TNFA, TAFA). Deterministic finite automata that have k markers or pebbles that they may put onto cells of their two-dimensional input and pick up later are called DMA(k).

The two-dimensional Turing machines that we investigate operate on square input-tapes and form the classes DTM^s, NTM^s, and ATM^s (in [4, 5] rectangular input tapes are considered. This leads to a more complicated notation which will not be used here.)

We are interested in space bounds $s(n)$ that map non-negative integers to positive integers, i.e. $s(n) \geq 1$. The Turing machines operating in space bounded by $s(n)$ where n is the number of rows resp. columns of the input are called $\mathrm{DTM}^s(s(n))$, $\mathrm{NTM}^s(s(n))$, and $\mathrm{ATM}^s(s(n))$.

If X is some class of automata then $\mathcal{L}[\mathrm{X}]$ is the class of languages accepted by these automata. We also consider the restriction of the automata in X to single-letter input alphabets. This will be denoted by X(0).

We write '$\subset$' for proper inclusion.

2 Links Between One- and Two-Dimensional Machines

Throughout the paper we will need the following operation ρ mapping one-dimensional words over an alphabet Σ to square tapes over $\Sigma \times \Sigma$. Let $w = a_1 a_2 \cdots a_n$ be a word of length n. Then $\rho(w) = z$ where $z(i,j) = (a_i, a_j)$ for $1 \leq i \leq n$ and $1 \leq j \leq n$. Thus a symbol of z in a certain row and column has the corresponding symbol of w in the first resp. second component. A word $w = a_1 a_2 \cdots a_n$ is mapped to

$$\begin{matrix} (a_1, a_1) & (a_1, a_2) & \cdots & (a_1, a_{n-1}) & (a_1, a_n) \\ (a_2, a_1) & (a_2, a_2) & \cdots & (a_2, a_{n-1}) & (a_2, a_n) \\ \vdots & \vdots & & \vdots & \vdots \\ (a_{n-1}, a_1) & (a_{n-1}, a_2) & \cdots & (a_{n-1}, a_{n-1}) & (a_{n-1}, a_n) \\ (a_n, a_1) & (a_n, a_2) & \cdots & (a_n, a_{n-1}) & (a_n, a_n) \end{matrix}$$

This operation is extended in the usual way to languages. Note that for a single-letter alphabet Σ ($|\Sigma| = 1$) we have $|\Sigma \times \Sigma| = 1$.

Lemma 1 *A one-dimensional language L is accepted by an alternating two-way two-head finite automaton if and only if $\rho(L) \in \mathcal{L}[AFA]$.*

Proof: The one-dimensional automaton simulates the two-dimensional device by storing the row and column in its head positions. It assembles the pair from the symbols read by the heads.

Conversely the two-dimensional automaton first checks that its input is a square by moving diagonally. In another universal branch it verifies that the first components of the input symbols within every row agree and that within every column the second components are identical. Note that this phase requires only down and right moves since the tests can be done in parallel.

Then the two-dimensional automaton starts a step by step simulation by storing one head-position as the current row number and the other position as the column number. The currently scanned symbols are available as the components of the symbol from Σ read by the head. □

Without modifying the simulation we get:

Lemma 2 *A one-dimensional language L is accepted by an alternating one-way two-head finite automaton if and only if $\rho(L) \in \mathcal{L}[AFA]$ for a two-dimensional automaton that moves its head only down and to the right.*

Lemma 3 *A one-dimensional language L is accepted by a nondeterministic (deterministic) two-way two-head finite automaton if and only if $\rho(L) \in \mathcal{L}[NFA]$ ($\mathcal{L}[DFA]$).*

Proof: The simulations are the same as in the proof of Lemma 1. The only difference is that now the two-dimensional automaton sequentially checks that its input is of an appropriate form, i.e. is in $\rho(\Sigma^*)$. □

Lemma 4 *Let $s(n) \geq \log n$ be a space bound. A language L is in XSPACE($s(n)$) if and only if $\rho(L) \in \mathcal{L}[XTM^s(s(n))]$ for $X \in \{A, N, D\}$.*

Proof: The two-dimensional machine first checks (without using any storage) the form of its input as in the proof of Lemma 3. Then it ignores the first components and uses the first row as the input when simulating the one-dimensional machine.

For the reverse implication the one-dimensional machine allocates two areas of size $O(\log n)$ for storing a row and column position of the two-dimensional machine and simulates it step by step. □

We will now present some applications of the above operation.

Concerning problem (3) of Section 2.2 in [4, 5] asking if there is a set in $\mathcal{L}$[NFA] not in $\mathcal{L}$[DMA(1)] it is claimed in [14] that the problem is at least as hard as the question whether DSPACE($\log n$) = NSPACE($\log n$). While the arguments given that the answer "no" implies DSPACE($\log n$) = NSPACE($\log n$) are correct and we can even slightly strengthen this statement below, an answer "yes" does not tell us anything about the relation between DSPACE($\log n$) and NSPACE($\log n$). This is because it may happen that *one* marker is not sufficient for simulating every NFA. An analogous situation can be observed in the case of DFA and NFA where it is known that DFA cannot accept the set of those square tapes with an odd side length that have a certain central symbol while this is possible for NFA, see Theorem 4.3.4 in [12].

Theorem 1 *If $\mathcal{L}[NFA] \subseteq \bigcup_{k\geq 0} \mathcal{L}[DMA(k)]$ then deterministic and nondeterministic logarithmic space bounded one-dimensional Turing machines are equivalent ($DSPACE(\log n) = NSPACE(\log n)$) and conversely.*

Proof: The graph accessibility problem GAP is complete for NSPACE($\log n$). It is easy to see that GAP can be accepted by a nondeterministic two-way two-head finite automaton. Hence by Lemma 3 ρ(GAP) is in $\mathcal{L}$[NFA]. It is clear that a DMA(k) working on a word from $\rho(\Sigma^*)$ can be simulated in logarithmic space which concludes the proof of the forward implication.

Now assume that DSPACE($\log n$) = NSPACE($\log n$). An NFA can be simulated by a nondeterministic one-dimensional Turing machine working in logarithmic space that receives the rows of the input tape separated by a special symbol. By the assumption this machine can be converted into a deterministic one which in turn is simulated by a one-dimensional marker automaton (see [15]). Finally the input can be replaced with a rectangular tape again. □

If the input alphabet is restricted to one letter we obtain a somewhat weaker result.

Theorem 2 *If $\mathcal{L}[NFA(0)] \subseteq \bigcup_{k\geq 0} \mathcal{L}[DMA(k)]$ then deterministic and nondeterministic linear space bounded one-dimensional Turing machines are equivalent ($DSPACE(n) = NSPACE(n)$) and conversely.*

Proof: In [9] it has been shown that DSPACE(n) = NSPACE(n) if and only if every language over a single-letter alphabet that can be accepted by a nondeterministic two-way one-counter automaton whose counter is bounded by the input length can be accepted by a deterministic logarithmic space bounded Turing machine. Such a counter automaton can be simulated on square inputs by Lemma 1 (obviously a head is at least as good as a counter). Conversely a Turing machine can within logarithmic space count the input length and then ignore its input tape. This latter computation can—assuming equality of the complexity classes–be done by a deterministic machine which is in turn simulated by a marker automaton. □

3 Decision Problems

We start with the positive solution of some open problems from Section 2.6 in [4, 5].

Theorem 3 *The emptiness problem for nondeterministic three-way finite automata on two-dimensional tapes (TNFA) is decidable.*

Proof: We effectively reduce the emptiness-problem for TNFA to the corresponding problem for TNFA(0). The latter is known to be decidable [3].

We recall from [13] that the behaviour of a deterministic two-way automaton entering a portion of tape can be thoroughly described by a mapping from states to states augmented by an element marking that the automaton will never leave this portion again. We will modify this concept in several ways. First we are dealing with nondeterministic automata. Therefore the mapping is to *sets* of states. Further other possibilities are to advance to the next row or accept the input (in [13] acceptance could only occur at the right border of the input).

Now we will describe how a TNFA M with k states is transformed into a TNFA(0) M' such that M' accepts an input tape of size $n \times m$ if and only if M accepts *some* input of the same size. Suppose M enters a row with the inscription wxy where x is the symbol on which M's head rests. Then M may enter w in any of the k states and for every state there are $k+2$ ways of continuing the computation. M may return to x in one of k states, accept, or go to the next row (we are not interested in rejecting or looping computations here). The total number of possible behaviours is $2^{(k+2)k}$. Generalizing this one step further we consider all tapes of length $i = |w|$ and obtain a set s_i describing all possibilities of maximum size $2^{2^{(k+2)k}}$. Note that these values are independent of the actual input size. Now given a set associated to a length $i = |w|$ it is clear that the set for $i+1$ can be generated by considering all symbols that can be attached to some input of length i. This may even be done by a deterministic one-way automaton A_L operating on a unary string of length $i+1$. The last property allows us to also compute s_{i-1} from s_i by using a technique due to Hopcroft and Ullman that is sketched in Lemma 4 of [2]. Essentially M' starts a backward simulation of the deterministic automaton until either the state becomes unique or the endmarker is reached. Then the automaton is run forward until it returns to its old position.

In a totally symmetric way the right portion of the input y is handled by a deterministic automaton A_R that works from right to left.

Initially M' determines the state that A_R is in by running it on the entire first row and resets A_L to its initial state. Then it simulates M by moving its head in one direction, updating A_L and A_R, guessing a sequence of input symbols and a sequence of matching crossing sequences compatible with the symbols as well as with the sets determined by A_L and A_R until it either reaches an accepting state of M or M goes to the next row. In the latter case the simulation continues in that row. For a computation without loops the length of the crossing sequences is bounded and therfore each sequence can be kept in the finite control of M'.

We note that M' never visits a tape cell twice although M in its computation may do so. □

Corollary 1 *The emptiness problem and the universe problem for deterministic three-way finite automata on two-dimensional tapes (TDFA) are decidable.*

Proof: The first statement is immediate. The second claim follows from the effective complementation of TDFA given in Theorem 1 of [14]. □

The following result solves another problem from Section 2.6 in [4, 5].

Theorem 4 *The emptiness problem for alternating three-way finite automata on two-dimensional tapes over a single-letter alphabet (TAFA(0)) is undecidable.*

Proof: In [11] it is shown that the emptiness problem for one-way two-head alternating finite automata over single-letter alphabets is undecidable in the one-dimensional case. By Lemma 2 every one-dimensional automaton A can be transformed into a TAFA(0) B such that A accepts some word w if and only if b accepts $\rho(w)$. This proves the theorem.

In order to make the method more transparent and prepare one of the corollaries we will briefly sketch the method in [11]. Essentially it consists of a reduction of the halting problem for two-counter automata to the emptiness problem. The counters are encoded by a single number that is stored as the distance of one of the heads to the right endmarker. The contents are stored as exponents of two primes, e.g. 2 and 3. Decrement is performed by dividing this number by the appropriate prime. Increment seems to be impossible since the head cannot move back. Therefore we initially multiply the number stored by a sufficiently large power of 5. Then a multiplication by 2 or 3 can be combined with a division by 5. As a result the head never has to move left. □

Note: In the above construction the head of the TAFA(0) never moves left. The result and the following corollaries therefore still hold for two-way automata with down and right movements only.

The emptiness problem above can be reduced to an inclusion resp. equivalence test with the empty set or disjointness with the set of all tapes over the given alphabet. Therefore the preceding result also solves those problems from [4, 5].

Corollary 2 *For TAFA(0) inclusion, equivalence, and disjointness are undecidable.*

In the proof from [11] sketched above no use was made of infinite computations. Hence it is possible to exchange accepting and rejecting states and obtain the

Corollary 3 *The universe problem is undecidable for TAFA(0).*

4 Hierarchy Results

From the translational technique of Lemma 4 for alternating Turing machines we get a hierarchy result analogous to Theorem 2.16 of [4, 5] searched for in the open problem of Section 2.4 of [4, 5].

Theorem 5 *Let $s_2(m)$ be a one-dimensionally space constructable function. Suppose $s_1(m) \in o(s_2(m))$ and $s_2(m) \geq \log m$. Then there exists a set in $\mathcal{L}[ATM^s(s_2(m))]$ but not in $\mathcal{L}[ATM^s(s_1(m))]$*

Proof: It is well-known that a corresponding space hierarchy exists for one-dimensional machines, see e.g. Exercise 3.6.5 in [1]. By Lemma 4 we obtain the hierarchy for two-dimensional machines. □

Remark: Since the diagonal argument above can be based on the input length we obtain a hierarchy result for languages over single-letter alphabets if the bound s_2 is fully two-dimensionally space constructable.

We recall the notion of strong and weak $s(n)$ space bounded computations. In the weak mode of operation it is sufficient that one of the (possibly many) accepting computation trees of an alternating machine obeys the bound while no assumption is made concerning the rejected inputs. For the strong mode we require that *every* reachable configuration respects the bound.

The following result can be shown for nondeterministic, deterministic, and alternating machines (the constructability notion has to match the dimension of the input tape), see Theorem 8.1.1 in [15].

Fact 1 *If a function $s(n)$ is fully space constructable then weak and strong mode of acceptance coincide for $s(n)$ space-bounded machines.*

For two-dimensional Turing machines the following has been shown in [7]:

Fact 2 *If $s_1(n)$ is space constructable by a two-dimensional Turing, $s_1(n) \leq \log n$ and $s_2(n) = o(s_1(n))$ then $\mathcal{L}[DTM^s(s_1(n))] \setminus \mathcal{L}[ATM^s(s_2(n))]$ is nonempty if the strong mode of acceptance is assumed for the complexity classes.*

A third important step is the full constructability on two-dimensional machines of a rich family of functions growing more slowly than $\log\log n$ as shown in [10]. As a special case of this result we have

Fact 3 *The following functions are fully constructable for all $k \geq 1$:*

$$f_k(n) = \begin{cases} \log^{(k)}(\sqrt{n/4}) & \text{if } n = 2^{2m} \text{ for some } m \geq 0 \\ 1 & \text{otherwise} \end{cases}$$

where $\log^{(k)}$ is the k-fold iteration of the logarithm.

These facts imply an infinite hierarchy of language classes for two-dimensional alternating Turing machines below $\log n$ analogous to Corollary 2.2 of [4, 5]:

Theorem 6 *For any constant $c > 0$ and each $k \geq 1$*

$$\mathcal{L}[AFA^s] = \mathcal{L}[ATM^s(c)] \subset \ldots \subset \mathcal{L}[ATM^s(f_{k+1}] \subset \mathcal{L}[ATM^s(f_k)] \subset \ldots$$

We summarize the hierarchy results for two-dimensional Turing machines

- Above $\log n$ the known hierarchies of the one-dimensional case can be transferred to two-dimensional machines with the help of the translational operation ρ (for deterministic and nondeterministic machines these hierarchies were known [4, 5]). The hierarchies are witnessed by languages over single-letter alphabets.
- Below $\log n$ the situation depends on the mode of operation:
  - Deterministic machines with a strong space-bound can be transformed into halting ones by Sipser's simulation (see e.g. [15]). As pointed out in [10] the full constructability of arbitrary slowly growing functions gives rise to a corresponding hierarchy of classes of languages over single-letter alphabets.

- For nondeterministic and alternating two-dimensional Turing machines we could show above that there are hierarchies of language classes provided that no restriction is imposed on the size of the alphabet.

Thus it remains open whether the hierarchies for nondeterministic and alternating machines can be based on single-letter alphabets. It would also be interesting to know if the notions of constructability and *full* constructability can be separated.

References

[1] J.Balcazár, J.Diáz, J.Gabárro: Structural Complexity II, Springer (1990).

[2] O.H.Ibarra, S.M.Kim, L.E.Rosier: *Some characterizations of multihead finite automata,* Inform. and Control 67 (1985) 114–125.

[3] K.Inoue, I.Takanami: *A note on decision problems for three-way two-dimensional finite automata,* IPL 10 (1980) 245-248.

[4] K.Inoue, I.Takanami: *A survey of two-dimensional automata theory,* Proc. Machines, Languages, and Complexity (J.Dassow, J.Kelemen eds.), Springer LNCS 381 (1989) 72–91.

[5] K.Inoue, I.Takanami: *A survey of two-dimensional automata theory,* Inf. Sci. 55 (1991) 99–121.

[6] T.Jiang, O.H.Ibarra, H.Wang: *Some results concerning 2-d on-line tesselation acceptors and 2-d alternating finite automata,* Theor. Comput. Sci. 125 (1994) 243–257.

[7] T.Jiang, O.H.Ibarra, H.Wang, Q.Zheng: *A hierarchy result for 2-dimensional TM's operating in small space,* Inf. Sci. 64 (1992) 49–56.

[8] K.N.King: *Alternating multihead finite automata,* Theor. Comput. Sci. 61 (1988) 149–174.

[9] B.Monien: *The LBA-problem and the deterministic tape complexity of two-way one-counter languages over a one-letter alphabet,* Acta Informatica 8 (1977) 371–382.

[10] H.Petersen: *On space functions fully constructed by two-dimensional Turing machines,* IPL 54 (1995) 9–10.

[11] H.Petersen: *Alternation in simple devices,* Proc. ICALP95.

[12] A.Rosenfeld: *Picture Languages,* Academic Press, New York (1979).

[13] J.C.Shepherdson: *The reduction of two-way automata to one-way automata,* IBM Journal of Res. and Dev. April (1959) 198–200.

[14] A.Szepietowski: *Some remarks on two-dimensional finite automata,* Inf. Sci. 63 (1992) 183–189.

[15] A.Szepietowski: *Turing machines with sublogarithmic space,* Springer LNCS 843 (1994).

How Hard is to Compute the Edit Distance*

Giovanni Pighizzini

Dipartimento di Scienze dell'Informazione
Università degli Studi di Milano
via Comelico, 39 – 20135 Milano – Italy
pighizzi@ghost.dsi.unimi.it

Abstract. The notion of *edit distance* arises from very different fields, such as self–correcting codes, parsing theory, speech recognition and molecular biology. The edit distance between an input string and a language L is the minimum number of *edit operations* (*substitution* of a symbol in another incorrect symbol, *insertion* of an extraneous symbol, *deletion* of a symbol) needed to change the input string into a sentence of L.
In this paper we study the complexity of computing the edit distance, discovering sharp boundaries between classes of languages for which this function can be efficiently evaluated and classes of languages for which it seems to be difficult to compute. Our main result is a parallel algorithm for computing the edit distance for the class of languages accepted by one–way nondeterministic auxiliary pushdown automata working in polynomial time, a class that strictly contains context–free languages. Moreover, we show that this algorithm can be extended in order to find a sentence of the language with minimum distance with respect to the input string.

1 Introduction

Consider a language $L \subseteq \Sigma^*$ and suppose to get strings which represent instances of L affected by errors due, for example, to typing process, to noise in the communication channel, *etc.* We want to find the original sentences of L from which, more likely, the strings we have received were originated. In order to find such sentences, we adopt a minimum distance criterion, i.e., given an input string x, we are interested in determining a sentence $y \in L$ with minimal *edit distance* from x.

The *edit* or *Levenshtein distance* [AP72, WF74] between two strings is the minimum number of edit operations (i.e., the *substitution* of a symbol by an incorrect symbol, the *insertion* of an extraneous symbol, and the *deletion* of a symbol), that transform a string into another. Thus, fixed a language L, the edit distance of L is the function which associates to any input string x the minimal distance between x and strings belonging to L.

* This work was supported in part by the ESPRIT Basic Research Action No. 6317: "Algebraic and Syntactic Methods in Computer Science (ASMICS 2)" and by the Ministero dell'Università e della Ricerca Scientifica e Tecnologica (MURST).

Edit distances were initially studied in code theory [Lev66], parsing theory [AP72, Pig92], and speech recognition [OTK76, Ack80]. Interesting combinatorial problems on strings, as for example the computation of longest common subsequences [AG87], can be reduced to edit distances. Very recently they regained a certain popularity due to their connections with problems in molecular biology [EGG90, Kar93].

In this paper, we are interested in establishing general results about the complexity of computing edit distances. In particular, we want to delineate the boundaries among classes of languages for which at least a strong evidence can be given about the "hardness" of edit distances and classes for which edit distances can be efficiently computed. The results we obtain are quite tight. More precisely, in Section 3, we show that the edit distance of any language accepted in logarithmic space by a deterministic Turing machine is computable in polynomial time if and only if P = NP. On the other hand, if we restrict our attention to one–way devices, edit distance becomes feasible. In fact, our main result, proved in Section 4, states that the edit distance of languages accepted by one–way nondeterministic auxiliary pushdown automata (i.e., logspace bounded Turing machines augmented with a pushdown store) working in polynomial time (a class of languages wider than context–free) can be efficiently computed.

In general people is not really interested in computing edit distances, but mainly in finding a sentence with minimal edit distance from the input string. In Section 5, we consider such a problem and we show that in the case of languages accepted by one–way nondeterministic auxiliary pushdown automata working in polynomial time, it can be solved in the same bounds given in Section 4 for computing the edit distance.

This paper is a continuation of a stream of research whose aim is that of identifying properties other than the membership which are easily computable for certain classes of languages (see, e.g., [Huy90, Huy91, ABP93, ÀJ93b]).

For brevity reasons, some of the proofs have been omitted from this version of the paper.

2 Preliminary definitions and results

Given a finite alphabet Σ and the monoid Σ^* of strings over Σ, the binary relation $\vdash$, representing the three kinds of syntax errors considered in this paper can be defined as follows. For x and y in Σ^*, $x \vdash y$ if and only if there exist $v, w \in \Sigma^*$, $a, b \in \Sigma$ such that $x = vaw$, $y = vbw$ and $a \neq b$ (*substitution* of a symbol), or $y = vaw$ and $x = vw$ (*deletion*), or $y = vw$ and $x = vaw$ (*insertion*). Observe that $\vdash$ is symmetric. By $\overset{k}{\vdash}$ we denote the composition of $\vdash$ with itself k times. Then $x \overset{k}{\vdash} y$ if and only if y can be obtained from x with k errors. The *edit distance* between two strings $x, y \in \Sigma^*$, denoted by $d(x, y)$, is defined as the minimum integer k such that $x \overset{k}{\vdash} y$.

For example given $\Sigma = \{a, b, +, *\}$, we have:

$$a + a * a \vdash aa + a * a \vdash aa + ba * a \vdash aa + bb * a \vdash aa + bb * .$$

Then $a + a * a \vdash^4 aa + bb*$. Moreover, it is possible to verify that this is optimal, i.e., $d(aa + bb*, a + a * a) = 4$.

The following properties of the edit distance can be easily verified:

- $\forall u, v, w, z \in \Sigma^*$: $d(uv, wz) \leq d(u, w) + d(v, z)$;
- $\forall x', x'', y \in \Sigma^*$: $d(x'x'', y) = \min\{d(x', y') + d(x'', y'') \mid y = y'y''\}$.

Given a language $L \subseteq \Sigma^*$ and a string $x \in \Sigma^*$, we define the edit distance $d(x, L)$ between x and L as the minimum of the distances between x and strings belonging to L, i.e., $d(x, L) = \min\{d(x, y) \mid y \in L\}$, with the convention that $d(x, \emptyset) = \infty$. We point out that if L is not empty then $d(x, L)$ is at most $|x| + m_L$, where m_L denotes the length of the shortest string belonging to L. In fact, we can assume that all $|x|$ symbols of x are insertions and all m_L symbols of the shortest string of L have been deleted. From this fact and the trivial inequality $d(x, y) \geq |(|x| - |y|)|$, it turns out that if $y \in L$ is a string with minimal distance from x, i.e., $d(x, L) = d(x, y) = \min\{d(x, z) \mid z \in L\}$, then $|y| \leq 2|x| + m_L$.

When we deal with Turing machines, we will denote by Σ the input alphabet, by Γ the worktape alphabet and by Q the set of states. We always suppose that these sets are pairwise disjoint.

We briefly recall that a *one–way nondeterministic auxiliary pushdown automaton* (1–NAuxPDA) [Bra77] is a nondeterministic Turing machine having a one–way, end–marked, read–only input tape, a pushdown tape, and a two–way, read/write work tape *with a logarithmic space bound.* (For more formal definitions see, e.g, [HU79].) "Space" on an 1–NAuxPDA means space on the work tape only (excluding the pushdown). Without loss of generality, we make the following assumptions about 1–NAuxPDAs: (1) at the start of the computation the pushdown store contains only one symbol Z_0; this symbol is never pushed or popped on the stack; (2) there is only one final state q_f; when the automaton reaches the state q_f the computation stops; (3) the input is accepted if and only if the automaton reaches q_f, the pushdown store contains only Z_0, all the input has been scanned and the worktape is empty; (4) if the automaton moves the input head, then no operations are performed on the stack; (5) every push adds exactly one symbol on the stack; and (6) input head can be kept stationary only after moves that do not depend on the content of currently scanned input square.

We recall that a *surface configuration* of a 1–NAuxPDA M [Coo71] on an input string of length n is a 5–tuple (q, w, j, γ, i) where q is the state of M, w is a string of worktape symbols (the *worktape contents*), j is an integer, $1 \leq j \leq |w|$ (the *worktape head position*), γ is a pushdown symbol (the *stack top*), and i is an integer $1 \leq i \leq n + 1$ (the *input head position*). Throughout the paper, we will denote by $\mathcal{S}_n$ the set of all surface configurations on inputs of length n, and, for any surface configuration A, by i_A the input head position in A. Observe that the size of any surface configuration belonging to $\mathcal{S}_n$ is at most $O(\log n)$.

The initial surface configuration, denoted by A_0, is $(q_0, \sharp, 1, Z_0, 1)$ where q_0 is the initial state, and by our assumptions on 1–NAuxPDAs the only *accepting surface configuration* on input of length n is the tuple $A_f^n = (q_f, \sharp, 1, Z_0, n + 1)$ where $\sharp$ represents the empty worktape.

Given a string $y \in \Sigma^*$ and an integer $t \geq 0$, a pair (A, B) of surface configurations is said to be *realizable on y within t steps* if M from A can reach in at most t steps B ending with its stack at the same height as in A, without popping it below its level in A at any point in this computation, and consuming the input substring y.[2] Of course, if (A, C) is realizable on y' within t' steps and (C, B) is realizable on y'' within t'' steps then (A, B) is realizable on $y'y''$ within $t' + t''$ steps. Moreover, for any surface configuration A, the pair (A, A) is realizable on ϵ in zero steps, and it cannot be realizable on nonempty strings.

Given two surface configurations A, B and an integer $t \geq 0$, by L^t_{AB} we denote the set of all strings $y \in \Sigma^*$ such that the pair (A, B) is realizable on y within t steps. Note that $|y| = i_B - i_A$. Thus, $L^{n^s}_{A_0 A^n_f}$ is the set of strings of length n accepted by a 1–NAuxPDA working in time n^s.

A *family of circuits* is a set $\{C_n \mid n \in N\}$ where C_n is a circuit for inputs of size n. $\{C_n\}$ is logspace uniform if the function $n \rightarrow C_n$ is computable on a deterministic Turing machine in logarithmic space. NC^k (AC^k) denotes the class of problems solvable by logspace uniform families of bounded (unbounded) fan–in Boolean circuits of polynomial size and $O(\log^k n)$ depth.

Unbounded fan–in circuits are equivalent to parallel random access machines with both concurrent read and concurrent write (CRCW PRAM). In particular the class AC^1 coincides with the class of functions computed by CRCW PRAM in $O(\log n)$ time, using a polynomial number of processors.

An *arithmetic circuit* is a circuit where the OR (addition) gates and the AND (multiplication) gates are interpreted over a suitable semi–ring. For further notions of parallel computation and arithmetic circuits the reader is referred to [Coo85],[MRK88].

3 Languages whose edit distance is unlike to be easy

The first problem we afford is that of identifying the smallest class of languages for which edit distances are hard to compute. In this section, we give strong evidence of the fact that this class is LOGSPACE. In fact, we prove that if LOGSPACE has polynomial time computable edit distances then P = NP.

Theorem 1. *The following statements are equivalent:*
(1) P = NP;
(2) for all languages in P *the edit distance is computable in polynomial time;*
(3) for all languages in LOGSPACE *the edit distance is computable in polynomial time.*

Proof. (1)⇒(2) For any language $L \in$ P, we consider the problem Π_L of deciding, given a string $x \in \Sigma^*$ and an integer $k \geq 0$, whether or not $d(x, L) \leq k$. Since the edit distance among two strings is computable in time proportional to the product of their lengths [WF74], it is not difficult to see that Π_L can be solved using the following nondeterministic polynomial time algorithm:

[2] This implies that the length of the input substring y is $i_B - i_A$.

input x, k
guess $y \in \Sigma^*$, with $|y| \leq 2|x| + m_L$
if $y \in L$ and $d(x,y) \leq k$ then accept
else reject

At this point, it is possible to devise an algorithm that computes $d(x, L)$ by calling such a machine on input (x, k) for increasing values of k. If P = NP this algorithm works in deterministic polynomial time.

(2)$\Rightarrow$(3) Obvious.

(3)$\Rightarrow$(1) Given a language L accepted by a nondeterministic Turing machine M in polynomial time $p(n)$, we consider the language L_{acc} whose strings represent accepting computations of M. More precisely, L_{acc} contains all strings of the form:

$$x \underbrace{\sharp\omega_0\sharp\omega_1\sharp\ldots\sharp\omega_{p(n)}}_{p(n)+1}, \tag{1}$$

where:

- $x \in \Sigma^*$, $n = |x|$, $\sharp \notin \Sigma \cup \Gamma \cup Q$;
- for all $i = 0, \ldots, p(n)$, ω_i is a string over $\Gamma \cup Q$ of polynomial length $p'(n)$ which encodes a configuration of M;
- ω_0 is the initial configuration of M;
- for all $j = 1, \ldots, p(n)$, ω_{j-1} yields ω_j by a move of M (on input x);
- $\omega_{p(n)}$ is an accepting configuration of M.

It is not difficult to verify that $L_{acc} \in$ LOGSPACE.

Now, we show how to decide membership to L by computing the edit distance for L_{acc}. With every input string $x \in \Sigma^*$ we associate the string

$$\phi(x) = x \underbrace{\sharp\overbrace{\$\ldots\$}^{p'(n)}\sharp\overbrace{\$\ldots\$}^{p'(n)}\sharp\ldots\sharp\overbrace{\$\ldots\$}^{p'(n)}}_{p(n)+1},$$

where $\$ \notin \Sigma \cup \Gamma \cup Q \cup \{\sharp\}$. Thus, given a string $x \in \Sigma^*$ and a string $I_y \in L_{acc}$ representing as in (1) an accepting computation of M on $y \in L$, using properties of the edit distance recalled in Section 2, it turns out that

$$d(\phi(x), I_y) = \min\{d(x, I') + d(\tilde{x}, I'') \mid I'I'' = I_y\},$$

where $\tilde{x} = (\sharp\$^{p'(n)})^{p(n)+1}$, i.e., $\phi(x) = x\tilde{x}$.

Since all \$s in $\tilde{x}$ are not in I'', we have $d(\tilde{x}, I'') \geq p'(n) \cdot (p(n)+1)$, for any I''. Moreover, I_x can be obtained from $\phi(x)$ substituting each block of $p'(n)$ \$s with a string representing a suitable configuration of M. Thus, it turns out that for $x = y$ the equality holds, and then $d(\phi(x), I_x) = p'(n) \cdot (p(n)+1)$. On the other hand, if $x \neq y$ then $d(x, I')$ is zero only when I' coincide with x and it is a prefix of y. In this case, it turns out that $I'' = y'\sharp\omega_0\sharp\ldots\sharp\omega_{p(|y|)}$, for some

$y' \in \Sigma^*$, $y' \neq \epsilon$, and the distance $d(\tilde{x}, I'')$ can be computed observing that all symbols of y', $\omega_0, \ldots, \omega_{p(|y|)}$ are not symbols of $\tilde{x}$. Since w.l.o.g. $p(|y|) \geq p(x)$, we have: $d(\tilde{x}, I'') \geq |y'| + \sum_{i=0}^{p(|y|)} |w_i| \geq 1 + p'(n) \cdot (p(n)+1) > p'(n) \cdot (p(n)+1)$. At this point, we can easily conclude that:

$$x \in L \quad \text{if and only if} \quad d(\phi(x), L_{acc}) = p'(n) \cdot (p(n)+1).$$

Thus, membership to L can be decided computing $d(\phi(x), L_{acc})$. □

4 Efficient computation of the edit distance

In [AP72] it is shown that the edit distance of context–free languages can be computed in polynomial time. Moreover, as proved in [Pig92], it can be efficiently computed using a parallel algorithm. In this section we consider a wider class of languages, namely languages accepted by 1–NAuxPDA working in polynomial time, and we exhibit a parallel algorithm for CRCW PRAM that computes edit distances in $O(\log n)$ time, using a polynomial number of processors. So, it turns out that the class of languages accepted in polynomial time by 1–NAuxPDA is the biggest class of languages we know for which edit distances can be efficiently computed.

The algorithm we present consists of defining and evaluating an arithmetic circuit over the commutative semiring $(\mathbb{N} \cup \{\infty\}, \min, +, \infty, 0)$ (in the following $(\mathbb{N}, \min, +)$). More precisely, given a 1–NAuxPDA M working in polynomial time n^s, and an input string x of length n, let $H = 2n + m_L$ be an upper bound on the length of the string accepted by M with minimal distance from x, and $\mathcal{S}_H$ the set of surface configurations on inputs of length H. We define an arithmetic circuit C_n with three types of nodes:

- *input nodes,* with associated an input value from $\mathbb{N} \cup \{\infty\}$:

$$N_l = \{(A, B, i, j, t) \mid 0 \leq i \leq j \leq n,\ A, B \in \mathcal{S}_H,\ 0 \leq t \leq 1\};$$

- *min nodes,* which compute the minimum among the values of their sons:

$$N_{\min} = \{(A, B, i, j, t) \mid 0 \leq i \leq j \leq n,\ A, B \in \mathcal{S}_H,\ 1 < t \leq H^s\};$$

- *plus nodes,* which computes the sums of the values of their sons:

$$N_+ = \{(A, C, B, i, k, j, t', t'') \mid 0 \leq i \leq k \leq j \leq n,\ A, B, C \in \mathcal{S}_H,\ 0 < t', t'' \leq H^s\}.$$

Connections among nodes are defined as follows. Node $(A, C, B, i, k, j, t', t'') \in N_+$ has exactly two sons, that is nodes $(A, C, i, k, t'), (C, B, k, j, t'') \in N_l \cup N_{\min}$. Node $(A, B, i, j, t) \in N_{\min}$ has three kinds of sons: (1) all nodes of the form $(A, C, B, i, k, j, t', t'') \in N_+$ such that $t' + t'' = t$; (2) all nodes $(A', B', i, j, t-2) \in N_l \cup N_{\min}$ such that A' can be reached from A via a push and B can be reached from B' via a pop of the same symbol; and (3) all nodes $(A, B, i, j, t-1) \in N_l \cup N_{\min}$.

Now, we define the value which will be associated with input nodes of the circuit, in such a way that each node $(A, B, i, j, t) \in N_{\min} \cup N_l$ of the circuit computes the minimum distance between the input substring $x_{i+1} \dots x_j$ and the language L^t_{AB}. So, the solution of the problem is constructed bottom–up from optimal solutions of subproblems, according to the optimality principle of dynamic programming [Bel57].

For input nodes of the form $(A, B, i, j, 0)$, observing that $L^0_{AB} = \{\epsilon\}$ if $A = B$, and $L^0_{AB} = \emptyset$ otherwise, we define:

$$value(A, B, i, j, 0) = \begin{cases} j - i & \text{if } A = B \\ \infty & \text{otherwise.} \end{cases}$$

The definition for input nodes of the form $(A, B, i, j, 1)$ is more complicated; in fact, L^1_{AB} can be either empty or it can contain the empty string or some element of Σ. For instance, if L^1_{AB} contains one symbol $\sigma \in \{x_{i+1}, \dots, x_j\}$, then $x_{i+1} \dots x_j$ can be obtained from L^1_{AB} with $j - i - 1$ insertions. On the other hand, if L^1_{AB} contains some element of Σ, but it does not contain any element of $\{x_{i+1}, \dots, x_j\}$, then $x_{i+1} \dots x_j$ can be obtained from L^1_{AB} with one substitution and $j - i - 1$ insertions. More precisely, the value associated with a node $(A, B, i, j, 1)$ is defined as:

$$value(A, B, i, j, 1) = \begin{cases} j - i - 1 & \text{if } A \text{ yields } B \text{ consuming an input symbol } \sigma \in \{x_{i+1}, \dots, x_j\} \\ j - i & \text{if } (A = B) \vee (A \text{ yields } B \text{ without consuming any input symbol belonging to } \{x_{i+1}, \dots, x_j\})^3 \\ \infty & \text{otherwise.} \end{cases}$$

The following lemma, whose proof can be given by induction, states the main property of the values computed by the gates of C_n with such inputs:

Lemma 2. *For any $A, B \in S_H$, $0 \le i \le j \le n$, $0 \le t \le H^s$, it holds that:*

$$value(A, B, i, j, t) = d(x_{i+1} \dots x_j, L^t_{AB}).$$

Finally, recalling that the length of the string of L with minimal distance from x is at most $H = 2n + m_L$, we have:

$$\begin{aligned} d(x, L) &= \min_{m=0,\dots,H} d(x, L^{m^s}_{A_0 A^m_f}) \\ &= \min_{m=0,\dots,H} d(x, L^{H^s}_{A_0 A^m_f}) \\ &= \min_{m=0,\dots,H} value(A_0, A^m_f, 0, n, H^s). \end{aligned}$$

Thus, we can extend circuit C_n with a node performing this operation and producing the output.

[3] This corresponds to two different situations: the pair (A, B) is realizable on the empty string, i.e., A yields B without consuming input symbols, or the pair (A, B) is realizable on some symbol $\sigma \in \Sigma$, but it is not realizable on symbols belonging to the set $\{x_{i+1}, \dots, x_j\}$.

As a consequence, we are able to prove our main result:

Theorem 3. *The edit distance of languages accepted by one-way nondeterministic auxiliary pushdown automata working in polynomial time is in* AC^1.

Proof. (outline) The edit distance can be computed using the arithmetic circuit C_n above defined. This circuit has polynomial degree and it is defined over a *commutative* semiring. Thus, it can be efficiently evaluated using the technique presented in [MRK88], which consists of $O(\log n)$ applications of a routine called *Phase*, where a single application of *Phase* consists of nothing more complicated than matrix multiplication over the semiring $(\mathbb{N}, \min, +)$. Such a multiplication can be done in constant depth using unbounded fan–in circuits [SV81]. So, it turns out that the whole computation is in AC^1. □

5 How to find the string with minimal edit distance

In Section 4 we showed how to compute the edit distance $d(x, L)$ between a string $x \in \Sigma^*$ and a language $L \subseteq \Sigma^*$ accepted by a 1–NAuxPDA working in polynomial time. Often, however, it is required to exhibit a string $y \in L$ with minimal distance from x. In this section we show how this problem can be solved: we give an algorithm for the computation of the string with minimal distance, which, as that presented in Section 4, can be executed by a CRCW PRAM in $O(\log n)$ time, using a polynomial number of processors.

The idea is that of modifying the evaluation function for gates of C_n, in order to compute optimal substrings during the evaluation process. To achieve this aim, we adopt the following scheme. With each node (A, B, i, j, t) of the circuit C_n we associate a pair (m, w), where $m \in \mathbb{N}$ represents the minimum distance between the input substring $x_{i+1} \dots x_j$ and the language L^t_{AB}, and $w \in \Sigma^*$ is a string of L^t_{AB} with such a distance, i.e., $d(x_{i+1} \dots x_j, L^t_{AB}) = d(x_{i+1} \dots x_j, w)$.

Subsequently, we evaluate circuit C_n over the semiring $R = ((\mathbb{N} \times \Sigma^*) \cup \{\infty\}, \oplus, \otimes, \infty, (0, \epsilon))$, where the operation $\oplus$ selects between two elements of $\mathbb{N} \times \Sigma^*$ the element whose first component is minimal and the operation $\otimes$ adds the first components and concatenates the second ones, i.e., $(m', w') \otimes (m'', w'') = (m' + m'', w'w'')$.

Leaves of C_n are evaluated as follows:

$$value(A, B, i, j, 0) = \begin{cases} (j - i, \epsilon) & \text{if } A = B \\ \infty & \text{otherwise,} \end{cases}$$

$$value(A, B, i, j, 1) = \begin{cases} (j - i - 1, \sigma) & \text{if } A \text{ yields } B \text{ consuming a symbol } \sigma \in \{x_{i+1}, \dots, x_j\} \\ (j - i, \epsilon) & \text{if } A = B \text{ or } A \text{ yields } B \text{ without consuming input symbols} \\ (j - i, \sigma) & \text{if } A \text{ yields } B \text{ consuming } \sigma \in \Sigma, \text{ but } A \text{ cannot yield } B \text{ consuming a symbol in } \{x_{i+1}, \dots, x_j\} \\ \infty & \text{otherwise.} \end{cases}$$

However semiring R is not commutative, and so it is not possible to apply the algorithm of [MRK88] to evaluate C_n gates. We show now how to overcome this problem in the case of the alphabet $\Sigma = \{0,1\}$. The extension to other alphabets is trivial.

We recall that every surface configuration A contains the position i_A of the input head; thus, given a node (A, B, i, j, t) of the circuit, each string $w \in L^t_{AB}$ can contribute to the output string from position i_A to position $i_B - 1$. Moreover, the maximal length $H = 2n + m_L$ of the output, for inputs of length n, is *a priori* known. This will permit us to represent elements of R over a commutative semiring. More precisely, if the value associated with the node (A, B, i, j, t) is the pair (m, w), then we represent it as the integer $h = m \cdot 2^H + w \cdot 2^{i_B}$ (where w here is interpreted as the integer number whose binary representation is the string w). Observe that the binary representation of h is obtained concatenating the binary representation of m and the string $0^{i_A - 1} w 0^{H - i_B + 1}$, whose length is H. In this way, elements of R are represented as integers. It is not difficult to see that, in this representation, semiring operations $\oplus$ and $\otimes$ are substituted by min and $+$ respectively. Thus, the noncommutative semiring R in the case under consideration can be represented by the *commutative* semiring $(\mathbb{N}, \min, +)$, on which we are able to efficiently evaluate C_n as explained in Section 4.

Thus, the following result holds:

Theorem 4. *For any language L accepted by one–way nondeterministic auxiliary pushdown automata in polynomial time, the edit distance and the function which associates with any input string x a sentence of L with minimal distance from x are in* AC^1.

6 Concluding remarks

In this paper we showed that the edit distance of languages accepted in polynomial time by 1–NAuxPDA can be efficiently computed. Our algorithm is given assuming that all edit operations have the same cost. However, it can be easily extended to work also for other cost functions (see [WF74] for the definition of *weighted* edit distance).

The evaluation of the edit distance is an example of optimization problem over the semiring $(\mathbb{N}, \min, +)$. In particular, our results show that edit distances for languages accepted by 1–NAuxPDA working in polynomial time are in the class of functions computable by arithmetic circuits of polynomial size and polynomial degree over the semiring $(\mathbb{N}, \min, +)$. By results of [MRK88] this class is contained in AC^1 and, on the other hand, it can be shown (coding string on integers as outlined in Section 5) that its maximization version contains OptSAC^1 [Vin91, AJ93a], i.e., the class of functions computed by arithmetic circuits of polynomial size and polynomial degree over $(\Sigma^*, \max, \mathrm{concat})$.

References

[ABP93] E. Allender, D. Bruschi, and G. Pighizzini. The complexity of computing maximal word functions. *Computational Complexity*, 3:368–391, 1993.

[Ack80] M. Ackroyd. Isolated word recognition using the weighted Levenshtein distance. *IEEE Transactions on Acoustics, Speech and Signal Processing*, 28:243–244, 1980.

[AG87] A. Apostolico and C. Guerra. The longest common subsequence problem revisited. *Algorithmica*, 2:315–336, 1987.

[AJ93a] E. Allender and J. Jiao. Depth reduction for noncommutative arithmetic circuits. In *Proc. 25th ACM Symposium on Theory of Computing*, pages 515–522, 1993.

[ÀJ93b] C. Àlvarez and B. Jenner. A very hard log–space counting class. *Theoretical Computer Science*, 107:3–30, 1993.

[AP72] A. Aho and T. Peterson. A minimum distance error–correcting parser for context–free languages. *SIAM J. Computing*, 1:305–312, 1972.

[Bel57] R. Bellman. *Dynamic Programming*. Princeton Univ. Press, 1957.

[Bra77] F. Brandenburg. On one–way auxiliary pushdown automata. In *Proc. 3rd GI Conference*, Lecture Notes in Computer Science 48, pages 133–144, 1977.

[Coo71] S. Cook. Characterization of pushdown machines in terms of time–bounded computers. *Journal of the ACM*, 18:4–18, 1971.

[Coo85] S. Cook. A taxonomy of problems with fast parallel algorithms. *Information and Control*, 64:2–22, 1985.

[EGG90] D. Eppstein, Z. Galil, and R. Giancarlo. Efficient algorithms with applications to molecular biology. In R. Capocelli, editor, *Sequences*, pages 59–74. Springer–Verlag, 1990.

[HU79] J. Hopcroft and J. Ullman. *Introduction to automata theory, languages, and computations*. Addison–Wesley, Reading, MA, 1979.

[Huy90] D. Huynh. The complexity of ranking simple languages. *Mathematical Systems Theory*, 23:1–19, 1990.

[Huy91] D. Huynh. Efficient detectors and constructors for simple languages. *International Journal of Foundations of Computer Science*, 2:183–205, 1991.

[Kar93] R. Karp. Mapping the genome: some combinatorial problems arising in molecular biology. In *Proc. 25th ACM Symposium on Theory of Computing*, pages 278–285, 1993.

[Lev66] V. Levenshtein. Binary codes capable of correcting deletions, insertions and reversals. *Soviet Physics–Doklady*, 10:707–710, 1966.

[MRK88] G. Miller, V. Ramachandran, and E. Kaltofen. Efficient parallel evaluation of straight–line code and arithmetic circuits. *SIAM J. Computing*, 17:687–695, 1988.

[OTK76] T. Okuda, E. Tanaka, and T. Kasai. A method for the correction of garbled words based on the Levenshtein metric. *IEEE Transactions on Computers*, 25:172–178, 1976.

[Pig92] G. Pighizzini. A parallel minimum distance error–correcting context–free parser. In *Theoretical Computer Science – Proceedings of the Fourth Italian Conference*, pages 305–316. World Scientific, 1992.

[SV81] Y. Shiloach and U. Vishkin. Finding the maximum, merging and sorting in a parallel computation model. *Journal of Algorithms*, 2:88–102, 1981.

[Vin91] V. Vinaj. Counting auxiliary pushdown automata and semi–unbounded arithmetic circuits. In *Proc. 6th Structure in Complexity Theory*, pages 270–284, 1991.

[WF74] R. Wagner and M. Fisher. The string–to–string correction problem. *Journal of the ACM*, 21:168–173, 1974.

On the synchronization of semi-traces*

Klaus Reinhardt

Wilhelm-Schickhard Institut für Informatik, Universität Tübingen
Sand 13, D-72076 Tübingen, Germany
e-mail: reinhard@informatik.uni-tuebingen.de

Abstract. The synchronization of two or more semi-traces describes the possible evaluation of a concurrent system, which consists of two or more concurrent subsystems in a modular way, where communication between the subsystems restricts the order of the actions. In this paper we give criteria, which tell us for given semi-traces in given semi-commutation systems, whether they are synchronizable and whether the synchronization is again a semi-trace; and criteria, which tell us for given semi-commutation systems, whether all semi-traces have this property. We prove that deciding these criteria is **NLOGSPACE**-complete for given semi-traces. The same holds for the synchronizability of all semi-traces for given semi-commutation systems. On the other hand the question, whether for given semi-commutation systems the synchronization of synchronizable semi-traces is a semi-trace is **co-NP**-complete. Furthermore we give a **co-NP**-complete condition for being able to decide synchronizability locally in $\mathbf{TC}^0$.

1 Introduction

Traces were introduced by Mazurkiewicz in [Maz87] to describe the behavior of concurrent systems. Traces are equivalence classes of words under partial commutations, which allow only to describe symmetric dependencies between actions. In order to enhance the possibilities of descriptions, M.Clerbout introduced the notion of semi-commutation in her thesis [Cle84] as a generalization of partial commutation, see also M. Clerbout and M. Latteux [CL87]. In a broader context it was observed in [HK89], [Och90] and [Och92], that semi-commutations are very useful for modeling behaviors of Petri-nets. For instance we are able to describe a producer/consumer system. For more information see [CGL+92], [OW93], [DR95].

If we want to describe a distributed system and we have described the single components by semi-dependence alphabets, where the alphabets can partially overlap at those actions, where the components interact, we need the operation of *synchronization* to get the consistent evaluations of the entire system.

The *partial traces* in [Die94] are closed under synchronization; but this notion gives up the efficient description by just one representing word. Semi-traces are

* this research has been supported by the EBRA working group No. 3166 ASMICS.

not closed under synchronization and hence we have to distinguish between *stable* and *unstable* synchronizations of semi-traces.
In this paper we classify the complexity of problems and operations on semi-commutation systems, which are semantically relevant for such concurrent systems. Since it is important for practical applications to be able to design algorithms with low complexity and particularly efficient parallel algorithms, it is a good news, that some of the problems are in such low complexity classes as **NLOGSPACE** and $\mathbf{TC}^0$.

2 Preliminaries

Let $\mathbf{TC}^0$ be the class of problems being recognized by a uniform circuit family of constant depth and polynomial size with threshold-gates or computable in constant time for the enlarged PRAM model in [Par90], respectively. $\mathbf{NC}^k$ ($\mathbf{AC}^k$) is the class of problems being recognized by a circuit family of $O(log^k n)$ depth and polynomial size with bounded (unbounded) fan-in gates [Ruz81]. The problems in $\mathbf{NC} = \bigcup_k \mathbf{NC}^k$ are regarded as the efficiently parallelizable problems. **NLOGSPACE** (**NP**) is the class of problems being recognized by a logarithmic space-bounded (polynomial time-bounded) nondeterministic Turing machine. For completeness we use **LOGSPACE**-reducibility. co-**NP** is the set of the complements of problems in **NP**. The following inclusions are know [Joh90]: $\mathbf{AC}^0 \subset \mathbf{TC}^0 \subseteq \mathbf{NC}^1 \subseteq \mathbf{LOGSPACE} \subseteq \mathbf{NLOGSPACE} \subseteq \mathbf{AC}^1 \subseteq \mathbf{TC}^1 \subseteq \mathbf{NC}^2 \subseteq \mathbf{NC} \subseteq \mathbf{P} \subseteq \mathbf{NP} \subseteq \Sigma_2^P \subseteq \mathbf{PSPACE}$. Although nearly no separations are known (e.g., $\mathbf{TC}^0 \neq \mathbf{NP}$? is not known), proper inclusions are conjectured, which motivates the difference for local checking in $\mathbf{TC}^0$ or **NLOGSPACE** regarded in this paper.
To describe semi-commutation, we use *semi-dependence alphabets*, of the form (A, SD) where $SD \subseteq A \times A$ is reflexive but possibly asymmetric. This defines the associated *semi-commutation system* $SC = \{ ab \Longrightarrow ba \mid (a,b) \notin SD \}$. A *semi-trace* over (A, SD) is by definition the set of words $[u\rangle = \{ w \in A^* \mid u \overset{*}{\underset{SC}{\Longrightarrow}} w \}$, which can be derived from a word $u \in A^*$ by applying semi-commutation rules from SC. This means that we can use one possible evaluation of the concurrent system expressed by the word u to describe all possible evaluations. For several semi-dependence alphabets (A_i, SD_i) the corresponding semi-traces $[u\rangle_i$ get the same index.

2.1 Graph representation

We represent a semi-trace $[a_1...a_n\rangle$ over (A, SD) by a graph with nodes labeled by $a_1...a_n$ and two kinds of edges: The *hard arcs* $a_i \rightarrow a_j$ with $i < j$ and $(a_i, a_j) \in SD$ and the *soft arcs*[2] $a_i \dashrightarrow a_j$ with $i < j$ and $(a_i, a_j) \in SD^{-1} \setminus SD$, which emphasize the semi-dependence structure (see also [DOR94]). The graph examples are restricted to the Hasse diagram, that means we do not show arcs,

[2] An asymmetric commutation can change the order described by a soft arc but then the outcome is not representing the complete semi-trace. $(SD^{-1} = \{(a,b) \mid (b,a) \in SD\})$

which are in the transitive closure of shown arcs.

Example: Consider the following semi-dependence alphabets

$(A_1, SD_1) =$ 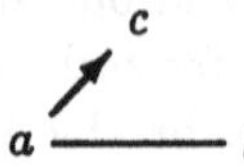

and $(A_2, SD_2) =$ 

and the following semi-traces together with their graph representation:

$[cacba\rangle_1 =$ 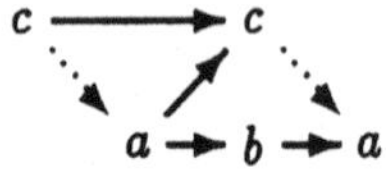

$[dbada\rangle_2 =$ 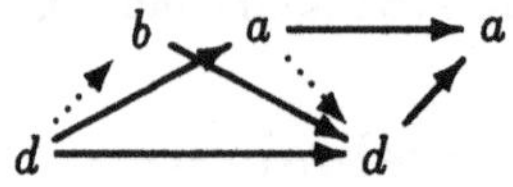

$[bdada\rangle_2 =$ 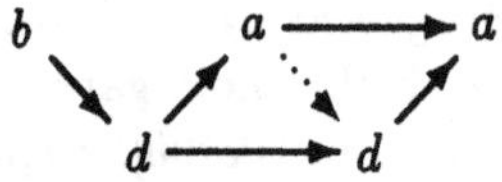

$[daadb\rangle_2 =$ 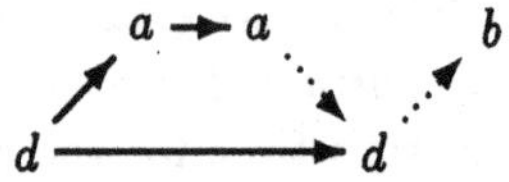

2.2 Synchronization

In order to construct complex systems in a modular way, we need a notion of synchronization for semi-commutation as it was introduced by Mazurkiewicz in [Maz87] and [DV89] for the symmetric case of partial commutation.

The synchronization $(A_1, SD_1) \parallel ... \parallel (A_m, SD_m)$ of semi-dependence alphabets is simply their union $(A, SD) = (\bigcup_{i \leq m} A_i, \bigcup_{i \leq m} SD_i)$. The synchronization[3] of semi-traces is $[u_1\rangle_1 \parallel [u_2\rangle_2 \parallel \ldots [u_m\rangle_m = \{w \in A^* \mid \Pi^A_{A_i}(w) \in [u_i\rangle_i \, \forall 0 < i \leq m\}$, where $\Pi^A_{A_i}(w)$ is the *projection* of the word w to the letters in A_i with $\Pi^A_{A_i}(a) = a$ for all $a \in A_i$ and $\Pi^A_{A_i}(a) = \lambda$ otherwise. We call semi-traces *synchronizable*, if their synchronization is not the empty set and *stably synchronizable* if their synchronization is a semi-trace.

On the graph representation level the synchronization is the union of the graphs of the semi-traces, where we have to identify nodes with the same labeling according to their order preserving all arcs describing dependencies. Of course $\forall i, j \leq m \, \forall a \in A_i \cap A_j \mid u_i \mid_a = \mid u_j \mid_a$ is a necessary condition for synchronizability; together with the non-existence of a cycle of hard arcs, the condition becomes also sufficient. If there is a soft arc (in either direction) in a cycle with hard arcs having the same direction, the soft arc can be deleted.

Continuing the above example we get the new semi-dependence alphabet $(A, SD) = (A_1 \cup A_2, SD_1 \cup SD_2) =$ 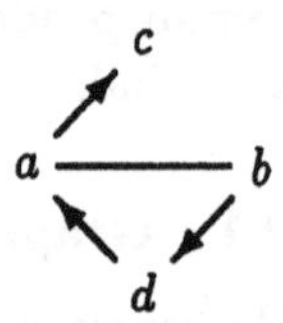

and for $[cacba\rangle_1 \parallel [dbada\rangle_2$ we get the stable synchronization

$[cdacbda\rangle =$ 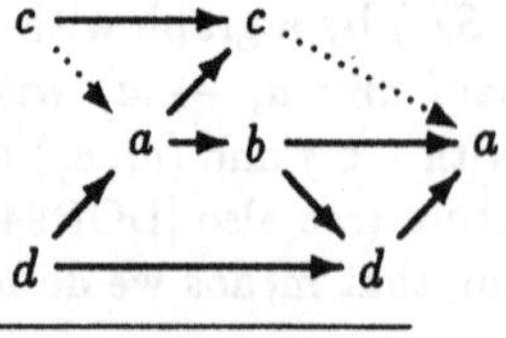

over (A, SD).

[3] It is easy to see that the synchronization on an identical alphabet is simply the intersection: $[u\rangle_i \parallel [v\rangle_i = [u\rangle_i \cap [v\rangle_i$

On the other hand we get $[cacba\rangle_1 \parallel [bdada\rangle_2 =$ which has a cycle of hard arcs and therefore $[cacba\rangle_1 \parallel [bdada\rangle_2 = \emptyset$.

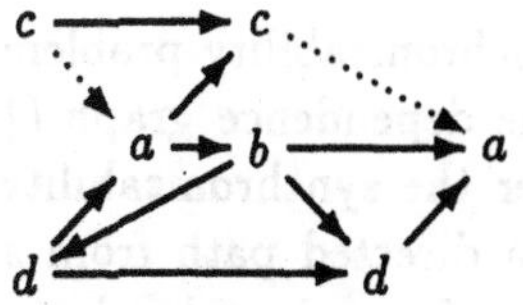

Another example is $[cacba\rangle_1 \parallel [daadb\rangle_2$. Here we get the graph

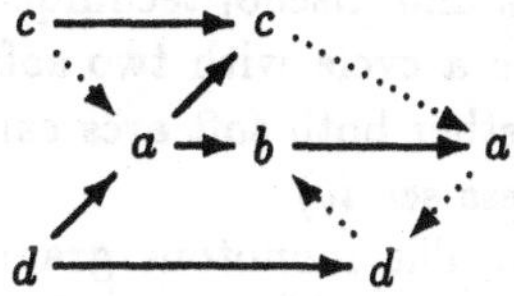

having a directed cycle containing two soft arcs, which means, that either the soft arc $a \dashrightarrow d$ or the soft arc $d \dashrightarrow b$ must be turned around to get a word describing a possible evaluation. Therefore $[cacba\rangle_1 \parallel [daadb\rangle_2 = [cdacbad\rangle_1 \cup [cdacdba\rangle_2$ is not a stable synchronization.

2.3 Deciding the synchronizability of semi-traces

The above observations lead to the following theorem:

Theorem 1. *The following problems are* **NLOGSPACE**-*complete:*

i) Given m semi-dependence alphabets (A_i, SD_i) with $0 < i \leq m$ and the semi-traces $[u_1\rangle_1, [u_2\rangle_2, \ldots [u_m\rangle_m$.
Are the traces synchronizable, that means $[u_1\rangle_1 \parallel [u_2\rangle_2 \parallel \ldots [u_m\rangle_m \neq \emptyset$?

ii) Given a semi-dependence alphabet (A, SD) and the semi-traces $[u\rangle, [v\rangle$ with $\forall a \in A \; |u_i|_a = |u_j|_a$.
Are the two traces synchronizable, that means $[u\rangle \cap [v\rangle \neq \emptyset$?

iii) Given m semi-dependence alphabets (A_i, SD_i) with $0 < i \leq m$ and the semi-traces $[u_1\rangle_1, [u_2\rangle_2, \ldots [u_m\rangle_m$.
Are the traces stably synchronizable, that means is there a word $w \in A^*$ with $[u_1\rangle_1 \parallel [u_2\rangle_2 \parallel \ldots [u_m\rangle_m = [w\rangle$?

iv) Given a semi-dependence alphabet (A, SD) and the semi-traces $[u\rangle, [v\rangle$ with $[u\rangle \cap [v\rangle \neq \emptyset$.
Are the two traces stably synchronizable, that means is there a word $w \in A^*$ with $[u\rangle \cap [v\rangle = [w\rangle$?

Proof. *i)*: A nondeterministic logarithmic space-bounded Turing machine first deterministically checks $\forall i, j \leq m \; \forall a \in A_i \cap A_j \mid u_i \mid_a = \mid u_j \mid_a$, then it guesses a cycle and tests each connection by testing precedence and dependence of the pair in each semi-trace. According to [Imm88] and [Sze88] the non-existence of the cycle can also be tested in **NLOGSPACE**. For hardness see *ii*).
ii): The problem is in **NLOGSPACE** because of *i*). The monotone graph reachability problem for directed graphs is well known to be **NLOGSPACE**-complete; for a given graph

$$G = (\{s = a_1, a_2, \ldots a_n = e\}, R)$$

with the property $(a_i, a_j) \in R \Rightarrow i < j$ the question is whether there exists a directed path from s to e. This can be reduced to the complementary of

the synchronizability problem by adding the edge (e, s) to the graph, which is now the dependence graph $(\{s, a_2, a_3 \dots a_{n-1}, e\}, R)$ $(SD := R \cup \{(e, s)\})$ and consider the synchronizability of $[esa_2a_3 \dots a_{n-1}\rangle$ and $[a_2a_3 \dots a_{n-1}es\rangle$. There exists a directed path from s to e, iff the semi-traces are not synchronizable, because of a cycle of hard arcs.
iii): A nondeterministic logarithmic space-bounded Turing machine first checks the synchronizability like in *i*), then is uses the [Imm88] and [Sze88] technique to do the opposite of the following: It guesses and tests a cycle with two soft arcs. Then again using [Imm88] and [Sze88] it tests, whether both soft arcs can be turned around without producing a cycle. For hardness see *iv*).
iv): The problem is in **NLOGSPACE** because of *iii*). The monotone graph reachability problem for directed graphs can be reduced to the problem by adding the additional letters b and c and the edges $(b, s), (b, c), (e, c)$ and (e, s) to the graph, which is now the dependence graph $(\{b, c, s, a_2, a_3, \dots a_{n-1}, e\}, SD)$ and consider, whether the synchronization of the semi-traces
$[cbs(a_2a_3 \dots a_{n-1})^n e\rangle$ and $[s(a_2a_3 \dots a_{n-1})^n ecb\rangle$ is a semi-trace.

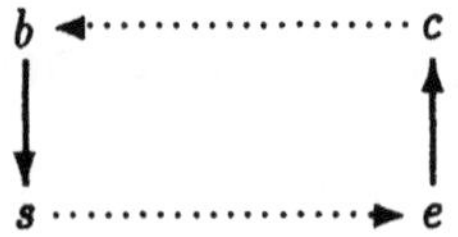

$(a_2 \dots a_{n-1})^n$

The synchronization is the semi-trace $[bs(a_2a_3 \dots a_{n-1})^n ec\rangle$, iff there is a path of hard arcs from s to e. □

For the existence of a non-confluent situation we get the same complexity:

Theorem 2. *[Rei94] The following problem is* **NLOGSPACE**-*complete: Given a semi-dependence alphabet* (A, SD) *and the semi-trace* $[u\rangle$. *Are there words* $v, w \in [u\rangle$ *with* $[v\rangle \parallel [w\rangle = \emptyset$?

3 The synchronizability in semi-commutation systems

It is easy to see that only a cycle of hard arcs from at least two semi-traces can be responsible for non-synchronizability of semi-traces, where $\forall i, j \leq m\ \forall a \in A_i \cap A_j \mid u_i \mid_a = \mid u_j \mid_a$. (The cycle may even consist of a symmetric dependence.) Of course this cycle must be also in the dependence graph and must be composed of edges from at least two dependence alphabets.

Theorem 3. *Given* m *semi-dependence alphabets* (A_i, SD_i) *with* $0 < i \leq m$. *The following assertions are equivalent:*

i) $\forall u_1 \in A_1^*, \dots u_m \in A_m^*, (\forall i, j \leq m\ \forall a \in A_i \cap A_j \mid u_i \mid_a = \mid u_j \mid_a) \Rightarrow [u_1\rangle_1 \parallel [u_2\rangle_2 \parallel \dots [u_m\rangle_m \neq \emptyset$

ii) $\neg \exists 1 < k, i \neq j, C = \{(x_1, x_2), \dots (x_{k-1}, x_k), (x_k, x_1)\}$ *with* $C \subseteq SD \wedge C \cap SD_i \neq \emptyset \wedge C \cap SD_j \neq \emptyset$

Again this property is **NLOGSPACE**-complete.
The following theorem says that semi-dependence alphabets have unstable synchronization (that means it can happen that the synchronization is neither empty nor a semi-trace), iff there exists a cycle in the union of the semi-dependence alphabets where

- all nodes are different,
- the two directed arcs (x_1, x_k) and (x_{j+1}, x_j) have reverse direction,
- there is no chord which separates them,
- there is no chord in backward direction and
- in each semi-dependence alphabet the cycle is either interrupted or has a directed arc in the opposite direction at some place, but then there is a another semi-dependence alphabet, having an arc at this place and is either interrupted or has a directed arc in the opposite direction at another place.

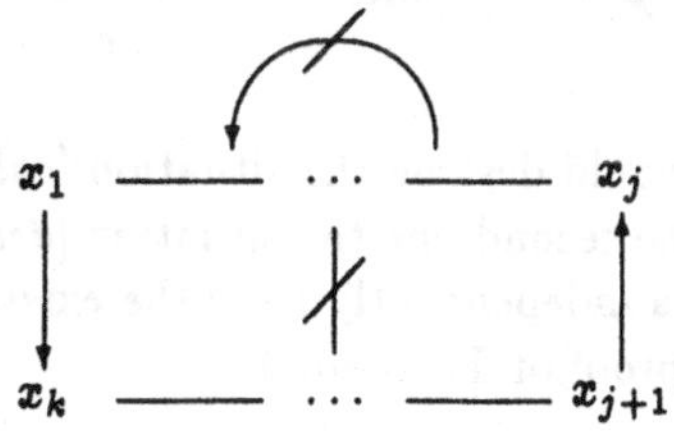

Theorem 4. *Given m semi-dependence alphabets (A_i, SD_i) with $0 < i \leq m$. The following assertions are equivalent:*

i) $\exists u_1 \in A_1^*, \ldots u_m \in A_m^* \; \forall w \in A^* \quad [w\rangle \neq [u_1\rangle_1 \parallel [u_2\rangle_2 \parallel \cdots [u_m\rangle_m \neq \emptyset$

ii) $\exists j \in \{1, \ldots k-1\}, (x_1, x_2), \ldots, (x_{k-1}, x_k) \in SD \cup SD^{-1}$ *with*

$$\begin{aligned}
&\forall n \neq p \;\; x_n \neq x_p \wedge \\
&(x_1, x_k), (x_{j+1}, x_j) \in SD \setminus SD^{-1} \wedge \\
&\forall (n, p) \in \{1, \ldots j\} \times \{j+1, \ldots k\} \setminus \{(1, k), (j, j+1)\} \; (x_n, x_p) \notin SD \cup SD^{-1} \wedge \\
&\forall n \in \{1, \ldots k\} \; \forall p \in \{n+2, \ldots k\} \;\; (x_p, x_n) \notin SD \wedge \\
&\forall i \leq m \; \exists n \leq k, ((x_{n+1}, x_n) \notin SD_i \wedge \\
&\qquad ((x_n, x_{n+1}) \notin SD_i \vee \\
&\qquad \exists l \in \{1, \ldots m\} \setminus \{i\} (x_n, x_{n+1}) \in SD_l \cup SD_l^{-1} \wedge \\
&\qquad\qquad \exists p \in \{1, \ldots k\} \setminus \{n\} \; (x_{p+1}, x_p) \notin SD_l))
\end{aligned}$$

The simplest example for this is $(A_1, SD_1) = (\{a, b\}, \{(a, b)\})$ and $(A_2, SD_2) = (\{a, b\}, \{(b, a)\})$ then $[ba\rangle_1 \parallel [ab\rangle_2 = \{ab, ba\} \neq [u\rangle$ for any u.
If we only regard semi-traces over one semi-dependence alphabet, then it follows, that it has unstable synchronization, iff there exists a cycle of different nodes with two directed arcs in both directions and only directed chords from the part of the cycle where two of these arcs start (in different direction) to the part of the cycle where the other two arcs end.

Corollary 5. *Given a semi-dependence alphabet (A, SD). The following assertions are equivalent:*

i) $\exists u, v \in A^* \ \forall w \in A^* \quad [w\rangle \neq [u\rangle \parallel [v\rangle \neq \emptyset$
ii) $\exists j \neq q \neq r \neq j \in \{1, ...k-1\}, (x_1, x_2), \dots (x_{k-1}, x_k) \in SD \cup SD^{-1}$ *with*
$\forall n \neq p \ \ x_n \neq x_p \ \wedge$
$(x_1, x_k), (x_{j+1}, x_j) \in SD \setminus SD^{-1} \ \wedge$
$(x_{q+1}, x_q), (x_{r+1}, x_r) \in SD^{-1} \setminus SD \ \wedge$
$\forall (n, p) \in \{1, ...j\} \times \{j+1, ...k\} \setminus \{(1, k), (j, j+1)\} \ (x_n, x_p) \notin SD \cup SD^{-1} \wedge$
$\forall n \in \{1, ...k\} \ \forall p \in \{n+2, ...k\} \ \ (x_p, x_n) \notin SD$

So there are only two such basic examples for semi-dependence alphabets (all other cases can be reduced to one of them by contraction of nodes in the graph):

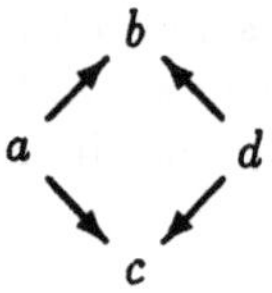

and

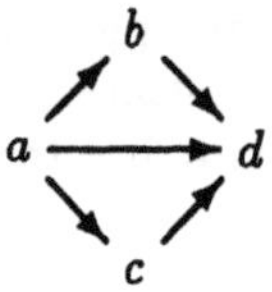

In the first one any chord would destroy the situation $[cabd\rangle \parallel [bdca\rangle = [abdc\rangle \cup [dcab\rangle \neq [u\rangle$ for any u. In the second one the situation $[dcab\rangle \parallel [bdca\rangle = [abdc\rangle \cup [cdba\rangle \neq [u\rangle$ for any u works independently from the existence of the arc from a to d. Now we come to the proof of Theorem 4:

Proof. $ii \Rightarrow i$: Take the shortest cycle of this kind and choose $u_i = x_{n+1} \dots x_k x_1 x_2 \dots x_n$ such that $(x_{n+1}, x_n) \notin SD_i$ and

$$(x_n, x_{n+1}) \notin SD_i \vee \exists l \leq m, \neq i, (x_n, x_{n+1}) \in (SD_l \cup SD_l^{-1}) \cap [u_l\rangle_l.$$

So the synchronization has a cycle with the two soft arcs (x_k, x_1) and (x_j, x_{j+1}) and no cycle of hard arcs.
$i \Rightarrow ii$: Choose v, v', u_i (length-)minimal with $\forall i \Pi^A_{A_i}(v) = u_i, \forall i \Pi^A_{A_i}(v') = u_i$ and $\neg\exists w$ with $\forall i \ \Pi^A_{A_i}(w) = u_i, v, v' \in [w\rangle$. Construct G as the union of all (graphs of) $[u_i\rangle_i$. Construct G' from G by replacing every $b \dashrightarrow a$ by $a \longrightarrow b$ if $a = x_1 \longrightarrow x_2 \longrightarrow \dots x_j = b$ in G. Because of v, v' the graph G' contains no cycle of hard arcs. So G' must have a cycle with at least two soft arcs

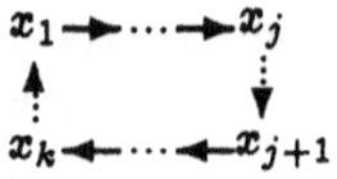

because otherwise G' would be the graph for $[w\rangle$ which must contain every hard arc of G anyway because of $\forall i \ \Pi^A_{A_i}(w) = u_i$ and cycles with one soft arc are deleted in the construction of G'. The shortest cycle of this kind has no chord. Now we can find a cycle with at least two soft arcs consisting only of original arcs from G and having only hard new chords in the direction of the cycle, which do not separate the two soft arcs with the following construction: We replace a new arc in cycle by the path of original hard arcs, which caused its existence. This may lead to a forbidden chord, but then we can find a shorting of the cycle with at least two soft arcs and no forbidden chord, where either the number of

soft arcs or the number of new hard arcs is reduced, so the construction must terminate.
Because of the minimality of the u_i there are no other vertices in G' except those in the cycle and no element of A appears twice. Furthermore the cycle can not contain soft chords and therefore there can be no arcs in the opposite direction. Now every arc in the cycle must come from one of the $[u_i\rangle_i$'s. But for every $[u_i\rangle_i$ the cycle must be open at a certain arc from x_n to x_{n+1} and it could be that $(x_n, x_{n+1}) \in SD_i$, then an arc at this place must be delivered from some other $[u_l\rangle_l$. But this $[u_l\rangle_l$ must be open at another arc from x_p to x_{p+1}. This is, what the last part of the formula in *ii* says. □

Theorem 6. *The assertions of Theorem 4 as well as the assertions of Corollary 5 are* **co-NP**-*complete to decide for semi-commutation systems.*

It is easy to see that the problem is in **co-NP**. The proof of hardness in [Rei94] uses a similar graph construction as in [DOR94] with the only difference, that we have to replace the two directed arcs by two pairs of directed arc to two additional nodes.
The complexity class Σ_2^P contains those languages which are accepted by an NP-machine with access to an NP-oracle. (For more background see e. g. [BDG88].) In analogy to another result in [DOR94] the following holds:

Theorem 7. *The following problem is* Σ_2^P*-complete: Given two dependence alphabets* (A, D) *and* (A, D') *such that* $D' \subseteq D$.
Does there exist a semi-commutation system SC such that

$$D = \{(a,b) \in A \times A \mid ab \Longrightarrow ba \notin SC \cap SC^{-1}\} \text{ and}$$
$$D' = \{(a,b) \in A \times A \mid ab \Longrightarrow ba \notin SC \cup SC^{-1}\} \text{ and}$$
$$\forall u, v \in A^* \ \exists w \in A^* \text{ with } [u\rangle \parallel [v\rangle = [w\rangle \text{ or } = \emptyset \ ?$$

This means can unstable synchronization be avoided by choosing the direction of asymmetric dependencies?

The same result holds for $2m$ dependence alphabets. Because the direction of the most arcs in the cycle is of no influence, we can not use the same graph construction as in [DOR94] for the proof; instead a different construction in [Rei94], which makes use of the direction of chords, works in a similar way to reduce the truth of quantified boolean formulae (see [Sto77], [Wra77]) to the problem.

Remark. Given $2m$ dependence alphabets (A_i, D_i) and (A_i, D_i') such that $D_i' \subseteq D_i$ for all $0 < i \leq m$. The direction of asymmetric arcs can be chosen for m semi-dependence alphabets in such a way that semi-traces having the same letters in the intersection are always synchronizable, that means the direction of asymmetric arcs can be changed in a way such that there is no directed cycle consisting of arcs from at least two of the dependencies iff there is no such cycle having at most one asymmetric arc. The property is hereby in **NLOGSPACE**.

4 The inclusion of semi-traces

For a valid representing evaluation of a concurrent system it is a basic question, whether another evaluation is also valid. If the system is described by partial commutation, it is the following trace-equivalence problem: Given a symmetric semi-commutation system C and two words w, w'; is it true that $w \overset{*}{\underset{C}{\Longrightarrow}} w'$?

The trace-equivalence problem is proved to be in $\mathbf{TC}^0$ in [AG91] by describing in first order logic enlarged by majority quantifiers, whether for each dependent pair (a, b) for every position in w there is a corresponding position in w' with the same number of preceding a's and b's. By a result of [BIS90] this follows to be in $\mathbf{TC}^0$ (but not in AC^0). The same is applied to asymmetric semi-commutation systems in [Rei94]:

Theorem 8. *The following semi-trace-inclusion problem is in* $\mathbf{TC}^0$*: Given an asymmetric semi-commutation system SC by the semi-dependence alphabet* (A, SD) *and two words* w, w'*; is it true that* $w \overset{*}{\underset{SC}{\Longrightarrow}} w'$ *respectively* $[w'\rangle \subseteq [w\rangle$*?*

5 Local checking of synchronizability of semi-traces

In [DV89] an easy testable ($\mathbf{TC}^0$), necessary condition for the synchronizability of traces was described. The *local checking property* is a criterion for the sufficiency of this condition. It is co-NP-complete [DOR94]. The necessary condition for the synchronizability of two semi-traces $[u_1\rangle_1, [u_2\rangle_2$ is the following:

$$\Pi^{A_1}_{A_1\cap A_2}(u_1) \overset{*}{\underset{SC'}{\Longrightarrow}} \Pi^{A_2}_{A_1\cap A_2}(u_2) \text{ for } SC' \text{ defined by } SD' = SD_1 \cap SD_2^{-1}$$

According to Theorem 8 this condition can be tested in $\mathbf{TC}^0$.

Lemma 9. *The above condition is sufficient, iff the* composition [RW91] *of the semi-commutation systems* SC_1 *and* SC_2^{-1} *is a semi-commutation system*[4].

Proof. If the composition of the semi-commutation systems SC_1 and SC_2^{-1} is a semi-commutation system, the composition SC' is determined by $SD' = SD_1 \cap SD_2^{-1}$ over $A_1\dot\cup\{a_1, \ldots a_n\} = A_2\dot\cup\{b_1, \ldots b_m\}$. If $\Pi^{A_1}_{A_1\cap A_2}(u_1) \overset{*}{\underset{SC'}{\Longrightarrow}} \Pi^{A_2}_{A_1\cap A_2}(u_2)$ then $u_1 a_1^{|u_2|_{a_1}} \ldots a_n^{|u_2|_{a_n}} \overset{*}{\underset{SC'}{\Longrightarrow}} u_2 b_1^{|u_1|_{b_1}} \ldots b_m^{|u_1|_{b_m}}$, since $(A_1 \times \{a_1, \ldots a_n\}) \cap SD_1 = \emptyset$ and $(A_2 \times \{b_1, \ldots b_m\}) \cap SD_2 = \emptyset$. Thus we have

$$u_1 a_1^{|u_2|_{a_1}}, \ldots a_n^{|u_2|_{a_n}} \overset{*}{\underset{SC_1}{\Longrightarrow}} s \overset{*}{\underset{SC_2^{-1}}{\Longrightarrow}} u_2 b_1^{|u_1|_{b_1}}, \ldots b_m^{|u_1|_{b_m}}$$

and s is in the synchronization of $[u_1\rangle_1$ and $[u_2\rangle_2$.
If the condition is sufficient, then for every $w_1 \overset{*}{\underset{SC'}{\Longrightarrow}} w_2$ there is a synchronization $s \in [\Pi_{A_1}(w_1)\rangle_1 \parallel [\Pi_{A_2}(w_2)\rangle_2$ and the derivation $w_1 \overset{*}{\underset{SC'}{\Longrightarrow}} w_2$ is composed of $w_1 \overset{*}{\underset{SC_1}{\Longrightarrow}} s \overset{*}{\underset{SC_2^{-1}}{\Longrightarrow}} w_2$. □

[4] this means that $f_{SC_1} \circ f_{SC_2^{-1}} = f_{SC'}$ for a semi-commutation system SC'.

Theorem 10. *The local checking property for synchronizability of two semi-traces is* **co-NP**-*complete.*

Proof. A criterion describing, whether the composition of two semi-commutation systems is a semi-commutation system, is given in [RW91]. It is easy to see that this criterion is in co-**NP** and since the criterion for confluence, which was shown to be co-**NP**-complete in [DOR94], is a special case for this, the problem is co-**NP**-complete, too. □

Remark. If we use a pairwise application of the condition for the local checking for m semi-commutation systems, we get the same result with an immediate generalization of [RW91]. We conjecture that this is also the case for conditions regarding projections to sub-alphabets of constant size.

Remark. It is easy to see that for regular string languages R_1, R_2 the synchronization is also regular but even in the special case for symmetric dependencies it is undecidable whether $f_{SC}(R_1) \parallel f_{SC}(R_2) \neq \emptyset$ according to [AH89]; for more information see [Rei94].

Acknowledgment: *I thank Volker Diekert, Klaus-Jörn Lange and an anonymous referee for helpful remarks.*

References

[AG91] C. Àlvarez and J. Gabarró. The parallel complexity of two problems on concurrency. *Information Processing Letters*, 38:61–70, 1991.

[AH89] IJ. J. Aalbersberg and H. J. Hoogeboom. Characterizations of the decidability of some problems for regular trace languages. *Mathematical Systems Theory*, 22:1–19, 1989.

[BDG88] J. L. Balcázar, J. Díaz, and J. Gabarró. *Structural Complexity* I. Number 11 in EATCS Monographs on Theoretical Computer Science. Springer, Berlin-Heidelberg-New York, 1988.

[BIS90] D.M. Barrington, N. Immerman, and H. Straubing. On uniformity within NC^1. *J. of Comp. and Syst. Sciences*, 41:274–306, 1990.

[CGL+92] M. Clerbout, D. Gonzalez, M. Latteux, E. Ochmanski, Y. Roos, and P.A. Wacrenier. Recognizable morphisms on semi commutations. Tech. Rep. LIFL I.T.-238, Université des Sciences et Technologies de Lille (France), 1992.

[CL87] M. Clerbout and M. Latteux. Semi-Commutations. *Information and Computation*, 73:59–74, 1987.

[Cle84] M. Clerbout. *Commutations Partielles et Familles de Langages.* Thèse, Université des Sciences et Technologies de Lille (France), 1984.

[Die94] V. Diekert. A partial trace semantics for Petri nets. *Theoretical Computer Science*, 134:87–105, 1994. Special issue of ICWLC 92, Kyoto (Japan).

[DOR94] V. Diekert, E. Ochmański, and K. Reinhardt. On confluent semi-commutation systems – decidability and complexity results. *Information and Computation*, 110:164–182, 1994. A preliminary version was presented at ICALP'91, Lecture Notes in Computer Science 510 (1991).

[DR95] V. Diekert and G. Rozenberg, editors. *The Book of Traces.* World Scientific, Singapore, 1995.

[DV89] V. Diekert and W. Vogler. On the synchronization of traces. *Mathematical Systems Theory*, 22:161–175, 1989. A preliminary extended abstract appeared at MFCS 88, Lecture Notes in Computer Science 324 (1988) 271-279.

[HK89] D. V. Hung and E. Knuth. Semi-commutations and Petri nets. *Theoretical Computer Science*, 64:67–81, 1989.

[Imm88] N. Immerman. Nondeterministic space is closed under complement. *SIAM Journal on Computing*, 17(5):935–938, 1988.

[Joh90] D. S. Johnson. A catalog of complexity classes. In J. van Leeuwen, editor, *Algorithms and Complexity*, volume A of *Handbook of Theoretical Computer Science*, chapter 2, pages 67–161. Elsevier, 1990.

[Maz87] A. Mazurkiewicz. Trace theory. In W. Brauer et al., editors, *Petri Nets, Applications and Relationship to other Models of Concurrency*, number 255 in Lecture Notes in Computer Science, pages 279–324, Berlin-Heidelberg-New York, 1987. Springer.

[Och90] E. Ochmański. Semi-Commutation and Petri Nets. In V. Diekert, editor, *Proceedings of the ASMICS workshop Free Partially Commutative Monoids, Kochel am See 1989*, Report TUM-I9002, Technical University of Munich, pages 151–166, 1990.

[Och92] E. Ochmański. Modelling concurrency with semi-commutations. In I. M. Havel and V. Koubek, editors, *Proceedings of the 17th Symposium on Mathematical Foundations of Computer Science (MFCS'92), Prague, (Czechoslovakia), 1992*, number 629 in Lecture Notes in Computer Science, pages 412–420, Berlin-Heidelberg-New York, 1992. Springer.

[OW93] E. Ochmański and P. A. Wacrenier. On regular compatibility of semi-commutations. In Andrzej Lingas, Rolf Karlsson, and Svante Carlsson, editors, *Proceedings of the 20th International Colloquium on Automata, Languages and Programming (ICALP'93), Lund (Sweden) 1993*, number 700 in Lecture Notes in Computer Science, pages 445–456, Berlin-Heidelberg-New York, 1993. Springer.

[Par90] Ian Parberry. A primer on the complexity theory of neural networks. In R.B. Banerji, editor, *Formal Techniques in Artificial Intelligence*, Amsterdam, 1990. North-Holland.

[Rei94] K. Reinhardt. *Prioritätszählerautomaten und die Synchronisation von Halbspursprachen.* Dissertation, Institut für Informatik, Universität Stuttgart, 1994.

[Ruz81] W. L. Ruzzo. On uniform circuit complexity. *Journal of Computer and System Sciences*, 22:365–383, 1981.

[RW91] Y. Roos and P. A. Wacrenier. Composition of two semi commutations. In A. Tarlecki, editor, *Proceedings of the 16th Symposium on Mathematical Foundations of Computer Science (MFCS'91), Kazimierz Dolny (Poland) 1991*, number 520 in Lecture Notes in Computer Science, pages 406–414, Berlin-Heidelberg-New York, 1991. Springer.

[Sto77] L. J. Stockmeyer. The polynomial time hierarchy. *Theoret. Comput. Sci.*, 3:1–22, 1977.

[Sze88] R. Szelepcsényi. The method of forced enumeration for nondeterministic automata. *Acta Informatica*, 26:279–284, 1988.

[Wra77] C. Wrathall. Complete sets and the polynomial hierarchie. *Theoret. Comput. Sci.*, 3:23–33, 1977.

Tiling with bars and satisfaction of boolean formulas

Eric Rémila

Université Jean Monnet Saint-Etienne
Institut Universitaire de Technologie de Roanne.
12 Avenue de Paris, 42300 Roanne, France.

Abstract. Let F be a figure formed from a finite set of cells of the planar square lattice. We prove that the problem of tiling such a figure with bars formed from 2 or 3 cells can be reduced to the the logic problem 2-SAT and we deduce a linear-time algorithm of tiling with these bars. Afterwards, we prove that the similar problem im the triangular lattice can be reduced to two successive matching problems and we deduce a polynomial algorithm of tiling.

1 Introduction

The problem of tiling a finite figure of the plane is at the junction of different topics of mathematics : group theory, graph theory, geometry. J. H. Conway and J.C.Lagarias [JCL90] have obtained important results using Cayley graphs of "tiling groups". Their work has been prolonged by W. P. Thurston [Thu90], which has produced algorithms to tile a figure without hole with lozenges, or with dominoes. W.PThurston' s ideas have been taken again by C. and R. Kenyon [RK92], which have obtained an algorithm to tile a simply connected figure with h_m tiles (horizontal rectangles of length m and width 1) and v_n tiles (vertical rectangles of length n and width 1), where m and n are fixed integers.
On the other hand, the problem of tiling a figure with h_m and v_n tiles has been studied with purely combinatorial methods. D. Beauquier and M. Nivat [MN91] have given a very easy characterization of simply connected figures admitting a unique tiling with h_m and v_n tiles. J. M. Robson [Rob91] has proved that, the problem of the existence of such a tiling is NP-complete when $m \geq 3$ or $n \geq 3$.
The subject of this paper is the problem of tiling a figure with bars of length 2 (i.e. h_2 tiles and v_2 tiles) or length 3 (i.e. h_3 tiles and v_3 tiles). A previous paper [Rem] has been devoted to this subject, but only the case of figures without holes had been treated in this paper.
The tile group of these tiles is isomorphic to $\mathbb{Z}^2$, thus it gives us no information. Hence, we have to use only combinatorial methods. Our main result is the following :

Main result : the problem of tiling a figure with bars of length 2 or 3 can be reduced to the logic problem 2-SAT.

Our reduction permits to obtain a linear algorithm of tiling with bars of length 2 or 3.
Afterwards,we study the similar problem in the triangular lattice, where the tiles are formed from 2 or 3 triangular cells. The same reduction cannot be applied but the use of the theory of matchings in graphs yields to a polynomial time algorithm of tiling.

2 Definitions and notations

A cell is a (closed) unit square of the plane $\mathbb{R}^2$ whose center has integer coordinates. The cell (x, y) denotes the cell whose center is point (x, y). Integer x (respectively y) is called the horizontal (respectively vertical) coordinate of (x,y).
Cell C' is a neighbor of cell C if C and C' have a common edge. One canonically define the left, right, upper and lower neighbor of a cell.
A figure is a finite union of cells. The area of a figure F is the number of cells included in F.
An isolated cell of a figure F is a cell of F with no neighbor in F.
A peak of a figure F is a cell of F which has exactly one neighbor in F. A peak C of F is said vertical (respectively horizontal, left, right, lower, upper) if C is the vertical (respectively horizontal, left, right , lower, upper) neighbor of a cell of F.
A vertical (respectively horizontal) bridge of a figure F is a cell of F which has exactly two neighbors in F which are its vertical (respectively horizontal) neighbors.
Let m be an integer such that $m \geq 2$. An m-bar is a rectangle of length m and unit width, formed from the union of m neighboring cells. An m-bar B is vertical (respectively horizontal) if all the cells of B have the same horizontal (respectively vertical) coordinate.

3 Different formulations of the problem

A tiling Φ of a figure F is a set of 2-bars and 3-bars included in F such that each cell of F is included in exactly one element of set Φ.
A packing Π of a figure F is a set of 2-bars included in F such that each cell of F is included in at most one element of set Π.
A default of a packing of F is a cell of F which is not element of a bar of this packing. A default C of a packing Π of F is pointed if there exists a 2-bar B of Π such that $B \cup C$ is a 3-bar.

Proposition 1. *Let F be a finite figure. The following propositions are equivalent (see figure 1):*

- *there exists a tiling Φ of F,*
- *there exists a packing Π of F all defaults of whose are pointed.*

Moreover, a tiling of F can be obtained from a packing of F in linear time (in the area of the figure) and a packing of F can be obtained from a tiling of F in linear time.

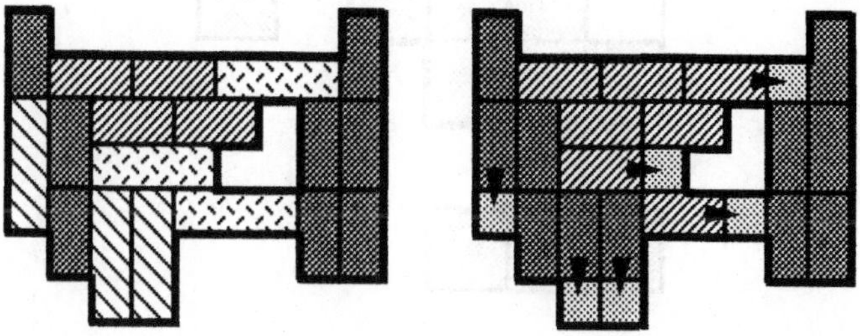

Fig. 1. tiling and packing with pointed defaults of the same figure.

4 Necessary conditions for a figure to be tileable

For the rest of this paper, F denotes a finite figure with no isolated cell and α (respectively β) denotes the set of the edges e of the peaks (respectively bridges) of F which are not edges of the frontier of F.

Definition 2. Let e and e' be two edges of $\alpha \cup \beta$. We say that e and e' create a conflict if there exists a non-empty sequence $(e_1, e_2, ...e_{2k})$ of edges of the frontier of F and a vector $u = (a, b)$, with $a^2 = b^2 = 1$, such that (see figure 2) :

- for each integer i, with $0 \leq i < 2k$, edges e_i and e_{i+1} have a common extremity,
- for each integer i, with $0 \leq i < 2k - 1$, $e_{i+2} = t_u(e_i)$, where t_u denotes the translation of vector u.
- edge e is vertical if and only if edge e_1 is vertical and e and e_1 have a common extremity which is not an extremity of e_2,
- edge e' is vertical if and only if edge e_{2k} is vertical and e' and e_{2k} have a common extremity which is not an extremity of e_{2k-1}.

conflict of edges.

For each edge e of $\alpha \cup \beta$, we define a boolean variable x_e. We state the following rules :

Rules of compatibility :

- if e is element of α, then $x_e = 1$,
- if e and e' are edges of the same bridge, then $x_e \vee x_{e'} = 1$,
- if e and e' are in conflict in F, then $\neg x_e \vee \neg x_{e'} = 1$.

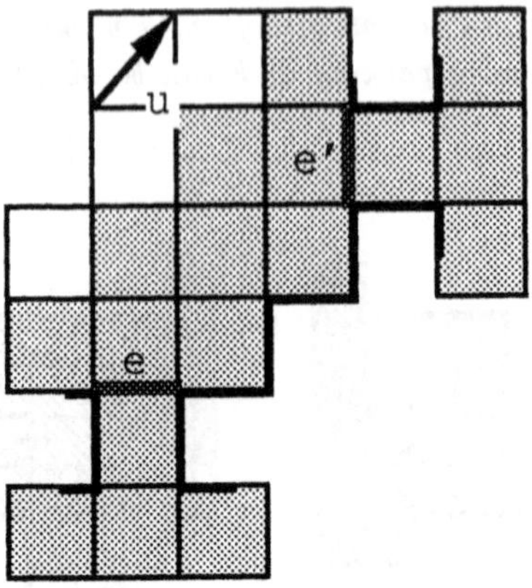

Fig. 2. conflict of edges.

Remark. For each each edge e of a b, variable x_e appears in at most 3 rules of compatibility.

Proposition 3. *If figure F can be tiled with bars, then the conjunction of rules of compatibility can be satisfied.*

5 Sufficiency of the conditions

Theorem 4. *let F be a figure (with no isolated cell) such that the rules of compatibility of F can be simultaneously satisfied. There exists a packing of F such that each default is pointed.*

Proof. assume that, for each edge e of $\alpha \cup \beta$, a value of x_e is given in such a way that all the rules of compatibility are satisfied. Consider the following algorithm (see figure 3) :

Algorithm of packing

initialization : construct a list λ of the vertical bridges and the vertical peaks of F, mark the edges of the frontier of F and construct a list Δ of cells with two vertical marked edges and a horizontal marked edge. Moreover, Π denotes a set of 2-bars. For initialization, Π is empty.
step 1 : successively take each cell A of λ. If A is either pointed or covered by a 2-bar which has been previously put, take the successive cell. Otherwise, let A' and A" be respectively the upper and lower neighbor of A and let e' and e" be respectively the upper and lower edges of A. If e' is an edge of $\alpha \cup \beta$ and $x_{e'} = 1$, put the vertical 2-bar $B_1 = A \cup A'$ in set Π and mark the vertical edges of A'. Otherwise, put bar $B_2 = A \cup A''$ in set Π, and mark the vertical edges of A". In the two cases, update list Δ.
step 2 : take the first cell C of Δ. If C is either pointed or covered by a 2-bar

which has been previously put, delete C from list Δ. Otherwise, let e denote the unmarked edge of C. Put vertical 2-bar $B = C \cup C'$ in set Π, where C and C' are the cells of F which share edge e, mark the vertical edges of C' and update list Δ. This instruction is repeated until list Δ is empty.
step 3 : successively take each cell D of F. If C is either pointed or covered by a 2-bar which has been previously put, take the successive cell. Otherwise let L and R respectively denote the left and right neighbors of D. If L is a cell of F which has not been previously covered by a 2-bar, put bar $B' = D \cup L$ in set Π, otherwise put bar $B'' = D \cup R$.

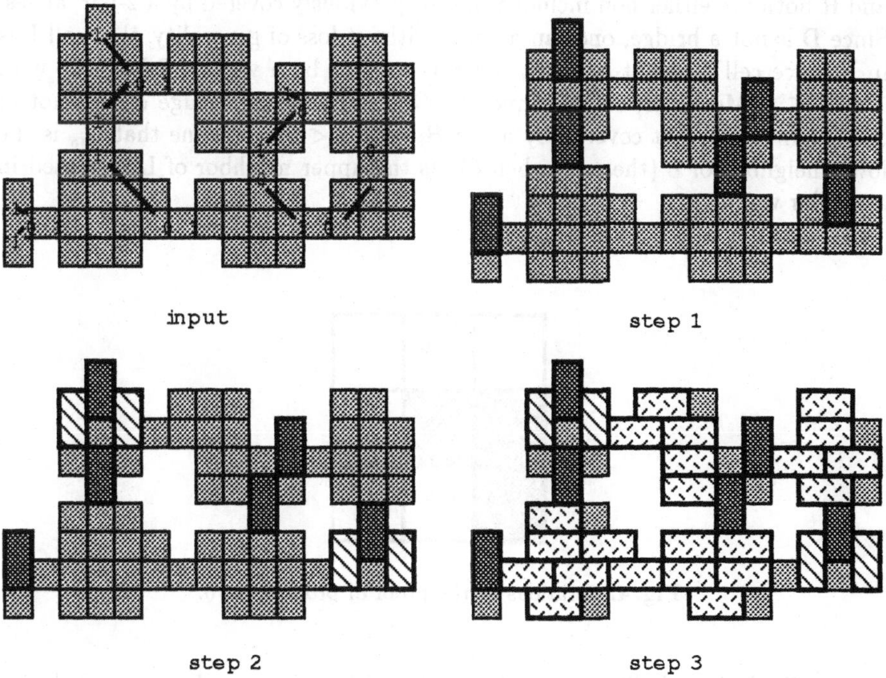

Fig. 3. Algorithm of packing

Proposition 5. *the algorithm above gives a packing Π of F with 2-bars such that each default is pointed.*

Proof. we have to prove that :

- in step 1, cells A' and A" have not been previously covered by a 2-bar. This is obvious, since A is not pointed,
- in step 2, cell C' is a cell of F which has not been previously covered by a 2-bar. This is obvious since edge e is unmarked (which proves that C' is in F) and C is not pointed.

– in step 3, at least one of the cells L and R is in F and has not been previously covered by a 2-bar. This needs much attention.

Let $(B_1, B_2, ..., B_q)$ denote the sequence of the vertical 2-bars used in the execution of step 1 and $(B_{q+1}, B_{q+2}, ..., B_p)$ denote the sequence of the vertical 2-bars used in the execution of step 2, in the order of placing. For each integer i such that $1 \leq i \leq p$, let S_i denote the set of marked edges after having put bars B_j, with $j \leq i$ (let S_0 denote the edges of the frontier of F), and let C_i denote the cell which has forced to put B_i (i. e. the bridge for step 1, the cell with three marked edges, one horizontal and two vertical for step 2). Assume that, during the execution of step 3, a non-pointed cell D of F whose horizontal neighbors L and R both are either non included in F or previously covered by a 2-bar, arises. Since D is not a bridge, one can assume without loss of generality, that cell L is in F Since cell D is not pointed, cell L is covered by a vertical 2-bar B_{i_0}, with $1 \leq i_0 \leq p$. Moreover, cell L cannot be C_{i_0}, since the right edge of L is not in S_{i_0-1} (since D is not covered by a bar B_j with $j < i_0$). Assume that C_{i_0} is the lower neighbor of L (the case when C_{i_0} is the upper neighbor of L is treated in a similar way).

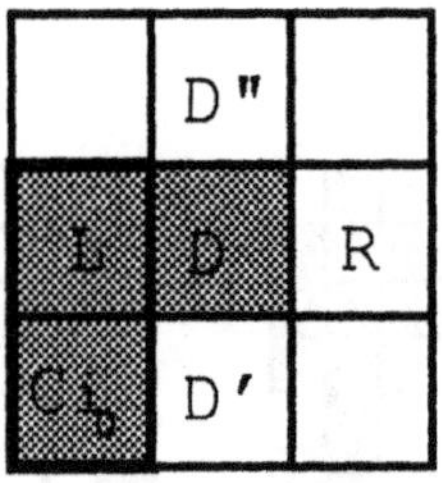

Fig. 4. notations of the proof of proposition 5.

Let D' denote the lower neighbor of D ; if D' is in F and is covered during step 1 or step 2, then D is pointed, which is a contradiction ; if D is in F and is not covered after step 2, then the right edge of C_{i_0} is not in S_{i_0-1}, which is a contradiction. Thus, necessarily, cell D' is not in F.
Let D" denote the upper neighbor of D. If D" is in F and is covered during step 1 or step 2, then D is pointed, which is a contradiction ; if D" is in F and is not covered after step 2, then remark that the right edge of D is necessarily marked at the end of step 2 (otherwise, R is in F and is covered by a horizontal 2-bar, which yields that D is not pointed). This is a contradiction, since D has three marked edges, which contradicts the emptiness of list L step 2. Thus cell D" is not in F. Let e and e' respectively denote denote the left and right edge of D. The remarks above prove that e is an edge of $\alpha \cup \beta$.

Lemma 6. *the equality :* $x_e = 0$ *holds.*

The proof of this lemma is given below. From this lemma, cell D is a horizontal bridge of F and e' is an edge of $\alpha \cup \beta$. As for x_e, one can obtain : $x_{e'} = 0$, which contradicts ii) of the rules of compatibility. Thus the existence of D is impossible, which proves the validity of the algorithm.

Proof. assume that $x_e = 1$.
If $1 \leq i_0 \leq q$, then C_{i_0} is either a peak or bridge of F. Let e_{i_0} be the upper edge of C_{i_0}. Edge e_{i_0} is an edge of $\alpha \cup \beta$, and $x_{e_{i_0}} = 1$. Thus edges e and e_{i_0} create a conflict, which is a contradiction. Thus we have $q < i_0 \leq p$.
This yields that the lower edge of C_{i_0} is marked, thus the lower neighbor of C_{i_0} is not in F (remark that each horizontal marked edge is an edge of the frontier of F). Moreover, the left neighbor L_{i_0} of C_{i_0} is in F (since C_{i_0} is not a peak of F) and the left edge e'_{i_0} of C_{i-0} is element of S_{i_0-1}, thus there exists a vertical 2-bar B_{i_1}, with $i_1 \leq i_0$, such that B_{i_1} contains L_{i_0}.
Edge e'_{i_0} is not element of S_{i_1-1}, since neither L_{i_0} nor C_{i_0} are covered by a 2-bar B_j such that $j \leq i_1 - 1$. Thus, we have $L_{i_0} \neq C_{i_1}$. Moreover, C_{i_1} is not the upper neighbor of L_{i_0}, since L is not covered by a 2-bar B_j such that $j \leq i_1 - 1$. Thus, cell C_{i_1} is the lower neighbor of L_{i_0}. As for i_0, we have $i_1 > q$, since otherwise the upper edge e_{i_1} of cell C_{i_1} is such that $x_{e_{i_1}} = 1$ and e and e_{i_0} create a conflict. Thus, as for C_{i_0}, the lower neighbor of C_{i_1} is not in F and the left neighbor L_{i_1} of C_{i_1} is covered by a 2-bar B_{i_2}, with $i_2 < i_1$, and C_{i_2} is the lower neighbor of L_{i_1}. This kind of argument can be infinitely repeated, thus we have an infinite sequence $(C_{i_j})_{j \in \mathbb{N}}$ of cells of F such that, for each integer j, $C_{i_{j+1}}$ is the lower neighbor of the left neighbor of C_{i_j}, which is a contradiction. Thus the assumption: $x_e = 1$ is false.

6 The algorithm

We recall all the steps of an algorithm to tile a figure F with 2-bars or 3 bars :

step 1 : verify that F has no isolated cell.
step 2 : construct the list of conditions of compatibility.
step 3 : solve, if possible, the instance of the logic problem 2-SAT, given by the rules of compatibility.
step 4 : execution of the algorithm of packing previously seen .
step 5 : construct a tiling from the packing.

Let m (respectively n) denote the number of cells (respectively edges) of F. The time complexity of the above algorithm is at most linear (in m or n) since 2-SAT has a complexity in O(n' + m'), where n' (respectively m') is the number of literals (respectively m') used (this is a classical result of complexity theory (see [Asp-Pla-Tar] for details)) and each cell appears at most once in each list used for the algorithm of packing.

7 Extension to the triangular lattice

In the triangular lattice, a 2-bar (respectively a 3-bar)) is a simply connected figure formed from two (respectively three) triangular cells (see figure).The main tool used is the theory of the matchings in a graph. We recall that a matching of a graph is a set of edges of this graph with pairwise disjoint extremities. Thus a packing (defined as in the square lattice) is a special case of matching.

Theorem 7. *let F be a finite figure of the triangular lattice. The following propositions are equivalent :*

i) there exists a tiling of F,
ii) there exists a packing Π of F such that there exists an injective mapping f from the set D_Π of the defaults of Π to the tiles of Π such that, for each default d, a side of d is also a side of f(d),
iii) there exists a maximal packing Πmax of F such that there exists an injective mapping f from the set $D_{\Pi max}$ of the defaults of Πmax to the tiles of Πmax such that, for each default d, a side of d is also a side of f(d).
iv) for each maximal (for the number of tiles) packing Πmax of F, there exists an injective mapping f from the set $D_{\Pi max}$ of the defaults of Πmax to the tiles of Πmax such that, for each default d, a side of d is also a side of f(d).

Proof. i) ⇒ ii) is obvious.
ii) ⇒ iii) : if Π is not a maximal packing, then there exists an alternating chain C which join two defaults (this is a classical result of matching theory (see [Ber73])). Let Π' be the packing obtained from Π by an "exchange" on the alternating chain C. For each default d of Π', let f'(d) be the tile of Π' which contains the side of d covers which is on the boundary of f(d). Hence f(d) = f'(d) if and only if f(d) does not cover some cells of C. Moreover, one easily verifies that for each tile t of Π' which covers two cells of C, there exists at most two defaults d and d' of Π' such that f'(d) = f'(d') = t, and when we have exactly two defaults, the two sides of t which are sides of the defaults are not on the boundary of the same cell covered by t. Thus a local transformation involving cells d, d', and the cells of t permits to delete defaults d and d' (see figure 5). After all those transformations are done, a packing Π" satisfying condition iii) is constructed. This argument is repeated until a maximal packing arises.
iii ⇒ iv) is done by the same kind of argument as the previous implication, since a sequence of "exchanges" on alternating cycles or alternating chains an extremity of which is default permits to pass from any maximal matching to any other one.
iv ⇒ i) is obvious.

The theorem above guarantees the validity of the following algorithm of tiling:
step 1 : construct a maximal packing Πmax of the figure
step 2 : construct the bipartite graph G whose vertices are the defaults and the tiles of Πmax such that the neighbors of a default d are the tiles which have a

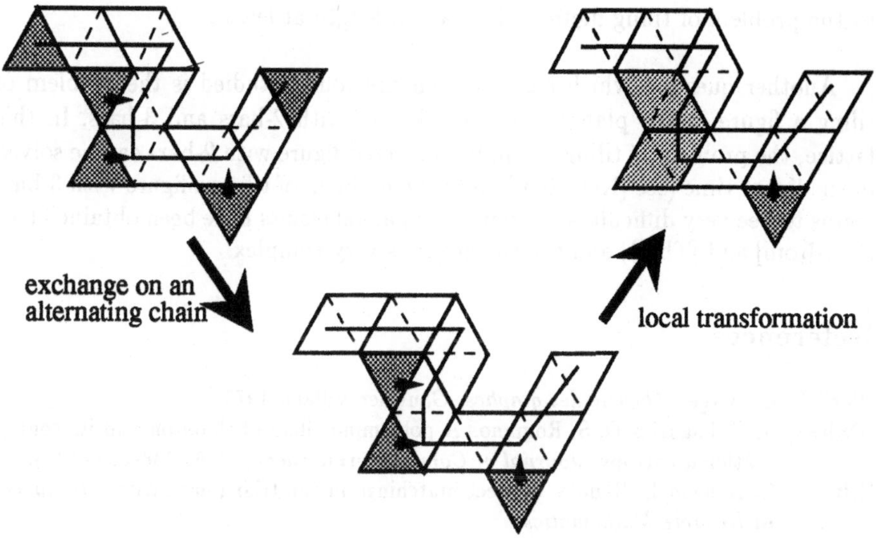

Fig. 5. proof of theorem 7 (*ii*) ⇒ *iii*))

common side with d and the neighbors of a tile t are the defaults which have a common edge with t.
step 3 : construct a maximal matching of G. If each default is matched, a tiling of the figure is canonically deduced . Otherwise, there exists no tiling.

This algorithm takes at most a polynomial time since there exists some polynomial algorithms for finding a maximal matching in a graph.

8 Open problems

All the results of complexity obtained about tiling with bars give a strange situation :

- the problem of tiling a figure with 2-bars is a matching problem in a graph and consequently can be solved in a polynomial time.,
- the problem of tiling a figure with 3-bars (or with horizontal 3-bars and vertical 2-bars) can reduce the logic problem 3-SAT and, consequently, is NP-complete (see [Rob]), even though the same problem limited to figures without hole is linear (see [Ken-Ken]).
- the problem of tiling a figure with 3-bars and 2-bars can be solved in a linear time.
- the problem of tiling a figure without hole with horizontal 3-bars and (vertical and horizontal) 2-bars (see [Ung]), but the complexity of tiling a figure (which may have holes) with the same bars is still unknown.

A direction of search is the problem of tiling figures with bars of length at least m (integer m being fixed). The problem treated in this paper is clearly equivalent to the problem of tiling figures with bars of length at least 2.

to the problem of tiling figures with bars of length at least 2.

Another question which has not been previously studied is the problem of tiling a figure of the planar hexagonal lattice with 2-bars and 3-bars. In this lattice, the problem of tiling a simply connected figure with 2-bars can be solved in the linear time (see [Ken-Rém]), and the problem of tiling a figure with 3-bars seems to bee very difficult, since only very partial results have been obtained (see [Lag-Rom] and [Thu]), and the tile group is very complex.

References

[Ber73] C. Berge. *Theorie des graphes.* Gauthier Villars, 1973.

[DSR93] J. C. Lagarias D. S. Romano. A polyomino tiling of thurston and its configurational entropy. *Journal of Combinatorial Theory,* A 63:338–358, 1993.

[ER] C. Kenyon E. Remila. Perfect matchings in the triangular lattice. *to appear in Discrete Mathematics.*

[JCL90] J. H. Conway J. C. Lagarias. Tiling with polyominoes and combinatorial group theory. *Journal of Combinatorial Theory,* A 53:183–208, 1990.

[MFP79] B. Aspvall M. F. Plass, R. E. Tarjan. A linear-time algorithm for testing the truth of certain quantified boolean formulas. *Information Processing Letters,* 8(3):121–123, 1979.

[MN91] D. Beauquier M. Nivat. Tiling pictures of the plane with two bars, a horizontal and a vertical one. pages 37–43, Universite de Creteil, 1991. seminaire Polyominos et pavages, D. Beauquier.

[Rem] E. Remila. Tiling a figure with bars of length 2 or 3. *to appear in Discrete Mathematics.*

[RK92] C. Kenyon R. Kenyon. Tiling a polygon with rectangles. pages 610–619. FOCS, 1992.

[Rob91] J. M. Robson. Le recouvrement d'une figure par hm et vn est np-complet. pages 95–103, Universit de Creteil, 1991. seminaire Polyominos et pavages, D. Beauquier.

[Thu90] W. P. Thurston. Conway's tiling group. *American Mathematical Monthly,* pages 757–773, 1990.

[Ung93] V. Unger. Pavage d'une figure par h2, v2, h3. rapport de stage de D. E. A. Laboratoire de l'Informatique du Parallelisme. Ecole Normale Superieure de Lyon, 1993.

Axiomatizing Petri Net Concatenable Processes

Vladimiro Sassone

BRICS* – Computer Science Dept., University of Aarhus

Abstract. The concatenable processes of a Petri net N can be characterized abstractly as the arrows of a symmetric monoidal category $\mathcal{P}[N]$. Yet, this is only a partial axiomatization, since $\mathcal{P}[N]$ is built on a concrete, ad hoc chosen, category of symmetries. In this paper we give a fully equational description of the category of concatenable processes of N, thus yielding an axiomatic theory of the noninterleaving behaviour of Petri nets.

Introduction

Concatenable processes of Petri nets have been introduced in [3] to account, as their name indicates, for the issue of process concatenation. Let us briefly reconsider the ideas which led to their definition.

The development of theory Petri nets, focusing on the noninterleaving aspects of concurrency, brought to the foreground various notions of process, e.g. [14, 5, 2, 12, 3]. Generally speaking, Petri net processes—whose standard version is given by the Goltz-Reisig *non-sequential processes* [5]—are structures needed to account for the *causal relationships* which rule the occurrence of events in computations. Thus, ideally, processes are simply computations in which explicit information about such causal connections is added. More precisely, since it is a well-established idea that, as far as the theory of computation is concerned, causality can be faithfully described by means of partial orderings—though interesting 'heretic' ideas appear sometimes—abstractly, the processes of a net N are ordered sets whose elements are labelled by transitions of N. Concretely, in order to describe exactly which multisets of transitions are processes, one defines a process of N to be a map $\pi\colon \Theta \to N$ which maps transitions to transitions and places to places respecting the 'bipartite graph structure' of nets. Here Θ is a *finite deterministic occurrence net*, i.e., roughly speaking, a finite conflict-free 1-safe acyclic net such that the minimal and maximal elements of the partial ordering $\preccurlyeq$ naturally induced by the 'flow relation' on the elements of Θ are places. The role of π is to 'label' the places and the (partially ordered) transitions of Θ with places and transitions of N compatibly with the structure of N.

Given this definition, one can assign the natural *source* and *target* states to a process $\pi\colon \Theta \to N$ by considering the multisets of places of N which are the image via π of, respectively, the minimal and maximal (wrt. $\preccurlyeq$) places of Θ. Now, the simple minded attempt to concatenate a process $\pi_1\colon \Theta_1 \to N$ with source u to a process $\pi_0\colon \Theta_0 \to N$ with target u by merging the maximal places of Θ_0 with the minimal places of Θ_1 in a way which preserves the labellings fails immediately. In fact, if more than one place of u is labelled by a single place of N, there are many ways to put in one-to-one correspondence the maximal places of Θ_0

* Basic Research in Computer Science, Centre of the Danish National Research Foundation. The author was supported by EU Human Capital and Mobility grant ERBCHBGCT920005. Work partly carried out during the author's doctorate at Università di Pisa, Italy.

and the minimal places of Θ_1 preserving the labels, i.e., there are many possible concatenations of π_0 and π_1, each of which gives a possibly different process of N. In other words, as the above argument shows, process concatenation has to do with *merging tokens*, i.e., instances of places, rather than *merging places.*

Therefore, any attempt to deal with process concatenation must disambiguate the *identity* of each token in a process. This is exactly the idea of *concatenable processes,* which are simply Goltz-Reisig processes in which the minimal and maximal places carrying the same label are linearly ordered. This yields immediately an operation of concatenation, since the ambiguity about the identity of tokens is resolved using the additional information given by the orderings. Moreover, the existence of concatenation leads easily to the definition of the category of concatenable processes of N. It turns out that such a category is a *symmetric monoidal category* whose tensor product is provided by the parallel composition of processes [3]. The relevance of this result is that it describes Petri net behaviours as *algebras* in a remarkably smooth way.

Naturally linked to the fact that they are algebraic structures, concatenable processes are amenable to abstract descriptions. In [3] the authors deal with this by associating to each net N a symmetric monoidal category $\mathcal{P}[N]$ isomorphic to the category of concatenable processes of N; such a characterization, however, is not completely abstract and it provides only a partial axiomatization of the algebra of concatenable processes of N, since in the cited work $\mathcal{P}[N]$ is built on a concrete, ad hoc constructed, category Sym_N. In this paper we show that Sym_N can be characterized abstractly, thus yielding a *purely algebraic* and *completely abstract* axiomatization of the category of concatenable processes of N. Namely, we shall prove that $\mathcal{P}[N]$ is the *free symmetric strict monoidal* category on the net N modulo two simple additional axioms.[1] This result complements the investigation of [3] on the structure of net computations by showing that they can be described by an *essentially algebraic theory* (whose models are symmetric monoidal categories), which, in our opinion, is a remarkable fact. In addition, our axiomatization of $\mathcal{P}[N]$ naturally provides a *term algebra* and an *equational theory* of concatenable processes of N, by means of which one can 'compute' with and 'reason' about them. The relevance of this is evident when one thinks of N as modelling a complex system whose behaviour is to be analysed.

Concerning the organization of the paper, Section 1 recalls the needed definitions; the reader acquainted with [12, 3] and with monoidal categories can safely skip it. In Section 2 we sketch the proof of our result. The present paper intends to be an extended abstract; therefore, most of the proofs are omitted and those remaining are just sketched. Full expositions can be found in [15, 16].

1 Background

The notion of *monoidal category* dates back to [1] (see [11] for an easy thorough introduction and [4] for advanced topics). In this paper we shall be concerned only with a particular kind of symmetric monoidal categories, name-

[1] We remark that the existence of a similar axiomatization was conjectured also in [6].

ly those which are *strict monoidal* and whose objects form a *free commutative monoid.* Remarkably, a very similar kind of categories have appeared as distinguished algebraic structures also in [10], where thery are called PROP's (for Product and Permutation categories), and in [8].

A *symmetric strict monoidal category* (SSMC in the following) is a structure $(\underline{C}, \otimes, e, \gamma)$, where $\underline{C}$ is a category, e is an object of $\underline{C}$, called the *unit* object, $\otimes: \underline{C} \times \underline{C} \to \underline{C}$ is a functor, called the *tensor product*, subject to the equations

$$\otimes \circ \langle \otimes \times 1_{\underline{C}} \rangle = \otimes \circ \langle 1_{\underline{C}} \times \otimes \rangle, \tag{1}$$
$$\otimes \circ \langle \underline{e}, 1_{\underline{C}} \rangle = 1_{\underline{C}}, \tag{2}$$
$$\otimes \circ \langle 1_{\underline{C}}, \underline{e} \rangle = 1_{\underline{C}}, \tag{3}$$

where $\underline{e}: \underline{C} \to \underline{C}$ is the constant functor which associate e and id_e respectively to each object and each morphism of $\underline{C}$, $\langle \_, \_ \rangle$ is the pairing of functors induced by the cartesian product, and $\gamma: {}_{-1} \otimes {}_{-2} \xrightarrow{\sim} {}_{-2} \otimes {}_{-1}$ is a natural isomorphism, called the *symmetry* of $\underline{C}$, subject to the Kelly-MacLane *coherence axioms* [9, 7]:

$$(\gamma_{x,z} \otimes id_y) \circ (id_x \otimes \gamma_{y,z}) = \gamma_{x \otimes y, z}, \tag{4}$$
$$\gamma_{y,x} \circ \gamma_{x,y} = id_{x \otimes y}. \tag{5}$$

Equation (1) states that the tensor is associative on both objects and arrows, while (2) and (3) state that e and id_e are, respectively, the unit object and the unit arrow for $\otimes$. Concerning the coherence axioms, axiom (5) says that $\gamma_{y,x}$ is the inverse of $\gamma_{x,y}$, while (4), the *real key* of symmetric monoidal categories, links the symmetry at composed objects to the symmetry at the components. A *symmetry* s in a symmetric monoidal category $\underline{C}$ is any arrow obtained as composition and tensor of *identities* and *components* of γ. We use $Sym_{\underline{C}}$ to denote the subcategory of the symmetries of $\underline{C}$.

A *symmetric strict monoidal functor* from $(\underline{C}, \otimes, e, \gamma)$ to $(\underline{D}, \otimes', e', \gamma')$, is a functor $\mathsf{F}: \underline{C} \to \underline{D}$ such that

$$\mathsf{F}(e) = e', \tag{6}$$
$$\mathsf{F}(x \otimes y) = \mathsf{F}(x) \otimes' \mathsf{F}(y), \tag{7}$$
$$\mathsf{F}(\gamma_{x,y}) = \gamma'_{\mathsf{F}x, \mathsf{F}y}. \tag{8}$$

Let $\underline{\mathsf{SSMC}}$ be the category of SSMC's and symmetric strict monoidal functors and let $\underline{\mathsf{SSMC}}^{\oplus}$ be the full subcategory consisting of the monoidal categories whose objects form *free* commutative monoids.

Next, we recall the definition of Petri nets formulated in [12].

Notation. We denote by $S^{\oplus}$ the *free commutative monoid* on S, i.e., the monoid of *finite multisets* of S. Recall that a finite multiset is a functions from S to ω which yields nonzero values at most on finitely many elements of S. As usual, we represent $u \in S^{\oplus}$ as a formal sum $\bigoplus_{i \in I} u(s_i) \cdot a_i$ where only the $a_i \in S$ such that $u(a_i) > 0$ appear.

A *Petri net* is a structure $N = (\partial_N^0, \partial_N^1 : T_N \to S_N^{\oplus})$, where T_N is a set of *transitions*, S is a set of *places*, and ∂_N^0 and ∂_N^1 are functions assigning to each transition, respectively, a source and a target multiset. A *morphism* of PT nets

$f: N_0 \to N_1$ is a pair $\langle f_t, f_p\rangle$, where $f_t: T_{N_0} \to T_{N_1}$, the transition component, is a *function* and $f_p: S^{\oplus}_{N_0} \to S^{\oplus}_{N_1}$, the place component, is a *monoid homomorphism* which respect source and target, i.e., the two diagrams below commute.

$$
\begin{array}{ccc}
T_{N_0} & \xrightarrow{\partial^0_{N_0}} & S^{\oplus}_{N_0} \\
{\scriptstyle f_t}\downarrow & & \downarrow{\scriptstyle f_p} \\
T_{N_1} & \xrightarrow[\partial^0_{N_1}]{} & S^{\oplus}_{N_1}
\end{array}
\qquad\qquad
\begin{array}{ccc}
T_{N_0} & \xrightarrow{\partial^1_{N_0}} & S^{\oplus}_{N_0} \\
{\scriptstyle f_t}\downarrow & & \downarrow{\scriptstyle f_p} \\
T_{N_1} & \xrightarrow[\partial^1_{N_1}]{} & S^{\oplus}_{N_1}
\end{array}
$$

The data above define the category Petri of PT nets.

Let N be a net. We recall now the construction of the symmetric strict monoidal category $\mathcal{P}[N]$. We start by introducing the *vectors of permutations* (*vperms*) of N,[2] which from the categorical viewpoint play the role of the symmetry isomorphism for $\mathcal{P}[N]$.

Notation. We denote by $\Pi(n)$ the group of permutations of n elements and we write $|\sigma| = n$ when $\sigma \in \Pi(n)$. To simplify notation, we shall assume that the empty function $\varnothing: \varnothing \to \varnothing$ is the (unique) permutation of zero elements.

For $u \in S^{\oplus}$, a *vperm* $s: u \to u$ is a function which assigns to each $a \in S$ a permutation $s(a) \in \Pi(u(a))$. Given $u = n_1 \cdot a_1 \oplus \ldots \oplus n_k \cdot a_k$ in $S^{\oplus}_N$, we shall represent a vperm s on u as a vector of permutations, $\langle \sigma_{a_1}, \ldots, \sigma_{a_k}\rangle$, where $s(a_j) = \sigma_{a_j}$, whence their name. One can define the operations of sequential and parallel composition of vperms, so that they can be organized as the arrows of a SSMC. The details follow (see also Figure 1).

Given the vperms $s = \langle \sigma_{a_1}, \ldots, \sigma_{a_k}\rangle: u \to u$ and $s' = \langle \sigma'_{a_1}, \ldots, \sigma'_{a_k}\rangle: u \to u$ their *sequential composition* $s; s': u \to u$ is the vperm $\langle \sigma_{a_1}; \sigma'_{a_1}, \ldots, \sigma_{a_k}; \sigma'_{a_k}\rangle$, where $\sigma; \sigma'$ is the composition of permutation which we write in the diagrammatic order from left to right. Given the vperms $s = \langle \sigma_{a_1}, \ldots, \sigma_{a_k}\rangle: u \to u$ and $s' = \langle \sigma'_{a_1}, \ldots, \sigma'_{a_k}\rangle: v \to v$ (where possibly $\sigma_{a_j} = \varnothing$ for some j), their *parallel composition* $s \otimes s': u \oplus v \to u \oplus v$ is the vperm $\langle \sigma_{a_1} \otimes \sigma'_{a_1}, \ldots, \sigma_{a_k} \otimes \sigma'_{a_k}\rangle$, where

$$(\sigma \otimes \sigma')(x) = \begin{cases} \sigma(x), & \text{if } 0 < x \le |\sigma|, \\ \sigma'(x - |\sigma|) + |\sigma|, & \text{if } |\sigma| < x \le |\sigma| + |\sigma'|. \end{cases}$$

Let γ be $\{1 \leftrightarrows 2\} \in \Pi(2)$ and consider $u_i = n^i_1 \cdot a_1 \oplus \ldots \oplus n^i_k \cdot a_k$, $i = 1, 2$, in $S^{\oplus}$, the *interchange vperm* $\gamma(u_1, u_2)$ is the vperm $\langle \sigma_{a_1}, \ldots, \sigma_{a_k}\rangle: u_1 \oplus u_2 \to u_1 \oplus u_2$ where

$$\sigma_{a_j}(x) = \begin{cases} x + n^2_j, & \text{if } 0 < x \le n^1_j, \\ x - n^1_j, & \text{if } n^1_j < x \le n^1_j + n^2_j. \end{cases}$$

It is now immediate to see that $\_;\_$ is associative. Moreover, for each $u \in S^{\oplus}$ the vperm $u = \langle id_{a_1}, \ldots, id_{a_n}\rangle: u \to u$, where id_{a_j} is the identity permutation, is an identity for sequential composition. Let 0 be the empty multiset on S. Then, the vperm $s: 0 \to 0$ is a unit for parallel composition. Now, given a net N, let

[2] Vperms are called *symmetries* in [3]. Here, in order to avoid confusion with the general notion of symmetry in a symmetric monoidal category, we prefer to use another term.

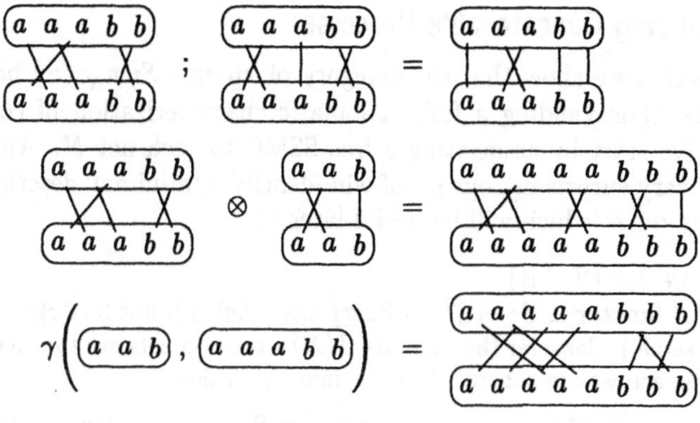

Figure 1: The monoidal structure of vperms

Sym_N be the category whose objects are the elements of $S_N^{\oplus}$ and whose arrows are the vperms $s: u \to u$ for $u \in S_N^{\oplus}$. Then, it is easy to show that Sym_N is a SSMC with respect to the given composition and tensor product, with identities and unit element as explained above and the symmetry natural isomorphism given by the collection $\gamma = \{\gamma(u,v)\}_{u,v \in Sym_N}$ of the interchange vperms.

We can now define $\mathcal{P}[N]$ as the category which includes Sym_N as subcategory and has as additional arrows those defined by the following inference rules:

$$\frac{t: u \to v \text{ in } T_N}{t: u \to v \text{ in } \mathcal{P}[N]}$$

$$\frac{\alpha: u \to v \text{ and } \beta: u' \to v' \text{ in } \mathcal{P}[N]}{\alpha \otimes \beta: u \oplus u' \to v \oplus v' \text{ in } \mathcal{P}[N]} \qquad \frac{\alpha: u \to v \text{ and } \beta: v \to w \text{ in } \mathcal{P}[N]}{\alpha; \beta: u \to w \text{ in } \mathcal{P}[N]}$$

plus axioms expressing the fact that $\mathcal{P}[N]$ is a SSMC with composition $\_;\_$, tensor $\_\otimes\_$ (extending those already defined on vperms) and symmetry isomorphism γ, and the following axioms

$$\begin{array}{ll} t; s = t, & \text{where } t: u \to v \text{ in } T_N \text{ and } s: v \to v \text{ in } Sym_N, \\ s; t = t, & \text{where } t: u \to v \text{ in } T_N \text{ and } s: u \to u \text{ in } Sym_N. \end{array} \qquad (\Psi)$$

In other words, $\mathcal{P}[N]$ is built on the category Sym_N by adding the transitions of N and freely closing with respect to sequential and parallel composition of arrows, so that $\mathcal{P}[N]$ is made symmetric strict monoidal and the axioms (Ψ) hold. The relevant fact about $\mathcal{P}[N]$ is that its arrows can be interpreted precisely as concatenable processes of N, i.e., $\mathcal{P}[N]$ represents exactly the noninterleaving behaviour of N, including its algebraic structure. (See [3] for the details.)

THEOREM 1.1 ($\mathcal{P}[N]$ vs. *Concatenable Processes [3]*)
Let N be a net. Then there exists a one-to-one correspondence between the arrows of $\mathcal{P}[N]$ and the concatenable processes of N such that, for each $u, v \in S_N^{\oplus}$, the arrows of the kind $u \to v$ correspond to the processes enabled by u and producing v, and such that sequential and parallel composition (tensor product) of processes (arrows) are respected.

2 Axiomatizing Concatenable Processes

In this section we show that the category of vperms Sym_N can be described abstractly, thus yielding a fully axiomatic characterization of concatenable processes. We start by associating a free SSMC to each net N. Although this may not be very surprising, our proof will identify a 'minimal' description of the free category on N which will be useful later on.

PROPOSITION 2.1 ($\mathcal{F} \dashv \mathcal{U}$)
The forgetful functor $\mathcal{U}: \underline{\mathsf{SSMC}}^{\oplus} \to \underline{\mathsf{Petri}}$ *has a left adjoint* $\mathcal{F}: \underline{\mathsf{Petri}} \to \underline{\mathsf{SSMC}}^{\oplus}$.

Proof. (Sketch.) Consider the category $\mathcal{F}(N)$ whose objects are the elements of $S_N^{\oplus}$ and whose arrows are generated by the inference rules

$$\frac{u \in S_N^{\oplus}}{id_u: u \to u \text{ in } \mathcal{F}(N)} \qquad \frac{a \text{ and } b \text{ in } S_N}{c_{a,b}: a \oplus b \to a \oplus b \text{ in } \mathcal{F}(N)} \qquad \frac{t: u \to v \text{ in } T_N}{t: u \to v \text{ in } \mathcal{F}(N)}$$

$$\frac{\alpha: u \to v \text{ and } \beta: u' \to v' \text{ in } \mathcal{F}(N)}{\alpha \otimes \beta: u \oplus u' \to v \oplus v' \text{ in } \mathcal{F}(N)} \qquad \frac{\alpha: u \to v \text{ and } \beta: v \to w \text{ in } \mathcal{F}(N)}{\alpha; \beta: u \to w \text{ in } \mathcal{F}(N)}$$

modulo the axioms expressing that $\mathcal{F}(N)$ is a strict monoidal category, namely,

$$\alpha; id_v = \alpha = id_u; \alpha \quad \text{and} \quad (\alpha;\beta);\gamma = \alpha;(\beta;\gamma),$$
$$(\alpha \otimes \beta) \otimes \gamma = \alpha \otimes (\beta \otimes \gamma) \quad \text{and} \quad id_0 \otimes \alpha = \alpha = \alpha \otimes id_0,$$
$$id_u \otimes id_v = id_{u \oplus v} \quad \text{and} \quad (\alpha \otimes \alpha');(\beta \otimes \beta') = (\alpha;\beta) \otimes (\alpha';\beta'),$$

the latter whenever the righthand term is defined, and the following axioms

$$\begin{aligned} c_{a,b}; c_{b,a} &= id_{a \oplus b}, \\ c_{u,u'}; (\beta \otimes \alpha) &= (\alpha \otimes \beta); c_{v,v'}, \quad \text{for } \alpha: u \to v,\ \beta: u' \to v', \end{aligned} \tag{9}$$

where $c_{u,v}$ for $u, v \in S_N^{\oplus}$ denote *any* term obtained from $c_{a,b}$ for $a, b \in S_N$ by applying recursively the following rules (compare with axiom (4)):

$$\begin{aligned} c_{0,u} &= id_u = c_{u,0}, \\ c_{a \oplus u, v} &= (id_a \otimes c_{u,v}); (c_{a,v} \otimes id_u), \\ c_{u, v \oplus a} &= (c_{u,v} \otimes id_a); (id_v \otimes c_{u,a}). \end{aligned} \tag{10}$$

Observe that equation (9), in particular, equalizes all the terms obtained from (10) for fixed u and v. In fact, let $c_{u,v}$ and $c'_{u,v}$ be two such terms and take α and β to be, respectively, the identities of u and v. Now, since $id_u \otimes id_v = id_{u \oplus v} = id_v \otimes id_u$, from (9) we have that $c_{u,v} = c'_{u,v}$ in $\mathcal{F}(N)$. Then, it can be shown that the collection $\{c_{u,v}\}_{u,v \in S_N^{\oplus}}$ is a symmetry natural isomorphism which makes $\mathcal{F}(N)$ into a SSMC which is free on N. This means that $\mathcal{F}$ extends to a functor left adjoint to $\mathcal{U}$. ✓

Thus, establishing the adjunction $\underline{\mathsf{Petri}} \rightharpoonup \underline{\mathsf{SSMC}}^{\oplus}$, we have identified the free SSMC on N as a category generated, modulo appropriate equations, from the net N viewed as a graph enriched with formal arrows id_u, which play the role of the identities, and $c_{a,b}$ for $a, b \in S_N$, which generate all the needed symmetries. In the following, we speak of the *free* SSMC on N to mean $\mathcal{F}(N)$ as constructed above.

The following is the adaptation to SSMC's of the usual notion of quotient algebras characterized, as usual, by a universal property.

Proposition 2.2 (*Monoidal Quotient Categories*)
For a given SSMC $\underline{C}$, let $\mathcal{R}$ be a function which assigns to each pair of objects a and b of $\underline{C}$ a binary relation $\mathcal{R}_{a,b}$ on the homset $\underline{C}(a,b)$. Then, there exist a SSMC $\underline{C}/\mathcal{R}$ and a symmetric strict monoidal functor $Q_{\mathcal{R}}: \underline{C} \to \underline{C}/\mathcal{R}$ such that

i) If $f\mathcal{R}_{a,b}f'$ then $Q_{\mathcal{R}}(f) = Q_{\mathcal{R}}(f')$;

ii) For each symmetric strict monoidal $H: \underline{C} \to \underline{D}$ such that $H(f) = H(f')$ whenever $f\mathcal{R}_{a,b}f'$, there exists a unique functor $K: \underline{C}/\mathcal{R} \to \underline{D}$, which is necessarily symmetric strict monoidal, such that $K \circ Q_{\mathcal{R}} = H$.

Proof. (Sketch.) Say that $\mathcal{R}$ is a $\otimes$-*congruence* if $\mathcal{R}_{a,b}$ is an equivalence for each a and b and if $\mathcal{R}$ respects composition and tensor, i.e., whenever $f\mathcal{R}_{a,b}f'$ then, for all $h: a' \to a$ and $k: b \to b'$, we have $(k \circ f \circ h)\mathcal{R}_{a',b'}(k \circ f' \circ h)$ and for all $h: a' \to b'$ and $k: a'' \to b''$, we have $(h \otimes f \otimes k)\mathcal{R}_{a' \otimes a \otimes a'', b' \otimes b \otimes b''}(h \otimes f' \otimes k)$. Clearly, if $\mathcal{R}$ is a $\otimes$-congruence, the following definition is well-given: $\underline{C}/\mathcal{R}$ is the category whose objects are those of $\underline{C}$, whose homset $\underline{C}/\mathcal{R}(a,b)$ is $\underline{C}(a,b)/\mathcal{R}_{a,b}$, i.e., the quotient of the corresponding homset of $\underline{C}$ modulo the appropriate component of $\mathcal{R}$, and whose arrow composition and tensor product are given by $[g]_{\mathcal{R}} \circ [f]_{\mathcal{R}} = [g \circ f]_{\mathcal{R}}$ and $[f]_{\mathcal{R}} \otimes [g]_{\mathcal{R}} = [f \otimes g]_{\mathcal{R}}$, respectively. Moreover, it is easy to check that $\underline{C}/\mathcal{R}$ is a SSMC with symmetry isomorphism given by the natural transformation whose component at (u,v) is $[\gamma_{u,v}]_{\mathcal{R}}$ and unit object e.

Observe now that, given $\mathcal{R}$ as in the hypothesis, it always possible to find the least $\otimes$-congruence $\mathcal{R}'$ which includes (componentwise) $\mathcal{R}$. Then, take $\underline{C}/\mathcal{R}$ to be $\underline{C}/\mathcal{R}'$ and $Q_{\mathcal{R}}$ to be the obvious projection of $\underline{C}$ into $\underline{C}/\mathcal{R}$. Clearly, $Q_{\mathcal{R}}$ is a symmetric strict monoidal functor. Moreover, it is not difficult to show that it enjoys the properties *(i)* and *(ii)* above.

Our next step is to show that $\mathcal{P}[N]$ is the quotient of $\mathcal{F}(N)$ modulo two simple additional axioms. In order to show this, we need the following lemma.

Lemma 2.3 (*Axiomatizing Sym_N*)
The arrows of Sym_N are generated via sequential composition by the vperms of the kind $id_u \otimes \gamma(a,a) \otimes id_v: u \oplus 2 \cdot a \oplus v \to u \oplus 2 \cdot a \oplus v$. Moreover, two such compositions yield the same vperm if and only if this can be shown by using the axioms

$$((id_{u\oplus a} \otimes \gamma(a,a) \otimes id_v)\,;\,(id_u \otimes \gamma(a,a) \otimes id_{a\oplus v}))^3 = id_{u\oplus 3\cdot a\oplus v},$$
$$((id_u \otimes \gamma(a,a) \otimes id_{2\cdot b\oplus v})\,;\,(id_{u\oplus 2\cdot a} \otimes \gamma(b,b) \otimes id_v))^2 = id_{u\oplus 2\cdot a\oplus 2\cdot b\oplus v}, \qquad (11)$$
$$(id_u \otimes \gamma(a,a) \otimes id_v)^2 = id_{u\oplus 2\cdot a\oplus v}.$$

where f^n indicates the composition of f with itself n times.

Proof. (Sketch.) Concerning the first claim, a vperm $p = \langle \sigma_{a_1}, \ldots, \sigma_{a_n} \rangle$ coincides with the tensor $\sigma_{a_1} \otimes \cdots \otimes \sigma_{a_n}$ which, exploiting the functoriality of $\otimes$, can be written as $(\sigma_{a_1} \otimes \cdots \otimes id_{u_n}); \cdots; (id_{u_1} \otimes \cdots \otimes \sigma_{a_n})$. Now, since σ_{a_i} is a permutation, it is a composition of transpositions of adjacent elements, and since the transposition $\tau_i: n{\cdot}a \to n{\cdot}a$, $1 \le i < n$, can be written as $id_{(i-1)\cdot a} \otimes \gamma(a,a) \otimes id_{(n-i-1)\cdot a}$ in Sym_N, we have that $\sigma_{a_i} = (id_{u'_1} \otimes \gamma(a_i,a_i) \otimes id_{u''_1}); \cdots; (id_{u'_k} \otimes \gamma(a_i,a_i) \otimes id_{u''_k})$. Therefore, the vperms $id_u \otimes \gamma(a,a) \otimes id_v$ generate via composition all the vperms of Sym_N. Concerning the axiomatization, it is easy to verify that the equations (11) hold in Sym_N. On the other hand, the completeness of axioms (11) follows non trivially from a non-trivial axiomatization of the groups of permutations [13]. ✓

PROPOSITION 2.4 (*Axiomating* $\mathcal{P}[N]$)
$\mathcal{P}[N]$ is the monoidal quotient of the free SSMC *on N modulo the axioms*

$$c_{a,b} = id_{a\oplus b}, \quad \text{if } a,b \in S_N \text{ and } a \neq b, \tag{12}$$

$$s;t;s' = t, \quad \text{if } t \in T_N \text{ and } s,s' \text{ are symmetries.} \tag{13}$$

Proof. (Sketch.) We show that $\mathcal{P}[N]$ enjoys the universal property of $\mathcal{F}(N)/\mathcal{R}$ stated in Proposition 2.2, where $\mathcal{R}$ is the congruence generated from equations (12) and (13). It follows then from general facts about universal constructions that $\mathcal{P}[N]$ is isomorphic to $\mathcal{F}(N)/\mathcal{R}$.

First of all observe that $\mathcal{P}[N]$ belongs to $\underline{\mathsf{SSMC}}^{\oplus}$. Therefore, corresponding to the Petri net *inclusion* morphism $N \to \mathcal{UP}[N]$, there is a symmetric strict monoidal functor $\mathsf{Q}\colon \mathcal{F}(N) \to \mathcal{P}[N]$ which is the identity on the places and on the transitions of N. It follows easily from the definition of $\mathcal{P}[N]$ that Q equalizes the pairs in $\mathcal{R}$. Then, we have to show that Q is universal among such functors.

We start by observing that Sym_N can be embedded in $Sym_{\mathcal{F}(N)}$ via a monoidal functor. Consider the mapping G of objects and arrows of Sym_N to, respectively, objects and arrows of $Sym_{\mathcal{F}(N)}$ which is the identity on the objects and such that

$$\begin{aligned} \mathsf{G}(id_u \otimes \gamma(a,a) \otimes id_v) &= id_u \otimes c_{a,a} \otimes id_v, \\ \mathsf{G}(p;q) &= \mathsf{G}(p);\mathsf{G}(q), \\ \mathsf{G}(id_u) &= id_u. \end{aligned}$$

It follows from Lemma 2.3 that the equations above define G on all vperms. Thus, to conclude that G is a functor we only need to show that it is well-defined; exploiting Lemma 2.3, this can be seen by showing that it respects axioms (11). Clearly, G is not symmetric strict monoidal, since $\mathsf{G}(\gamma(a,b)) = id_{a\oplus b} \neq c_{a,b}$, i.e., axiom (8) does not hold. However, G is strict monoidal in the sense that (6) and (7) hold.

Let $\underline{C} = (\underline{C}, \otimes, e, \gamma)$ be a SSMC and suppose that there exists a symmetric strict monoidal functor $\mathsf{H}\colon \mathcal{F}(N) \to \underline{C}$ such that, for any pair $a \neq b \in S_N$ and for any symmetries s and s', $\mathsf{H}(c_{a,b}) = \mathsf{H}(id_{a\oplus b})$ and $\mathsf{H}(s;t;s') = \mathsf{H}(t)$. We have to show that there exists a unique $\mathsf{K}\colon \mathcal{P}[N] \to \underline{C}$ such that $\mathsf{H} = \mathsf{KQ}$. We consider the following definition of K on objects and generators

$$\begin{aligned} \mathsf{K}(u) &= \mathsf{H}(u), && \text{if } u \in S_N^{\oplus}, \\ \mathsf{K}(s) &= \mathsf{H}(\mathsf{G}(s)), && \text{if } s \text{ is a symmetry} \\ \mathsf{K}(t) &= \mathsf{H}(t), && \text{if } t \in T_N, \end{aligned}$$

extendend to $\mathcal{P}[N]$ by $\mathsf{K}(\alpha;\beta) = \mathsf{K}(\alpha);\mathsf{K}(\beta)$ and $\mathsf{K}(\alpha \otimes \beta) = \mathsf{K}(\alpha) \otimes \mathsf{K}(\beta)$.

First of all, we have to show that K is well-defined, i.e., that the equations which hold in $\mathcal{P}[N]$ are preserved by K. Since H and G are strict monoidal, it follows that the functoriality of $\otimes$, axioms (1)–(3) and (Ψ) are preserved. The key to show that the same holds for the naturality of the symmetry, for (4) and for (5) is to show that $K(\gamma(u,v)) = \gamma_{K(u),K(v)}$, which can be done by induction on the least of the sizes of u and v. Once this fact is established, the aforesaid points follow from fact that $\underline{C}$ is a SSMC.

The next task is to show that $H = KQ$. It follows from the fact that Q is symmetric strict monoidal and from the definition of the symmetries of $\mathcal{F}(N)$ that H and KQ coincide on $Sym_{\mathcal{F}(N)}$. Then, one proves that $H = KQ$ by proving, by easy induction on the structure of the terms, that each arrow of $\mathcal{F}(N)$ can be written as the composition of symmetries and arrows of the kind $id_u \otimes t \otimes id_v$, for $t \in T_N$. Finally, concerning the uniqueness condition on K, observe that it must necessarily be $K(id_u \otimes \gamma(a,a) \otimes id_v) = id_{H(u)} \otimes \gamma_{H(a),H(a)} \otimes id_{H(v)}$, which, by Lemma 2.3, defines K uniquely on Sym_N. Moreover, the behaviour of K on the arrows formed as composition and tensor of transitions is uniquely determined by H. ✓

The next corollary gives an alternative form for axiom (13).

COROLLARY 2.5 (*Axiom (13) revisited*)
Axiom (13) in Proposition 2.4 can be replaced by the axioms

$$\begin{aligned} t; (id_u \otimes c_{a,a} \otimes id_v) &= t \quad \text{if } t \in T_N \text{ and } a \in S_N, \\ (id_u \otimes c_{a,a} \otimes id_v); t &= t \quad \text{if } t \in T_N \text{ and } a \in S_N. \end{aligned}$$

Proof. Since $(id_u \otimes \gamma_{a,a} \otimes id_v)$ and all the identities are symmetries, axiom (13) implies the present ones. It is easy to see that, on the contrary, the axioms above, together with axiom (12), imply (13). ✓

Finally, in the next corollary, we sum up the purely algebraic characterization of the category of concatenable processes that we have proved in the paper.

COROLLARY 2.6 (*Axiomatizing Concatenable Processes*)
The category $\mathcal{P}[N]$ of concatenable processes of N is the category whose objects are the elements of $S_N^{\oplus}$ and whose arrows are generated by the inference rules

$$\frac{u \in S_N^{\oplus}}{id_u: u \to u \text{ in } \mathcal{P}[N]} \qquad \frac{a \text{ in } S_N}{c_{a,a}: a \oplus a \to a \oplus a \text{ in } \mathcal{P}[N]} \qquad \frac{t: u \to v \text{ in } T_N}{t: u \to v \text{ in } \mathcal{P}[N]}$$

$$\frac{\alpha: u \to v \text{ and } \beta: u' \to v' \text{ in } \mathcal{P}[N]}{\alpha \otimes \beta: u \oplus u' \to v \oplus v' \text{ in } \mathcal{P}[N]} \qquad \frac{\alpha: u \to v \text{ and } \beta: v \to w \text{ in } \mathcal{P}[N]}{\alpha; \beta: u \to w \text{ in } \mathcal{P}[N]}$$

modulo the axioms expressing that $\mathcal{P}[N]$ is a strict monoidal category, namely,

$$\begin{aligned} \alpha; id_v = \alpha = id_u; \alpha \quad &\text{and} \quad (\alpha;\beta);\gamma = \alpha;(\beta;\gamma), \\ (\alpha \otimes \beta) \otimes \gamma = \alpha \otimes (\beta \otimes \gamma) \quad &\text{and} \quad id_0 \otimes \alpha = \alpha = \alpha \otimes id_0, \\ id_u \otimes id_v = id_{u \oplus v} \quad &\text{and} \quad (\alpha \otimes \alpha');(\beta \otimes \beta') = (\alpha;\beta) \otimes (\alpha';\beta'), \end{aligned}$$

the latter whenever the righthand term is defined, and the following axioms

$$\begin{aligned} c_{a,a}; c_{a,a} &= id_{a \oplus a}, \\ t; (id_u \otimes c_{a,a} \otimes id_v) &= t, \quad \text{if } t \in T_N, \\ (id_u \otimes c_{a,a} \otimes id_v); t &= t, \quad \text{if } t \in T_N, \\ c_{u,u'}; (\beta \otimes \alpha) &= (\alpha \otimes \beta); c_{v,v'}, \quad \text{for } \alpha: u \to v,\ \beta: u' \to v', \end{aligned}$$

where $c_{u,v}$ for $u, v \in S_N^{\oplus}$ is obtained by repeatedly applying the following rules:

$$\begin{aligned} c_{a,b} &= id_{a\oplus b}, \quad \text{if } a = 0 \text{ or } b = 0 \text{ or } (a, b \in S_N \text{ and } a \neq b), \\ c_{a\oplus u,v} &= (id_a \otimes c_{u,v}); (c_{a,v} \otimes id_u), \\ c_{u,v\oplus a} &= (c_{u,v} \otimes id_a); (id_v \otimes c_{u,a}). \end{aligned}$$

Proof. Easy from Proposition 2.1, Proposition 2.4 and Corollary 2.5. ✓

Acknowledgements. I wish to thank José Meseguer and Ugo Montanari to whom I am greatly indebted for introducing me to this subject and for many helpful discussions. Many thanks to Mogens Nielsen for his comments on the exposition of this paper.

References

[1] J. BÉNABOU. Categories with Multiplication. *Comptes Rendue Académie Science Paris*, n. 256, pp. 1887–1890, 1963.

[2] E. BEST, AND R. DEVILLERS. Sequential and Concurrent Behaviour in Petri Net Theory. *Theoretical Computer Science*, n. 55, pp. 87–136, 1987.

[3] P. DEGANO, J. MESEGUER, AND U. MONTANARI. Axiomatizing Net Computations and Processes. In *Proceedings of the 4th LICS Symposium*, pp. 175–185, IEEE, 1989.

[4] S. EILENBERG, AND G.M. KELLY. Closed Categories. In *Proceedings of the Conference on Categorical Algebra*, La Jolla, S. Eilenberg *et al.*, Eds., pp. 421–562, Springer, 1966.

[5] U. GOLTZ, AND W. REISIG. The Non-Sequential Behaviour of Petri Nets. *Information and Computation*, n. 57, pp. 125–147, 1983.

[6] R. GORRIERI, AND U. MONTANARI. Scone: A Simple Calculus of Nets. In *Proceedings of CONCUR* '90, LNCS n. 458, pp. 2–31, 1990.

[7] G.M. KELLY. On MacLane's Conditions for Coherence of Natural Associativities, Commutativities, etc. *Journal of Algebra*, n. 1, pp. 397–402, 1964.

[8] W. LAWVERE. *Functorial Semantics of Algebraic Theories*. PhD Thesis, Columbia University, New York, 1963. An abstract appears in *Proceedings of the National Academy of Science*, n. 50, pp. 869–872, 1963.

[9] S. MACLANE. Natural Associativity and Commutativity. *Rice University Studies*, n. 49, pp. 28–46, 1963.

[10] S. MACLANE. Categorical Algebra. *Bulletin American Mathematical Society*, n. 71, pp. 40–106, 1965.

[11] S. MACLANE. *Categories for the Working Mathematician*. Springer-Verlag, 1971.

[12] J. MESEGUER, AND U. MONTANARI. Petri Nets are Monoids. *Information and Computation*, n. 88, pp. 105–154, Academic Press, 1990.

[13] E.H. MOORE. Concerning the abstract group of order *k*! isomorphic with the symmetric substitution group on *k* letters. *Proceedings of the London Mathematical Society*, n. 28, pp. 357–366, 1897.

[14] C.A. PETRI. *Non-Sequential Processes*. Interner Bericht ISF–77–5, Gesellschaft für Mathematik und Datenverarbeitung, Bonn, Germany, 1977.

[15] V. SASSONE. *On the Semantics of Petri Nets: Processes, Unfoldings and Infinite Computations*. PhD Thesis TD 6/94, Dipartimento di Informatica, Università di Pisa, 1994.

[16] V. SASSONE. *Some Remarks on Concatenable Processes*. Technical Report TR 6/94, Dipartimento di Informatica, Università di Pisa, 1994.

Functional Sorts in Data Type Specifications

A Geometric Approach to Semantics

Klaus-Dieter Schewe

Technical University of Clausthal, Computer Science Institute
Erzstr. 1, D-38678 Clausthal-Zellerfeld, FRG
schewe@informatik.tu-clausthal.de

Abstract. There are several notions of type with different semantics in computer science. The approach in this paper considers an extension of algebraic type specifications with respect to functional sorts and tries to give a suitable semantics for them.
The basic constituent of the theory is an extended notion of *signature*, which now consists of sorts, constructors and axioms. For sorts and constructors the semantics is defined by coherent Grothendieck topoi. Then it can be shown that initial topoi always exist.
Since each topos defines a canonical theory of a higher-order (intuitionistic) logic, the axioms in the signature define a theory. It is known that models of such theories are uniquely defined by logical functors, which define the models of type specifications in general.

1 Motivation and General Framework

There are several notions of types in computer science. Types occur in modern programming languages to classify values. The overall goal of this work is to contribute to a suitable semantics for *typed program specifications*. In an ADT sense this corresponds to "types over types".

Some approaches to type semantics are based on set theory. Especially, standard formal specification languages [2, 7, 31] use specifications with sets, but there are several problems with such an approach:

- The algebraic community [13, 14, 16, 34] points out the missing integration of sets of values and the operations on them.
- Researchers working on semantics on the basis of λ-calculi point out the insufficiency with respect to higher-order structures — there is no set D with a bijection to the set of functions D^D.
- It is known that there is no set-theoretic model e. g. for the important Girard-Reynolds polymorphism [27].

Moreover, looking at a type as as set or algebra is only one part of the problem. More generally, we are interested in a suitable program logic that includes the type theory. E. g., Dijkstra's approach [12, 25] assures that program properties can be expressed logically via associated predicate transformers. One problem is then to assure the existence of these predicate transformers in the program logic. Classically, a suitable logic is infinitary first-order logic $\mathcal{L}^{\omega}_{\omega\infty}$.

Another problem in type semantics concerns "subtyping" which is in general more than set inclusion [3, 24]. E. g., $NAT \times NAT$ is usually considered a subtype of NAT (via projection), and the functions from A to C are considered a subtype of the functions from B to C (via restriction), where B is a subtype of A. Consequently, subtyping need not preserve algebraic structures. We need a suitable semantics of subtyping that overcomes the purely syntactical approach via type-checking rules.

In database systems, two more problems occur. Firstly, a theory of type parameterization (sets, lists, ...) is required. Such a theory exists for algebraic specifications [13] and should be preserved. Secondly, infinite data structures defined e. g. by rational trees and higher-order functions occur naturally, when generic updates are considered in object-oriented databases schemata, or more generally in the case of inherent cyclic inclusion constraints [28, 29].

The approach in this paper is part of a project which tries to approach general type semantics such that the following properties are satisfied [30]:

- the theory should preserve the coupling of values and operations known from the algebraic approach,
- the theory should be able to support higher-order structures, even for the case of $A \cong A^A$,
- infinite (rational tree) structures should be supported,
- the theory should allow type parameterization and instantiation,
- the theory should give a semantics for subtyping as an additional relation between types that need not be structure preserving, and
- it should be possible to associate a canonical program logic $\mathcal{L}$ to type specifications such that predicate transformers exist in $\mathcal{L}$.

In this paper we approach part of these problems by reconsidering algebraic type specifications with a semantics in coherent Grothendieck topoi. Roughly speaking, a Grothendieck topos generalizes the category of sets and preserves most of its properties, especially those used in the algebraic theory of types [14, 34]. In general, a topos is a cartesian closed, finitely complete category with a subobject classifier, and a coherent Grothendieck topos is a category of sheaves on a site that can be constructed from finite covering families.

In particular, this suggests to generalize the algebraic approach taking functional sorts into consideration. Such extensions to signatures form the starting point in Section this piece of work.

Then it is known that there is a close connection between topos theory and higher-order intuitionistic logic in that every topos corresponds to a theory T of such a logic as the ***topos*** $\mathbb{E}(T)$ ***of definable types***. Moreover, each model of T in an arbitrary topos F corresponds to a ***logical functor*** $\mathbb{E}(T) \rightarrow F$ [15]. It is rather easy to see that the logic of a topos extends $\mathcal{L}^{\omega}_{\omega\infty}$ [23].

Due to space limitations we can not give an introduction to topoi nor to category theory. For category theory consider standard literature [5, 21], for topos theory one of [6, 18, 19, 22] and [8, 15, 20, 22, 23] for the logic of topoi. Other examples for the application of topoi or sheaves in computer science can be found in [26, 32].

2 Assigning Grothendieck Topoi to Signatures

The basic constituent of algebraic specifications [14, 34] is the *signature* that consists of sorts and operators. Based hereon, (conditional) equations can be added to define data-type specifications. The most common semantics for such specifications is based on initial algebras.

A Σ-algebra for a given signature Σ associates sets with the sorts and functions with the operators. The major idea underlying this piece of work is to vary the underlying category, i. e. to switch from *Set* to arbitrary Grothendieck topoi.

The first reason for this switch is purely technical in nature. To the author's best knowledge nearly all the proofs in the theory of algebraic specifications do not depend on specific properties of *Set*, but generally hold in topoi. Next, Grothendieck topoi are always equivalent to categories of sheaves over a site, hence in a certain intuitive sense can be constructed from some underlying small category, the site. Thirdly, signatures may give rise to sketches [6] and in this way allow to construct suitable Grothendieck topoi. On the other hand, topoi are cartesian closed, which suggests to take also functional sorts into consideration, but then set semantics usually causes problems.

Definition 1. A *signature* Σ consists of a finite set *Sorts* of sorts and a finite set *Cons* of constructors such that each constructor $c \in Cons$ has an *arity* $sig(c) = dom(c) \rightarrow codom(c)$ with $codom(c) \in Sorts$ and $dom(c) \in \overline{Sorts}$, where the set $\overline{Sorts}$ of *higher-order sorts* is the smallest set satisfying

- $\omega \in \overline{Sorts} - Sorts$,
- $Sorts \subseteq \overline{Sorts}$,
- if $s, t \in \overline{Sorts}$, then also $(s \rightarrow t) \in \overline{Sorts}$ and
- if $s_1, \ldots, s_n \in \overline{Sorts}$, then $s_1 \times \ldots \times s_n \in \overline{Sorts}$.

The semantics of such signatures will be defined on the basis of Grothendieck topoi that are determined by the constructors in *Cons*. The underlying idea just mimiques the classical approach. We let the higher-order sorts and the constructors correspond to objects and morphisms respectively in a Grothendieck topos E such that there is a site $(C, \mathcal{J})$ with a full small subcategory C of E and a topology $\mathcal{J}$ "generated" by the morphisms associated with the constructors.

Definition 2. A *Σ-topos* of a signature $\Sigma = (Sorts, Cons)$ is a Grothendieck topos $E \cong Sh(C, \mathcal{J})$ with the following properties:

(i) For each higher-order sort $s \in \overline{Sorts}$ there is an object A_s in E such that $A_\omega \cong \Omega$ for the truth value object Ω in E, $A_{(s \rightarrow t)} \cong A_t^{A_s}$ and $A_{s_1 \times \ldots \times s_n} \cong A_{s_1} \times \ldots \times A_{s_n}$ hold.

(ii) For each constructor $c : dom(c) \rightarrow s$ in *Cons* there is a morphism $\bar{c} : A_{dom(c)} \rightarrow A_s$ in E.

(iii) C is a full small subcategory of E containing $\{A_s \mid s \in \overline{Sorts}\}$ such that $\{\bar{c} \mid c \in Cons, codom(c) = s\}$ forms a *subbasis* of the topology $\mathcal{J}$, i. e. they define a pretopology on C via composition and pullback diagrams and $\mathcal{J}$ is the topology generated by this pretopology.

(iv) The set $G = \{A_s \mid s \in Sorts\}$ is a set of generators for E.

In algebraic specifications we know that Σ-algebras for a signature Σ form a category. Therefore, we look for suitable morphisms between Σ-topoi. In general, we have defined two kinds of "morphisms" for topoi. Logical morphisms allow to preserve all of the structure of a topos, whereas geometric morphisms abstract from continuous functions. In order to combine both notions we use *logical-geometric* (or *atomic*) morphisms, i. e. geometric morphisms $f : E \to F$ such that the inverse image functor f^* is logical.

Definition 3. Let E, E' be Σ-topoi with generator sets $G = \{A_s \mid s \in Sorts\}$ and $G' = \{A'_s \mid s \in Sorts\}$ respectively. A *Σ-topos-morphism* $f : E \to E'$ is a logical-geometric morphism with $f^*(A'_s) = A_s$.

It is easy to see that Σ-topoi and Σ-topos-morphisms define a category mod_Σ. Moreover, note that f^* being logical implies that $f^*(A'_s) = A_s$ also holds for all higher-order sorts $s \in \overline{Sorts}$.

3 Initial Topoi for Signatures

Assume to be given a signature Σ. Since our goal is to generalize the algebraic approach to data type specifications we may ask for initial Σ-topoi in the category mod_Σ. Let us sketch their construction. For this we assume that nested product sorts do not occur in $\overline{Sorts}$.

The idea is to define a *sketch* [5, 6]. First we define a directed graph $\mathcal{G}$ with $\overline{Sorts} \cup \{\bot\}$ as its set of vertices.

- For each vertex s there are edges id_s from s to itself and edges $\bot_s$ from $\bot$ to s.
- For each product sort $s_1 \times \ldots \times s_n$ containing $dom(c)$ for a constructor $c \in Cons$ we have an edge $c_{s_1 \times \ldots \times s_n}$ from this vertex to $codom(c)$.
- For each product sort $s_1 \times \ldots \times s_n$ and each subsequence $i_1, \ldots, i_k$ of indices there is an edge $\pi^{s_1 \times \ldots \times s_n}_{i_1,\ldots,i_k}$ from $s_1 \times \ldots \times s_n$ to $s_{i_1} \times \ldots \times s_{i_k}$.
- For each pair of sorts (s, x) there is an edge η^s_x from x to $(s \to x \times s)$. Similarly, for each pair (s, t) there is an edge $ev_{s \to t}$ from $(s \to t) \times s$ to t.
- For each edge g from y to x and each vertex s there is an edge $l_s g$ from $y \times s$ to $x \times x$. Similarly, for each edge f from t to u there is an edge $r_s f$ from $(s \to t)$ to $(s \to u)$.
- For each edge f from $x \times s$ to t there is an edge $\hat{f}$ from x to $(s \to t)$. Similarly, for each edge g from x to $(s \to t)$ there is an edge $\check{g}$ from $x \times s$ to t.
- There is an edge w from the empty product sort to ω.

All these vertices and edges in $\mathcal{G}$ shall become objects and morphisms in our Σ-topos. They correspond to the identity morphisms, the constructors, the projections given by products and to the counit and unit morphisms associated with the exponential adjoints.

Next we define a set $\mathcal{D}$ of diagrams in $\mathcal{G}$. In the Σ-topos to be constructed these diagram shall be commuting.

- The first diagrams contain three vertices $s_1 \times \ldots \times s_n$, $s_{i_1} \times \ldots \times s_{i_k}$ for a subsequence of indices and $codom(c)$ for some $c \in Cons$ and three edges $c_{s_1\times\ldots\times s_n}$, $c_{s_{i_1}\times\ldots\times s_{i_k}}$ and $\pi^{s_1\times\ldots\times s_n}_{i_1,\ldots,i_k}$.
- Next, we have diagrams with three vertices $s_1 \times \ldots \times s_n$, $s_{i_1} \times \ldots \times s_{i_k}$ for a subsequence of indices and s_{i_j} for some $j \in \{1,\ldots,k\}$ and three edges $\pi^{s_1\times\ldots\times s_n}_{i_1,\ldots,i_k}$, $\pi^{s_1\times\ldots\times s_n}_{i_j}$ and $\pi^{s_{i_1}\times\ldots\times s_{i_k}}_{i_j}$.
- Diagrams of the third kind contain six vertices s (twice), x, $x \times s$, $(s \to t)$ and $(s \to t) \times s$. The seven edges are id_s, $\pi_1^{x\times s}$, $\pi_2^{x\times s}$, $\pi_1^{(s\to t)\times s}$, $\pi_2^{(s\to t)\times s}$, $\hat{f}$ and $l_s\hat{f}$.
- The next kind of diagrams contains three vertices x, $(s \to t)$ and $(s \to x \times s)$ and three edges $\hat{f}$, $r_s f$ and η_x^s.
- Similarly, we have diagrams with three vertices $x \times s$, t and $(s \to t) \times s$ and edges $\check{g}$, $l_s g$ and $ev_{s\to t}$.
- The sixth kind of diagram just contains two vertices $x \times s$ and t and two edges f and $\check{\hat{f}}$ from the first vertex to the second one.
- Similarly, we have diagrams with two vertices x and $(s \to t)$ and two edges g and $\hat{\check{g}}$ from the first vertex to the second one.
- Next consider diagrams with seven vertices $y \times s$, $x \times s$ (twice), $(s \to t) \times s$, $(s \to u) \times s$, t and u. The seven edges are $l_s g$ (twice), $l_s f$, $l_s r_s h$, $\check{f}$, h and $ev_{s\to u}$.
- Finally, take similar diagrams with seven vertices y, x, $(s \to t)$ (twice), $(s \to u)$, $(s \to y \times s)$ and $(s \to x \times s)$ and seven edges g, $\hat{f}$, η_y^s, $r_s l_s g$, $r_s f$ and $r_s h$ (twice).

The next constituent is a set $\mathcal{K}$ of cones in $\mathcal{G}$. These will become limits in the Σ-topos.

- Elements of the first kind of cones have a discrete base diagram with vertices $s_1, \ldots, s_n$, a root $s_1 \times \ldots \times s_n$ and edges $\pi^{s_1\times\ldots\times s_n}_{s_j}$ for all $j = 1, \ldots, n$. For the special case of an empty base $(n = 0)$ the root is the empty product sort.
- The second kind of cones has a base with three vertices s, $dom(c)$ and $dom(c')$ and two vertices $c_{dom(c)}$ and $c_{dom(c')}$ for some $c, c' \in Cons$ with $codom(c) = codom(c') = s$. For $c = c'$ the root is $dom(c)$ and the three edges of the cone are $c_{dom(c)}$ and $id_{dom(c)}$ (twice). For $c \neq c'$ the root is $\perp$ and the three edges of the cone are $\perp_{dom(c)}$, $\perp_{dom(c')}$ and $\perp_s$.

The last constituent of a sketch is a set C of cocones to become colimits in topos to be constructed. First we have the cocone with an empty base and the coroot $\perp$, which will turn $\perp$ into an initial object.

A second kind of cocones is given by a discrete base consisting of all $dom(c)$ for $c \in Cons$ with $codom(c) = s$. This s then gives the coroot and the edges $c_{dom(c)}$ complete the cocone.

Hence $(\mathcal{G}, \mathcal{D}, \mathcal{K}, C)$ defines a sketch and there exists a small category C with objects A_s for all $s \in \overline{Sorts}$. In particular, all diagrams in $\mathcal{D}$ give rise to commuting diagrams in C, all cones in $\mathcal{K}$ define finite limits in C, and all cocones in C define colimits in C.

In particular, we have $A_{s_1\times\ldots\times s_n} \cong A_{s_1} \times \ldots \times A_{s_n}$ and $A_{(s\to t)} \cong A_t^{A_s}$. Moreover, A_s is the coproduct $\coprod_{c\in Cons, codom(c)=s} A_{dom(c)}$.

Next, we define the Grothendieck pretopology on C. For this let $\bar{c}$ for $c \in Cons$ denote the morphism in c corresponding to the edge $c_{dom(c)}$. Let $\{\bar{c} \mid c \in$

Cons, $codom(c) = s\}$ be a cover of A_s and let all other non-trivial covers be generated by composition and pullbacks. Due to the second kind of cones in $\mathcal{K}$, there are no non-trivial covers defined by pullbacks.

Then let $\mathcal{J}$ be the Grothendiecj topology on C generated from the pretopology. Then $E = Sh(C, \mathcal{J})$ is a Grothendieck topos and C can be embedded into E by $L \circ y$, where y is the Yoneda embedding into $Set^{C^{op}}$ and L is the associated sheaf functor.

Since $\{A_s \mid s \in Sorts\}$ is a generator set for C, $\{(L \circ y)(A_s) \mid s \in Sorts\}$ must be a generator set for E. It can be shown that the chosen construction allows L to preserve exponents.

Finally, there is a morphism $(L \circ y)(A_\omega) \to \Omega$, which results as the classifying morphism of $(L \circ y)(\bar{w}) : \mathbb{1} \to (L \circ y)(A_\omega)$ for the edge w in $\mathcal{G}$. We may force this to be an isomorphism. Thus, E will be a Σ-topos and it can be shown that it is initial in mod_Σ.

Theorem 4. *For any signature Σ there exists an initial Σ-topos.* □

4 General Type Specifications

Let us extend the theory by adding restricting formulae to signatures as in algebraic specifications. When we are given a Σ-topos E for a signature, then E determines a higher-order language $\mathcal{L}(E)$ [15]. Recall that the sorts in this language correspond to the objects A in E, the constants of sort A are the morphisms $\mathbb{1} \to \tilde{A}$ and the power sort mapping is given by $[A_1, \dots, A_n] = \Omega^{\Pi_{i=1}^{n} A_i}$.

Moreover, since $E \cong \mathbb{E}(T(E))$ — the topos of definbable types — for the theory associated with E, we may use predicative restrictions and λ-abstraction for definable types and definable total functions. Finally, we write $f(x)$ instead of $ev(f, x)$. Then the morphisms corresponding to constructors and operators can be used as usual to define terms.

However, in order to do so, we need the topos E. In order not be fixed a priori to initial topoi we use a sublanguage $\mathcal{L}(\Sigma)$ that contains just those expressions that can be written in terms of constructors and operators.

After these preliminary remarks concerning the language $\mathcal{L}(\Sigma)$ we are now prepared to define general specifications and their semantics.

Definition 5. A ***specification*** consists of a signature Σ and a finite set Ax of formulae in $\mathcal{L}(\Sigma)$.

If $\Xi = (\Sigma, Ax)$ is a specification and E is a Σ-topos, then Ax defines a theory $T_E(\Xi)$ over $\mathcal{L}(E)$.

Definition 6. Let $\Xi = (\Sigma, Ax)$ be a specification. Then a ***Ξ-type*** consists of a Σ-topos E and a model $\mathcal{F}$ of $T_E(\Xi)$ or equivalently a logical morphism $\mathbb{E}(T_E(\Xi)) \to F_\mathcal{F}$ into some topos $F_\mathcal{F}$.

A ***Ξ-type-morphism*** from $(E, \mathcal{F})$ to $(F, \mathcal{G})$ consists of a Σ-topos-morphism $f : E \to F$ and a logical morphism $k : F_\mathcal{F} \to F_\mathcal{G}$ such that $\mathcal{G} = k \circ \mathcal{F} \circ f\text{-}int^*$, where $f\text{-}int : \mathbb{E}(T_E(\Xi)) \to \mathbb{E}(T_F(\Xi))$ is the unique logical-geometric morphism induced by f.

Again Ξ-types and Ξ-type-morphisms define a category $\mathcal{M}od_{\Xi}$. Let us now address the problem how to relate different specifications. Analogously to algebraic specifications we consider signature morphisms and then extend them to specification morphisms.

Definition 7. A *signature-morphism* σ from Σ to Σ' is a pair of mappings σ_S : $Sorts \to Sorts'$ and $\sigma_C : Cons \to Cons'$ such that we have (with $\overline{\sigma_S}$ being the canonical extension of σ_S to $\overline{Sorts}$) for all $c \in Cons$

$$dom(\sigma_C(c)) = \overline{\sigma_S}(dom(c)) \text{ and}$$
$$codom(\sigma_C(c)) = \sigma_S(codom(c))$$

If $\sigma : \Sigma \to \Sigma'$ is a signature-morphism and E' is a Σ'-topos, then just as in the case of algebraic specifications we may construct a canonical Σ-topos E in the following way:

- Take $A_s = A'_{\sigma_S(s)}$ for $s \in Sorts$, $\bar{c} = \overline{\sigma_C(c)}$ for $c \in Cons$ and let C be a small finitely complete subcategory of E' containing all the objects A_s and all morphisms $\bar{c}$.
- Let $\mathcal{J}$ be the Grothendieck topology on C generated by the morphisms $\bar{c}$ as a subbasis.
- Let $E = Sh(C, \mathcal{J})$.

Then there is a canonical logical-geometric morphism $\bar{\sigma} : E \to E'$. We write σ-$mod(E')$ for the Σ-model E. Then each interpretation $\mathcal{F}$ of $\mathcal{L}(\Sigma')$ in some topos F induces an interpretation of $\mathcal{L}(\Sigma)$ in F denoted by σ-$int(\mathcal{F})$. Hence the following definition.

Definition 8. A *specification-morphism* σ from $\Xi = (\Sigma, Ax)$ to $\Theta = (\Sigma', Ax')$ consists of a signature-morphism $\sigma : \Sigma \to \Sigma'$ such that for all Θ-models $E', \mathcal{F})$ we have σ-$int(\mathcal{F}) \models T_{\sigma\text{-}mod(E')}(\Xi)$.

Then the category of specifications with specification morphisms will be denoted by *G-spec*. A *specification-morphism* $\sigma : \Xi \to \Theta$ induces a functor $\mathcal{M}od_{\Theta} \to \mathcal{M}od_{\Xi}$ denoted by σ-*mod*.

Definition 9. A *parameterized specification* is an injective specification morphism $\sigma : \Theta \to \Xi$ (i.e. the underlying signature morphism is injective) together with a functor $\tau : \mathcal{M}od_{\Theta} \to \mathcal{M}od_{\Xi}$ such that for all constructors c in Ξ, but not in Θ, we have $codom(c) \in Sorts_{\Xi} - Sorts_{\Theta}$, and σ-$mod \circ \tau$ is equivalent to the identity on $\mathcal{M}od_{\Theta}$.

We remark that this definition includes the persistence requirement that is common in algebraic specifications. As in the algebraic case there is a natural choice for the functor τ given by the left-adjoint σ-*free* of σ-*mod*. The existence of this left-adjoint functor is due to the existence of an initial Ξ-type.

Proposition 10. *Let $\sigma : \Theta \to \Xi$ be a parameterized specification. Then σ-mod : $\mathcal{M}od_{\Xi} \to \mathcal{M}od_{\Theta}$ has a left-adjoint σ-$free : \mathcal{M}od_{\Theta} \to \mathcal{M}od_{\Xi}$ such that σ-$mod \circ$ σ-$free$ is equivalent to the identity functor on $\mathcal{M}od_{\Theta}$.* □

5 Examples

Example 1. Let us first look at the (trivial) example of natural numbers that is well known in algebraic specifications. We use the syntax of the specification language SAMT [30].

Type NAT
  Sort nat
  Constructor $0 :\rightarrow nat,$
    $succ : nat \rightarrow nat,$
    $\_+\_ : nat \times nat \rightarrow nat,$
    $\_\leq\_ : nat \times nat \rightarrow \omega$
  Axiom With $x :: nat.\, 0 + x = x,$
    With $x, y :: nat.\, succ(x) + y = succ(x + y),$
    With $x, y :: nat.\, 0 \leq x = true \wedge succ(x) \leq succ(y) = x \leq y$
End NAT □

Example 2. As a second simple example let us see how exponential sorts $s \rightarrow t$ come into play. Here we use the keyword `BasedOn` to denote that the specification NAT is included. This may always be regarded as an abbreviation for an injective specification morphism.

Type FUN_NAT
  BasedOn NAT
  Sort fun_nat
  Constructor $fun : (nat \rightarrow fun\_nat) \rightarrow fun\_nat,$
    $consfun : nat \rightarrow fun\_nat,$
    $arity : fun\_nat \rightarrow nat$
  Axiom With $f :: nat.\, arity(consfun(f)) = 0,$
  Axiom With $f :: (nat \rightarrow fun\_nat), x :: nat.\, arity(\lambda x.f) = succ(arity(f(x)))$
End FUN_NAT □

Example 3. Let us finally look at a trivial example concerning finite sets.

Type $FSET[X :: TRIV]$
  BasedOn NAT
  Rename $X.s$ **As** elt
  Sort $fset$
  Constructor $\emptyset :\rightarrow fset,$
    $single : elt \rightarrow fset,$
    $union : fset \times fset \rightarrow fset,$
    $card : fset \rightarrow nat$
  Axiom ...
End $FSET$

Here $TRIV$ denotes a collection of restrictions on the possible instantiations. In this case we assume that there is at least one sort s. Such restrictions always look like type specifications. □

6 Concluding Remarks

In this paper a first introduction to type semantics on the basis of topos theory was given. It could be shown that initial topoi exist. The use of topos theory was motivated by an extension to functional sorts corresponding to exponential objects and an integration of the algebraic [9, 34] and the functional approach [10, 24] to type semantics.

Though the result seems to be reasonable, it still has to be explicitly shown that the algebraic results on the coupling of structures and operations [14, 16, 17, 33], parameterization and domain equations [13] can be preserved.

Moreover, the close connection with higher-order categorical logic [8, 15, 20] guarantees the existence of predicate transformers [12, 25] and can hence be used for defining the semantics of typed programs [11].

To define a uniform semantical framework with these features is fundamental for the mathematical foundation of recent developments especially in database and concept theory [1, 3, 4, 26, 28, 29, 32].

References

1. S. Abiteboul, M. Vardi, V. Vianu: *Computing with Infinitary Logic*, in J. Biskup, R. Hull (Eds.): *Proc. 4th Int. Conf. on Database Theory*, Springer LNCS 646, 1992, 113-123
2. J. R. Abrial: *A Formal Approach to Large Software Construction*, in J. L. A. Van de Snepscheut (Ed.), *Mathematics of Program Construction*, Springer LNCS, vol. 375, 1989, 1-20
3. R. M. Amadio, L. Cardelli: *Subtyping Recursive Types*, ACM TOPLAS, vol. 15 (4), 1993, 575-631
4. M. P. Atkinson, P. Buneman: *Types and Persistence in Database Programming Languages*, ACM Comp. Surveys, vol. 19 (2), 1987, 105-190
5. M. Barr, C. Wells: *Category Theory for Computing Science*, Prentice-Hall 1990
6. M. Barr, C. Wells: *Toposes, Triples and Theories*, Springer Grundlehren der mathematischen Wissenschaften 278, 1985
7. D. Bjørner, C. B. Jones: *Formal Specification and Software Development*, Prentice Hall, 1982
8. A. Boileau, A. Joyal: *La Logique des Topos*, Journal of Symbolic Logic, vol. 46 (1), 1981, 6-16
9. M. Broy: *Equational Specification of Partial Higher Order Algebras*, in M. Broy (Ed.): *Logic of Programming and Calculi of Discrete Design*, Springer, NATO ASI Series F, vol. 36, 1986, 185-242
10. K. B. Bruce, A. R. Meyer: *The Semantics of Second Order Polymorphic Lambda Calculus*, in G. Kahn, D. B. MacQueen, G. Plotkin (Eds.): *Semantics of Data Types*, Springer LNCS 173, 1984, pp. 131-144
11. P. Cousot: *Methods and Logics for Proving Programs*, in J. van Leeuwen (Ed.): *The Handbook of Theoretical Computer Science*, vol B: "Formal Models and Semantics", Elsevier, 1990, 841-993
12. E. W. Dijkstra, C. S. Scholten: *Predicate Calculus and Program Semantics*, Springer Texts and Monographs in Computer Science, 1989
13. H.-D. Ehrich, U. Lipeck: *Algebraic Domain Equations*, Theoretical Computer Science, vol. 27, 1983, 167-196

14. H. Ehrig, B. Mahr: *Fundamentals of Algebraic Specification 1*, Springer EATCS Monographs, vol. 6, 1985
15. M. P. Fourman: *The Logic of Topoi*, in J. Barwise (Ed.): *Handbook of Mathematical Logic*, North-Holland Studies in Logic, vol. 90, 1977, 1053-1090
16. J. A. Goguen: *Types as Theories*, Oxford University, 1990
17. J. A. Goguen, R. M. Burstall: *Institutions: Abstract Model Theory for Specification and Programming*, J.ACM, vol. 39 (1), 1992, 95-146
18. R. Goldblatt: *Topoi - The Categorial Analysis of Logic*, North-Holland, Studies in Logic, vol. 98, 1984
19. P. Johnstone: *Topos Theory*, Academic Press, 1977
20. A. Kock, G. Reyes: *Doctrines in Categorial Logic*, in J. Barwise (Ed.): *Handbook of Mathematical Logic*, North-Holland Studies in Logic, vol. 90, 1977, 283-313
21. S. Mac Lane: *Categories for the Working Mathematician*, Springer GTM, vol. 5, 1972
22. S. Mac Lane, I. Moerdijk: *Sheaves in Geometry and Logic - A First Introduction to Topos Theory*, Springer Universitext, 1992
23. M. Makkai, R. Paré: *Accessible Categories: The Foundations of Categorial Model Theory*, Contemporary Mathematics, vol. 104, AMS, Providence (Rhode Island), 1989
24. J. C. Mitchell: *Type Systems for Programming Languages*, in J. van Leeuwen (Ed.): *The Handbook of Theoretical Computer Science*, vol B: "Formal Models and Semantics", Elsevier, 1990, 365-458
25. G. Nelson: *A Generalization of Dijkstra's Calculus*, ACM TOPLAS, vol. 11 (4), 1989, 517-561
26. J. Palomäki: *Towards a Foundation of Concept Theory*, in H. Kangassalo, H. Jaakkola: *Information Modelling and Knowledge Bases V*, IOS Press, Amsterdam, 1994
27. J. C. Reynolds: *Polymorphism is not Set-Theoretic*, in G. Kahn, D. B. MacQueen, G. Plotkin (Eds.): *Semantics of Data Types*, Springer LNCS 173, 1984, 145-156
28. K.-D. Schewe, B. Thalheim: *Fundamental Concepts of Object Oriented Databases*, *Acta Cybernetica*, vol. 11 (4), 1993, 49-85
29. K.-D. Schewe, J. W. Schmidt, I. Wetzel: *Identification, Genericity and Consistency in Object-Oriented Databases*, in J. Biskup, R. Hull (Eds.): *Proc. 4th Int. Conf. on Database Theory*, Springer LNCS 646, 1992, 341-356
30. K.-D. Schewe: *Specification of Data-Intensive Application Systems*, Habilitationsschrift, TU Cottbus, 1994
31. J. M. Spivey: *Understanding Z, A Specification Language and its Formal Semantics*, Cambridge University Press, 1988
32. C. Tuijn, M. Gyssens: *Views and Decompositions of Databases from a Categorial Perspective*, in J. Biskup, R. Hull (Eds.): *Proc. 4th Int. Conf. on Database Theory*, Springer LNCS 646, 1992, 99-112
33. W. Wechler: *Universal Algebra for Computer Scientists*, Springer EATCS Monographs, vol. 25, 1992
34. M. Wirsing: *Algebraic Specification*, in in J. van Leeuwen (Ed.): *The Handbook of Theoretical Computer Science*, vol B: "Formal Models and Semantics", Elsevier, 1990, pp. 675-788

Springer-Verlag and the Environment

We at Springer-Verlag firmly believe that an international science publisher has a special obligation to the environment, and our corporate policies consistently reflect this conviction.

We also expect our business partners – paper mills, printers, packaging manufacturers, etc. – to commit themselves to using environmentally friendly materials and production processes.

The paper in this book is made from low- or no-chlorine pulp and is acid free, in conformance with international standards for paper permanency.

Lecture Notes in Computer Science

For information about Vols. 1–899

please contact your bookseller or Springer-Verlag

Vol. 900: E. W. Mayr, C. Puech (Eds.), STACS 95. Proceedings, 1995. XIII, 654 pages. 1995.

Vol. 901: R. Kumar, T. Kropf (Eds.), Theorem Provers in Circuit Design. Proceedings, 1994. VIII, 303 pages. 1995.

Vol. 902: M. Dezani-Ciancaglini, G. Plotkin (Eds.), Typed Lambda Calculi and Applications. Proceedings, 1995. VIII, 443 pages. 1995.

Vol. 903: E. W. Mayr, G. Schmidt, G. Tinhofer (Eds.), Graph-Theoretic Concepts in Computer Science. Proceedings, 1994. IX, 414 pages. 1995.

Vol. 904: P. Vitányi (Ed.), Computational Learning Theory. EuroCOLT'95. Proceedings, 1995. XVII, 415 pages. 1995. (Subseries LNAI).

Vol. 905: N. Ayache (Ed.), Computer Vision, Virtual Reality and Robotics in Medicine. Proceedings, 1995. XIV,

Vol. 906: E. Astesiano, G. Reggio, A. Tarlecki (Eds.), Recent Trends in Data Type Specification. Proceedings, 1995. VIII, 523 pages. 1995.

Vol. 907: T. Ito, A. Yonezawa (Eds.), Theory and Practice of Parallel Programming. Proceedings, 1995. VIII, 485 pages. 1995.

Vol. 908: J. R. Rao Extensions of the UNITY Methodology: Compositionality, Fairness and Probability in Parallelism. XI, 178 pages. 1995.

Vol. 909: H. Comon, J.-P. Jouannaud (Eds.), Term Rewriting. Proceedings, 1993. VIII, 221 pages. 1995.

Vol. 910: A. Podelski (Ed.), Constraint Programming: Basics and Trends. Proceedings, 1995. XI, 315 pages. 1995.

Vol. 911: R. Baeza-Yates, E. Goles, P. V. Poblete (Eds.), LATIN '95: Theoretical Informatics. Proceedings, 1995. IX, 525 pages. 1995.

Vol. 912: N. Lavrac, S. Wrobel (Eds.), Machine Learning: ECML – 95. Proceedings, 1995. XI, 370 pages. 1995. (Subseries LNAI).

Vol. 913: W. Schäfer (Ed.), Software Process Technology. Proceedings, 1995. IX, 261 pages. 1995.

Vol. 914: J. Hsiang (Ed.), Rewriting Techniques and Applications. Proceedings, 1995. XII, 473 pages. 1995.

Vol. 915: P. D. Mosses, M. Nielsen, M. I. Schwartzbach (Eds.), TAPSOFT '95: Theory and Practice of Software Development. Proceedings, 1995. XV, 810 pages. 1995.

Vol. 916: N. R. Adam, B. K. Bhargava, Y. Yesha (Eds.), Digital Libraries. Proceedings, 1994. XIII, 321 pages. 1995.

Vol. 917: J. Pieprzyk, R. Safavi-Naini (Eds.), Advances in Cryptology - ASIACRYPT '94. Proceedings, 1994. XII, 431 pages. 1995.

Vol. 918: P. Baumgartner, R. Hähnle, J. Posegga (Eds.), Theorem Proving with Analytic Tableaux and Related Methods. Proceedings, 1995. X, 352 pages. 1995. (Subseries LNAI).

Vol. 919: B. Hertzberger, G. Serazzi (Eds.), High-Performance Computing and Networking. Proceedings, 1995. XXIV, 957 pages. 1995.

Vol. 920: E. Balas, J. Clausen (Eds.), Integer Programming and Combinatorial Optimization. Proceedings, 1995. IX, 436 pages. 1995.

Vol. 921: L. C. Guillou, J.-J. Quisquater (Eds.), Advances in Cryptology – EUROCRYPT '95. Proceedings, 1995. XIV, 417 pages. 1995.

Vol. 922: H. Dörr, Efficient Graph Rewriting and Its Implementation. IX, 266 pages. 1995.

Vol. 923: M. Meyer (Ed.), Constraint Processing. IV, 289 pages. 1995.

Vol. 924: P. Ciancarini, O. Nierstrasz, A. Yonezawa (Eds.), Object-Based Models and Languages for Concurrent Systems. Proceedings, 1994. VII, 193 pages. 1995.

Vol. 925: J. Jeuring, E. Meijer (Eds.), Advanced Functional Programming. Proceedings, 1995. VII, 331 pages. 1995.

Vol. 926: P. Nesi (Ed.), Objective Software Quality. Proceedings, 1995. VIII, 249 pages. 1995.

Vol. 927: J. Dix, L. Moniz Pereira, T. C. Przymusinski (Eds.), Non-Monotonic Extensions of Logic Programming. Proceedings, 1994. IX, 229 pages. 1995. (Subseries LNAI).

Vol. 928: V.W. Marek, A. Nerode, M. Truszczynski (Eds.), Logic Programming and Nonmonotonic Reasoning. Proceedings, 1995. VIII, 417 pages. 1995. (Subseries LNAI).

Vol. 929: F. Morán, A. Moreno, J.J. Merelo, P. Chacón (Eds.), Advances in Artificial Life. Proceedings, 1995. XIII, 960 pages. 1995 (Subseries LNAI).

Vol. 930: J. Mira, F. Sandoval (Eds.), From Natural to Artificial Neural Computation. Proceedings, 1995. XVIII, 1150 pages. 1995.

Vol. 931: P.J. Braspenning, F. Thuijsman, A.J.M.M. Weijters (Eds.), Artificial Neural Networks. IX, 295 pages. 1995.

Vol. 932: J. Iivari, K. Lyytinen, M. Rossi (Eds.), Advanced Information Systems Engineering. Proceedings, 1995. XI, 388 pages. 1995.

Vol. 933: L. Pacholski, J. Tiuryn (Eds.), Computer Science Logic. Proceedings, 1994. IX, 543 pages. 1995.

Vol. 934: P. Barahona, M. Stefanelli, J. Wyatt (Eds.), Artificial Intelligence in Medicine. Proceedings, 1995. XI, 449 pages. 1995. (Subseries LNAI).

Vol. 935: G. De Michelis, M. Diaz (Eds.), Application and Theory of Petri Nets 1995. Proceedings, 1995. VIII, 511 pages. 1995.

Vol. 936: V.S. Alagar, M. Nivat (Eds.), Algebraic Methodology and Software Technology. Proceedings, 1995. XIV, 591 pages. 1995.

Vol. 937: Z. Galil, E. Ukkonen (Eds.), Combinatorial Pattern Matching. Proceedings, 1995. VIII, 409 pages. 1995.

Vol. 938: K.P. Birman, F. Mattern, A. Schiper (Eds.), Theory and Practice in Distributed Systems. Proceedings,1994. X, 263 pages. 1995.

Vol. 939: P. Wolper (Ed.), Computer Aided Verification. Proceedings, 1995. X, 451 pages. 1995.

Vol. 940: C. Goble, J. Keane (Eds.), Advances in Databases. Proceedings, 1995. X, 277 pages. 1995.

Vol. 941: M. Cadoli, Tractable Reasoning in Artificial Intelligence. XVII, 247 pages. 1995. (Subseries LNAI).

Vol. 942: G. Böckle, Exploitation of Fine-Grain Parallelism. IX, 188 pages. 1995.

Vol. 943: W. Klas, M. Schrefl, Metaclasses and Their Application. IX, 201 pages. 1995.

Vol. 944: Z. Fülöp, F. Gécseg (Eds.), Automata, Languages and Programming. Proceedings, 1995. XIII, 686 pages. 1995.

Vol. 945: B. Bouchon-Meunier, R.R. Yager, L.A. Zadeh (Eds.), Advances in Intelligent Computing - IPMU '94. Proceedings, 1994. XII, 628 pages.1995.

Vol. 946: C. Froidevaux, J. Kohlas (Eds.), Symbolic and Quantitative Approaches to Reasoning and Uncertainty. Proceedings, 1995. X, 420 pages. 1995. (Subseries LNAI).

Vol. 947: B. Möller (Ed.), Mathematics of Program Construction. Proceedings, 1995. VIII, 472 pages. 1995.

Vol. 948: G. Cohen, M. Giusti, T. Mora (Eds.), Applied Algebra, Algebraic Algorithms and Error-Correcting Codes. Proceedings, 1995. XI, 485 pages. 1995.

Vol. 949: D.G. Feitelson, L. Rudolph (Eds.), Job Scheduling Strategies for Parallel Processing. Proceedings, 1995. VIII, 361 pages. 1995.

Vol. 950: A. De Santis (Ed.), Advances in Cryptology - EUROCRYPT '94. Proceedings, 1994. XIII, 473 pages. 1995.

Vol. 951: M.J. Egenhofer, J.R. Herring (Eds.), Advances in Spatial Databases. Proceedings, 1995. XI, 405 pages. 1995.

Vol. 952: W. Olthoff (Ed.), ECOOP '95 - Object-Oriented Programming. Proceedings, 1995. XI, 471 pages. 1995.

Vol. 953: D. Pitt, D.E. Rydeheard, P. Johnstone (Eds.), Category Theory and Computer Science. Proceedings, 1995. VII, 252 pages. 1995.

Vol. 954: G. Ellis, R. Levinson, W. Rich. J.F. Sowa (Eds.), Conceptual Structures: Applications, Implementation and Theory. Proceedings, 1995. IX, 353 pages. 1995. (Subseries LNAI).

VOL. 955: S.G. Akl, F. Dehne, J.-R. Sack, N. Santoro (Eds.), Algorithms and Data Structures. Proceedings, 1995. IX, 519 pages. 1995.

Vol. 956: X. Yao (Ed.), Progress in Evolutionary Computation. Proceedings, 1993, 1994. VIII, 314 pages. 1995. (Subseries LNAI).

Vol. 957: C. Castelfranchi, J.-P. Müller (Eds.), From Reaction to Cognition. Proceedings, 1993. VI, 252 pages. 1995. (Subseries LNAI).

Vol. 958: J. Calmet, J.A. Campbell (Eds.), Integrating Symbolic Mathematical Computation and Artificial Intelligence. Proceedings, 1994. X, 275 pages. 1995.

Vol. 959: D.-Z. Du, M. Li (Eds.), Computing and Combinatorics. Proceedings, 1995. XIII, 654 pages. 1995.

Vol. 960: D. Leivant (Ed.), Logic and Computational Complexity. Proceedings, 1994. VIII, 514 pages. 1995.

Vol. 961: K.P. Jantke, S. Lange (Eds.), Algorithmic Learning for Knowledge-Based Systems. X, 511 pages. 1995. (Subseries LNAI).

Vol. 962: I. Lee, S.A. Smolka (Eds.), CONCUR '95: Concurrency Theory. Proceedings, 1995. X, 547 pages. 1995.

Vol. 963: D. Coppersmith (Ed.), Advances in Cryptology -CRYPTO '95. Proceedings, 1995. XII, 467 pages. 1995.

Vol. 964: V. Malyshkin (Ed.), Parallel Computing Technologies. Proceedings, 1995. XII, 497 pages. 1995.

Vol. 965: H. Reichel (Ed.), Fundamentals of Computation Theory. Proceedings, 1995. IX, 433 pages. 1995.

Vol. 966: S. Haridi, K. Ali, P. Magnusson (Eds.), EURO-PAR '95: Parallel Processing. Proceedings, 1995. XV, 734 pages. 1995.

Vol. 967: J.P. Bowen, M.G. Hinchey (Eds.), ZUM '95: The Z Formal Specification Notation. Proceedings, 1995. XI, 571 pages. 1995.

Vol. 969: J. Wiedermann, P. Hájek (Eds.), Mathematical Foundations of Computer Science 1995. Proceedings, 1995. XIII, 588 pages. 1995.

Vol. 970: V. Hlaváč, R. Šára (Eds.), Computer Analysis of Images and Patterns. Proceedings, 1995. XVIII, 960 pages. 1995.

Vol. 971: T.E. Schubert, P.J. Windley, J. Alves-Foss (Eds.), Higher Order Logic Theorem Proving and Its Applications. Proceedings, 1995. VIII, 400 pages. 1995.

Vol. 972: J.-M. Hélary, M. Raynal (Eds.), Distributed Algorithms. Proceedings, 1995. XI, 333 pages. 1995.

Vol. 973: H.H. Adelsberger, J. Lažanský, V. Mařík (Eds.), Information Management in Computer Integrated Manufacturing. IX, 665 pages. 1995.

Vol. 974: C. Braccini, L. DeFloriani, G. Vernazza (Eds.), Image Analysis and Processing. Proceedings, 1995. XIX, 757 pages. 1995.

Vol. 975: W. Moore, W. Luk (Eds.), Field-Programming Logic and Applications. Proceedings, 1995. XI, 448 pages. 1995.

Vol. 978: N. Revell, A M. Tjoa (Eds.), Database and Expert Systems Applications. Proceedings, 1995. XV, 654 pages. 1995.